RENEWABLE ENERGY
TECHNOLOGY AND THE ENVIRONMENT

Proceedings of the 2nd World Renewable Energy Congress
Reading, UK, 13-18 September 1992

Edited by
A. A. M. SAYIGH
Department of Engineering, University of Reading, UK

Organized by

WORLD RENEWABLE ENERGY COMPANY LTD.
Volume 5

PERGAMON PRESS
OXFORD · NEW YORK · SEOUL · TOKYO

U.K.	Pergamon Press Ltd, Headington Hill Hall, Oxford OX3 0BW, England
U.S.A.	Pergamon Press, Inc., 660 White Plains Road, Tarrytown, New York 10591-5153, U.S.A.
KOREA	Pergamon Press Korea, KPO Box 315, Seoul 110-603, Korea
JAPAN	Pergamon Press Japan, Tsunashima Building Annex, 3-20-12 Yushima, Bunkyo-ku, Tokyo 113, Japan

Copyright © 1992 Pergamon Press Ltd

All Rights Reserved. No part of this publication may be reproduced, stored in a retrieval system or transmitted in any form or by any means: electronic, electrostatic, magnetic tape, mechanical, photocopying, recording or otherwise, without permission in writing from the publishers.

First edition 1992

Library of Congress Cataloging in Publication Data
World Renewable Energy Congress (2nd : 1992 : Reading, England)
Renewable energy technology and the environment : proceedings of the 2nd World Renewable Energy Congress, Reading, UK, 13-18 September 1992 / edited by A. A. M. Sayigh ; organized by World Renewable Energy Company Ltd.
v. <1 >
Includes bibliographical references and index.
1. Renewable energy sources--Environmental aspects--Congresses.
I. Sayigh, A. A. M. II. World Renewable Energy Congress Company.
III. Title
TD 195.E49W68 1992 333.79'4--dc20 92-20446

British Library Cataloguing in Publication Data
A catalogue record for this book is available from the British Library.

ISBN 0 08 041268 8 (5 volume set)

*The WORLD RENEWABLE ENERGY COMPANY LTD.
operates as a registered charity, No. 1009879*

In order to make this volume available as economically and as rapidly as possible the authors' typescripts have been reproduced in their original form. This method unfortunately has its typographical limitations but it is hoped that they in no way distract the reader.

Printed and bound in Great Britain by BPCC Wheatons Ltd, Exeter

CONTENTS OF VOLUME 5

Related Topics

* Papers were not received in time to be included in the proceeding.

Related Topics

INTEGRATED RESOURCE PLANNING

Frank Kreith

National Conference of State Legislatures
1560 Broadway, Suite 700
Denver, Colorado 80202, USA

ABSTRACT

This article presents the essential features of an Integrated Resource Planning (IRP) process designed to provide energy for societal and industrial needs at least cost. Use of renewable energy sources and energy conservation measures, as well as consideration of social costs, are described. Available data on societal costs and estimates for energy cost of conservation measures and renewable energy systems are included.

KEYWORDS

Renewable Energy; Conservation; Integrated Resource Planning; Societal Costs of Energy; Demand Side Management.

INTRODUCTION

Energy is a mainstay of industrial society. It is an essential input to run the world's factories and provide many of the comforts such as mobility, heat and light. But, using energy requires use of finite natural resources, generates pollution, and creates health problems. It is therefore important that energy be generated and used efficiently. Opportunities for improving energy use efficiency exist all over the world. For example, western Europe uses only 57% of the energy and Japan only 44% of the amount of energy used in the United States to produce a unit of GNP. In centrally planned economies, such as the former USSR, it has been estimated by Cooper and Schipper (1992) and Siuyak (1991) that the amount of energy used to generate one unit of GNP is much higher than in the U.S., but no data for developing countries could be found.

Electric power is important for industrial nations and for the past thirty years electricity use in the United States has been growing at a faster rate than the economy. During the same period (1973-1990) the overall ratio of energy consumption to GNP declined by almost 30%. The increasing use of electric power is a result of continuing electrification in the U.S. economy, which according to some economists, improves industrial productivity. However, to provide one unit of electric energy requires three to four times the amount of primary energy compared to direct burning of fossil fuels. Consequently, if a given task can be performed by direct application of heat from the combustion of oil, natural gas or coal, or can be produced from renewable energy, such as solar thermal or wind, the societal need can be satisfied with less input of primary energy and less adverse environmental impact. Furthermore, the societal cost of using electric power from fossil or nuclear sources is considerably larger than the cost of using primary energy directly through fossil combustion or solar systems. Because, at present, external costs such as environmental degradation and health impacts are not properly represented in the prices of energy in the marketplace, increased use of electric power can add to the socio-economic burden and increase the energy to GNP ratio.

Since more than one third of the primary energy consumed in the U.S. is used for generating electricity, considerable emphasis has recently been placed by electric utilities on new and more efficient ways to meet the energy needs for which heretofore electricity was the preferred and, in some cases, the sole energy source. This process, called integrated resource management (IRP) or least cost planning (LCP), will be described.

INTEGRATED RESOURCE PLANNING (IRP)

Traditionally, the planning process of electric utilities consisted of comparing the electric production capacity with the projected demand and building the additional production capacity needed to meet the expected demand in compliance with safety regulations and environmental standards. The utility selected the types of fuels, power plants, distribution systems and power purchases that would meet its objectives while optimizing its profits. Energy demand was taken as a "given" that could not be altered and only supply options were considered. No efforts were made to reduce or shape the demand and no attempts were made to integrate supply and demand-side options.

This process was satisfactory as long as energy resources were plentiful and cheap. Recently, however, the cost of energy resources has increased and the public has become concerned about environmental degradation. Hence, many utilities realize that the traditional way of planning for the future needs to be modified. The modification consists primarily of introducing demand-side management (DSM), a process designed to reduce the amount and influence the timing of the customers' energy use. DSM affects the system energy and total capacity that an electric utility must provide to meet the demand. DSM is a resource option complementary to supplying power and provides an important component in a modern utility's energy resource mix.

IRP is the process of simultaneously examining side by side all energy savings and energy producing options to optimize the mixture of resources and minimize the total costs while including consideration of environmental and health concerns. There is no unique method for IRP, but an extensive study conducted by Schweitzer *et al.* (1990) showed that the following sequence of steps is generally used:

1. Develop a load forecast.
2. Inventory existing resources.
3. Identify future electricity needs not being met by existing resources.
4. Identify potential resource options, including DSM programs.
5. Screen all options to identify those that are feasible and economic.
6. Identify and quantify environmental and social costs of these options.
7. Perform some form of uncertainty analysis.
8. Select a preferred mix of resources, including conservation measures and load shaping, which are treated as synonymous to supply options.
9. Implement least cost mix of supply and conservation options.

Figure 1. Schematic Diagram for IRP Process

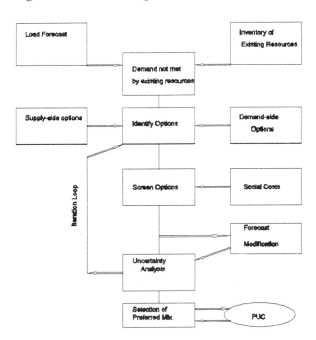

Figure 1 shows a schematic diagram for an IRP process that includes externalities. It gives a systematic procedure to evaluate demand-side options, compare these to supply-side options and develop an energy policy that will integrate environmental and social costs. The goal is to develop a long term energy strategy that will acquire the most inexpensive resources first and internalize social costs in the rate structure. When externality costs are incorporated into the IRP process Eto (1990) suggests it be called least cost planning (LCP), but both IRP and LCP use the same methodology. The IRP process is generally carried out by computer models. These models vary in complexity from screening tools for PCs to sophisticated integrated planning models. A summary of these tools has been prepared by Eto (1990). The new features in this process, DSM and the social costs of energy (SCE) will now be considered in more detail.

DEMAND-SIDE MANAGEMENT (DSM)

Demand-side management is a broad term that encompasses the planning, implementation and evaluation of utility-sponsored programs to influence the amount or timing of customers' energy use. This in turn affects the system energy (kWh) and total capacity (kW) that the electrical utility must provide to meet the demand. DSM is a resource option complementing power supply. It is used to reshape and reduce customer energy use and demand, thus providing an important component of a modern utility's energy resource mix. Four basic techniques for influencing and reducing energy use in demand-side management are:

Peak Clipping is the reduction of the system peak loads. It uses direct load control, commonly practiced by direct utility control of customers' appliances. While many utilities use this mainly to reduce peaking capacity or capacity purchases during the most probable days of system peak, direct load control can also be used to reduce operating cost and dependence on critical fuels.

Valley Filling builds off-peak loads. This may be particularly desirable when the long-run incremental cost is less than the average price of electricity because adding lower-priced, off-peak load under those circumstances decreases the average price. A popular valley filling methods is to use thermal energy storage for industrial water or space heating.

Load Shifting involves shifting load from on-peak to off-peak periods. Examples include use of storage for water or space heating, cold storage, and customer load shifts. The load shift from storage devices displaces peak loads which would have existed if conventional appliances without storage had been installed.

Strategic Conservation is the load shape change that results from utility-stimulated conservation programs designed to reduce end use consumption. Conservation was not always considered load management because in the past it reduced power sales and profits, as well as changed the use pattern. In employing energy conservation, the utility planner must consider what conservation actions would occur naturally and then evaluate the cost-effectiveness of utility programs to accelerate or stimulate more action. Examples include weatherization and appliance efficiency improvement.

Conservation technologies in both the residential and industrial sectors can be used for load management. For example, improved insulation for a building reduces energy consumption and is therefore classified as strategic conservation. Using a high-efficiency compressor in an air conditioning system reduces consumption during peak load and therefore achieves peak clipping as well as strategic conservation. To entice utilities to

2291

implement conservation programs they must be allowed to earn a fair profit on "saved" as well as on "sold" energy.

In addition to the four energy conservation and load shaping programs used by utilities to influence the amount or timing of customers' energy use, utilities that have excess capacity often attempt to increase their sale of power. This process is called strategic load growth and may involve incentives to switch from gas to electric appliances or rebates for the installation of electric devices.

Table 1 shows the energy cost and payback time for some typical DSM measures estimated by Kreith (1992) from data supplied by the Western Area Power Administration (1991). The cost of saved energy in the last column was calculated with a real discount of 3%, a long term average of the difference between the interest rate and the inflation in the U.S. according to Goldstein *et al.* (1990).

Table 1. Estimated Payback and Energy Cost for Conservation Technologies*

Technology	Life (yr)	Payback (yr)	Cost of Saved Energy (c/kWh)
Bldg. Insulation	20	1-4	1.5-1.9
Storm Windows	20	5-10	3.5-7.0
Solar Films	3-15	~5	3.2-13
Weather Stripping	2.5	1.6	5.2
Heat Pumps	15	3-4	2-3
Evaporative Cooling	5-20	--	1.1-3.3
Efficient Motors	~7	1.3**	~5
Heater Insulation	10	1.1	0.9
Low-Flow Shower Head	10	0.4	0.4
High Effic. Refrig.	20	1.3**	0.7
High Effic. Fluorescent Lighting	20,000 (hr)	~1**	~2.4

*   Abstracted from Kreith (1992)
** Payback is based on incremental cost

SOCIAL COSTS OF ENERGY AND VALUE OF ENERGY SAVINGS

For a realistic integrated resource planning process, it is necessary to include the cost of externalities, usually called "social costs", in the planning process. Externality costs are the result of prices in the marketplace not reflecting the full costs of resources, particularly those borne by society as environmental and health related costs. For example, damage from air pollution or acid rain is an externality cost not included in the energy production costs and must therefore be paid for by society. Some of the most important environmental and health damages from energy production include air and water pollution, land use, health effects and disposal of ash or radioactive waste.

SCE must be based on a common unit of service, usually the kWh for electricity or the MBtu (million Btu) for natural gas or oil combustion. The social cost of an adverse environmental or health impact can be estimated from the relation

Social Cost = (Size of Impact) x (Damage Cost per Unit of Impact)   (1)

The social cost is the total cost in dollars per kWh of electricity produced that is borne by society because it is not included in the market cost. The size of the impact is expressed in physical units per kWh such as pounds of air pollutants emitted per kWh or number of people likely to contract respiratory diseases from air pollution per kWh. The damage cost per unit of impact is the economic effect of the adverse impact in dollars per unit of impact, e.g. the average cost of the respiratory disease cure per person or the cost of mitigating the air pollution emission per pound of pollutant.

Externality costs are difficult to estimate and vary from place to place. One of the pioneers was the New York Public Service Commission who estimated in 1989 the economic cost of mitigating the residual air emission from a "base" coal power plant that barely meets federal New Source Performance Standards (NSPS) and used that figure as the externality cost as shown in Table 2. An overview of how other states in the U.S. incorporate externalities in the IRP process is presented by Kreith (1992). It was found that the externality estimates for the New York bidding process are less than estimates presented by Koomey (1990).

Table 2. New York Externality Cost Estimates*

Externality	Emission from NSPS X Coal Plant (lbs/MWh)	Control Cost = ($/lb)	Mitigation Cost (c/kWh)
Air Emission			
SO_2	6.0	0.416	0.25
NO_2	6.0	0.92	0.55
CO_2	1820.0	0.00055	0.10
Particulates	0.3	0.167	0.005
Water Impacts	NA	NA	0.100
Land Use	NA	NA	0.400
Total			1.405

*Calculated from data given by Putta (1990) and Foley and Lee (1990).

The cost (or value) of conserved energy (CCE) by installation of a conservation measure such as a high efficiency motor must also be expressed as c/kWh. It is common practice to use the levelized cost over the lifetime of the system. As shown by Kreider and Kreith (1982), the cost of energy from a conservation system is:

$$\text{CCE} = \frac{\text{Initial Cost of Conservation Device x CRF}}{\text{energy saved per year}} \qquad (2)$$

CRF in the above relation is the capital recovery factor which accounts for the time value of money invested initially. Its numerical value depends on the lifetime of the conservation device, t, and the discount rate, r, or:

$$\text{CRF } (r,t) = \frac{r}{1-(1+r)^{-t}} \qquad (3)$$

Specifically, the capital recovery factor is the ratio of the annual payments to the total sum that must be repaid. For example, if the lifetime is 10 years and the effective discount rate is 0.03 or 3%, CRF = 0.117 and an initial investment of $1,000 costs $117/yr to repay in 10 years. If this device can save 10,000 kWh per year the cost of energy saved will be $117/10,000 kWh or 1.17 c/kWh. Essentially, the same basic approach can be used to determine the cost of energy from a solar system, such as a photovoltaic power system, or a passive system, such as movable window insulation. A more complete analysis that includes the cost of a backup system, the effects of fuel cost escalation and possible tax credit is given by Kreider and Kreith (1982). Ranges of the energy cost of some existing solar options for use in IRP programs according to Howard and Sheinkopf (1991) and Kreith (1991) are given in Table 3.

Table 3.   Estimated Cost Ranges of Some Solar Options for IRP

System	Cost of Energy in $/Kwh
Advanced Windows (H&S)	0.011 - 0.05
Daylighting and Controls (H&S)	0.02 - 0.04
Solar Domestic Hot Water (H&S)	0.04 - 0.16
Solar Process Heat (H&S)	0.015 - 0.052
Photovoltaic DC (1990) (H&S)	0.25 - 0.35
Solar Thermal (Kreith)	0.09 - 0.15
Wind (Kreith)	0.047 - 0.072

EXTERNALITY COSTS FOR ELECTRICITY FROM FOSSIL FUELS

Koomey (1990) recently surveyed available studies on the external costs of electric power from fossil power plants in the U.S. Excluding CO_2 costs, the externality costs of existing coal fired power plants was found to range from 1.93 to 3.54 c/kWh, excluding results for California which were three times as high (California Energy Commission, 1989) and an early EPRI study which gave only about half of the above values (EPRI, 1987). For new coal fired power plants that meet current emission standards (NSPS), the externality costs, excluding CO_2, ranged from 0.83 to 1.53 c/kWh if the values from California and EPRI are omitted. For an average cost of electric power of 6.6 c/kWh, externality costs are about 18% for new plants and 42% for older plants without state-of-the-art pollution control equipment. The results of the survey for coal power plants are fairly close to previous estimates by Hohmeyer (1988) for externalities in Germany. Excluding California, Table 4 gives Koomey's averaged values of the externality costs for gas, oil, and coal fired electric power plants and combustion of natural gas, including estimates of CO_2 effects. To obtain the total energy cost, the social costs must be added to the market price of the energy. The estimates of the California Energy Commission (1990) for the cost of electric power from various fuel sources are shown in Table 5. It is apparent that the externality costs for fossil fuels are substantial, but externality cost per kWh of heat from direct use of natural gas is considerably less than from electrical heating. No data of the social cost of nuclear power in the U.S. could be found in the open literature.

Table 4.   Summary of Externality Costs of Energy for the United States

Technology	Average Delivered Cost (1989 cents/kWh)	Average Externality Cost (1989 cents/kWh)	Externality Cost as % of Delivered Cost
Existing Steam Plants:			
Natural Gas	6.6	0.78	12%
Oil	6.6	1.67	25%
Coal	6.6	2.94	45%
Direct Use of Natural Gas	1.61	0.46	29%
New NSPS Plants:			
Coal Steam (base load)	8.3	1.51	18%
CT Gas (peak load)	5.5	0.95	17%

Table 5.   Estimated* Cost of New Electrical Generation in 1987 Dollars

	Low	High
		(in cents/kWh)
Solar Thermal Hybrid	6.0	7.8
Nuclear	5.3	9.3
Natural Gas (Intermediate)	5.3	7.5
Hydro	5.2	18.9
Wind	4.7	7.2
Coal Boiler	4.5	7.0
Natural Gas Combined Cycle	4.4	5.0
Geothermal Flash Steam	4.3	6.8
Biomass Combustion	4.2	7.9

*Note: These estimates do not include social costs.

SUMMARY

There exist many opportunities for conserving energy at costs below the market price of electric power. The IRP process, which includes demand side management, offers the means to provide for future energy needs at less cost than the conventional planning process of utilities. Since social costs of energy are a substantial fraction of current market prices, they should be integrated into future economic planning and utility rate structure. When social costs are added to the current market price for power, many renewable technologies will become cost effective because their externality costs are below those of fossil fuel power.

REFERENCES

Cooper, R.C. and L. Schipper (1992). The Efficiency of Energy Use in the USSR-an International Perspective. *Energy, the Int. Journal*, 17, 1-24.

Eto, J.H. (1990). An Overview of Analysis Tools for Integrated Resource Planning. Report No. 28692, Lawrence Berkeley Lab., Berkeley, CA.

Foley, L.O. and A.D. Lee (1990). Scratching the Surface of the New Planning: A Selective Look. *The Electricity Journal*,, 3, 48-55.

Goldstein, D. *et al.* (1990). Initiating Least Cost Energy Planning in California-Preliminary Methodology and Analysis. Presentation of NRDC and Sierra Club to the California Energy Resources, Conservation and Development Assoc., Sacramento, CA.

Hohmeyer, O. (1988). Social Costs of Energy, Springer Verlag, New York.

Howard, B.D. and K.G. Sheinkopf (1991). Solar Building Options for Demand Side Management. In: *1991 Solar World Congress*, Vol. 1, Part II, pp. 685-690. Pergamon Press, Oxford.

Koomey, J. (1990). Comparative Analysis of Monetary Estimates of External Costs Associated with Combustion of Fossil Fuels. Report No. 28313, Lawrence Berkeley Lab., Berkeley, CA.

Krause, F. and J.H. Eto (1988). Least Cost Utility Planning Handbook for Public Utility Commissioners. Vol. 2, *The Demand Side: Conceptual and Methodological Issues*. Nat. Assoc. of Reg. Utility Comm., Washington, D.C.

Kreider, J.F. and F. Kreith (1982). Solar Heating and Cooling - Active and Passive Design. McGraw Hill Book Comp., New York.

Kreith, F. (1991). Solar Thermal Energy - Current Status and Future Potential. Energy and the Environment Proc. First World Renewable Energy Cong., Reading, UK.

Kreith, F. (1992). Energy Management and Conservation - An Agenda for State Action. National Conference of State Legis., Denver CO (in press).

Putta, S. (1990). Valuing Externalities in Bidding in New York. *The Electricity Journal*, 3, 42-47.

Sinyak, Y. (1991). U.S.S.R.: Energy Efficiency and Prospects. *Energy, the Int. Journal*, 16, 791-816.

Western Area Power Association, Energy Services (1991). *DSM Pocket Guidebook*, v. 1 Residential Technologies and v. 2 Commercial Technologies, Golden CO.

ELECTRICITY FROM GEOTHERMAL ENERGY: 88 YEARS OF PRODUCTION

E. BARBIER

CNR-International Institute for Geothermal
Research, Piazza Solferino 2, 56126 Pisa, Italy

ABSTRACT

Eighty-eight years ago, in 1904, at Larderello in Tuscany, electricity was generated for the first time from geothermal steam. The industrial history of Larderello had however begun in 1818 with the manufacture of boric acid extracted from hot waters of natural pools in that area. In 1913 a 250 kW plant generating electric energy from geothermal steam went into operation in Larderello providing for the first time electricity on an industrial basis. Years later, other countries in the world followed the Italian example, and today 38 milliard kWh are generated yearly with an installed geothermal electric capacity of around 6000 MW.

KEYWORDS

Geothermal energy; geothermal steam; geothermal waters

Electricity was produced for the first time 88 years ago from geothermal energy. The first successful experiment was conducted on 4 July 1904 at Larderello, a hundred kilometres south of Florence, when Prince Piero Ginori Conti lit 5 bulbs with the electricity produced from a small dynamo driven by geothermal steam. The steam came from a nearby very shallow well.

The few watts generated that day have, 88 years later, increased to the roughly 6000 megawatts that are currently being produced throughout the world from the geothermal source.

The tiny Tuscan village of Larderello was the cradle of the industrial development of geothermal energy, both for electricity generation and the so-called "non-electric uses". Let us now take a closer look at the history of Larderello, which began with the manufacture of boric acid (Mascagni, 1779; Pilla, 1845; Jervis, 1874).

Boric acid (H_3BO_3) was obtained for the first time by W. Homberg in 1702, after treating borax ($Na_2B_4O_7$) with sulfuric acid. It was assumed at the time that borax might have antispastic healing properties, and it was prescribed under the brand name "Homberg sedative salts".

In 1777 the German chemist Uberto Francesco Hoefer, Chief Apothecary of the Grand Duke of Tuscany, discovered boric acid in the hot waters of the natural pools which, at that time, covered a vast area, the present site of Larderello and its surroundings, near the Etruscan town of Volterra (Fig.1). For centuries the inhabitants of the area had referred to these small natural craters, filled with muddy water and kept on a fierce boil by underground springs of boron-enriched steam, as "lagoni". Jets of hissing steam often shot violently into the air from fissures in the ground. Bubbling pools and hissing steam, parched hot earth, yellowish sulfur and reddish ferrous oxide deposits, and the sound effects accompanying these phenomena created a wild infernal landscape. Many hot water springs in the vicinity of these pseudo-volcanic manifestations had been used for centuries as "baths" and their healing properties extolled by the Etruscans, whose settling in the area dates back to 1500 B.C., by the Romans and, more recently, by Medieval and Renaissance "physicians".

Fig.1. A bubbling hot pool in the Larderello area.

The first cartographic document recording the "lagoni" of Larderello is the *Tabula Itineraria Peutingeriana* (Peutinger map, Fig.2), dating to the 3rd century A.D., which shows the main routes and military roads of the Roman empire. Many illustrious personages took the waters there, including Lorenzo the Magnificent, Lord of Florence, who vowed that it did wonders for his gout.

Hoefer, the chemist, began his experiments by concentrating the natural hot waters to obtain a substance that was identical in all respects to the "Homberg sedative salts", i.e. boric acid. Judging from the accurate description given in 1779 by the physiologist and anatomist Paolo Mascagni, Hoefer's discovery was the first step in developing what was to become a very florid industry, and a borax factory was built in the area. Due to a lack of adequate facilities, however, this enterprise failed and all activity was abandoned for a time.

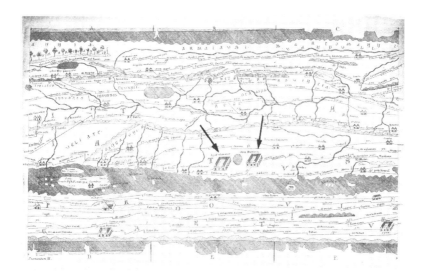

Fig.2. Part of the *Tabula Itineraria Peutingeriana* 3rd century A.D. showing the main roads of the Roman Empire and the spas in the Larderello area called *Aquas Volaternas* and *Aque Populaniae*. (Courtesy of ENEL).

The first serious attempt at producing borax in this way was postponed until 1818, when Francesco Larderel, a French emigré who had settled in Tuscany, overcame difficulties of all kinds and launched a flourishing industry that was to remain one of the leading manufacturers of boric acid worldwide for more than 100 years.

Exploitation of the energy content of the geothermal fluids was to follow, and not so very long after. In 1827 Francesco Larderel had the ingenious idea of utilizing the natural steam to heat the boron-enriched waters of the pools, and collect the boric acid left after their evaporation. Until then the local woods and forests had provided the fuel required to concentrate these waters, but wood supplies were becoming scarcer and more expensive. Long flat sheets of lead, called Adrian Boilers after their inventor's name Adriano de Larderel, were used to evaporate the water and concentrate its boron content. Geothermal steam flowing beneath these evaporators heated the overlying waters. The boric acid concentration increased from 2-4 g/l on entry to 150-160 g/l. After cooling and refining, the boric acid was ground to a fine powder or to minute particles. This was the first *industrial utilization* of geothermal energy.

The business was administered by the Larderels until 1925, although as early as 1899 they had nominated Prince Doctor Piero Ginori Conti, son-in-law of the last Larderel male descendent, as General Manager of the entire works.

In 1903 Prince Ginori Conti began experimenting with the utilization of natural steam to produce mechanical energy. A steam jet was directed against a bladed wheel, which drove a machine tool as it rotated. The next step was to generate electricity, which was achieved on 4 July 1904, using a piston engine coupled to a dynamo, with which Ginori Conti lit his famous five bulbs (Fig.3). Encouraged by his success, in 1905 the Prince decided to widen his experiment. Using an old Cail steam-engine, he managed to generate about 40 hp, which was used to light the factory and to drive some small electric engines. The steam was taken from one of the biggest wells for the time, with a flowrate of 4 t/h at 165°C and a pressure of 3.5 atm abs. By then drilling for steam at Larderello had become a routine operation, although somewhat risky for the men on the drill-site.

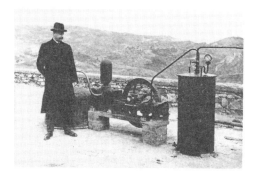

Fig.3. Prince Piero Ginori Conti and his steam engine used in 1904 in the first experiment at generating electricity from geothermal steam. (Courtesy of ENEL).

Right from the start of the boric acid industry the natural steam jets were integrated with steam from wells drilled near natural manifestations. The oldest drilling technique, still in use when the first experiments in electricity generation were being conducted (Fig.4), consisted of mounting a rock bit on a rigid column of drillpipes. This device was hauled upwards by a winch, initially driven manually, and then left to drop into the ground. The winches were later operated by small electric engines or steam-engines driven by the steam from productive wells. The steam was found at varying depths, but never below 200 m, as this was the absolute limit for the equipment available at that time in the region.

Prince Ginori Conti had divided the wells into two types, which he called wet and dry. The former produced water and steam, and were used to extract boric acid. The dry wells produced dry steam, and were used to generate electricity. At the beginning of this century the wells had a diameter of 20 - 40 cm; as the bit gradually descended a protective casing of riveted iron tubes was lowered into the well. The riveted part tended to corrode easily and defects in the joints caused steam to escape and reduction in flowrate. Later, oxyacetylene welding solved this serious problem.

The quantity of steam produced by the well usually increased as drilling proceeded, and if all the water in the well had evaporated, the well blew out spontaneously. As the steam flowrate gradually increased the debris produced by the bit was also carried to the surface along with the steam. If water remained in the well there was no spontaneous production; when this happened the skilled workmen in charge went on drilling until they reached what they considered the steam "vein" and then lowered a piston, called the "serpent", which was later extracted rapidly by means of a winch. Release of the pressure of the water column, which was a few tens of metres high and which balanced the steam pressure, brought about an eruption of mud, steam and rock fragments. At times this was so violent that the upper part of the derrick was destroyed. Despite the shallow depths reached by drilling, 100 m on average, flowrates in the early 1900s were between 6 and 20 t/h, with shut-in pressures of as much as 5 atm abs. It took from 2 to 6 months to drill a well, provided that no accidents occurred, such as walls caving in or deformation of the iron casings.

Fig.4. Larderello, the boric acid industry, 1828.
Hand-driven drilling rig for steam wells.
(Courtesy of ENEL).

Studies and experiments on the generation of electricity continued until, in 1912, the decision was taken to construct an electric power-station to supply electricity to the borax factory buildings, the salt works at Saline di Volterra, the village of Pomarance and the town of Volterra. A steam turbine was chosen instead, as a low pressure piston engine of a few hundred kW capacity would have taken up too much space. The obvious choice for feeding the turbine was natural steam, but this project had to be abandoned. Although Prince Ginori Conti's staff had years of experience with the Cail engine, they had no idea what would happen with such a complicated and delicate machine as a steam turbine, nor could they predict how the metals would react in the presence of natural steam. But this was not this their only problem. The successful operation of a low pressure steam engine depends greatly on the efficiency of the condenser. As they needed a large steady flow of steam, the geothermal well had to be allowed to work completely open, but its pressure dropped too low. Consequently the condenser had to guarantee a fairly good vacuum. However, if they fed geothermal steam directly into the turbine, non-condensable gases would also enter the condenser. Considering the quantity of these gases present in the steam (about 60 g/kg of fluid), their extraction with a pump would have drained much of the turbine power. They were thus forced to use the natural steam to heat and evaporate a secondary fluid: fresh water. Four Proche & Bouillon tube nest heat exchangers were used to generate pure steam, with geothermal steam circulating around them. Each nest of tubes was sheathed in iron. Natural steam entered through a sheet iron tube, after passing through a separator, where condensation water and impurities were collected. The geothermal steam had a pressure of about 2 atm abs at a flowrate of 3 t/h. The pure secondary steam had a pressure of 1.5 atm abs and fed a 250 kW, 3000 r.p.m. low pressure, action-reaction turbine constructed by the Italian company Franco Tosi. The turbine was coupled to a 4000 volts, 50 periods Gans alternator. Aluminium was widely used for the electric circuits, as it was almost totally resistant to hydrogen sulfide. A step cooler supplied circulating cool water to the condenser

This experimental plant operated from 1913 to 1916; in 1914 it supplied electricity to Volterra and Pomarance, and in 1915 to Saline di Volterra as planned. The distribution network was rated at 16,000 volts for the lines leaving Larderello, and at 220 volts for use within the works.

This prototype provided some valuable experience which was later used to design and construct a much larger and more complex power-plant, with three Franco Tosi turboalternators of 2500 kW each; this plant began operations in 1916 (Luiggi, 1917; Ginori Conti, 1917, 1924, 1925; Anon. 1926; Società Boracifera 1928; Nasini, 1939). Industrial exploitation of electricity from geothermal energy was thus a "fait accompli" (Figs.5, 6).

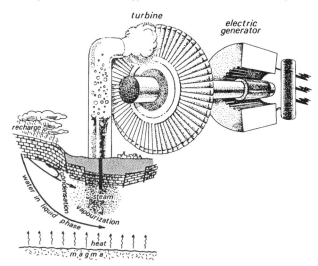

Fig.5. Electricity from geothermal steam. Sketch
of a geothermal field on the left.

Fig.6. Larderello (Tuscany, Italy), nowadays and its 400 MW power plants.

The example set by Italy was followed by other countries in which considerable surface manifestations denoted the probable existence underground of large quantities of high-temperature fluids. By 1919 the Japanese had successfully drilled some wells at Beppu in search of steam, in 1924 electric energy of geothermal origin was produced in small quantities though they started producing geothermal electricity industrially only in 1966. Although the production of geothermal electric energy in the United States had only begun in 1960, the first well was drilled in California in 1921 in The Geysers area. In Iceland, the heat of geothermal fluids has been used since 1925 for heating homes and greenhouses. Since 1949 thermal waters in China have been used for industrial and agricultural purposes and since 1958 electric power has been produced from natural steam. In 1950, a large research programme was initiated in New Zealand and despite a relatively late start this effort led to the achievement of significant industrial results and made a substantial contribution to the knowledge of geothermal fields.

The 1990 status of worldwide geothermal electric power is shown in Table 1, along with the figures for 1982 showing that the electric capacity has doubled in the past 8 years. The installed electric capacity in 1990 was 5838 MW with the generation of 38 milliard kWh.

In industrialized countries, where the installed electric power reaches high figures (tens or up to hundreds of thousands of MW) geothermal energy is unlikely to account for more than 1%, at most, of the total in the next decade. In developing countries, on the contrary, where electrical consumption is still limited but geothermal prospects are good, electric energy of geothermal origin could make quite a significant contribution to the total. At the moment, for instance, 14% of the electricity in the Philippines, 19% in El Salvador, and 8% in Kenya, come from geothermal sources. The future contribution of geothermal energy to the generation of electricity in the world can be estimated at about 9000 MW in 1995 (Table 1) and 12,000 MW in the year 2000 (Barbier, 1991) with a generation cost of the kWh in the range of 4-6 US cents of dollar (every cost included). This range takes into account different geological situations, quality of steam, output of wells and size of power plants.

Country	Megawatts		
	1982	1990	1995
United States	936	2770	3200
Philippines	570	894	2164
Mexico	180	700	950
Italy	440	548	885
Japan	215	215	270
New Zealand	202	283	342
Indonesia	30	142	380
El Salvador	95	95	180
Costarica			110
Kenya	30	45	105
Iceland	41	45	110
Nicaragua	35	35	100
Turkey	0.5	20	40
China	4	21	50
Soviet Union	11	11	70
France (Guadal.)		4	4
Portugal (Azores)	3	3	3
Guatemala		2	15
Greece		2	12
Romania		1.5	1.5
St.Lucia			10
Argentina		0.6	0.6
Thailand		0.3	3.3
Zambia		0.2	0.2
TOTAL	2793	5838	9006

Table 1. Geothermal electric capacity in the world and forecast for 1995

REFERENCES

Anon. (1926) Convegno minerario di Larderello (20 giugno 1926). (Mining Conference at Larderello, 20 June 1926) Bollettino dell'Associazione Mineraria Italiana, May-June 1926.

Barbier,E. (1991). Geothermal energy: its role in the generation of electricity and its environmental impact. In: Electricity and the environment. Background papers. International Atomic Energy Agency, Vienna, IAEA-TECDOC-624, 163-176.

Ginori Conti, P. (1917). L'impianto di Larderello. (Larderello power-plant) L'Elettrotecnica, 15-25 September 1917, n.26-27, 1 - 11.

Ginori Conti, P. (1924). The natural steam power-plant of Larderello. World Power Conf., Wembley, July 1924.

Ginori Conti, P. (1925). The manufacture of boric acid in Tuscany. J. Soc. chem. Ind., 17 July 1925, XLIV, 29, 343-345.

Jervis, G. (1874). I tesori sotterranei dell'Italia. Repertorio di informazioni utili ad uso delle Amministrazioni Provinciali e Comunali, dei Capitalisti, degli Istituti Tecnici ed in genere di tutti i Cultori delle Scienze Mineralogiche. Parte seconda: Regione dell'Appennino e vulcani attivi e spenti dipendentivi. (Italy's underground treasures. Catalogue of useful information for provincial and communal administrations, businessmen, technical institutes and all scholars of the mineralogical sciences. Second part: The Apennine region and its active and extinct volcanoes). Ermanno Loescher editore in Torino, Vol. 2 (in 4 Vols.).

Luiggi, L. (1917) La centrale termo-elettrica di Larderello. (Larderello thermal electric power-plant). *Giornale del Genio Civile*, Rome, May 1917, 1 - 12.

Mascagni, P. (1779). *Dei Lagoni del Senese e del Volterrano. Commentario di Paolo Mascagni al Signor Francesco Caluri, professore nella Regia Universita di Siena.* (On the pools of Siena and Volterra regions. Report of Paolo Mascagni to Signor Francesco Caluri, Professor at the Royal University of Siena), Stamperia Vinc. Pazzini Carli & Figli, Siena.

Nasini, R. (1939). *I soffioni e i lagoni della Toscana e l'industria boracifera.* (The steam vents and pools of Tuscany and the boraciferous industry) Associazione Italiana di Chimica, Rome.

Pilla, L. (1845). *Breve cenno sulla ricchezza minerale della Toscana.* Di Leopoldo Pilla, professore di Geologia nella Imperiale Regia Universita di Pisa. (Brief note on Tuscany's mineral riches. By Leopoldo Pilla, Professor of Geology at the Royal Imperial University of Pisa). Presso Rocco Vannucchi, Pisa.

Societa Boracifera di Larderello (1928). I primi cento anni di una grande conquista industriale, 1827-1927. (The first hundred years of a great industrial conquest, 1827-1927).

WAVE ENERGY: PROSPECTS AND PROTOTYPES

L J DUCKERS

COVENTRY UNIVERSITY
PRIORY STREET
COVENTRY
CV1 5FB

ABSTRACT

The average energy content of ocean waves in some parts of the World is very large. Extracting some of this energy and converting it to mechanical, thermal, or more usually, electrical energy is an attractive proposition, partly because of the economic benefits but especially because of the extremely low environmental impact of wave converter schemes. Around the world a large number of converter concepts have been developed theoretically, at model scale or tested as prototypes. This paper considers those concepts which are at, or close to, prototype testing.

KEYWORDS

Wave energy; prototypes; economics; environmental impact.

INTRODUCTION

The possibility of extracting energy from ocean waves has intrigued man for centuries, and although there are a few early examples over 100 years old, it is only in the past two decades that technically suitable devices have been proposed. In general these devices have few environmental drawbacks. The economic projections for some devices look extremely promising and especially so in areas of the world where the wave climate is energetic.

WAVE ENERGY AND WAVE POWER

Ocean waves are generated by wind passing over extensive stretches of water. Because the wind is originally derived from solar energy we may consider waves to be stored, moderately high density, form of solar energy.

In a typical 'sea state' a variety of wavelengths, or periods, of the constituent waves are observed and these form the wave spectrum. The power per unit frontage is then given by

$$P_s = \alpha_s H_s^2 T_s \quad \text{kWm}^{-1}$$

H_s = significant wave height
= 4 x rms wave height
= average of highest $\frac{1}{3}$ of waves
T_s = zero up crossing period
α_s = 0.49 kWs^{-1} m^{-2}

for example, then, a sea with a significant wave height of 2.5m and zero

2306

crossing period of 9 seconds would have a mean power of 0.49 x 2.5² x 9 = 27.6 kW m⁻¹.

Figure 1 shows estimates for the wave power density around the world and the largest resources are found in the regions receiving rather constant wind due to their climatic conditions. For example the north east Atlantic is subjected to the air stream from the Gulf of Mexico which consequently generates a substantial wave climate off the European Atlantic coast.

Figure 1 is adapted from reference 1 and the estimates shown are average wave power values in kW per metre of wave front in deep water. The total World wide wave energy resource at any one time is of the order of 2 TW (reference 2). In the North Atlantic 50 kW per metre is typical, whereas around Japan 15 kW per metre is more usual. Energy is lost as waves run into shallower water and so shore mounted devices are subjected to lower power wave climates.

The wave climate is not steady, indeed seas vary on a minute by minute basis as well as seasonally. It is important to note that generally the most energetic Atlantic seas occur during the winter when the demand for electricity is greatest. The variation in wave height, period and power with time means that devices have to be carefully designed for optimal energy capture and have also to be able to withstand the considerable loadings that result from the largest storms.

THE TECHNOLOGY

In order to capture energy from sea waves it is necessary to intercept the waves with a structure which can respond in an appropriate manner to the forces applied to it by the waves. If the structure is fixed to the seabed or seashore then it is easy to see that some part of the structure may be allowed to move with respect to the fixed structure and hence convert the wave energy into some mechanical energy (which is probably subsequently converted into electricity). Floating structures can be employed, but then a stable frame of reference must be established so that the 'active' part of the device moves relative to the main structure. This can be achieved by the application of inertia or by making the structure so large that it spans several wave crests and hence is reasonably stable in most sea states.

FIXED DEVICES

An important group of wave energy devices are the bed and shore mounted ones since, excepting the Japanese vessel, Kaimei, they are the only ones so far tested as prototypes at sea. As a fixed frame of reference and with good access for maintenance they have obvious advantages over the floating devices, but do operate in reduced power levels and may ultimately have limited sites for extensive future deployment.

Probably the majority of devices tested and planned are of the oscillating water column (OWC) type. An air chamber pierces the surface and the contained air is forced out of and then into the chamber by the approaching crests and troughs. On its passage from and to the chamber the air passes through an air turbine generator and so produces electricity. A novel air turbine, the Wells, which is self rectifying and has aerodynamic characteristics particularly suitable for wave application, is proposed for many OWCs.

Oscillating water columns have been built in Norway, Japan, India and Scotland (see references 3, 5, 7, 8 and 11) and are proposed for the Azores by the Portuguese (reference 4). Kaimei was a floating collection of OWCs which was first tested in 1977. A further four (fixed) OWC type devices have been tested as prototypes in Japan. The first of these was constructed and tested at Sanze on the north west coast of Japan in 1983. The front of the column was 8.1m wide and 5m high and a tandem Wells turbine was employed to extract the energy. The average output was only eleven kW at a cost of £0.16/kWh. The device has been decommissioned. The most recently installed (December 1989) Japanese OWC is at Sakata, also on the north west coast. Here an extension to the harbour wall has had one 20m section constructed as a wave energy converter, again incorporating a

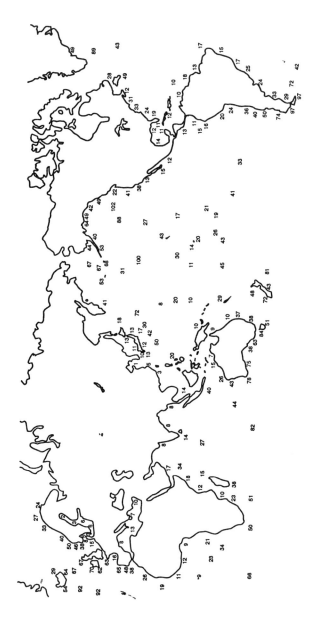

Figure 1   average wave power in kW per metre

tandem Wells turbine rated at 60kW. By functioning both as breakwater and energy generator the Japanese believe that the system is cost effective and they will work at further exploitation of such schemes when the results of this prototype are firmly established. Other prototypes have been tested in Japan, several having mechanical linkage between a moving component, such as a hinged flap, and the fixed part of the device. An example of this is the pendular, two examples of which have been operating on the northern island of Hokkaido since 1983 and 1985. It has hinged pendulum which is positioned one quarter wavelength from the rear caisson wall and has a nominal output of 5kW. Further examples are proposed.

In Norway a device for capturing water at an elevated level as a result of waves running up a tapered channel has proved to be very successful. TAPCHAN, as it is called, needs to be very carefully located as it is susceptible to tides and wave direction. TAPCHAN could reasonably be regarded as the most successful wave energy scheme in the World, the demonstration prototype has been operating on the coast 40km north west of Bergan since 1985. A 40m wide horn shaped collector is designed to harvest the energy from a range of incident wave frequencies and directions. Waves entering the collector are fed into the wide end of the tapered channel, which has a wall height of 10m (from -7m to +3m), where they propogate towards the narrow end with increasing wave height. The wave height is amplified until the crests spill over the walls into the resevoir at a level of 3m above the mean sea level. The wave energy has then been converted into potential energy and is then to be converted to electricity by allowing the water to return to the sea via a low head hydroelectric Kaplan system where a 350kW induction generator delivers electricity into the grid.

A new demonstration prototype is being considered for Indonesia, and a commerical scheme of 1.0 to 1.5MW is being costed for King Island, Tasmania as an alternative to a new diesel power station.

The second Norwegian device is the Multi resonant Oscillating Water Column (MOWC) designed and manufactured in 1985 by Kvaerner Brug. The oscillating water column chamber is set back into a cliff face which falls vertically to a water depth of 60m. The set back of the column produces two harbour walls which broaden the frequency response curve for the water column allowing the system to absorb energy over a wide frequency band. The oscillating air flow is fed through a 2m diameter Wells turbine rotating within the speed range 1000 - 1500 rev/min. The turbine is directly coupled to a 600kVA generator, and the output passed through a frequency converter before being fed to the grid. The performance exceeded predictions and provided energy at about £0.04/kWh. Two severe storms in December 1988 tore the column from the cliff and to date the scheme has not been replaced, although future designs could be much more robust.

FLOATING DEVICES

Floating devices, such as the Clam and Duck from the United Kingdom, and the floating OWC°, such as the WHALE, and Backward Bent Duct Buoy, from Japan are under active consideration. They would be able to harvest more energy since the wave power density is greater offshore than in shallow water and since there is little restriction to the deployment of large arrays of such devices.

The floating OWC° designs have been tested at model scale and the Japanese would like to take the Whale version to a full scale prototype. A rather massive structure is required to provide frame of reference, but since the concept incorporates uses such as a breakwater, leisure provision etc in addition to the generation of electricity the research team believes that the Whale will be cost effective.

The Clam is a floating rigid toroid. Twelve air cells are arranged around the circumference of the toroid and these cells are all coupled together by an air ducting which contains twelve Wells turbines. Thus the air forced from one cell will pass through at least one turbine on route to other cells. Each cell is sealed against the water by a flexible rubber membrane. Performance measurements, together with mathematical modelling and outline full scale design and costings, lead to a cost of delivered electricity of about £0.05/kWh. More details will be given in another paper at this congress (reference 9).

The Edinburgh Duck was originally envisaged as many cam-shaped bodies linked together on a long flexible floating spine which was to span several kilometres of the sea. More recently interest has centered on the case of a single Duck which would demonstrate the technology at full scale and because of point absorber effects would produce significant amounts of energy.

RESEARCH ACTIVITY AROUND THE WORLD

As indicated above there is, or has been, significant research activity in the UK, Norway and Japan. In fact Japan probably has the most substantial current research programme with many teams working on a variety of projects of which only the most important have been mentioned. Several hundred wave powered navigation buoys are also deployed around the Japanese coastline. Further details of much of the work listed below can be found in references 5 and 8.

JAPAN

Apart from the extensive work mentioned already there are also some fundamental developments taking place in Japan. The most interesting are the focussing devices, shaped plates 2m beneath the water surface which concentrate the waves at the centre of a circle. A sea trial is proposed.

NORWAY

Important work is continuing on phase controlled latching to improve energy capture and on the theory of rotating cylinders which have a larger or smaller image size depending upon the direction of rotation as a means of optimising the capture of a Bristol Cylinder type wave energy converter.

UK

During the 1980's the number of device teams in the UK declined due to lack of funding. The surviving teams are engaged on the following projects. The Bristol Cylinder is a submerged cylinder which follows the orbital water paths of the waves but is constrained by mooring cables attached to the sea floor. Energy is extracted at the mooring cables. The pitching and surging FROG is a reactionless wave energy converter which achieves energy absorbing behaviour by the movement of internal inertial mass. A team at the National Engineering Laboratory have considered hydrodynamics and control of OWCs. Studies of wave resource and wave ray tracing are also in hand.

CHINA

There appears to be some wave energy activity developing in China; several papers were presented at the Japanese Symposium (reference 5). Interestingly, much of the Chinese work linked to Japan, either in concept or by the exchange of ideas and staff (to Japan). Some of the work concentrates on navigation buoys, some on theoretical modelling but one group has deployed a small shoremounted OWC of about 8kW installed capacity in the Pearl River estuary. A 5.3W navigation buoy based on the backward bent buoy has also been tested on the River Pearl.

KOREA

Some theoretical modelling of OWCs is taking place at the University of Ulsan and at the Korea Institute of Ships and Ocean Engineering. It seems likely that the work will extend to model tests and sea trials in the future.

INDIA

A sea trial of a 150kW multiresonant OWC device has commenced off the Trivandrum coast. If the cost of the breakwater is shared between the harbour wall and the power plant the electrical production is calculated at Re 1/kWh. A 2.0m diameter Wells turbine and 150kW rated induction generator were installed. The device is expected to deliver an average of 75kW from April to November and 25kW from December to March.

Since the average wave power density along the Indian coast is only 5 to 10kW/m it is again remarkable to see such research and development activity. However many more harbours are planned on the Indian coastline and the potential application of OWC wave energy converters will therefore be considered.

DENMARK
There has been a research effort in Denmark based upon a tethered buoy. The large floating buoy responds to wave activity by pulling a piston in a sea bed unit. This piston pumps water through a submersed turbine. An array of these buoys could be deployed and arranged to have an integrated, and hence smoothed, output. There have been some difficulties with seals on the prototype but these should be overcome with further development.

SWEDEN
A similar concept to the Danish buoys was investigated but using reinforced rubber hose as the tether and pumping mechanism. The research appears to have ceased due to lack of funding.

PORTUGAL
A 350kW OWC is planned for the island of Pico, part of the Azones in the North Atlantic. This will be located on the sea bottom, close to the rocky shoreline. A Wells turbine will be incorporated into the column. Future developments might include air chamber flow latching and variable pitch turbine blades in order to improve overall performance.

EIRE
The West coast of Eire is particularly suitable for the deployment of both shore mounted and offshore wave energy converters. Research in Eire has concentrated on OWC^s and self rectifying air turbines as alternatives to the Wells turbine

USA
A small amount of work has been carried out in the United States. Government support has been modest, but commercial organisations have promoted several concepts to preliminary design and model testing. These have included a scheme based on the OWC with the McCormick counter rotating turbine and the SEAMILL concept which resembles an OWC but has a float on top of the internal water surface. The motion of this float moves a turbine through a bio degradeable oil working fluid and hence generates electricity. Tank tests are being conducted and a 200kW prototype is planned (reference 6).

ECONOMICS

Wave energy, like many other renewable technologies, has high capital costs but low operating costs. The high capital costs arise from the need to build and deploy large structures to capture small amounts of energy as the "power density" of wave environments is quite low at around 50 kW per metre. On the other hand the operating costs are low because one has to consider only operational, repair and maintenance costs, which together might only amount to a few percent per annum of the capital cost, and there is no cost associated with the fuel, the waves - unless governments impose an abstraction tax!

The consequence of high capital cost, but low operating cost, is generally a long pay back period, and this seems to be a major drawback as far as government and commercial investors are concerned. The fact is, though, that some wave energy concepts are already looking economically attractive. The long term financial returns can be extremely high. Being an environmentally clean technology it may be that the value of the output should be enhanced with respect to electricity derived from some of the conventional sources.

The value of wave energy converters is very dependent upon discount rates, the resource power density, the local cost of conventional energy and the possibilities for secondary uses such as breakwaters or leisure activities. Clearly these parameters vary from country to country and perhaps even within a country. The method of assessment of economical viability is therefore likely to be very different from site to site.

PROTOTYPES

Wave energy is a long term technology, it will take some further years of research and development to produce prototypes of some devices and refine the design of others. Further optimisation of the cost effectiveness of the designs should accompany these R&D programmes.

Table 1 shows that a considerable number of prototypes have been already tested and that the output rating of these varied from the 50 W of the navigation buoys to a 500kW OWC. Some of the prototypes have suffered setbacks whilst others have been very successful.

The author is aware of a number of future prototypes and these are listed in table 2., clearly, though, it is quite likely that there may be other schemes at an early planning stage which are not known to the author.

Wave energy is already being utilised in some parts of the world. Where a remote island has expensive conventional energy and a reasonable wave climate it is likely that prototype devices may be economically competitive. We should not close our minds to the possibilities for other devices, emerging in the future with enhanced cost effectiveness.

CONCLUSION

Wave energy converters are being developed and tested in as many as ten countries. The author believes that the TAPCHAN concept, and the shore mounted OWC's will be economically attractive in many locations around the world. These devices are simple and easily maintained. In the longer term a major contribution from wave energy will probably arise from the deployment of arrays of floating offshore or near shore devices. Urgent research and development is needed to bring these to the prototype stage.

REFERENCES
(1)   'Energi från havets vågor', Claeson L (in Swedish) Published by Energiforskningsnämnden (Efn) Stockholm, Sweden (1987)
(2)   'Wave Energy' Evaluation for C.E.C, Lewis A, published by Graham & Trotman Ltd (1985)
(3)   'Wave Energy Devices' Ed. Duckers L J Meeting C57, Coventry, (1989) The Solar Energy Society
(4)   Wave Energy Project in Portugal OWC demonstration plant. Falcao A F, Gato, L M C Teresa Pontes, M, Sarmento AJNA ISES solar World Congress, (1989), Kobe, Japan.
(5)   3rd Symposium on Ocean Wave Energy Utilization. (Largely in Japanese) Ed, Miyazaki, T and Hotta, H. Tokyo, Japan (1991)
(6)   'Project Seamill'. Bueker R A in Oceans "91" Symposium, Honolulu, Hawaii.
(7)   'State of the Art in Wave Power Recovery', Carmichael A D and Falnes J in 'Ocean Energy Recovery: The State of the Art' Ed Seymour R J to be published.
(8)   'Wave Energy Research and Development in Japan' Migazaki T in 'Oceans "91" Symposium', Honolulu, Hawaii.
(9)   "Towards a Prototype Floating Circular Clam Energy Converter" Duckers L J, Lockett F P, Loughridge B W, Peatfield A M, West M J and White P R S (to be presented at World Renewable Energy Congress II September 1992).
(10)   'Wave Energy' One day meeting S027 (1991), London Institution of Mechanical Engineers, 1 Birdcage Walk, London.
(11)   'Islay Gully Shoreline Wave Energy Device Phase 2: Device Construction and Pneumatic Power Monitoring' Whittaker T J T, Long A E, Thompson A E and McIlwaine S J, Contractor Report to ETSU (ETSU WV1680) (1991).

Table 1
Prototypes

Year	Type	Location	Owner	Installed Capacity kW	Comments
1965	Navigation Buoy OWC	Japan	Maritime Agency	0.05	Several hundred now deployed around the coastline of Japan
1978 to 1986	Kaimei	Japan	IEA	375 to 1000	Vessel motion compromised the system performance. No further interest in energy testing but fundamental data on moorings and materials.
1983	OWC	Sanze, Japan	Mitui+Fuji	40	Low output, discommissioned after one year.
1983	Pendular	Muroran, Japan	Muroran Institute of Technology	5	Still Operational.
1984	Kaiyo floating terminator	Okinowa, Japan	Institute of Ocean Environmental Technology	Not known	Research programme completed.
1985	OWC	Toftestallen Norway	Kvaerner Brug	500	Good performance.Destroyed by storms in December 1988.
1985	Tapchan	Toftestallen Norway	Norwave	350	Good performance, still operational.
1985	Pendular	Mashike, Japan	Mashike Port	5	Supplies Hot water, still operational.
1985	OWC	Neya, Japan	Taisei Corp	40	Wells Turbine drove/heat generating eddy current type device. Tests finished in 1988.

1988	OWC array	Kujukuri, Japan	Takenaka Komuten Co	30	Array of 10 OWC, with rectifying valves feeding a common high pressure reservior. Planned to continue until 1985.
1989	Hinged flap	Wakasa Bay, Japan	Kansai Electric Power Co	1	Under test.
1989	Tethered floats	Hanstholm Denmark	Danish wave power aps	45	Problems with rubber seals. Further trials are planned.
1989	OWC	Sakata, Japan	Port and Harbour Research Institute	60	The OWC is an integral part of a new harbour wall. Now operational.
1991	OWC	Islay, UK	UK Government Queens Univ. Belfast	75	Still operation. Over 1000 hours of testing.
1991	OWC	Trivandrum, India	IIT Madras	150	Latest available information in 1991 was than device was nearing completion.

Table 2
Future Prototypes

Type	Location	Owner	Installed Capacity kW	Comments
OWC	Pico Island Azores	Portugal	350-500	Currently at planning stages.
Whale floating OWC	Japan	Japan Marine Science and Technology centre.	100-400	Funding requested for Japanese authorities.
Backward Bent Duct Buoy	Japan and/or Hawaii	Ryokuseisha Corporation	330	1:10 model tested at sea.
Pendular	Yagishiri, Japan	Muroran Inst. of Technology	125	3 cells with a total frontage of 25m used to provide heat.
Clam	West Coast of Scotland	Coventry University	2000	1:15 model tested in Loch Ness. Component development and theoretical modelling proceeding.
Tapchan	Indonesia Tasmania	Norwave	1000 to 1500	Under consideration.
Seamill	California, USA	Hydropower Corp.	200	Early stages of planning.
OWC	Scotland	UK Government	500-1000	Early stages of planning.
Tapchan	Shetland, Scotland	Norwave/ACER/ Shetland	3000	Funding requested.

2315

A NEW POWER BASE:

RENEWABLE ENERGY POLICIES FOR THE NINETIES AND BEYOND[1]

Dr. Keith Lee Kozloff

World Resources Institute

I. CONTEXT FOR RENEWABLE ENERGY POLICY IN THE U.S.

Sufficient renewable energy flows are available in the aggregate to displace U.S. dependence on fossil fuels for electricity and building thermal end uses. The benefits potentially conferred by doing so include long term sustainability of energy-dependent economic activities, reduction in conventional fuel cycle environmental impacts as well as greenhouse gas emissions, avoidance of economic risks associated with reliance on fossil fuels for electric generation and other end uses, and more equitable distribution of the benefits and costs of energy production.

Achieving some of these benefits depends on the trajectory of market penetration. Time dependent benefits include avoidance of the ecological irreversibilities associated with greenhouse gas emissions, acid rain, and fossil fuel extraction and transportation activities; reduction of long term capital investments in conventional energy supply and use infrastructure;

---

[1] This paper summarizes a report of the same title forthcoming from World Resources Institute.

and avoidance of investments in other replacements for dwindling fossil fuel stocks. While these considerations do not converge to some absolute future date, they do imply that the benefits from renewables will be reduced absent a major contribution from them by the middle of the next century.

Energy markets tend not to reflect the relationship between long-term benefits and market penetration of renewables. Consequently, short term and often volatile market conditions drive the demand for renewables. For example, fluctuating natural prices were one cause of the 1991 bankruptcy of the major solar thermal electric developer in the U.S. Because of the long lead times for both technological and market development, however, the U.S. cannot afford to wait to commercialize renewables until its reserves of cheap natural gas as well as energy savings from inexpensive efficiency improvements are exhausted. Even when fully commercialized, it will take several decades for many renewable applications to saturate markets, starting from their initially small base.

The extent to which renewable energy flows are captured to displace fossil fuels depends equally on the ability to cost effectively match renewable energy flows with the location and scale of end use demands. While virtually all regions of the U.S. have the ability to supply some portion of their energy needs with renewables, there are major regional disparities between the magnitude of renewable energy flows and centers of energy demand, particularly those in metropolitan areas. The potential contribution of renewables to these high energy density areas will not be realized under current institutions governing energy transmission.

2317

Barriers that further inhibit the deployment of renewables stem largely from decision rules not accounting for full private benefits of renewables, private decision makers being unable to capture social benefits, and the current lack of competitiveness of many renewable energy applications. In other words, even if energy prices reflected social costs, many renewable energy applications are currently too expensive to compete with fossil fuels. And for some applications, even if the cost of energy from renewables is reduced, institutional barriers would still prevent decision makers from responding to price signals.

Despite the potential social benefits from renewables and technical improvements that have reduced their cost of energy, the rate at which renewables penetrate different end use markets is likely to be limited. This is due to the above barriers as well as projected low prices and availability of fossil fuels. The high avoided cost conditions that led to the first wave of renewable electric generation, for example, no longer exist. Existing public policies are unlikely to achieve rates of deployment consistent with realizing the potential benefits from renewables.

Developing Commercialization Strategies

A coordinated commercialization strategy is necessary to address the multiple barriers facing renewable energy development. Because diverse technical and economic characteristics cause different barriers to be binding, rapid market penetration of renewables is not amenable to a single "magic bullet" policy initiative. While individual policies implemented in the past have stimulated growth in many renewable energy applications, the effectiveness of these policies was hampered by the lack of an overall commercialization strategy.

Policy coordination would help identify opportunities for synergism, reduce the potential for redundant or offsetting policies, and improve the allocation of scarce fiscal resources and political capital. Coordination is important both among policies directed at different barriers as well as those aimed at the same barrier but implemented by different government entities. To fully achieve potential utility applications, for example, regulatory policies implemented by different agencies that influence energy capital investments (generation, transmission, distribution and storage), rate design, environmental compliance, and demand management should be consistent with each other.

While it is difficult to precisely quantify the optimal levels of public investment for and amounts of energy stimulated from different renewable sources, we can establish some rules for guiding public policy and investment decisions. Strategies for the commercialization of renewable energy technologies should have the following characteristics:

(1) Cost effectiveness should be a primary consideration in selecting from among policy options considered by different levels of government. To the extent that policy makers are faced with two or more instruments intended to promote the same objective, their cost effectiveness ranking should be considered along with other attributes such as equity, administrative feasibility, etc.

2319

Policies that merely correct distorted energy price signals and other resource acquisition decision criteria impose minimal short run social costs (other than administration and data analysis) relative to new public investments such as research, development, and demonstration projects. Such distortions should be corrected to the extent feasible prior to making major new investments. Policies that require government revenue should be implemented only when the potential gains outweigh the nonnegligible social costs of raising the necessary government funds.

(2) The specific characteristics of different renewable energy technologies and applications should be used to determine which policies are appropriate. Regulations, incentive levels, and investments need to be targeted to the widely varying commercial maturity levels of individual renewable technologies, as well their short and long-term energy contribution and operating characteristics.

(3) Commercialization strategies should actively involve key stakeholder groups. The public sector is hampered in picking winning technologies because it lacks market feedback. On the other hand, the private sector may underinvest in technologies that could ultimately become winners because it tends to focus on the short term. If both public and private sectors are involved in the allocation of research and development investments in precommercial technologies, for example, the likelihood is increased that such investments will yield high returns. Risk sharing may also reduce the potential for a commercialization policy to be coopted by interested parties.

(4) For renewable technologies that are not limited in their applicability to a subnational region, the federal government should assume a leadership role in coordinating commercialization activities. This is because market aggregation is most effective at the national level; markets in even large states may not be sufficient to achieve potential cost reductions. Also, energy consumers in any region potentially benefit from commercialization activities undertaken at the state or utility level that reduce the cost of energy from renewables. Many environmental effects mitigated by renewables are regional or national in scope. Finally, the federal government has unique advantages in sharing and spreading the risks associated with new technologies.

At the same time, national commercialization strategies must recognize that the value of renewable energy varies by location and application. States and localities have advantages over the federal government in promoting the identification and matching of renewable energy sources with high value demands. States also have the greatest leverage in determining the mix of generation resources acquired by private utilities.

(5) All aspects of commercialization policy should be subject to feedback and correction. The lack of a perceived energy crisis combined with fiscal constraints means that expensive "shotgun" approaches formerly used to develop policy are no longer feasible. Policy options (such as procurement, export promotion, and tax incentives) should be evaluated ex ante by their long run potential for building sustainable markets and reducing the cost of energy, rather than simply by the number of installations stimulated. The signals sent by a policy to renewable

energy manufacturers, installers/developers, and end users should thus induce reductions in the cost of renewably-produced energy.

Ex post evaluation should also be an integral component of public policies and programs that constitute renewable energy commercialization strategies. Inadequate evaluation has limited the lessons that can be learned about the relative cost effectiveness of past public policies in commercializing renewables. Because past initiatives were implemented over a period of concurrent changes in energy markets and varied considerably among states, causal relationships are tenuous between individual policies and the energy contribution from renewable sources. Much of the available evidence of policy cost effectiveness is anecdotal in nature. In contrast, the evolution of energy efficiency programs demonstrates the importance of systematic monitoring and evaluation as an input to subsequent program design.

(6) Policies should be crafted so as not shield renewable energy applications from market competition. Otherwise, the incentive for greater economic efficiency is dampened. For electric applications, resource acquisition should both be fair between renewables and nonrenewables and not skewed in favor of utility or independent ownership. Utility bidding schemes for resources to meet future demand should recognize that electricity is a bundle of services having characteristics related to load curve, location, reliability, power quality, social benefits, dispatchability and risk. This suggests the need for multi-attribute bidding schemes.

2322

For precommercial technologies, there may need to be an explicit public decision that the application needs a protected operating environment for a limited period of time to achieve commercial maturity. Precedent for such a declaration exists with nuclear fission for which continued federal involvement in virtually all phases of the fuel cycle is likely for the foreseeable future.

(7) Because capital markets perceive the competitiveness of renewables to depend on erratic public policy, capital availability for commercial development is constrained. Policies must be sufficiently consistent over time and predictable to allow strategic planning, especially in the face of volatile fossil fuel prices.

Policy consistency is necessary for building sustainable markets. One characteristic of a sustainable market for renewables is the presence of a sufficient number of firms in each technology to guarantee competition and innovation. At present, manufacturing capability is relatively concentrated for some renewable technologies; the bankruptcy of Luz meant the loss of the only significant solar trough developer in the US.

II. POLICY RECOMMENDATIONS

The potential benefits from renewables, barriers to their deployment and guidelines for policy coordination together constitute a framework for crafting comprehensive policy strategies for renewable energy technologies. These require implementation of both new initiatives as well as policies that already exist or have been tried, but need to be refined, modified, or enhanced. Based on several recent projections, a coordinated renewable energy strategy implemented now could result in 2-3 times the annual rate of market penetration of renewables that would otherwise prevail in the next decade.

Getting Prices Right

The criteria that private decision makers use in making energy investment decisions should be reformed to reduce or offset specific failures of the market to reflect social costs and benefits of energy options. This goal is distinguished from the popular concept of creating a "level playing field."[2]

Policies should cause private energy decisions to reflect the social (particularly environmental) costs associated with the fuel cycles of different energy sources. When these costs diverge sharply from market prices, energy choices are distorted. While the extent to which social cost internalization by itself stimulates renewable energy deployment is uncertain, in its absence, designing effective policy initiatives becomes more difficult.

---

[2] The premise of "creating a level playing field" is at best oversimplified, and at worst, misleading. What constitutes levelness is in the eye of the beholder. Different energy resources have characteristics that make them difficult to compare in terms of the energy services they provide. More importantly, there is little agreement over how energy decisions are affected by the overlapping or offsetting multitude of fiscal, regulatory, and other policies. For these reasons, it is unlikely that a level playing field could ever be identified, much less implemented.

The manner in which social costs are internalized should be consistent with the principle of efficiency. Taking into account those costs already implicit in existing environmental policy, private decision makers should face price signals that are consistent with a social accounting framework.

Applied to environmental costs, the efficiency principle would favor levying environmental taxes at the appropriate level. For example, national carbon taxes are preferred to state-level cents/kWh adders for carbon emissions that are imposed on utility resource acquisition decisions. Measurement, distributional, and effectiveness problems associated with state-imposed environmental adders further suggest that taxation or other economic approaches to environmental internalization are preferable for transboundary impacts. A patchwork of state-level adders may be effective, however, in inducing federal action.

Revising the Tax System

Tax incentives and expenditures can be used to achieve policy objectives, such as making private energy choice incentives consistent with social values. As a policy instrument, a tax incentive should have the following features. It should be a cost effective mechanism for achieving a specific policy objective relative to alternative policy instruments. It should send signals to manufacturers, developers, utilities, and other end users that are conducive to increasing the targeted technologies' commercial maturity. It should be targeted only to those technologies whose current stage of commercial maturity is sensitive to the level and type of incentive offered. Coordination with other policies is important to minimize adverse distributional effects and deadweight losses. Finally, the incentive should be consistent and predictable over time, with an announced gradual phase-out.

Public investment in renewables in the form of tax incentives or expenditures can be high and short-lived or low and long-lived, for the same budget exposure. Given that one of the barriers to investment in renewables is volatility in fossil fuel prices, an objective of tax policy should be to reduce this source of risk. In order for tax expenditures or incentives to support this objective, they should be structured to be relatively long-lived and low.

Past and proposed renewable energy tax incentives have also been justified on the basis rectifying existing biases in tax codes, offsetting past disproportionate government support for nonrenewable energy sources, decreasing the cost of energy from renewables through production economies, reducing the high upfront cost of renewables, and redressing uninternalized environmental effects from nonrenewables. Some of these justifications, such as environmental cost internalization, are well-grounded.

Other justifications, however, are only weakly supported. For example, the evidence is inconclusive that specific energy investment decisions are, on net, significantly biased against renewables due to current federal and state tax codes. To the extent that tax codes are found to bias energy technology choices, codes should be reformed to minimize distortions, rather than new incentives created to offset them.

Reforming Utility Resource Planning and Acquisition

When fully implemented, least cost planning (LCP) is perhaps the single most important state-level policy action that would promote renewable energy development. LCP provides a comprehensive and consistent framework for analyzing and incorporating the full range of benefits and costs associated with resource options. Full implementation of LCP:

(1) considers of resource-specific benefits such as savings in capital and operating costs related to the location of the resource in the utility system, reliability benefits, and benefits related to nongrid-connected applications;

(2) applies cost effective supply-side and demand-side measures to integrate intermittent/nondispatchable resources into the system;

(3) incorporates risks associated with resource options; and

(4) identifies appropriate regional resources for meeting end use needs.

To the extent that least cost planning and associated resource-specific quantification of avoided costs require data and analytical tools that are not commonly available to all utilities, information and training programs should be promoted.

Both a better accounting of differences in riskiness of energy options and a reallocation of how risks are shared should be incorporated in utility resource planning. Risks affecting net revenue streams tend to be concentrated in the near term for renewables relative to nonrenewables. Furthermore, regulation allows utility resource acquisition decisions to be insulated from many sources of risk associated with the lifecycle of energy resource options. One technique for addressing the former would assign different risk-adjusted discount rates to resource options. Examples of ways to reduce the latter would make utilities liable for the costs associated with predictable future environmental regulations and limit fuel cost pass throughs.

Utility resource acquisition processes such as competitive bidding should also be reformed to be unbiased toward both renewable versus nonrenewable and nonutility versus utility generation. Risks should be shared fairly between utilities and nonutility generators. Contracts should be standardized, long term, and allowed to be front-end loaded without penalty.

At the federal level, improving transmission access is critical for maximizing the use of renewable energy flows. To reduce transactions costs for independent renewable energy developers, utilities should be required to issue standard wheeling tariffs for "qualifying facilities" under the Public Utility Regulatory Policy Act. If transmission capacity is limited, utilities should be obligated to wheel power from QFs before other sources.

While few utilities have demonstrated strong interest in developing renewable energy to date, there do not appear to be overwhelming advantages to society from utility versus nonutility generation of renewable energy. Consequently, federal and state regulatory policy should cause the procedures and criteria used by utilities in energy resource decisions to treat all resource options equally, whether utility or independently owned.

Overcoming Institutional Barriers to Renewable Energy Use in Buildings

The building sector accounts for over a third of US energy consumption. Even when renewable applications in buildings are cost competitive (such as passive solar design), institutional barriers often prevent their being chosen. A combination of economic incentives, regulations, and information programs are needed to eliminate institutional barriers to matching diverse renewable resource flows with the location and timing of building energy demands.

As with energy efficiency measures, builders do not capture many of the benefits from building applications of renewables that accrue to building energy users, utility ratepayers, and society. Because of this inability, builders have little incentive to incorporate renewable energy features that increase the upfront costs of the building. Policies should be sought that allow builders to capture a portion of renewables' risk-reducing benefits to lenders and avoided cost benefits to utilities or that otherwise lower up-front capital costs of renewable buildings applications.

Gas and electric utilities should offer programs that lower the upfront cost of retrofitted thermal end use measures similarly to approaches used in other demand-side management programs. For new construction, hook up fees should reflect capacity and other costs imposed on the utility. Lenders should include consideration of building operating costs in determining the maximum amount of mortgage loans to be approved.

2329

Regulations are also called for in the building sector to complement economic incentives. The use of renewable energy in new building stock should be promoted by building, zoning, and subdivision code reforms that insure cost effective solar site orientation, guarantee solar access, and require energy efficiency standards that complement renewable energy applications.

Information programs should be geared to increasing builder, lender, and buyer confidence in the contribution of renewable energy design components and equipment by developing uniform standards and building energy rating systems. Information should be disseminated on operating characteristics, applications, and social benefits of renewables that is targeted to utilities, regulators, the building industry, and end users.

Making Research and Development Investments More Effective

R&D activities should be integrated into an overall commercialization strategy. While early federal R&D for renewables had both successes and failures, too often it was not driven by market needs and lacked adequate technology transfer. To move renewable energy technologies toward commercial maturity, sustained production levels are needed to provide market feedback to research and development activities. Mechanisms for allocating public funds among renewable technology projects and for sharing risk between public and private sectors should be established in the context of an overall commercialization plan. Recent DOE collaborations with the private sector reflect movement in this direction.

2330

Reallocation of the public energy R&D budget should be based on complementing (not replacing) private sector R&D such that aggregate public/private R&D spending promotes national objectives at minimum cost to taxpayers. Given that R&D is inherently risky, the federal energy R&D investment portfolio should be robust under uncertain future R&D outcomes and energy market conditions. A single large project (such as nuclear fusion) exposes the portfolio to more risk than an equal investment spread among many smaller projects. This suggests a mix of basic, applied, and cross-cutting research projects related to energy technologies at different stages of commercialization. The portfolio should include both short term high probability projects (such as wind machine blade design) and low probability, long term, high payoff projects (such as hot dry rock and ocean technologies). Applying these criteria consistently across energy supply options would likely suggest a very different budget allocation than currently exists.

Enhancing Access to Capital for Sustainable Markets

For several renewable technologies, the relationship between their cumulative production and cost of energy is sufficiently well understood to aggregate markets to achieve future production economies. Because of barriers in energy and capital markets, however, insufficient capital at acceptable cost is available to achieve potential cost reductions.

As suggested above, both federal and state governments have important roles in commercialization of renewable energy technologies. The federal government should exercise

2331

its leverage to overcome the constraints to achieving these cost reductions. Federal power marketing agencies should be given both the mandate and the financial resources required to meet a portion of their customers' future needs from renewable energy flows in their service areas. Besides serving national cost reduction objectives, such projects can demonstrate renewable energy applications to private utilities and the investment market. As well, the federal government should facilitate market aggregation of renewables through cooperative initiatives with private utilities and changes in its regulation of the wholesale power sector. Finally, procurement practices of nonpower agencies should be made more favorable to renewables.

For their part, states should participate in federally coordinated commercialization strategies as well as implement commercialization policies with state benefits. To achieve potential cost reductions, policy approaches can focus on quantities or prices. An example of the former approach is for states to require private utilities to competitively acquire some portion of projected resource requirements from renewable energy sources. The latter approach could take the form of allowing utilities a modest and time-limited earnings incentive to invest in renewable energy applications appropriate to their service area.

An emerging issue is the perceived tension between the common interpretation of least cost resource acquisition at the state level and the need to sustain national markets for renewables in the near term. This tension becomes expressed through state regulatory decisions that require utilities to acquire some renewable capacity, even when energy efficiency or natural gas generation appear to be cheaper options to meet projected near term requirements. The

2332

tension will likely diminish over time given demand growth and application of the least cost concept that accounts for risk, diversity, and the time scale of different resource options.

Redistributing Social Costs and Benefits

The discrepancy between the geographic incidence of social costs and benefits associated with renewables calls for specific policy responses. With respect to the environmental benefits of renewables relative to fossil fuels, the incidence of some important benefits are global (reduced carbon dioxide emissions) or multistate regional (reduced acid rain). The environmental impacts of renewables, however, tend to be relatively localized. To the extent feasible, policies should result in the environmental costs of different energy sources falling on those who cause them to be incurred.

Conflicts over siting renewable energy facilities of all types, which reflect that their social benefits do not accrue to local residents, are likely to hamper future renewable energy development as their contribution grows. Siting policies should seek a fair distribution of energy resource benefits and costs among affected groups such as through local rate breaks, other economic compensation, or in-kind benefits such as lands set aside for community parks.

Much of the renewable electric development in the 1980s enjoyed relatively high buy back rates based on projected fossil fuel costs. The spread between renewable energy production costs and buy back rates allowed ample funds for site mitigation. As electric generation becomes increasingly deregulated, pressure will grow among competing generators to discount the costs of site mitigation or compliance with other environmental regulations. Competitive bidding and other resource acquisition procedures should not give developers incentives to understate environmental costs.

RENEWABLE ENERGY TECHNOLOGIES IN DEVELOPING COUNTRIES: A NEW
APPROACH FOR SUCCESSFUL DIFFUSION

Federico M. Butera

Dipartimento di Programmazione, Progettazione e Produzione Edilizia, Facoltà di
Architettura, Politecnico di Milano - via Bonardi 3, 20133 Milano, Italy

ABSTRACT

Renewable energy technologies diffusion in developing countries has been very far
from the expectations, in spite of the clear evidence of the crucial importance they
have for the development of the poorer countries. The reason of this evidence must
come from the deep, since most of the renewable energy projects in the long run has
shown to be a failure. On the basis of a study carried out about the reasons of success
and failure of renewable energy technology projects in developing countries, and by
applying a new scientific paradigm (the so-called Science of Complexity) to the
analysis of social systems in the transient in which a new technology is going to
diffuse in it, a new methodology is proposed. The methodology can be applied to all
the phases of a project: identification, monitoring and ex-post evaluation, and allows
for both an holistic approach and semi-quantitative evaluations for decision making.
An example of application to a real case is also given.

KEYWORDS

Renewable Energy, Technology Transfer, Complexity, Entropy, Information, Organization

INTRODUCTION

In order to create conditions more favorable to the development of the developing countries
...The international community ...will have to rely increasingly on new and renewable sources
of energy, seeking to reserve hydrocarbons for non-energy and non-substitutable uses ...
(United Nations, 1981).

The above quoted statement was adopted by the United Nations Conference on New and Renewable Energy Sources, held in Nairobi in 1981. Since then, more than ten years later, little progress has been done, especially in developing countries. Actually, while some energy conservation measures have been promoted in most developed countries, bringing to a reduction of the Energy/GDP ratio in all of them, renewable energy sources take off in developing countries is still very far from being realized. Why?

Of course, many reasons can be identified, according to the point of view, and many approaches have been used to face the problem: none of them, however, has proved to be fully effective. This evidence, and the continuing difficulty encountered in managing successfully the diffusion process of renewable energy technologies either in developed and in developing countries, let us believe that conceptual models, methodologies and tools up to now used may be inadequate.

On the ground of this assumption a new conceptual model is here presented, developed in a different context and adapted to the case of technology transfer/diffusion, with the aim to make available a practical and effective tool for analysis and decision.

THEORETICAL BASIS OF THE MODEL

One key element stemming from the analysis of cases of success and failure in the transfer/diffusion process of renewable energy technologies is the difficult and complex relationship between the different conceptual models characterizing physical and human sciences (Butera, 1989). The problem arises from the fact that the worlds of physical and human sciences must necessarily interact in the transfer/diffusion process, but they do not share a common language. It is necessary, then, to develop a communication system allowing for the two different worlds to understand each other.

One of the most interesting and promising attempts to bridge the gap between the two worlds is the so-called *Science of Complexity*, dealing with self-organization processes and deterministic chaos (Prigogine and Nicolis, 1991).

In the framework of the Science of Complexity, social organizations may be defined as complex system, containing people and technological devices, interacting with the external environment made of other complex systems (Butera, 1991); within such systems, all components are mutually interacting through non linear relationships, whose parameters are subject to random fluctuations. Such a degree of complexity is found in natural living systems: biological and ecological, with which social organizations share some common features: i) they are open systems, i.e. capable of exchanging energy and matter with their environment; ii) they are made up of sub-systems linked by functional relationships; iii) they are capable of homeostasis, that is of reacting to changes of their environment in order to guarantee their survival; iv) they are capable of evolving, namely of transforming themselves in differently structured and organized systems.

As open dynamic systems, they can also be treated as thermodynamic systems far from equilibrium and the following balance equation can be applied:

$$dS = d_eS + d_iS$$

where dS is the entropy change in the system, d_eS is the transfer of entropy across the boundaries of the system, d_iS is the entropy produced within the system. According to the second law of thermodynamics, the entropy production inside the system is positive. Therefore, for a system keeping or improving its organization ($dS \leq 0$), d_eS must be negative; i.e. some negative entropy (neg-entropy, or organization flow) must be transferred from the environment to the system.

One of the most relevant aspects of social system's analysis in terms of entropy and neg-entropy flows is linked to the concept of resource. For a thermodynamic system far from equilibrium (or a system analyzed as such) resources are neg-entropy flows enabling the system itself to keep its organization or improve it. In this definition it is implicitly admitted that the concept of resource (or neg-entropy flow) is subjective: anything might be considered as a neg-entropy flow, provided it is usable and used by the system.

If we look into rapidly evolving complex systems such as human societies, we note that "progress" is always coupled with the "recognition" and "metabolization" of new resources by the system itself. Naturally the acknowledgment of new resources by the social system is coupled also with a transformation of its structure as far as both new components (technologies) and organization are concerned.

The terms "recognize" and "metabolize" indicating the capability of a system to use a neg-entropy flow have not been chosen by chance. They highlight the important links existing between fields such as thermodynamics, communication systems, biology and ecology, as suggested by Science of Complexity.

Let's recall that, according to the Information Theory, the information H emitted by a source of signals is calculated by:

$$H = -\sum_{i=1}^{i=N} p_i \log_2 p_i$$

where p_i is the probability of emission of a signal i. The maximum value H_{max} is obtained when all the signal are emitted with the same probability; this condition corresponds to the maximum uncertainty about the signal that will be emitted and, thus, to the maximum "surprise" of the receiver. The theory, consequently, states that the more a signal is unlikely, the higher is the information associated to it.

The Information Theory only takes into account the syntactic level of information, the organization of signs (Brillouin, 1962); it does by no means deal with the information's "content": the fact that signals embody some information (or intelligible message) about the weather or a declaration of war has no relevance whatsoever.

We may then distinguish between the information dealing with signals (the one treated by the Information Theory) and the one dealing with the meaning of these signals (i.e. of the information/message transmitted by means of the signals). The latter has been defined as *pragmatic information* (Jantsch, 1980). Pragmatic information, that is transmitted and received according to the principles established by the Theory of Information, is made of messages that generate reactions in the recipient and modify him. But, in order to do that, messages must be recognized by the recipients. A book written in Chinese is carrier of pragmatic information only if the reader can understand Chinese; otherwise is only a source of signs. In general, data are information only for those capable to give sense to them within a cognitive framework. This gives rise to the fact that pragmatic information must be recognized by the receiving system in order to be information.

Hence, both physical neg-entropy and information are resources only if the user (the recipient) is structurally prepared to use them.

NOVELTY, CONFIRMATION AND EVOLUTION

Systems evolve through accidental fluctuations of internal and/or external parameters that —in special conditions— give rise to processes synergically moving towards a demolition of the pre-existing system and the creation of a new structure (Prigogine, 1980).

Any physical system in dynamic equilibrium is always struggling against the surplus neg-entropy it is not capable of metabolizing. The neg-entropy flow, penetrating within a physical system is also a physical flow and, nearly always, never is a pure one. Namely, it is matter containing either elements which can be metabolized by the system and elements which cannot. Hence, the system has to continuously sift out the neg-entropy it recognizes from the rest, the "impurities" of the resource.

If the same reasoning is translated in terms of information, then the impurities become the so-called noise: a non-recognized message, i.e. a flow of meaningless signals.

We might then generalize and say that both information and neg-entropy reach the system in packages containing novelty and confirmation; by novelty we mean all that is not recognized and by confirmation all that is recognized. Systems tend to discard novelty and retain confirmation: it is a fundamental operation for survival.

More so, it is exactly the interplay, indeed the continuous fluctuation novelty-confirmation, that has made (and still does make) possible evolution in biological and ecological systems, and development in social systems (Jantsch, 1980). Evolution occurs if and when fluctuations within the system's structure are in phase with the novelty-confirmation fluctuations of the neg-entropy input, namely when the system is, accidentally, capable of recognizing a bit of novelty. At that point a very fast reaction may take place, involving all the system's components, re-arranging them according to the new and more advanced structure which is more complex and capable of metabolizing a new resource.

This is also the description of the learning process, which always implies a structural change (it is a common experience, when trying to solve a problem, to be suddenly enlightened by an

2337

idea and to have all the other concept connected to it rearranged, and a new reading of the problem).

This is the way ecological, biological and cognitive systems behave; the conceptual model here proposed is based on the hypothesis that the same rules stand also for social systems, in which the diffusion of new technical or technological innovations takes place.

Within this conceptual framework development is a learning process, in which social systems keep modifying their structure (keep "learning") and, with it, their "perception" of the neg-entropy/information flows coming from the surrounding environment.

SPECTRAL ANALYSIS OF A SOCIAL SYSTEM

Since social systems in which we try to introduce new techniques or technologies may be treated as complex physical systems and, at the same time, as cognitive systems subject to a learning process, they can be considered as a source of messages; they "talk" to the environment with a language that is as much rich and complex as rich and complex are the interactions system-environment. Each system, thus, according to the "perception" it has of the environment, communicates with it in a different way.

A transformation, deriving from the diffusion of an innovation reveals itself in a different communication pattern between the system and its environment, i.e. in a different perception that the system has of its environment: new interactions are activated, and many of the old ones are modified.

Since the interaction between a system and its environment is only determined by the subjective perception (recognized neg-entropy flows) that the system has of the environment's components, an effective way to analyze the system's structure is to evaluate this perception.

In order to do this, a social system (including technologies) at any scale (region, town, enterprise, tribe, family, etc.) interacts with its environment (all the other systems with which it is interconnected) through the processes that are going on in it.

The introduction of a new technology implies changes in one or more of these processes. Therefore, the analysis of the interactions between processes and environment can unveil the special way the system "talks" to the environment. Environment, on the other hand, is perceived by a social system in terms of actors, factors and mutual interactions, and this perception will be changed because of the introduction of new technologies.

Once identified the process or the processes that, within the social system, are involved in the technology transfer or diffusion, for each of them the following steps have to be followed:

a) identification of the components of the environment as perceived by the system analyzed: a detailed analysis of each process has to be made, step by step (or action by action), and for each action the actors (individual, social, institutional) and the factors (climate, transportation facilities, energy availability, etc.) potentially affecting it have to be identified;

2338

b) evaluation of the intensity of the interactions existing between each action of the process and the actors or factors involved. In order to execute this part of the methodology, a score S is given to each interaction (tab. 1). Then the values of the information H are calculated for each action and for each actor and each factor, according to the scheme shown in tab. 2.

Table 1

Actions	ACTORS OR FACTORS						
	A	B	...	i	...	Z	
1	$S_{1,A}$	$S_{1,B}$	...	$S_{1,i}$	...	$S_{1,Z}$	$S_1 = \Sigma_{(i=A..Z)} S_{1,i}$
2	$S_{2,A}$	$S_{2,B}$	...	$S_{2,i}$	...	$S_{2,Z}$	$S_2 = \Sigma_{(i=A..Z)} S_{2,i}$
...	...	...	...	...	...	...	...
j	$S_{j,A}$	$S_{j,B}$	...	$S_{j,i}$	...	$S_{j,Z}$	$S_j = \Sigma_{(i=A..Z)} S_{j,i}$
...	...	...	...	...	...	...	...
n	$S_{n,A}$	$S_{n,B}$	...	$S_{n,i}$	...	$S_{n,Z}$	$S_n = \Sigma_{(i=A..Z)} S_{n,i}$
	$S_A = \Sigma_{(j=1..n)} S_{j,A}$	$S_B = \Sigma_{(j=1..n)} S_{j,B}$	...	$S_i = \Sigma_{(j=1..n)} S_{j,i}$	...	$S_Z = \Sigma_{(j=1..n)} S_{n,Z}$	

Table 2

Actions	ACTORS OR FACTORS						H_{Action}
	A	B	...	i	...	Z	
1	$p_{1,A}$	$p_{1,B}$	...	$p_{1,i}$	...	$p_{1,Z}$	$(S_1/S^{Act})\Sigma p_{1,i} \log_2 p_{1,i}$
2	$p_{2,A}$	$p_{2,B}$	...	$p_{2,i}$	...	$p_{2,Z}$	$(S_2/S^{Act})\Sigma p_{2,i} \log_2 p_{2,i}$
...	...	...	...	...	...	...	...
j	$p_{j,A}$	$p_{j,B}$	...	$p_{j,i}$	...	$p_{j,Z}$	$(S_j/S^{Act})\Sigma p_{j,i} \log_2 p_{j,i}$
...	...	...	...	...	...	...	...
n	$p_{n,A}$	$p_{n,B}$	...	$p_{n,i}$	...	$p_{n,Z}$	$(S_n/S^{Act})\Sigma p_{n,i} \log_2 p_{n,i}$

Where: $p_{j,i} = \dfrac{S_{j,i}}{S_j}$ and $S^{Act} = \sum_{j=1}^{j=n} S_j$

With reference to tab. 2, H_{Action} is an index of the novelty of the action, i.e. it is an index able to take into account either the structure and the intensity of the interactions between each action and the perceived environment, defined as the whole universe of the actors and factors. High relative values of H_{Action} identify the most critical actions, the ones that have to be closely monitored in order to let the action develop as desired; They are actions requiring a great learning capability of the system and therefore they are the ones that need more support. H_{Action}, thus, is also an index of the complexity of the action.

$H_{A/F}$ is an index of novelty or complexity too, and indicates how much an actor or a factor is

Table 3

Actions	A	B	...	i	...	Z
1	$p_{1,A}$	$p_{1,B}$	...	$p_{1,i}$	...	$p_{1,Z}$
2	$p_{2,A}$	$p_{2,B}$	...	$p_{2,i}$	...	$p_{2,Z}$
...	...	...	...	...	...	...
j	$p_{j,A}$	$p_{j,B}$	...	$p_{j,i}$	...	$p_{j,Z}$
...	...	...	...	...	...	...
n	$p_{n,A}$	$p_{n,B}$	...	$p_{n,i}$	...	$p_{n,Z}$
$H_{A/F}$	$(S_A/S^{A/F})\Sigma p_{i,A}\log_2 p_{i,A}$	$(S_B/S^{A/F})\Sigma p_{i,B}\log_2 p_{i,B}$	...	$(S_i/S^{A/F})\Sigma p_{i,i}\log_2 p_{i,i}$	...	$(S_Z/S^{A/F})\Sigma p_{i,Z}\log_2 p_{i,Z}$

$$\text{Where:} \quad p_{j,i} = \frac{S_{j,i}}{S_i} \quad \text{and} \quad S^{A/F} = \sum_{i=A}^{i=Z} S_i$$

critical in the management of the productive process; a perturbation modifying the structure and the intensity of the interaction of an actor or factor showing a high relative value of $H_{A/F}$ will affect strongly the productive process, either in a positive or negative way. The values of $H_{A/F}$ rank the actors and the factors as more or less critical and, thus, needing more or less monitoring effort for a successful process execution.

If the calculated values of H are plotted in a bar diagram, as shown in fig. 1, the plot can be interpreted as a detailed picture of the system's structure, like the spectrum of a light or sound source. This sort of "spectral analysis" of the complexity of a productive system allow us to identify −at a glance− the actions, or actors or factors perceived as most critical for the process in order to work properly. When we compare the "spectrum" of the present process with the "spectrum" obtained by analyzing an improved, innovative process, then we can identify the subsystems (actions, actors, factors) with the highest "novelty" content; these are the subsystems on which we have to work harder in order to let the improved process work as expected.

When many subsystems show a high degree of novelty, it may be wise to reduce the scope of the innovation, since the probability of success of the improved process would be very low; in order to identify the most critical action, actor or factor, a further evaluation can be made, by analyzing the "sensitivity to perturbation" of each subsystem. This is made by "perturbating" the subsystem, i.e. modifying randomly the scores within a wide range, and analyzing the effect on the "spectrum" of the whole system.

The spectral distribution of the communication system/environment is also an index of the global complexity of the mutual interaction. A "flat" spectrum is an index of a very mature system, balanced and with a very complex structure. According to the Information Theory, such a system is also characterized by a low value of the redundancy (the redundancy R of a source of signals is defined as $R = 1 - H/H_{max}$). As evidenced by the by ecological systems, a system with low redundancy has little adaptation capability and requires, for its survival on the long run, to be maintained into a very rich environment. The relative height of the lines of the spectrum is an index of the organization level that supports each action. The higher is the line, the more complex is (or it should be) the structure and the organization that allows for the action to develop.

The spectral analysis of the communication system-environment may be used in several phases of the diffusion process of an innovation: from the feasibility study to the monitoring, to the ex-post evaluation.

AN APPLICATION OF THE METHODOLOGY

The methodology has been applied, on behalf of an international body, for both analyzing the unsuccessful performance of a south American institution –whose mission was the promotion of electricity and renewable energy sources for improving farms' productivity– and designing the characteristics of a new institution able to cope successfully with the assigned mission. The first step was to identify the present situation, by means of interviews to the personnel. The result of this preliminary survey was that the institution (lets call it EAA, Energy for Agriculture Agency, from here onwards) had very few interaction with its environment: its communication spectrum had few and low lines, i.e.:
a) the institution was insensitive to environment's changes and, conversely, didn't exert any influence on it;
b) the internal structure was very simple, with few internal interconnections, unable to produce or receive "unpredictable messages", i.e. novelty/information.
The second step consisted in the identification of the social and institutional actors that ideally should have interacted with a well working institution dealing with the promotion of energy use in rural areas (factors were not examined since they appeared far less important than actors in that particular context). These potential actors were identified by means of interviews carried on either with the managers of EAA and by involving all the reasonably possible counterparts. The main identified actors were: the Electric Utility (EU), the national Oil Company (OILCO), the Ministry of Agriculture (MA), the Regional Governments (RG), the Cooperatives (COOP), the Non Governmental Organizations (NGO), the Unions (UNIO), the Universities and other research institutions (UNIV) and the Small and Medium Industry organizations (SMIN).
By means of the interviews and allowing for both the national institutional framework and the national energy system, it was identified a sort of ideal EAA, i.e. an institution capable to satisfy need and wishes expressed by each of the above listed actors. The set of actions that the ideal EAA should have been able to cope with is reported in table 4, representing also the format-matrix used for carrying on the calculation procedure described in tables 1, 2 and 3. The scores to attribute to each interaction were given by a group of local experts. As a result of the calculations, the spectrum of the ideal EAA was found (fig.1).

Only four functions show a value of novelty exceeding 2, namely the functions 5), 6), 8) and 9); all the others are within the range 1.5 ÷ 2.0. These functions, therefore, show an excess novelty that can be faced in two ways: i) eliminate the function; ii) modify the attributes of

Table 4

ACTIONS	ACTORS								
	EU	OILC	MA	RG	COOP	NGO	UNIO	UNIV	SMIN
1) Monitoring of:									
a) executing projects									
b) executed projects									
c) proposed projects									
d) energy consumption									
e) perceived energy needs									
f) renewable energy potential									
g) training demand & needs									
h) R&D needs									
j) environment changes									
2) Diffusion of information on:									
a) new energy technologies									
b) data on renewable energy									
c) technological window									
3) Training as:									
a) executing structure									
b) support structure									
4) Projects approval-rejection-modification									
5) Identification of hi. priority integr. proj.									
6) Promotion of R&D									
7) Promotion of local industrial capability									
8) Promotion of the interaction with envir.									
9) Promotion of information exchange									
10) Prices and tariffs									
11) Promotion of rural energy investments									

the function, reducing the degree of interaction with the environment. The latter option be was chosen; in addition, also the function 7), *promotion of local industrial capability*, has been eliminated, even if its novelty value is <2, as a consequence of the modification and specification of the function 2): if the diffusion of information is successful, the function 7) becomes a necessary byproduct.

Similarly the actions *monitoring, diffusion and training* were analyzed in detail: also in these actions peaks of novelty were found, ad some of them were eliminated or aggregated or simplified.

Table 5

1)	Monitoring of:
	a) Rural Energy projects (executed, executing, proposed)
	b) energy potential & consumption
	c) training demand & needs
	d) national R&D status & needs
	e) international R&D status
	f) environment changes
2)	Diffusion of information
	a) Newsletter & audiovisuals
	b) Technological window
	c) National Conference on Rural Energy (CNER)
	d) Training on projects monitoring & evaluation and on management
3)	Projects approval/planning
4)	Rural Energy prices and tariffs fixing
5)	Promotion of Rural Energy investment

The final result of the analysis, i.e. the actions that the new EAA should carry on is shown in table 5. In fig. 2 the spectrum of the new institution is shown; from it derives that the most critical action, the one that must be kept more closely under control, is the action 2.

The degree of involvement of each actor included in the environment in which new EAA has to operate is shown in fig. 4. Most involved, by far, are NGOs and RDs, followed by Cooperatives; a lower but still appreciable value show both OILCO and the Universities. These actors, then, are the most critical, and if their full participation is not ensured, and kept active, it is very unlikely the fulfillment of the projected objectives. The analysis has been carried on in full detail, action by action, and it has been extended also to the internal organization of the new institution, allowing for the identification of the most suitable functional structure and the professional characteristics required for the personnel.

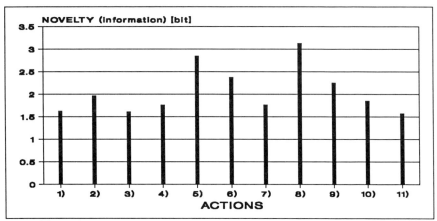

Figure 1

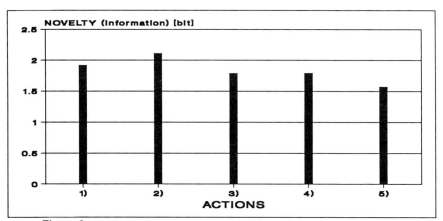

Figure 2

CONCLUSIONS

Either in the case here discussed and in other experiences, the methodology used has proved to be a valuable tool for decision making, given its simple use and its capability to highlight the most critical elements of a process.

The method, however, shouldn't be considered as an alternative to the other commonly used, that can effectively manage sectoral aspects of the whole problem, but as an "integrating" tool that gives the possibility to catch multiplicating effects of single actions apparently little or weakly connected among them. This method should be seen as a *dynamic guide* that, by highlighting the parameters that have to be kept under control, let us manage with higher success probability a very complex process such as the technology transfer and diffusion. Its usefulness goes behind the possibility to use the same tool for different phases of development projects: identification, appraisal, monitoring; since stochastic perturbations are allowed for and given the nature of its scientific background, the methodology is among the most suitable also for analyzing the mutual interaction between man-made and natural environment.

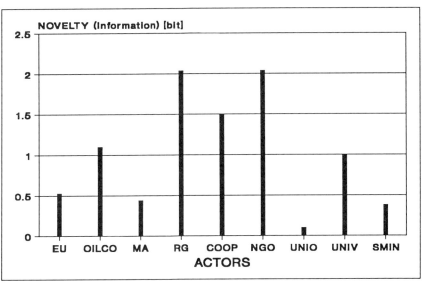

Figure 3

REFERENCES

Brillouin L. (1962). *Science and Information Theory*. Academic Press, New York.

Butera F. M. (1989). *Renewable Energy Sources in Developing Countries: Successes and failures in Technology Transfer and Diffusion*. LB-18, Progetto Finalizzato Energetica CNR/ENEA.

Butera F.M. (1991). Nuovi paradigmi scientifici e sviluppo eco-compatibile. *Oikos*, 3/91, 125-140.

Jantsch E. (1980). *The Self-Organizing Universe*. Pergamon Press, Oxford.

Prigogine I. (1980). *From Being to Becoming*. Freeman & Co., San Francisco.

Prigogine I., G. Nicolis (1991). *La complessità*. Einaudi, Torino. Original title: *Exploring Complexity. An Introduction*, published in 1987.

United Nations (1981). *Report of the United Nations Conference on New and Renewable Sources of Energy*. United Nations, New York.

HARNESSING THE TAX SYSTEM TO BENEFIT ALTERNATIVE ENERGY

R. A. WESTIN

University of Houston Law Center, Houston, Texas 77004

ABSTRACT

Tax credits, accelerated depreciation and other existing tax mechanisms can be used to encourage environmentally benign energy. Implementation is controversial.

KEYWORDS

Tax credits, depreciation, legislative process, carbon taxes, market failure.

LEGISLATIVE PROCESS

Tax legislation is a specialized process, calling for review by technicians who evaluate pending legislation to assure that any proposed new tax act does not conflict with other aspects of the tax law. Special interest groups watch pending tax legislation closely to assure that it does not injure their particular interest. However, tax legislative processes tend to ignore the impact that any new law might have on the environment. This is not surprising, given that the relationship between the two usually appears remote. The United States has recently added one environmental specialist to the staff of the House-Senate Joint Committee on Taxation. I expect that person to provide nonbinding commentary on the environmental implications of proposed tax legislation. In my view, all countries should take this step. Moreover, the activities of the staff member I just described should be reviewed periodically to assure that his or her role has in fact been useful.

NEED TO USE APPLIED ECONOMISTS FOR FOLLOW-UP WORK

It is a chronic problem in the USA and other countries that after tax initiatives are enacted, there is a failure by government to follow up on the impact of the legislation. Such follow-ups are not easy to perform because many responses will be biased; nevertheless, applied statistics and the computer revolution invite the effort.

ELIMINATION OF DESTRUCTIVE TAX PROVISIONS

Most tax systems are not normative. Instead, they are founded on the basic proposition that they exist to raise revenue in an economically efficient manner within the equitable values that the country's populace shares. Until recently, there has been almost no consideration of the relationship between the tax laws as a whole and environmental values, including the promotion of alternative energy sources.

Scope of Technical Conflicts

Each country has a unique bundle of tax laws. A study edited by this author (Gaines et al 1992) involving the US, UK, France, Germany and Sweden showed that each country's laws contain significant conflicts between tax inducements and express national environmental policies. Interestingly, it seems that the one subject that the tax laws have never been able to grapple with successfully is the personal automobile. As a first step, governments should review their laws for conflicting provisions and consider their elimination, placing the burden of retaining the conflicting provision on their proponents. Many of the conflicts are fairly subtle, such as the those encouraging extraction of natural resources over recycling.

ROLE OF MICROECONOMICS

Microeconomics has contributed to the understanding of the relationship between taxes and environmental concerns.

It is generally accepted in governmental planning circles that the price any particular good bears should include its full environmental cost. For example, the price of a kilowatt hour of electricity generated by a nuclear energy facility should include an appropriately prorated share of the cost of dismantling the facility. Economists have embodied the concept in the term "market failure", something that is considered to occur when the price of the output excludes the full range of inputs, including social costs associated with pollution, thereby causing excessive production and consumption (i.e., a misallocation of resources). This concept has earned two Nobel prizes.[1]

Technical discussion

Microeconomists assume firms and consumers are rational profit maximizers that will use up valuable but unpriced common resources such as clean air and water in their businesses. Moreover, firms will produce goods until the marginal cost of the last good produced equals the price it can command.

Outside the firm, there is a demand curve for each good, showing the quantity of a good that consumers will collectively purchase at any

---

[1] Kenneth J. Arrow (1972) and Gerard Debreu (1983).

given price per unit. Producers have a supply curve, which shows the quantity of a good they will collectively produce (supply) at any given price per unit. The supply curve is derived from the producers' marginal cost curves (i.e., cost of producing each incremental unit of output). Supply curves and demand curves vary over time with such factors as tastes and technological change.

Concept of a Tax on Output

The graph below illustrates a single firm's marginal cost curve and the demand curve for the product. The vertical axis is the price per unit. The horizontal axis is the number of units produced.

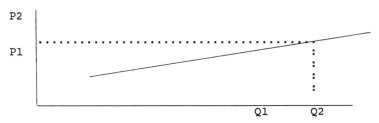

If the firm appropriated a common resource to the detriment of others, and if it were assessed damages for doing so, then its price per unit would rise. The following graph shows a second, higher curve that reflects such a full price, and is the "correct" level of output, in light of the appropriation of common resources such as clean air and water (i.e., pollution damages). The new equilibrium at Q1/P2 would correct the market failure.

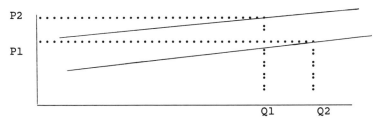

The second line is referred to as "social marginal cost" because it includes the full price to society.

If the remedy is a tax, as opposed to direct control, then the tax should be set at (P2 minus P1). The revenues equal (P1 minus P2) X Q1. The higher the social cost and the less "elastic" the demand curve, the higher the revenues. Elasticity relates to sensitivity of amount purchased to prices.

This new price level will choke back production and consumption. Economists have no clear answer as to what to so with the revenues of such a tax. Increasing the price without repairing the damage does not in itself assure significant environmental benefits, but it cures the "market failure". That is sufficient to traditional microeconomics, and it may be a genuinely valuable result. For example, it may move consumers to substitute new, environmentally benign sources of energy for current sources.

Setting the tax is practically impossible. For example, how does one measure the implicit cost of extinguishing a species? In short, this tax is an economist's dream, but it is nevertheless very powerful forensically, because it scrupulously illustrates the flaw in operating a tax system within a putatively free market economy wherein the regulatory mechanisms fail to account faithfully for environmental degradation.

Emissions Tax to Induce Restricted Pollution

Microeconomic concepts can be used to set a tax that would induce pollution reduction in lieu of direct government intervention by to making it is more attractive to pay to correct the problem at its source than to pay the tax. The following materials express one simple model. It involves one firm and entails a tax applicable only on part of the firm's emissions.

This model assumes government planners have developed a goal of a certain amount of emissions reduction, based on advice of experts.

The first step in this process is to construct a marginal cost curve, reflecting the additional cost the firm has to incur to remove the offending emission. In general, the cost per unit rises as the level of removal rises. For example, the government might conclude it would cost $1.00 to reduce the first unit, but $100 to remove the 100th unit of pollution. The following graph assumes the increase in cost for removing pollution rises steadily.

$/Unit

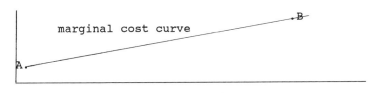

marginal cost curve

Units of Emission Reduction

The next step is to build a marginal benefit curve. It reveals the societal benefit in avoided damages of eliminating the emission. In general, the elimination of each incremental unit of emission has less value than the elimination of the prior unit of emission. In the following graph, the slope is very steep, indicating that the value of eliminating the initial units was very high, but declined rapidly.

2349

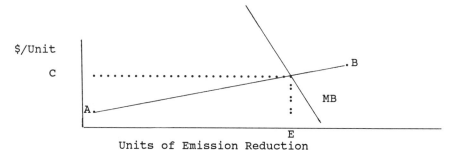

$/Unit

C

A

B

MB

E

Units of Emission Reduction

The inference of this graph is that the firm should stop trying to eliminate the emission at output level "E", at which point it costs "C" dollars to produce that last unit. Any further reduction is unduly expensive, i.e., it costs the firm more than society benefits. Conversely, if emissions reduction stopped short of point E, the result would be unfair to society as a whole. Alternatively, the tax could be structured on an industry-wide basis and it might be imposed on all offending emissions, or both combined. If the tax were imposed industry-wide, then one would have to construct an industry-wide cost curve.

It is known that the firm should reduce E units of emission, from the analysis so far. The level of the tax should be C dollars. The reasoning is that we know that it costs less than C/unit to eliminate emissions up to point E. Therefore, the first will prefer to avoid the tax and, e.g., install the scrubber, thereby eliminating the economically appropriate E units of emission. Thus, the tax will not be collected, but the problem will be cured and the firm will be able to select the most appropriate technology to eliminate the emission.

The major flaw is lack of information. Constructing marginal cost curves and marginal benefit curves for even one firm is hard. The sources of information may be imperfect and may be biased against providing honest answers. Also, technology can change; appraisals of benefits can change as a result of new health data, and so forth.

Implications of Defective Information

The tax may have been set wrong because the marginal costs turn out to have been understated. Accordingly, the real cost curve is reflected by MC1. Here, the tax will not cause the firm to avoid pollution at all. It always costs the firm over C dollars/unit to eliminate the emission.

Case 1: Government underestimates marginal costs

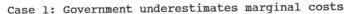
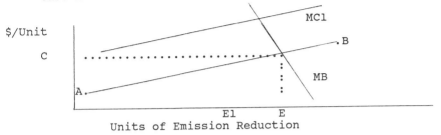

$/Unit

C

A

E1   E

Units of Emission Reduction

Case 2: Government overestimates marginal costs

If the real cost curve is in fact MC2, then the tax will induce the firm to eliminate _more_ emissions than is socially desirable in the sense of allocating excessive resources. The excess is E2 _minus_ E.

$/Unit

C

A

E   E2

Units of Emission Reduction

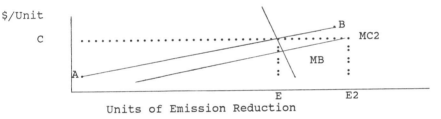

This problem takes its most exaggerated form when the marginal cost curves are fairly flat and the marginal benefit curve is steep. The most important policy implications are (i) these taxes are extremely difficult, if not impossible, to structure precisely and (ii) is that the choice of whether to use a tax or marketable rights, depends on the policy maker's view of the facts. Marketable rights assure that a particular level of pollution abatement will be achieved, but the cost is not known; taxes assure that the costs will be known, but the level of abatement will not be.

Taxes or pollution trading rights

Taxes are better than marketable rights because one knows the cost in advance of achieving a particular given level of compliance. Marketable rights are better because one knows the level of compliance. The choice between taxes and marketable rights depends upon estimates about how emissions and abatement costs are related to the damage caused by pollution. If pollution harm increases gradually as emissions increase but that abatement costs take a sharp jump upward at a certain abatement level, one might favor a tax, because a small mistake in tax level would not cause an enormous increase in pollution or impose enormous unknown costs. If pollution harm increases suddenly and dramatically when a threshold level is passed but that abatement costs rise gradually, one favors marketable rights. It would be important to assure that the threshold pollution level not be reached, while a mistaken estimate of the cost of reducing pollution to that level would not be very important. (S. Breyer 1982).

Implication of Microeconomic Analysis of Environmental Taxes and Alternative Energies

I believe a tax on output to correct market failure is a fantasy. Even in an extreme scenario, the business world will not stand for (and will, I believe, successfully argue against) a tax that is imposed in lieu of correcting the environmental deficiency solely in order to improve the workings of the marketplace. It does, however, create a foundation for arguing the case for benign sources of energy from an economically conservative vantage point by observing that failure to account for the full cost of "dirty" energy violates the rules of the laissez faire marketplace. As such, it is a powerful tool. The tax on pollution likewise makes the case for environmentallly clean energy in the obscure sense that it illustrates the great difficulty in establishing the socially acceptable level of pollution.

Moreover, emissions taxes are already here. So far, Sweden, the Netherlands, Finland and Denmark have all enacted carbon taxes. The taxes are diverse in nature and are subject to numerous technical criticisms and difficulties. Their fundamental impact, however, is to cause purchasers of carbon based fuels to internalize (however imprecisely) the environmental costs of the use of carbon based fuels. Public pressure to expand the scope of such taxes can only assist the development of alternative energy. In fact, in order to equalize competitive opportunities, the EC seems forced to either repudiate such taxes or enact a uniform carbon tax.

NEGATIVE TAXES

Economic theory does not inherently reject the concept of imposing mandatory pollution standards and accompanying them with direct subsidies so that the polluter can choose the technological change that is best adapted to its circumstances. (The OECD's "polluter pays" principle does, however, reject this approach.) In form, a subsidy can be structured as a tax credit for investments in changing technologies. Also, in a jurisdiction without a tax system, credits have no meaning; direct subsidies must instead be used.

There are policy reasons for not using credits or other tax incentives in the place of direct subsidies. The argument is that tax incentives can be enacted without the visibility (hence accountability) of direct grants. Once in place, the argument goes, they are unduly protected from repeal in the sense that they are not automatically subject to annual scrutiny. By contrast, budget appropriations require annual direct action. Finally, tax incentives tend to be regressive in the sense that if they depend on deductions in a progressive tax system, they are most beneficial to high income-tax bracket taxpayers. Credits are neutral in this respect, but they are useless to people who do not pay taxes (because they are poor), unless the credit can actually produce a refund in the absence of tax liability.

INITIATIVES TO ENCOURAGE ALTERNATIVE ENERGIES

The US has a history of using tax stimuli to encourage the development of alternative energies. The ideological argument for using tax benefits instead of direct grants is that the administrative costs of using taxes is lower than the administrative costs of direct grants. Former president Reagan's antipathy for alternative energy resulted in their demise, subject to a small handful of incentives for business energy credits, only one of which is meaningful. The following brief discussion outlines the US initiatives, most of which are extinct. They are traditional examples of using the tax system to provide individual taxpayers the incentive to change their behavior, and can be reproduced easily. There is uncertainty as to their effectiveness. Twelve reported GAO studies tend to show that only relatively large tax credits are effective. (BNA Daily Report for Executives, G-5 (March 18, 1982))

Tax-free financing for alternative energy facilities

Prior US law permitted the issuance of tax-exempt bonds whose proceeds were used to finance the construction of hydroelectric facilities. This offered a potentially major favoritism for hydro projects over fossil fuel projects. Holder of the bonds were not subject to federal income taxes on the interest income the bonds produced.

Residential energy credits

Prior law granted a generous credit, equal to 40% of the purchase price, of renewable energy devices used for heating, cooling, or producing hot water in the taxpayer's principal residence. The device could rely on wind, solar, or geothermal energy. Although hard data on the precise impact of the credit are apparently unavailable, it was clear at the time that the credit had an enormous impact on consumer behavior. It appears that the elimination of the credit almost wiped out the solar hot water industry. (Whatever Happened to Solar?, Garbage p.24 (Jan/Feb 1991)). It seems that the 1978 law (Public Utilities Regulatory Policies Act) forcing utilities to purchase electricity produced by alternative energies has at least partly reversed that trend.

In addition, there was a 15% credit for so-called "qualified energy conservation expenditures", meaning virtually any purchase that could increase the heating or cooling efficiency of the taxpayer's primary residence. The credit was limited to the first $2,000 of

2353

expenditures. Again, hard data is unavailable, but the consensus is that the credit had a major impact on fuel conservation.

Credit for producing fuels from nonconventional sources

Section 29 of the Internal Revenue Code grants a credit for getting fuel from difficult sources, especially coal seams, tar sands and geopressured brine. At first it appears to be a mere giveaway to particular industries, but it in fact serves an important function with respect to coal. Coal seams contain a vast amount of methane, a severe greenhouse gas. Most coal operators let it go. If there is a sufficient incentive, they will be encouraged to trap and sell the methane. There are significant "section 29 projects" in the USA and elsewhere. My discussions with business people convince me the credit is needed until natural gas prices rise significantly.

Business energy credits

At this point the US has only a small number of business-related energy incentives. The rate is 10% of the expenditures (i) for installing equipment to trap geothermal steam for generating electricity and (ii) is for solar power equipment used to heat, cool or produce electricity in a business setting. The credits are ringed with restrictions. Under prior law there was seemingly endless list of innovative technologies that could qualify for the credit.

Duration of credits

A major issue is how long to continue a credit that is designed to stimulate new technologies. In a sense, this is an old-fashioned question, because the debate arose in a setting that ignored the fundamental issue of internalizing costs, which produces an argument for setting taxes on other forms of energy sufficient to force their cost to the public to include the full measure of their implicit damage. Nevertheless, the accepted dogma in the US is that the credit should not be indefinite, but rather should be sufficiently extended (and predictable) to allow the necessary subsidy, but not long enough to be addictive. In the US experience, at the height of the windpower boom in the late 1970's and early 1980's there were numerous manufacturers of wind turbines. Now there are two viable companies. That may be the right answer in a laissez-faire world if the other energy sources are appropriately regulated.

STEPS TO AVOID

There are risks associated with touting tax incentives as optimal solutions that proponents of alternative energy should be wary of, based on US experience:

1. If the tax incentive is capable of being syndicated and sold to wealthy investors as a "tax shelter", the technology will lose credibility. The US windpower industry has suffered this fate.

2. Do not base the credit solely on the monetary investment; base it

at least on performance where possible. This will add credibility to the program and will help stimulate improved technology.

3. Be careful not to allow the production of energy from alternative sources to result in negative taxes to commercial producers. This will poison the attitude of legislators unless they advised of the de facto net subsidy in advance. This problem has recently arisen in the US with respect to extending tax credits for alternative energy.

4. Try to make the duration of incentives stable in advance, rather than getting an attractive incentive for an unknown period. Business people must be able to plan their projects. This calls for fiscal stability. If not, investors will require unduly high returns in order to account for risk of unanticipated withdrawal of tax benefits.

REFERENCES

1. S. Breyer, Regulation and Its Reform, p. 273 (Harvard 1982).
2. S. Gaines & R. Westin, Taxation for Environmental Protection: A Multinational Study, with Eriksson, Hertzog, Tiley, Williams and von Zezschwitz.
3. Whatever Happened to Solar?, Garbage p.24 (Jan/Feb 1991).

2355

PAPER WITHDRAWN AT PRESS STAGE

Pages 2356 - 2400 were reserved for the paper entitled "Towards a Fossil Free Energy Future" by Mr Stuart Boyle, Greenpeace House, London, U.K.

We regret that this paper was withdrawn at press stage due to a misunderstanding between the author and Greenpeace.

ENERGY-METEOROLOGY: A new discipline?

L.E. OLSSON

World Meteorological Organization, C.P. 2300
CH-1211 Geneva 2, Switzerland

ABSTRACT

The energy sector accounts for over 50% of the radiative forcing due to the anthropogenic emissions of greenhouse gases into the atmosphere. The Intergovernmental Panel on Climate Change (IPCC) is suggesting that the relative importance of the energy sector in accelerating climate change will increase. The most efficient ways in which to reduce greenhouse gas emissions are through energy conservation, including improved energy efficiency and through a change towards non-fossil fuel technologies. New and Renewable Sources of Energy, including hydropower, solar and wind, offer special advantages in this respect. Many of these energy systems draw on climate related resources. The World Meteorological Organization (WMO), together with its national Members, has a wealth of information and knowledge in the area of energy-meteorology available for service to the energy sector.

KEYWORDS

Energy-meteorology; climate; climate change; hydropower; solar energy; wind energy.

Introduction

At the First World Renewable Energy Congress in 1990, I presented the concept of Energy-meteorology. Within the World Meteorological Organization (WMO) and its Members, represented by the national Meteorological Services (NMS), activities related to energy have been in the foreground for many decades. Many so-called sustainable energy systems draw on a resource, which is observed, studied and described by professionals in climatology, hydrology and meteorology.

Within the family of the United Nations, UN-organizations have co-ordinated many of the activities relating to New and Renewable Sources of Energy (NRSE) in accordance with the recommendations made at the UN Conference on

NRSE held in Nairobi in 1981. At that conference WMO presented two basic publications relating to solar and wind energy respectively (WMO 1981). The Nairobi Programme of Action for Development and Utilization of NRSE has formed the basis for joint activities within the UN and thus also in the preparatory work for the UN Conference on Environment and Development (UNCED), held in June 1992. A major stepping stone in the development of energy strategies within WMO was the the Second World Climate Conference (SWCC) (WMO et al., 1990). The first assessment report of the Intergovernmental Panel on Climate Change (IPCC) was presented in 1990 and reviewed by the SWCC and a supplement was issued in 1992 in preparation for UNCED. WMO has been instrumental in co-ordinating the work in the area of energy-meteorology through the World Climate Programme. In the following paragraphs some of the most important developments relating to promotion and furthering the development of NRSE are discussed.

The Intergovernmental Panel on Climate Change (IPCC)

The IPCC was established by WMO and UNEP in 1988 in response to the growing concern for the potential adverse impacts of projected climate change. A major concern expressed in the IPCC reports is the increasing contribution from the energy sector to the so-called "radiative forcing" caused primarily by the emission of greenhouse gases. In the 1980's the energy sector accounted for over 50% of the radiative forcing caused by anthropogenic sources. Among the major conclusions forwarded to UNCED from IPCC (WMO/UNEP, 1992) may be noted:

"o emissions resulting from human activities are substantially increasing the atmospheric concentrations of the greenhouse gases: carbon dioxide, methane, chlorofluorocarbons, and nitrous oxide;

o the evidence from the modelling studies, from observations and the sensitivity analyses indicate that the sensitivity of global mean surface temperature to doubling CO_2 is unlikely to lie outside the range 1.5° to 4.5°C;

o there are many uncertainties in our predictions particularly with regard to the timing, magnitude and regional patterns of climate change due to our incomplete understanding;

o global mean surface air temperature has increased by 0.3° to 0.6°C over the last 100 years;

o the size of this warming is broadly consistent with predictions of climate models, but it is also of the same magnitude as natural climate variability. Thus the observed increase could be largely due to this natural variability; alternatively this variability and other human factors could have offset a still larger human-induced greenhouse warming;

o the unequivocal detection of the enhanced greenhouse effect from observations is not likely for a decade or more."

This implies a future warming at rates of about 0.3°C/decade and a corresponding sea level rise of between 2-4 cm/decade due to oceanic thermal expansion only.

2402

In attempting to make a comprehensive assessment of technological options for mitigating global warming, IPCC presents, among others, the following tentative findings:

"a)   Energy conservation and improved efficiency in the production, conversion distribution and end-use of energy is one of the most effective options available now and in the future. System restructuring, such as energy cascading and infrastructure improvement, has promising potential.

b)   The technologies to capture and sequester CO_2 from fossil fuel combustion deserve investigation, considering the expected continuing dependence on fossil fuels as primary energy sources.

c)   Nuclear power has the technological potential to be one of the major energy sources in the next century, but faces various socio-economic, security and safety constraints.

d)   There are various existing and promising non-fossil fuel technologies such as photovoltaics (PV), wind, hydropower, geothermal, biomass and solar thermal systems. PV may be first applied in a small scale on roofs, and then in larger scale on deserts and ocean surface if energy distribution technology can be advanced substantially.

e)   The physical potential of biomass for energy use is apparently high but in some regions, competition in land use for food may limit its production. Environmentally sound intensification of agriculture for more efficient production of food may be considered."

IPCC concludes that response strategies include:

"-   use of cleaner, more efficient sources with lower or no emissions of greenhouse gases."

Although NRSE are referred to several times by IPCC, it is often referred to in general terms such as "energy efficient, clean technologies, etc".

It should be obvious, however, that the results of the IPCC assessments, in general, support development of NRSE. This includes hydropower, solar energy, biomass, etc.

The Second World Climate Conference (SWCC)

The main objectives of the SWCC were:

"-   To create an awareness through specific case studies or examples drawn from experience in the World Climate Programme of the economic impact of climate and benefits from climate applications;

-     To assess the current state of knowledge on the global issues of climate change and greenhouse gases and requirements for future scientific activity and implications for public policy."

This was done considering the results of the IPCC assessments. The SWCC was organized in two parts, i.e. a Scientific and Technical Session, which led to a Conference Statement and a Ministerial Session which adopted a Ministerial Declaration. More than 1000 participants representing 137 countries met for six days in Geneva in the Fall of 1990. The most relevant outcome for the NRSE sector may be found in the proceedings and in the Conference Statement:

"In order to stabilize atmospheric concentrations of greenhouse gases while allowing for growth in emissions from developing countries, industrialized countries must implement reductions even greater than those required, on average, for the globe as a whole. However, even where very large technical and economic opportunities have been identified for reducing energy-related greenhouse gas emissions, and even where there are significant and multiple benefits associated with these measures, implementation is being slowed and sometimes prevented by a host of barriers. These barriers exist at all levels - at the level of consumers, energy equipment manufacturers and suppliers, industries, utilities, and governments. Overcoming the barriers obstructing least-cost approaches to meeting energy demands will require responses from all parts of society - individual consumers, industry, governments, and non-governmental organizations.

Developing countries also have an important role in limiting climate change. Maintaining development as a principal objective, energy and development paths can be chosen that have the additional benefit of minimizing radiative forcing."

In the Ministerial Declaration it is recommended that:

"Financial resources channelled to developing countries should, inter alia, be directed to:

(i)   Promoting efficient use of energy, development of lower and non-greenhouse gas emitting energy technologies and paying special attention to safe and clean new and renewable sources of energy;

(ii)   Arranging expeditious transfer of the best available environmentally sound technology on a fair and most favourable basis to developing countries and promoting rapid development of such technology in these countries."

Final remarks

WMO, together with many other UN agencies, is more or less directly involved in promoting the development of NRSE. The recent increased concern for adverse impact of climate change has led to an improved general awareness of the potentials for NRSE. Sustainable development implies development in harmony with the environment and that certainly includes the climate, be it on a global, regional or local scale. There are many links between the energy-sector and the different aspects of weather, climate and hydrology. WMO and its Members, represented by the national Meteorological Services, are involved in many activities which would have direct relation to the further development of NRSE. This includes provision of basic data

and analyses for assessing the potential for hydro- and wind-power, solar
energy and biomass production, as well as for providing supporting service
during the operational phase. These services need to be guided by the
requirements of the energy sector and this congress is an excellent attempt
along this line.

REFERENCES

WMO (1981). Meteorological aspects of the utilization of solar radiation
   as an energy source. WMO Technical Note No. 172 (available in English
   and French, Spanish version in preparation).
WMO (1981). Meteorological aspects of the utilization of wind as an energy
   source. WMO Technical Note No. 175 (available in English and Spanish,
   French version in preparation).
WMO (1990). Climate Change: Science, Impacts and Policy. Proceedings of
   the Second World Climate Conference (edited by J. Jäger and
   H.L. Ferguson). Cambridge University Press.
WMO/UNEP (1990). Climate Change: The IPCC Scientific Assessment. Report
   of Working Group I (edited by J.T. Houghton, G.J. Jenkins and
   J.J. Ephraums). Cambridge University Press.
WMO/UNEP (1990). Climate Change: The IPCC Impacts Assessment. Report
   of Working Group II (edited by W.J. McG. Tegart, G.W. Sheldon and
   D.C. Griffiths). Cambridge University Press.
WMO/UNEP (1990). Climate Change: The IPCC Response Strategies. Report
   of Working Group III. Cambridge University Press.

DEVELOPMENT, ENERGY AND ENVIRONMENT:
A LONG TERM PERSPECTIVE

Benjamin DESSUS* and François PHARABOD**

*PIRSEM, National Centre for Scientific Research, 4 rue Las Cases, 75007 Paris, France
**CPE, Ministry of Research and Technology, 1 rue Descartes, 75005 Paris, France

ABSTRACT

After decades of strong growth, the next century might be that in which the world population is stabilized at around 11 billion inhabitants. The development hoped for by the Third World and the probable consequencies on the world global environment are major issues for the next century. Toward 2100, how should we evaluate the evolution of energy needs and supplies. Is there any room for renewables in the energy mix ?

The proposed NOE scenario shows that it is possible to avoid both greenhouse effect and cumulation of nuclear waste threats; it requires a very important effort in energy efficiency and conservation for northern countries, an effort which can be achieved with the existing technologies, the massive transfer of efficient energy technologies for the development of southern countries, and, definitely, the mobilization of renewable energies.

An evaluation of the potential of these renewable energies has been performed; taking into account economically available technologies and local conditions make it possible to define reasonable annual energy amounts relevant for each technology. It appears that this potential can contribute significantly to the satisfaction of energy needs, and to the protection of the environment, both by decentralized applications as well as centralized equipments.

KEYWORDS

Energy, scenario, renewable energy, long term, environment, development.

INTRODUCTION

The dawning of the 1990s is just the right time for millenarian stirrings. The depletion of natural resources, the spread of pollution over the earth's surface, the rising of water levels, the swallowing up of new atlantides, and the climatic revolution each occupy a prime position among the numerous concerns.

The development hoped for by the Third World, whose runaway increase in population will continue for a long time to come, the energy risks which may be engendered by the possible dissemination of nuclear energy to every region of the globe, the probable consequences of the increase in the greenhouse effect due to man-produced gas emissions, all of the above lead us to raise three questions regarding the next century:

- At this future time, might we not find ourselves faced with an energy shortage?
- In order to avoid a possible energy shortage, might we not go from Charybde to Scylla by adopting modes of energy production that are dangerous for both nature and humanity?
- Does not the development of new countries threaten to throw the planet completely off-balance, from both an energy and ecological point of view?

1990 GLOBAL SITUATION AND TRENDS

From the point of view of energy and environment the present situation is clearly marked by different alarming statements.

Wasting and penury of energy

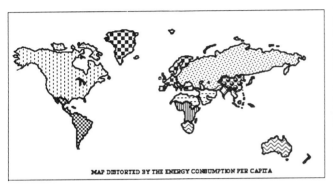

MAP DISTORTED BY THE ENERGY CONSUMPTION PER CAPITA

These issues are fully illustrated by Fig. 1 where the areas of the ten main regions of the world have been amended by their annual per capita energy consumption. In 1990, North American people consume 25 times more energy than Indian people, respectively 7.6 toe and .3 toe par capita (Pharabod, 1989); more than a third of the whole population of the world has no access to the basic services of electricity (lighting, cooling, water pumping).

Fig. 1. Disparities in energy consumptions for ten regions of the world

Global environment threats

Fossil fuel combustion increases each year the greenhouse gas concentration in the atmosphere. CO_2 emissions of 5.4 billion tons of carbon are reported for 1990 from fossil fuel combustion (80% from the North, 20% from the South); the deforestation of more than 10 million hectares each year, mainly due to agricultural needs in developing countries, adds 1.6 billion tons of carbon to CO_2 emissions; oil pollution spreads over the whole marine areas; the environment problems due to nuclear long life waste are not solved.

A rapid outlook of principal trends for the 60 next years, assuming a "business as usual" behaviour associated with a reasonable development and demographic prospectives for southern and eastern countries, shows a really alarming evolution (Frisch, 1989): very startling problems should appear for oil before 2050, as illustrated on Fig. 2, due to the depletion of the proven reserves everywhere but in the Middle-East region; greenhouse gas emissions should keep on growing up and their concentration in the atmosphere still more; the dissemination of nuclear plants will give rise to new risks both from the point of view of security (weapons and accidents) and of long term activity nuclear waste accumulation.

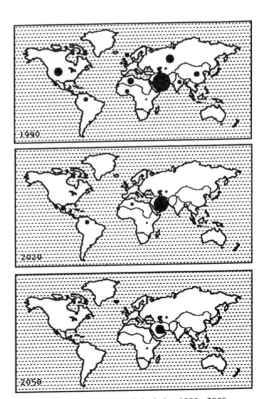

Fig. 2. World oil depletion 1990 - 2050

2407

THE PROSPECTIVE APPROACH

Faced with these threats, the aim of the prospective approach is to provide descriptions of diversified scenarios, to open new possibilities to get round the principal impediments and risks, in order to enlight strategic decisions for the 21st century.

At this point there are two very divergent ways of looking at energy consumption laws. The most classical one proceeds from the following line of arguments:

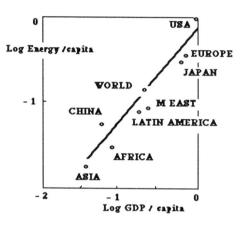

- The first one is the assertion of an experimental historical "law"; it says that the elasticity between energy consumption and Gross Domestic Product (GDP) is 1. The 1973-1986 period when that elasticity has been quite different (zero or negative for most OECD countries) has to be considered as an historical accident.

- The second step is to place on a graph the GDP per capita and the energy per capita of the different regions of the world at a given time. When using a Log Log scale (Fig. 3) the curve appears as linear: Log GDP= K.Log E. So, to reach a given GDP per capita each country has to "climb" that straight line. Finally, the energy consumption of a given country at a future horizon is simply given by a formula of the following type:

E= E$_0$.(1 + pop increase).(1 + GDP per cap increase) where E$_0$ is the initial consumption of the country.

Fig. 3. Energy and development: a somehow static vision

The problem with that timeless way of reasoning is that it is totally missing the technical progress which occurs with time. Indeed coming from the degree of development of India to the one of the United States would need time, perhaps 50 or 80 years. Meanwhile it is quite sure that technologies used will evolve. For example if Europe had developed its industry to the present extent with the 1900s technologies it would have probably required three times the energy consumption that we use to day!

When one considers the recent history of industrial countries it appears clearly that:

- First, the energy consumption of each of the services to the final consumer reduces quickly: from 1950 to 1990 the new household energy consumption in France has been divided by a factor 3.5, the oil specific consumption of equivalent cars by a factor 2.2, the lighting consumption by a factor 10, the one of TV monitors by a factor 20.

- Second, the manufacturing of the tools which deliver these services is less and less energy consuming: 6 toe were necessary to build an individual house of 110 square meters in 1950 in France, 4 toe were necessary for a much more confortable one of the same size in 1990; or .8 toe to manufacture a car (Peugeot 203) in 1950 versus .5 toe for the ECO 2000 prototype of the same company. This has been made possible with technological progress in materials manufacturing and implementation techniques.

- Third, these progress contribute to reduce the manufacturing costs; between 1972 and 1988 in France, mean prices of refrigerators have been reduced of 20% and their mean energy consumption of 33% (Barbier, 1991); from 1977 to 1990 price's index of house building in France has been reduced of 14% and energy consumption of 50% (Olive, 1991). This last statement is of special interest since it contradicts a trivial speech about the impossibility for developing countries to use energy efficient technologies because they are too expensive.

So it appears that a dynamic view of the technical progress has to be taken into account in order to make some projections on the energy consumption associated with development. The long term historical analysis (Martin, 1988) presented on Fig. 4 gives a confirmation to this thesis; the energy intensity (measured in toes per 1000 $ of GDP) of each country while it is developing begins to increase, goes through a maximum and then decreases. The heavy infrastructures setting explains this initial increase; the technical progress and the saturation of high

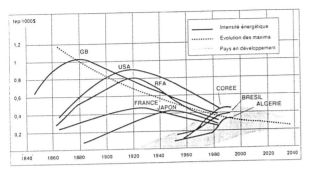

energy content goods which comes with the raising of the standards of living both contribute to the decreasing which follows.

Furthermore, the successive maxima find their place on a continuously decreasing curve; this means simply that the development of new countries occurs with more efficient technologies than the one of their predecessors.

Fig. 4. Energy intensity and development

This alternative analysis justifies the study of new scenarios including both a real development of the third world and low energetic profiles.

THE NOE SCENARIO, "NEW OPTIONS FOR ENERGY"

We have studied different scenarios (Dessus and Pharabod, 1990) in order to quantify the main energy issues for the next century: depletion of oil resources, development of road transportation, greenhouse effect, nuclear dissemination... Among those scenarios there is one, the NOE scenario, which strives to mitigate simultaneously two main environmental threats of the future, namely the increase of CO_2 in the atmosphere and the cumulation of nuclear waste on earth. There is no reason indeed, in the present state of our knowledge, to give preference to one of these two threats:

- Either we let the greenhouse effect increase in order to avoid the use of nuclear energy, by wagering that we will find solutions for eliminating CO_2, or that we will adapt ourselves to the probable climatic disruptions,
- Or we let long-life nuclear waste accumulate at a parabolic rhythm in order to avoid the greenhouse effect, by wagering that we or future generations will find solutions for the permanent decontamination of these waste.

The NOE scenario assumes a willful sustained effort towards energy conservation and diversification of energy resources. In particular, stress is placed on the use of renewable energy resources, in substitution for fossil fuel resources.

The set of constraining assumptions retained pertain to:

- The regional demographic evolutions, until 2100, according to United Nations estimations, which appear as the main factor of demand (Fig. 5);

- The estimation of fossil fuel and renewable resources; this estimation is based on the World Energy Council publications for fossil fuels and completed for renewable energy resources by our analysis (next page);

- The strong connection between the energy system and the environment which induces external constraints on the structure of world energy supplies.

POPULATION	1960	1985	2020	2060	2100
(in millions)					
North America	200	265	330	380	400
Europe	360	430	450	450	450
Japan Australia N Zeal	125	170	230	240	250
USSR Central Europe	310	390	490	530	550
All of the N countries	995	1255	1500	1600	1650
Latin America	215	410	710	1100	1250
N Africa Middle East	145	230	580	800	900
Africa	195	390	1140	1700	2000
India	420	760	1310	1700	1800
China	660	1040	1360	1600	1600
Asia Oceania	390	760	1400	1700	1800
All of the S countries	2025	3590	6500	8600	9350
World Total (UN source)	3020	4845	8000	10200	11000

Fig. 5. Evolution of the world population

Environmental constraints lead to the following decisions:
- the return, at the latest by 2100, of fossil fuels induced CO_2 emission levels lower or equal to the biosphere's maximal capacity of absorption supposed to be 3 gigatons of carbon per year;
- the halt of all nuclear fission energy production (at least with present technologies), at the latest by 2100.

The first phase of the explorative approach therefore consists in estimating the evolution of the demand for energy in each of the ten regions retained. Goldemberg *et al.* (1988) have shown how technological evolution and the spread of effective energy measures could make it possible to reduce by 2020 the consumption level per inhabitant in the northern countries by half. Their analysis also shows how the southern countries can afford a sustainable development which would give access to satisfactory living conditions, without their energy needs surpassing their equipment and financial capacities. We will retain the results of this study, but we shifted to the middle of next century an average reduction of one-half for the consumption level per inhabitant in industrialized countries (2.4 toe per capita) and an average rise to 1 toe per capita in the southern countries.

The average consumption values, per inhabitant, are then broken up, taking into account the specific characteristics of the ten regions considered. The total demand in primary energy per region is deduced by taking account of the anticipated growth of populations.

In these conditions the global evolution of the energy consumption of North and South between 1990 and 2100 is illustrated by Fig. 6. This figure is summing the results of the 4 regions of the North (North America, Europe, Japan Australia New Zeland, CEI) and 6 for the South (Latin America, North Africa Middle East, Africa, India, China, Asia Oceania)

It is interesting to note the future importance of the southern energy demand which overtakes the northern one in 2020 even with the modest energy per capita which has been assumed in that scenario. This is the consequence of the rapid growth of the southern population.

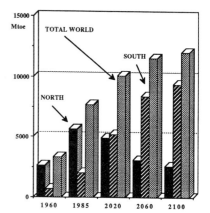

Fig. 6. NOE energy demand projection

In the second phase of this exploratory approach, we try to build a structure of supplies meeting the demand as it is defined. The iteration proceeds in three steps:

- Estimation of fossil fuel energy supplies using observed trends, account taken of a self-reduction in fossil-fuel energy consumption in industrialized countries which would occur, first, out of fear of the greenhouse effect (regulations and taxes), then because of the progressive depletion of crude oil and gas reserves;

- Sustained mobilization of renewable energy resources, first depending on their regional potential, with a dynamics which takes account of the values observed in the past, then with a partial energy exploitation of uninhabited regions (solar energy in the Sahara, hydraulics in the Himalayas, wind energy in Greenland);

- The balance of energy requirements is met by nuclear power, while assuring the feasibility of the resulting dynamics and a progressive reduction to zero after 2020.

In order to achieve this phase and assess the contribution of renewables we have developed a specific study of the world renewable energy potential (Dessus *et al.*, 1992).

WORLD POTENTIAL OF RENEWABLE ENERGIES

The role that renewable energies could play in the world's regional and global energy balances gives rise to varied and often quite different evaluations including almost unbounded ones. Reliable information is now available on the different technologies and their economic costs. But, as the commercial development of these renewables is generally very weak, the arguments for postponing their real possible world and regional impact are frequently confused with short term market type considerations. To overcome that difficulty we made a comprehensive analysis, region by region, of the actually accessible renewable energies at the horizon 2000 with the same methodology as the one employed to derive "proven fossil energy reserves" from "energy resources" in which resources are defined by quantitative information on physical potential, when reserves take into account technical and economical accessibility.

Apart from the unique renewable character of the reserves that we have to keep in mind, going from "resources" to "reserves" in renewable energy raises specific problems: first these resources are fluctuating with time, second they are diluted in space and not readily storable or transportable. Then it appears that, in addition to physical resources and economic considerations, it is necessary to take into account the presence of populations or activities near enough to be able to benefit of these diluted and somehow volatile energies, at least during the coming two decades.

Geographical, social and economic considerations led us to share the world in 22 sufficiently homogeneous regions. Ten principal technologies have been considered of global interest since they have achieved technical and economical demonstrations (at least in favourable conditions) and are presumably able to fit for needs of significative sectors in a number of regions of the world.

Those technologies are: solar water heating, decentralized photovoltaic electricity, grid connected solar electricity (PV and thermal), hydroelectricity, decentralized electricity from wind, grid connected wind electricity, energy from wood, energy from urban waste, energy from rural waste, biomass energy crops.

It is quite evident that other technologies should have been considered such as waves, offshore wind or tidal energy; but we have considered that, for the time being, energy from waves and offshore wind have not been industrially demonstrated and tidal, although demonstrated, is a very pin point resource through the world. This is not the case for a longer term projection. It stands to reason that each new technological development will implicate new classifications and new evaluations of the different estimated regional annual renewable "reserves" as well known for fossil fuels (for example if space power satellites or remote solar hydrogen plants are demonstrated).

Then again the population growth and the GDP increase of the considered regions will obviously have a direct influence on energy needs; this implies the possibility for renewable reserves, directly related to sectorial needs, to grow up with time.

It is important to notice that the estimation presented below is neither a market analysis nor a definitive evaluation of the whole energy reserve accessible in a long term perspective, but a realistic approach of the actually accessible renewables potential for 1990 - 2000, the "nineties".

Renewable theoretical resources can be defined as the amount of energy which can be produced from a natural flux of energy without any consideration of cost for the technological conversion.

Regional annual reserves of each renewable technology is derived from theoretical resources using the knowledge or the evaluation of different parameters. The first one is the competitiveness which is assumed for each technology with specified equipment costs, given discount rate, and comparative prices of fossil fuels. In these conditions the competitiveness is no more a limit to the concerned reserve. Other considerations appear which are specific of each of the technologies: local productivity, system efficiency, potentially concerned population, regional needs per capita, maximum tolerable fluctuating energy ratio on the grid...

Global results for the 1990s are illustrated by Fig. 7 for the ten main regions of the world. Accessible 1990s renewable energy annual reserves are never negligible whatever region is considered. 43% of the world 1985 primary energy supplies (7670 Mtoe) could be satisfied by these reserves: 24 % in Market Economy Industrialized Countries, 28 % in ex-Centrally Planned Industrialized Countries, 85 % in the Third World.

The major potential contributions come from wood: 1650 Mtoe compared with the 780 Mtoe consumption in 1985. We want to emphasize the fact that the increase of wood consumption is essentially due to efficient use of "commercial wood", which reaches 980 Mtoe versus 670 Mtoe for non commercial wood compared with figures for 1985, respectively 175 and 605 Mtoe. The share of non commercial wood has been choosen in continuity with 1985 figures to take into account the consumption practices of most developing countries. Waste takes the second place (500 Mtoe), far before solar water heating (100 Mtoe).

Primary electricity annual reserves come first from hydro (3990 TWh versus 2030 TWh production of 1985), second from grid connected wind and solar (440 TWh) and last from remote off grid wind and solar (260 TWh).

This last figure, even if it may appear as quite modest in the global energy balance is in fact most important since it corresponds to a unique chance of electricity access for more than two billion people which have practically no hope to be connected to the grid for the next 20 years. It must be remembered that small scale hydro and biomass-derived gas engines, already accounted for in the bulk of hydro and wood, would play an equivalent urgent role for isolated populations, but with the same modest impact on the global energy balance. Comparing this potential with the contribution of renewable in the 1985 world energy balance, it appears that the reserve existing for the nineties (3360 Mtoe) is 2.5 times the present contribution of renewable (1340 Mtoe).

2411

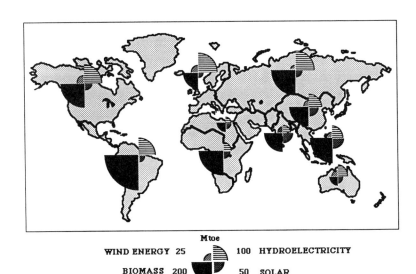

WIND ENERGY 25   Mtoe   100 HYDROELECTRICITY

BIOMASS 200   50 SOLAR

Fig. 7. Map of nineties renewable annual reserves

An extension to the year 2020 shows that the combination of technological progress, population growth and development of southern regions give the opportunity to extend the renewable potential from 3360 Mtoes to 4850 Mtoes without any technical breakthrough over the period. That potential is to compare with the energy demand in 2020, that is 10100 Mtoes in the scenario NOE. After 2050 we may take into account a partial equipment of uninhabited zones capable of furnishing the energy necessary for manufacturing materials or new storable and transportable energy carriers. In this approach we may envisage in a long term perspective an international trade of energy carriers derived from renewable energies, and first of all for biofuels and hydrogen. The renewable energy potential could then reach 8300 Mtoes in 2100, more than the two thirds of the needs identified in the NOE scenario at the end of the 21st century.

ENERGY SUPPLIES FOR A SUSTAINABLE DEVELOPMENT

With the NOE hypotheses and strategy, world energy supplies (Fig. 8) are established for each of the ten regions till 2100, leading to the following main elements:

- For the northern countries, the very large effort towards energy conservation (4800 Mtoe), the very rapid fall in the recourse to fossil fuel energy resources, which would be reserved as much as possible to the southern countries, a moderate growth (30%) of nuclear energy until 2020 before its continuous fall until 2100. For renewable energies, an increase of the contribution of biomass to supplies, especially in the form of fuel, sometimes imported from other production zones, the progressive use of the very important hydraulic reserves in the CEI (ex-USSR) and the rise in solar and wind energy, which would reach significant values after 2060.

- For the southern countries, a major initial increase in the recourse to fossil fuel energy resources, followed by the beginning of a decrease in 2050 and stabilization at approximately 3000 Mtep in 2100, a nuclear energy supply which remains confined to a

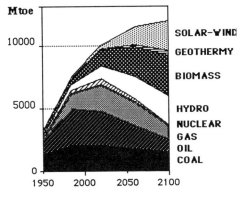

Fig. 8. Evolution of the world energy supplies

few countries in which it is already present before phasing out, an international effort made towards equipping the numerous hydraulic sites in Latin America, Asia and Africa. Nevertheless, it is biomass and solar energy that would play, in the long term, the major role in these countries reaching nearly half of the 2100 world energy supply, 27% for biomass and 20% for solar energy.

The NOE scenario's impact on the environment appears in Fig. 9. The concentration of carbon dioxide in the atmosphere would be stabilized around 470 ppmv (345 ppmv in 1985), without reaching the level at which it would be doubled. The stock of nuclear waste would reach a ceiling at .5 million tons before the end of the century. In case of the "business as usual" evolution, CO_2 concentration should double before 2050, cumulation of nuclear waste reach 2 million tons in 2100 and keep growing, nuclear plants spread over several tens of countries.

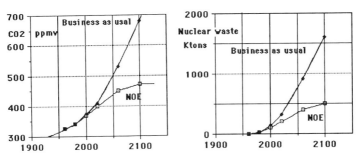

Fig. 9. CO_2 and nuclear waste cumulation

CONCLUSION

It has been demonstrated that a long term sustainable scenario for development, energy and environment, avoiding both grenhouse and nuclear threats is conceivable. Nevertheless it appears clearly:

- That this type of scenario imposes first, an important effort of energy conservation and second, a very determined mobilization of renewables,
- That it should not be realistic to imagine at the 2100 horizon a whole renewable civilisation,
- But that renewables have yet a very important role to play for the next century since they could supply more than 65% of the world's energy needs in 2100 and highly contribute to the preservation of our global environment.

In these conditions it's really time to begin with the development of energy conservation and of renewables on a world scale, including northern and southern countries. This development must be initialized by a large demonstrative effort which is mainly from the responsability of international financial bodies. On the other hand, these energy resources can now bring a significant improvment in living conditions of rural people of the South. In our mind it is one of the major political and scientific challenges to achieve the technological transfer and cooperation with southern countries on these items by the end of the 20th century.

REFERENCES

Barbier, C. (1991). *Evolution de la consommation d'énergie et du prix des réfrigérateurs sur le marché français depuis 20 ans*, Rapport DESS Economie et Politique de l'Energie, Université de Nanterre.
Dessus, B. and F. Pharabod (1990). Jérémie et Noé, deux scénarios énergétiques mondiaux à long terme. *Revue de l'Energie*, 421, 3-19.
Dessus, B., B. Devin and F. Pharabod (1992). World potential of renewable energies. *La Houille Blanche*, 1.
Frisch, J.R. (1989). *World Energy Horizons: 2000-2020*. Technip, Paris.
Goldemberg, J., T. Johansson, A.K.N. Reddy and R. Williams (1988). *Energy for a Sustainable World*. Wiley Eastern Ltd, New Delhi.
Martin, J.M. (1988). L'intensité énergétique de l'activité économique dans les pays industrialisés. *Economie et Société*, 41.
Olive, G. (1991). *Evolution de la consommation d'énergie et du prix des logements en France*. Rapport CNRS PIRSEM, Paris.
Pharabod, F. (1989). *Atlas Mondial de l'Energie - World Energy Atlas*. Aditech, Paris.

SOLAR ENERGY EDUCATION - AN IMPORTANT PART OF
WORLDWIDE SOLAR ENERGY ACTIVITIES

Lars Broman

Solar Energy Research Center
University College of Falun Borlänge
P. O. Box 10044, S-781 10 Borlänge, SWEDEN

ABSTRACT

Proper knowledge of solar energy is lacking on many levels of
society. Several groups and activities must be targetted to
overcome this major barrier. These include elementary and
secondary school education, vocational training, university
courses, educating decision makers, and educating the general
public. In making solar energy education understandable,
development of curricula and learning materials is important.
Curricula must have very different contents and approaches for
different target groups. Learning materials include textbooks,
laboratory equipment, video cassettes, exhibitions, demo.
installations, and even planetariums. International
co-operation in solar education is promoted by IASEE,
International Association for Solar Energy Education, which is
also ISES Working Group on Education.

KEYWORDS

Solar energy; education; laboratory; planetarium;
international co-operation.

INTRODUCTION

Fossil fuels have to be phased out both for environmental
reasons (carbon dioxide and greenhouse effect, sulfur and
nitrogen oxides, acid rain, etc.) and due to depletion of
non-renewable energy resources. Nuclear power, on the other
hand, is a highly controversial source of electric power,
including handling, safety and waste problems as well as high
costs. Furthermore, woodfuel shortage and desert growth is a
large problem in many parts of the world. Thus, renewable
energy sources are increasingly more interesting. Among these,
the direct utilization of energy from the sun is one of the
most promising outlooks for the future.

Advantages of solar energy:
* Very abundant and almost worldwide
* Environmentally benign
* Many applications are technologically mature
* Many applications are economically feasible
* Suitable for application in both developed and developing
  countries

Disadvantages of solar energy:
* It is a diluted energy source
* Some applications still need technical development
* Some applications are more expensive than competing energy
  sources
* Knowledge on solar energy is lacking on many levels in
  society
* It does not always fit easily into existing energy systems

These disadvantages are possible to overcome and educational
efforts will play a big part in this undertaking, especially
the fourth point. The problem that solar energy does not fit
into existing energy systems might in the future be overcome
by for example converting it into hydrogen.

The present paper is largely (but not totally) based on the
author's own experiences (Broman and Ott 1988 and 1991; Broman
1990a; Broman 1991a and b). Its purpose is to identify the
different groups to which educational activities need to be
directed as well as giving a comprehensive review of some
important methods that can be utilized.

ELEMENTARY AND SECONDARY SCHOOL EDUCATION

Basic knowledge of present and future energy systems and
energy conversion technologies has its place in school
curricula at many levels in physical science as well as social
science classes. Furthermore, some students will continue to
polytechnic schools and universities, and solar energy physics
(heat transfer, radiation, photovoltaic effects, etc.) is an
important prerequisite for these students (Ott 1990).

Appropriate science teaching requires not only textbooks but
also adequate laboratory equipment. This applies to solar
energy as well, and pedagogical laboratory experiments must be
included in the courses.

One example of how this can be achieved is the electricity
experiment kit developed at SERC (Broman, 1990b). Using a
board with twenty connection points, a large number of
different basic and medium-advanced electric experiments can
be performed. The kit is powered solely by PV charged NiCd
cells, and a number of photovoltaic experiments are included
in the manual. It is remarkable how much could be done within
a shoestring budget provided the teacher knows how! Other
examples are given below.

VOCATIONAL TRAINING

Skilled technicians are needed for manufacturing, installation and maintenance of various kinds of solar energy installations. Excellent results from training of PV equipment maintenance technicians on South Pacific islands have been reported by Wade and Lai (1989). While a mere 1/4 of the installations are funtioning on islands with no trained technician to take care of the equipment, 90 % are in working condition on islands where such personnel is available.

UNIVERSITY COURSES

There is a demand for university trained teachers, architects, mechanical engineers, electrical engineers, economists, and social scientists with advanced knowledge of relevant aspects on solar energy - and the demand is hopefully growing. Solar energy courses can be part of the regular undergraduate program as well as masters and doctorate courses of different lengths.

Examples of such solar energy education are the 1-year MSc course Principles of Renewable Energy Use that is given regularly at Univ. of Oldenburg (Blum et al, 1988) and the 1-year MSc course Energy Engineering with special emphasis on solar energy that is in the planning at Univ. College of Falun/Borlänge (Broman and Hådell, 1991). At this level, sometimes international experts can be persuded to lecture, as on the 8-credit course Solar Thermal Process Engineering held in Borlänge during the spring semester 1990 with J. A. Duffie as lecturer (Broman, Duffie and Lindberg, 1991).

In-service training of different professional groups, esp. teachers, is another important task. In particular science and technology teachers at all different school levels are important in this respect.

EDUCATING DECISION MAKERS

Decision makers are politicians, bureaucrats, and or company executives. While many are well-informed in their specialties, their knowledge about solar energy can however seldom be underestimated. On the other hand, if they are to be convinced about the benefits of a new technology, they have to be encountered and taught by top-ranking specialists in this field.

At SERC, we arrange so-called Solar Energy Days every year. Two or three consecutive days are devoted to special topics like "large solar heating plants with seasonal storage", "solar energy in the built environment", or "solar electricity for development". A typical Day consists of five lectures by different experts followed by a panel discussion. Companies manufacturing or selling solar energy equipment display their products in the lobby throughout the Days. The coffee breaks are long enough so the participants both have time to study the exhibition and to interact informally with the lecturers.

We use mostly Swedish specialists, but also invite one or two internationally reknown experts.

The Solar Energy Days are chiefly announced by sending letters to professionals all over Sweden who are working in energy-related positions. Mass media are always notified and have responded positively.

EDUCATING THE GENERAL PUBLIC

Ordinary people are the ultimate utilizers of energy from the sun and accordingly need basic knowledge in how to make use of this new technology and be motivated to use it. A number of ways to educate large populations are readily available. Some proven examples:

- Mass media. This newspapers, weekly magazines, radio, and TV. You address professional journalists, and if you manage to teach them some basic facts, they will frequently make a good job in popularizing what they have learned.

- Exhibitions. We have built both Science Center exhibitions (1986 and 1990 on solar measurements for the Futures' Museum in Borlänge) and travelling exhibitions (Alternative Energy 1976, Solar Energy Exhibition 1989; Broman and Gustafsson, 1991). The didactic value of an exhibition is greatly improved if it provides hands-on experiences.

Another kind of exhibition is the trade fair with commercial and institutional exhibitors. Such fairs can range in size from the one hundred sqm or so of exhibitions that accompany SERC's Solar Energy Days to the multi-acre outdoor exhibition of the UN Conference on New and Renewable Sources of Energy in Nairobi 1981. Such fairs contain up-to-date technological information for many categories of visitors and should be made available both to professionals and to the general public.

- Lectures, etc. General admission popular lectures sometimes attract good-size crowds, especially if arranged as debates or panel discussions, or if a well-known speaker is featured. Lectures can also be video-taped, and can, with appropriate solar powered equipment, be shown just about anywhere (Arafa, 1990).

- Community college courses. These are excellent in giving interested individuals more-than-basic knowledge. The aim of such courses can even be that every participant builds his own solar collector (Stenson, 1991).

MAKING SOLAR ENERGY UNDERSTANDABLE:
EXPERIMENTS AND HANDS-ON EXPERIENCE

I have found it imperative to mix lectures with laboratory exercises and provide the students with hands-on experience of the equation backgrounds. But before I've even come that far, I bring my students into the planetarium!

2417

Solar Paths in the Planetarium

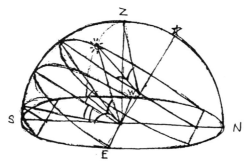

To draw pictures like this on the blackboard requires some
skill with the chalk. To sit in the classroom and comprehend
them requires a lot of imagination - for many students too
much imagination. It just isn't easy to convey understanding
of 3-dimensional relations using a 2-dimensional board.

Imagine how much easier it would be if you were able to bring
your student right into the center of the picture and let
them experience the sun's motion on the sky at different
times of the year and at different locations on earth; or to
put coordinate grids up in the sky and show what the sun's
declination, hour angle, azimuth and zenith distance really
are, and how these angle relate to one another.

This can be done if the students are brought to a planetarium,
a wonderful circular cinema where the projection screen is a
dome-shaped ceiling that covers the whole room. Alternatively,
you can rent an inflatable Starlab planetarium and have it set
up in your nearest gymnasium.

The facilities of planetariums vary, but you will always be
able to show the sun's path over the sky during different
seasons and for different latitudes. Other frequently
available features are the cardinal points, the zenith point,
the celestial pole and equator, the meridian, the ecliptic,
and so forth.

Once the students have been in the planetarium, they will
appreciate a nice litte computer program that was written by
Eva Lindberg (1991). It is called Solvej, which is both a
Nordic girl's name and a Nordic word for sun path - which is
just what Solvej draws on the screen or on paper for any
location on earth and any day of the year.

Measuring Meteorological Quantities

Some precision meteorological measurements are easier than
others, but most of them are important from a solar energy
point of view. While the temperature itself is easy to
measure accurately, it is valuable for a student to detect
herself e. g. that, on a sunny day, the temperature curve is
not symmetric around noon. The systematic changes in air
humidity require some thinking to grasp (as well as the
physics of a psychrometer), and the constantly varying

windspeed must be monitored to be fully appreciated. All
these are quantities that are used in solar energy
calculations, but the most important ones are trickier.

The basic insolation meter is a horizontally mounted
pyranometer that measures global radiation. In order to
separate beam and diffuse radiation, the latter is usually
measured with another pyranometer surrounded by a shadowing
ring. Especially at high latitudes, a sun-tracking
pyrheliometer is usually preferred. Properly calibrated,
these instruments should give the correct insolation values
within 5 % or less.

These instruments are quite expensive and therefore not
always available to the solar enegy educator. For student
laboratory purposes, other light meters will then have to do.
A good luxmeter, which measures visible light with a
wavelength response curve similar to the eye's, may be used
to estimate the insolation; the recommended unit conversion
ratio is 110 lux to 1 W/m^2. It is also possible to use a
solar cell connected to an amperemeter with low internal
resistance, providing you have been able to get it calibrated
against a precision pyranometer. With such simplified
devices, you would still be able to measure the daily
variations of global, direct and diffuse radiation with only
a few percent relative error, and absolutely maybe within
10 %.

Measuring Optical and Thermodynamic Properties

An efficient solar collector absorbs most of the incident
sunlight while at the same time the radiative, conductive and
convective losses are small. The two properties are usually
referred to as optical efficiency η_o and thermal losses,
respectively.

The optical efficiency of a flat plate collector equals
(approximately) the product of the glazing transmittance τ
and the absorber surface absorptance α. By means of a simple
optical bench, a collimated light beam, and the lux meter
mentioned above, it is easy to obtain approximate values of τ
for transparent layers, since the combination of the spectrum
from an incandescent lamp and the spectral response of the
luxmeter bears some resemblence to the solar spectrum. The
anglar dependence of τ is also easy to study, but ς is not.
In principle, of course, α equals unity minus reflectance _,
and ς is a measurable quantity. The optical bench can however
only be used for determining specular reflectance, and a
typical absorber does not reflect specularly.

More precise and complete optical measurements require a
spectrophotometer equipped with an integrating sphere. It is
very nice to be able to let student measure both τ and α,
since then they can calculate the optical efficiency of the
collector.

In order to give the students an idea of a solar collector's
thermal properties, we let them do k-value measuements of a
model collector. In this experiment, the absorber is heated

2419

electrically, and when stagnation temperature is reached, the
thermal losses for that specific difference between absorber
and ambient temperature equal the electric power $\underline{P}$.

The student who does several experiments with different $\underline{P}$'s
will find a (more or less) linear correlation between the
heat loss $\underline{P}$ and the temperature difference $\underline{T}$. The ratio $\underline{P/\Delta T}$
is called the collector's k-value, e. g. given in W/$^{\circ}$C.

In a solar collector, the rate of absorbed solar energy is
$\underline{I*\eta_o}$ and the useful collected power $\underline{I}$ equals the difference
between the absorbed power and the thermal losses. The
efficiency $\boldsymbol{\eta}$ = $\underline{I_u/I}$ at any given temperature is thus

$$\boldsymbol{\eta} = I /I = (I\,\boldsymbol{\eta_o} - P)/I,$$

where P = $\boldsymbol{\Delta}$T*k, giving

$$\boldsymbol{\eta} = \boldsymbol{\eta_o} - k*\boldsymbol{\Delta}T/I$$

This last equation is therefore a simple theoretical
expression with empirical constants: A solar collector's
efficiency curve determined indoors in the physics lab
without even measuring the intensity of the sunlight!

Direct Measurement of Collector Efficiency

Model solar collector: A plastic box with transparent
lid, half-filled with insulation and a black-surface
copper sheet. The sheet can also be heated electrically.

Now of course the students want to verify their efficiency
curve! The model collector they use does not have any cooling
water tubings, only a 1 mm thick black-surface copper sheet.
When brought out into the sunshine, all collected energy goes
into increased temperature of the copper sheet. Knowing the
specific heat of copper and the dimensions of the sheet,
monitoring the temperature increase gives the rate of energy
collected. Simultaneous measurement of the insolation then
yields the efficiency - and as a function of of $\underline{\Delta T/I}$, since
the absorber temperature rizes during the observations. Thus,
the students can draw an experimental efficiency curve and
compare it with the theoretical one.

2420

After having investigated the behaviour of the model collector, the students continue with measurements on real solar collectors. Since the sensors for insolation, inlet and outlet water temperature, and water flow are hooked up to a computer, the students won't get in direct contact with the physical quantities. Instead they will experience the contemporary way of doing technical measurements, which complements the previous experiments nicely. This part is of course specially important for the engineering students.

INTERNATIONAL CO-OPERATION THROUGH IASEE

The International Association for Solar Energy Education IASEE was founded in 1989 (Broman, 1990c). The aim of IASEE is, according to its by-laws,
"to promote solar energy education worldwide by means of
  * publication of a Newsletter,
  * organization of International Meetings,
  * topical activities, and
  * being ISES Working Group on Education.
Solar energy is here defined as direct use of the sun's energy as well as other renewable sources of energy."

IASEE has members from over 60 countries worldwide, features a Newsletter and arranges International Meetings - the first two International Symposiums on Renewable Energy Education were held in Borlänge, Sweden, 19-20 June 1991 and in Oldenburg, Germany, 11-12 June 1992.

IASEE's policy is to keep membership dues at a minimum and payable also in non-convertible currencies. As soon as there are several members in a country, it is assumed that a national section should be formed.

Further information is obtained from IASEE Secretariat, c/o SERC, P. O. Box 10044, S-781 10 Borlänge, SWEDEN, phone +46-2437 3757, fax 3750.

CONCLUDING REMARKS

The use of solar energy includes, for most people, an insight into a new technology, and even if many applications are techologically mature, the possibilities for direct utilization of the sun's energy is still unknown to many. Furthermore, much development of other applications need to be done by trained engineers and scientists. Thus, solar energy education is needed on many levels.

While much work is going on worldwide, and some examples have been given above, there is still need for development of curricula, laboratory equipment, suitable texts, and video tapes. On the other hand, such development may have been done by educators, but their results have not been communicated to others.

The recent establishment of IASEE has opened a new channel for international communication between solar energy educators. It is hoped that this channel will provide more momentum to the growth of high performance educational activities worldwide in this field.

ACKNOWLEDGEMENT

I want to express my deeply felt gratitude to Dr. Aadu Ott, University of Göteborg, for two decades of co-operation with solar energy education, which constitute tha basis for my knowledge in and commitment to the field.

REFERENCES

Arafa, Salah (1990). Solar powered video training system for village production. IASEE Newsletter No. 4.
Blum, Konrad, J. Luther, E. Naumann, and F. Neemann (1988). The postgraduate course "Principles of renewable energy use", Proc. Int. Conf. North Sun'88, Swedish Council for Building Research, pp 637-641.
Broman, Lars (1990a). Solar Energy Education (invited paper). Proc. National Solar Energy Convention of the Solar Energy Society of India, Tata McGraw-Hill, pp 455-459 (1990).
Broman, Lars (1990b). An electricity experiment kit for secondary schools using photovoltaic generated electricity. Proc. ISES Congress Clean and Safe Energy Forever, Pergamon Press, pp 427-431.
Broman, Lars (1990c). International Association for Solar Energy Education. Proc. World Renewable Energy Congress, Pergamon Press, pp 2835-2840.
Broman, Lars (1991a). Guest editorial, Sun at Work in Europe, Vol. 6, No. 4, p 2.
Broman, Lars (1991b). Making solar energy understandable, Sun at Work in Europe, Vol. 6, No. 4, p 17-18.
Broman, Lars, John A. Duffie and Eva Lindberg (1991). A concentrated course in solar thermal process engineering. Proc. 1991 Solar World Congress, Pergamon Press, pp 3815-3820.
Broman, Lars and Kjell Gustafsson (1991). An educational travelling exhibition on solar energy. Proc. 1991 Solar World Congress, Pergamon Press, pp 3849-3852.
Broman, Lars and Olle Hådell (1991). The new one year energy course at University College of Falun/Borlänge. Progress in Solar Energy Education No. 1.
Broman, Lars and Aadu Ott (1988). Solar education: The way forward. Sun at Work in Europe No.6, pp 24-25.
Broman, Lars and Aadu Ott (1991). On the need for solar energy education. Progress in Solar Energy Education No. 1.
Lindberg, Eva (1990). Solar Position Diagram. Report SERC/UCFB-90/0034 from University College of Falun/Borlänge. (The program is distributed by SERC.)
Ott, Aadu (1990). Solar Energy Education in Upper Secondary School. Proc. World Renewable Energy Congress, Pergamon Press, pp 2841-2844.
Stenson, Björn (1991). Building solar collectors on your own. Solsverige 1992, Larson, pp 65-71 (in Swedish).
Wade, Herbert A. and Henri Lai (1989). A photovoltaic training programme for the Pacific islands. Proc. 9th European PV Solar Energy Conference, Kluwer Acad. Publ., pp 1065-1068.

SOLAR ENERGY - EDUCATIONAL PATHWAYS

PROFESSOR W.W.S. CHARTERS

DEAN OF ENGINEERING
UNIVERSITY OF MELBOURNE
PARKVILLE VICTORIA AUSTRALIA   3052

Abstract,

Over recent decades considerable publicity has been given to the potential role of solar technologies in terms of decreasing reliance on fossil fuels in the world energy economy. This emphasis on clean renewable energy supply was triggered first by the post OPEC 'price hike' in the mid seventies and has been reinforced by concerns, in the public domain, regarding environmental degradation and the long term sustainability of the energy economy.

Considerable attention has been devoted to developing alternative energy technologies in a range of diverse forms for a multiplicity of domestic, commercial and industrial applications. Much less regard has been paid until recently to the role of education and training in gaining acceptance for use of these new technologies in replacing current equipment and practices in energy supply and demand.

In this paper attention is devoted to the various formal and informal educational pathways available in society today. A brief overview is provided of recent solar educational initiatives with particular reference to Australian projects in the hope that similar projects may be of use in many developed and developing nations of the world.

Key Words

Energy; solar; education; training; publicity.

INTRODUCTION.

In dealing with the question of the provision of education and training in solar techniques and solar technologies it is first necessary to define the scope of the educational objective. In a recent publication of Energy Policy {1} the editorial introduction by Tim Jackson contained the following statement regarding renewable energy technology -

'There is no common training ground, no universal terminology, no real transferability'.

In his editorial he stressed the fact that there is a more complex web of interrelated factors for renewable energy sources than for the conventional well defined fossil fuel types. The institutional complexity and variability associated with renewable energy conversion techniques stem from political, economic, social, environmental, and technical considerations.

If we wish to provide adequate education and training in the overall field of renewable energy utilisation we have to acknowledge the complexities alluded to above and to distinguish the individual client groups requiring formal or informal education.

Breadth of the educational task.

It is worth considering first the range of client groups who are either actively seeking knowledge on renewable resources or who should have as part of their professional armoury an understanding of these matters.

(i)   School education at the primary and secondary school level.

In recent years in Victoria considerable attention has been paid to providing adequate school educational material so that teachers can cope with the introduction of renewable energy material into formal course curriculum. {2}

Primary School Courses (Years 4 to 6)

For future generations it is essential that the basic facts of energy conversion and utilisation are acquired at an early stage in the formal educational process. It has been well demonstrated that language teaching can start at the primary school level and that basic concepts inculcated early on are more likely to be retained.

Special activities can be easily developed to show the forces which can be harnessed using wind, water or sun power by allowing children to build simple devices from cheap and readily available materials. {3}

This form of education requires careful planning and the provision of adequate and relevant resource kit material in audio visual and text form developed in conjunction with the school teachers.

Secondary school courses (Years 10 to 12)

If the ground work has been adequately covered at the primary school level
as outlined earlier then it is sensible to delay further formal education in renewables till the basic elements of science have been obtained. In this way the level of understanding is such that it is possible to explain in some degree the principles of operation of the various devices used in experimentation. These may range from simple flat plate or parabolic solar water or air heaters to elementary models of heliostat power towers.

With the recent upsurge of computing in the secondary school system it is also now possible to provide user friendly software for students which can be used to give them an insight into the solar design process. Such a programme was developed by the Victorian Solar Energy Council for inclusion in its secondary school kit {4}.

In any formal process at primary or secondary schools it is obviously important to 'educate the educators' and therefore an essential element of any such programme must be the provision of teacher training sessions.

(ii)   Technical and vocational training courses.

In parallel with the standard academic educational streams many countries run full time vocational training courses leading to technical qualifications appropriate to tradesmen such as plumbers, electricians, etc.

Specialist training is necessary for such tradesmen in solar energy technologies dealing with problems of manufacture of solar equipment and installation of small and large scale solar equipment. This training should cover the fields of solar thermal conversions, solar thermal/electric conversion, and solar photovoltaic technology.

Typical of such courses is one developed in Australia in recent years which is currently being taught both in Victoria and in Queensland {5}.

(iii) Post Secondary Educational courses.

At the advanced educational level there is room for incorporation of teaching in renewable energy technologies at both the undergraduate and postgraduate level.

It is my personal belief that, at the present state of the solar industry, there is insufficient demand to warrant specialist undergraduate degrees in solar science or solar engineering. However there is plenty of opportunity to use solar concepts in engineering and science subjects by using illustrative examples of heat and mass transfer processes in solar collectors, solar distillation units, solar thermal and power systems, etc.

Equally in non-technical subjects in Arts Commerce courses it is possible to select useful solar examples to illustrate the contribution of solar energy in social and economic development, and the benefits of clean renewable energy sources in lessening environmental degradation and in ensuring sustainable energy futures. In this context it is worth noting that much of the material developed for science at secondary school level may well be appropriate for non-technical subjects at post secondary school level.

At the postgraduate level solar energy provides ample material for generalist and specialist Masters courses of various types. Such courses are now offered as routine in many countries of the world including Australia, U.S.A., France, Germany amongst many others. {6} UNESCO has commissioned several studies of such educational programmes in an attempt to publicise this activity and disseminate information in a global sense {7}. In addition many universities and colleges world-wide have mounted full scale research studies leading to Masters and Doctoral qualifications based on many aspects of solar systems. From such research investigations have come a series of innovative solar developments ranging from advanced heat pipe collectors, through solar heat pumps to new high efficiency photovoltaic devices.

(iv)   Adult education courses.

One of the strengths of the solar energy movement has been its appeal at the 'grass-roots' level in society and there is ample opportunity to capitalise on this support by providing educational courses at the community level.

Material presented in such courses is of necessity of a non-specialist and non-technical nature but it is important to provide a clear and unbiassed view of solar options for public consideration.

Often such courses can be substantially assisted by the ready access of the public to Energy Information and Demonstration Centres supported by Government and/or Industry funding.   At such centres it is vital to have simple succinct 'fact-sheets' on various aspects of renewable technology as well as working examples of small scale solar equipment to illustrate the potential domestic, commercial, or industrial applications of such plant.

(v)   Publicity and promotional activities.

Although not education in the formal sense it is impossible to ignore the potential impact of the media on all aspects of modern life.   In both developed and developing nations the influence of newspapers, radios and television can be far reaching in spreading information on new developments.

Solar scientists, technologists, and activists need to recognise the importance of this information dissemination process, and to train themselves in the range of techniques suitable for the diverse media outlets. {8}   This includes the writing of press releases and simple factual solar articles, the preparation for short radio and television interviews and the techniques for lobbying successfully with politicians and bureaucrats who control the use of Government funds for renewable energy research and development.

Conclusion

Much has been done over the three to four decades, since the first solar energy meeting in Phoenix, Arizona in 1955, to spread the message on solar and renewable energy technologies.

Much remains to be done, in many countries, to convince Government and Industry leaders that renewable energy technologies can compete with conventional energy supplies and can be used as a pathway to a sustainable energy future.

All of us involved in the educational process 'at any level' bear a special responsibility to provide leadership in this educational role within society.

In this brief paper my intent has been to indicate the breadth of educational materials and courses necessary to provide an adequate infrastructure on which to build a solar energy future and to point out some examples of steps along the way already undertaken in various countries of the world.

REFERENCES.

{1}   Energy Policy Editorial
       Tim Jackson
       Energy Policy Vol. 19 No. 8   October '91.
       Butterworth-Heinemann Ltd., U.K.

{2}   Development of Primary and Secondary School Teaching
       Packages for Renewable Energy Education.
       W.W.S. Charters
       Proc. Energy & The Environment Conference, Reading, U.K.
       Pergamon Press U.K. 1980. pp 2348-2360.

{3}   The Sun, Energy and Us.
       Primary School Kit.
       Victorian Solar Energy Council
       Melbourne, Victoria, Australia.   1986.

{4}   Energy, Technology, & Society.
       Secondary School Kit.
       Victorian Solar Energy Council.
       Melbourne, Victoria, Australia. 1987.

{5}   Advanced Certificate in Renewable Energy Technology
       State Training Board of Victoria.
       Melbourne, Victoria, Australia, 1989.

{6}   Solar Energy Programmes in Universities in Malaysia.
       D.S. Chuah; K.S. Ong; Tran Van-Vi.
       Seminar on the Applications of Solar Energy in South &
       South East Asia.
       Bangkok, Thailand.   February, 1979.

{7}   Training in Solar Energy Curriculum Development
       Cleland McVeigh.
       UNESCO Report. Paris, France 1982.

{8}   Using the Media - A Practical Guide for Solar Advocates.
       Peter Fries.
       'Energy for a Sustainable World'.
       ANZSES Solar 91. Vol. 2 p. 648-653.
       Adelaide, Australia.   December '91.

RENEWABLE ENERGY EDUCATION IN DEVELOPING COUNTRIES : INDIAN SCENARIO

H.P. GARG AND T.C. KANDPAL
Centre for Energy Studies
Indian Institute of Technology, Delhi
Hauz Khas, New Delhi-110016 (INDIA)

ABSTRACT

The need for renewable energy education in developing countries, the relevant characteristics of developing areas of the world and desirable features of renewable energy education programmes in these areas are briefly described. A preliminary overview of the university level renewable energy teaching/training prorgrammes in India is also presented.

KEYWORDS

Renewable Energy Education; Developing Countries; Teaching Programme.

INTRODUCTION

Availability of cheap and abundant energy with minimum environmental and ecological hazards associated with its production and use is one of the important factors for the desired improvement in the quality of life of the people living in developing areas of the world. The growing scarcity of fossil fuels and the adverse ecological and environmental impacts associated with their excessive production and use have raised global interest in the harnessing of various renewable energy resources. In most of the developing countries, however, the various programmes in energy conservation, optimal utilization of fossil fuels and large scale utilization of renewable energy sources is very much handicapped by the unavailability of technical manpower in this field at all levels. A balanced and accelerated adoption of renewable energy technologies in the developing countries would, therefore, require an adequate number of well trained and competent manpower for resource assessment, technology development, planning and information processing etc. Education and training in this field at all levels and through all modes are, therefore, of prime importance (Badran 1990, Berkovski, 1990).

In the last two decades, several teaching institutions in the developing countries initiated efforts towards introducing renewable energy resources and technologies and courses on other relevant aspects in their programmes. Most of these courses are floated at the university level either as independent postgraduate level programmes on renewable energy technologies or as electives in conventional engineering/applied science curricula. Some preliminary efforts have also been made to introduce energy relative concepts in the school level curricula as well. The present paper deals with the various aspects of renewable energy education in developing countries, in general, and with the existing university level teaching/training programmes in India in particular.

OBJECTIVES OF RENEWABLE ENERGY EDUCATION

The specific objectives of any desirable programme of energy education may include (McVeigh, 1982; Shah, 1990)

i)   To develop an awareness among students about the nature and cause of present energy crisis.

ii)   To make the students aware of various types of nonrenewable and renewable sources of energy, their resource potential, existing technologies to harness them, economics and energetics of these technologies, and their socio-cultural and environmental aspects.

iii) To make the students appreciate the consequences of various energy related policy measures.

iv)   To enable the students to suggest alternative strategies towards solving the energy crisis and also to provide more energy for the desired improvement in the quality of life of the large population in developing countries.

CLASSIFICATION OF RENEWABLE ENERGY EDUCATION PROGRAMMES

Renewable Energy Education programme can be classified on the basis of (i) mode of education (ii) the audience to be educated (iii) geographical area covered and (iv) the duration of the programme.

To achieve the objectives of the renewable energy education programmes both the formal and nonformal modes of education should be used. Formal education includes instructions given in schools, colleges, universities etc. and is planned and well structured with identified objectives and needs. The non formal mode of education, on the other hand, involves learning from mass communication media and other informal means and primarily helps (i) those who have never gone to school (ii) the unemployed school leavers, and (iii) adults interested in acquiring new skills. While the renewable energy education through formal means is gaining momentum in many developing countries, the non-formal means are yet to be fully exploited for this purpose (Shah, 1990).

Regarding the target group of the renewable energy education programmes, it is now almost globally accepted that the education on new and renewable sources of energy should be provided (McVeigh, 1982)

i)   at various levels in schools, colleges and universities.

ii)   to technicians so as to enable them to deal with special needs and problems of relevant technologies.

iii) to the general public.

2431

In the developing countries the teaching/training programmes in the area of renewable sources of energy at the university level are now coming up. However, there is not sufficient progress in introducing the relevant concepts in the school curricula. Similarly there is a serious lack of specialized technician training programmes in this area. So far, the number of public awareness programmes for improving the understanding and acceptance of renewable sources of energy by the general public has also been very small in developing countries and calls for sincere efforts in this direction.

On the basis of the geographical area covered the renewable energy education programmes in developing countries can be classified as (i) international (ii) regional, and (iii) national level programmes. United Nations University (UNU) sponsored training programmes at IIT Delhi, India and at Bogota, Colubmia belong to the category of international training programmes which provide an opportunity to the personnel from ministries, universities, industries etc. of various developing countries to receive training on the relevant aspects. On a regional level a few regional programmes (e.g. at AIT, Bangkok, Thailand) exist while the national centres cater more to the local, specific needs in this area.

The short terms courses may either be designed to provide specialized training or for providing generalized orientation depending upon the needs and backgrounds of the trainees. The long term renewable energy education programmes on the other hand provide indepth coverage of all relevant aspect on a broader level.

RELEVANT CHARACTERISTICS OF DEVELOPING COUNTRIES

Some of the existing characteristics of developing countries of the world which may directly or indirectly affect the development and establishment of renewable energy education programmes are as following:

(i)   In order to improve the quality of life of most of their population it is imperative for the developing countries to provide more energy per capita to its people.

(2)   Most of the developing countries are oil-importing which necessitates developing a suitable infrastructure for harnessing of new and renewable sources of energy as early as possible. In many developing countries, however, the R&D programmes in this area are just being initiated.

(3)   The developing countries, in general, have good insolation characteristics and moderate to high wind speeds on a large portion of their geographical area. Similarly, most of these have quite a large amount of biomass feedstocks available for use as a renewable resource of energy.

(4)   Presently, only a few developing countries have their own teaching/training programmes in the field of renewable energy at the university level.

(5)   In most of the developing countries, a large fraction of the population is still illiterate and does not receive education through schools/colleges. Hence the need for informal education programme in renewable energy.

(6)   An acute lack of resources/funds is another common characteristic of
       most of the developing countries

(7)   Unemployment/underemployment prevails in almost all developing
       countries thus making it very necessary that renewable energy
       education is directly linked with employment opportunities.

(8)   At present, in most of the developing countries there is a serious
       lack of trained teachers, suitable text books, teaching aids in the
       area of renewable energy.

(9)   The local socio-cultural issues are quite detrimental in the mass
       level adoption of renewable energy technologies. In all the
       developing countries where efforts have been made to harness new and
       renewable sources of energy, there has not been many success
       stories.

(10)   In almost all developing countries the problem of energy is
       generally compounded by other important issues related to health,
       nutrition etc. It is therefore necessary that the renewable energy
       education is suitably linked with other relevant aspects as well.

DESIRABLE FEATURES OF RENEWABLE ENERGY EDUCATION PROGRAMMES IN DEVELOPING COUNTRIES

The desirable features of a renewable energy education programme in
developing countrie include:

(1)   It should include all renewable energy resources with particular
       emphasis on some specific ones depending upon the local needs and
       characteristics.

(2)   It should cover all aspects of renewable energy technologies such as
       (i) resource assessment (ii) technology (iii) economics and
       energetics (iv) socio-cultural issues and (v) ecological and
       environmental impacts.

(3)   Separate renewable energy education curricula for different levels
       (school, polytechnic, university) and for different audience be
       developed.

(4)   Although to some extent the renewable energy education programmes
       may cater to the local, site specific needs, they should at the same
       time fit in well with the national, regional and international
       programmes.

(5)   It should be flexible and dynamic thus allowing for improvement in
       future, if desired.

(6)   Renewable energy education preferably be provided in local language
       for better acceptance and effectiveness. Good quality text books,
       teaching aids etc. should also be available in the local language.

(7)   It should provide a balance between theory and practical aspects of
       renewable energy education and should cover all aspects of
       teaching/training including lecture, laboratory, demonstration,
       hand-on-skills training, design, manufacture, trouble shooting etc.

(8)   It should be capable of providing energy education to all the persons in minimum of time and should be economical so that maximum number of people may be educated within the existing financial resources.

(9)   Renewable energy programmes should also ensure employment/self employment to the concerned persons.

(10)   It should be compatiable with the global efforts in this direction and should allow effective and mutially beneficial experiences sharing and interaction.

RENEWABLE ENERGY TEACHING/TRAINING PROGRAMMES IN INDIA

Courses on renewable energy and for that matter on energy did not form part of normal undergraduate or postgraduate programme in any University in India till 1975. The first initiative in this direction was taken at the Indian Institute of Technology, Delhi by setting up an interdisciplinary school on energy studies with a special funding from the Government of India which was later given the status of a National Centre for Energy Studies. One of the objectives of this centre was to generate a cadre of scientists and engineers in this area for academic and research institutions, industry and the Government organisations. Master of Technology and Ph.D programmes in energy studies were therefore started in the centre.

With the growing interest in energy, other universities/institutes also initiated academic programmes in this area. The Indian Institute of Technology, Bombay started similar programme at the post-graduate level while the Indian Institute of Technology at Kharagpur floated an undergraduate programme in Energy Engineering. Some of the undergraduate/post-graduate programmes covering various renewable energy resources and technologies alongwith other energy related aspects is presented in the following paragraphs.

IIT DELHI

(a)   **Master of Technology in Energy Studies (M.Tech.)**

The two year Master of Technology/programme in Energy Studies was started at the Centre for Energy Studies of this Institute in 1981. This programme broadly covers promising new energy sources and offers an opportunity to the students to study one of them in detail and to carry out a suitable major project in that area. The duration of the programme has been reduced to one and a half years since July 1983.

The admission to all M.Tech. programme in the Institute is made on the basis of an all India competitive examination called GATE. Those who have completed M.Sc. in physics or bachelors degree in mechanical/electrical/chemical engineering are eligible for admission. Seats are offered on the basis of the students merit in GATE and his preference for the field of specialization. All selected students receive a scholarship and live in the student hostels of the Institute on the campus.

In July 1989 an evening M.Tech. programme in Energy Studies was started for sponsored candidate from different organisations in and around Delhi. The students doing this course normally take five semesters to successfully complete the course and some students of the first batch have just finished their training in December 1991

(b)   United Nations University, Tokyo (Japan) Sponsored
      Training Programme in Renewable Energy Systems

The United Nations University (UNU) has been sponsoring a training programme in the area of Renewable Energy Systems at the Indian Institute of Technology, Delhi since 1981. The trainees are drawn from other developing countries. The minimum qualification for admission is a bachelors degree in physical/biological sciences. The candidates should be engaged in energy planning, research or training in a government organisation. They should be sponsored by their employers and given leave for attending the course with salary paid during the training period.

The first four programmes were of 12 months duration but the duration was reduced to 8 months in 1989. The present 8-month course work is designed to give a strong foundation in the basic principles and their application for development of renewable energy technologies. Attention is also paid to the performance evaluation, cost-effectiveness and ecological and environmental impact of various technologies. Energy conservation techniques, energy resource assessment and planning form an integral part of the curriculum. Visits to industries and other centres of research in and around Delhi are arranged during the training period. The learning experience finally culminates in a project which puts to use the knowledge acquired by the trainees to solve one of the problems relevant to their needs back home. The UNU awards a certificate on the successful completion of the training.

IIT BOMBAY

This Institute offers a three semester M.Tech. programme in Energy Systems Engineering. The admission is open to engineering graduates in computer   science/mechanical/chemical/-electrical/aeronantical/civil/metalurgical engineering. The process of selection is the same as that followed at IIT Delhi.

IIT KHARAGPUR

An integrated undergraduate programme in energy leading to a B.Tech. degree in Energy Engineering is being offered at IIT Kharagpur. The course is open to students of Physical Sciences who have completed the final year school examination (after twelve years of schooling) and have qualified in an all India level joint entrance examination held for this purpose. The programme is designed to impart adequate appreciation of the nature and availability of resources, their economic exploitation commensurate with the preservation of our environment, interaction of energy and technology and finally, the imact of all these on the society at large.

MADURAI KAMRAJ UNIVERSITY, MADURAI

Two post graduate courses are offered at the Madurai Kamaraj University, Madurai (Tamilnadu). One of the courses is the M.Phil programme in Energy, Environment and Natural Resources and the second is an M.Sc. Course in Energy Science.

In the M.Phil programme the first year courses are very broad and the specialized courses are offered in the second year. In each course the student is expected to offer seminars and assignments on the thrust areas. The students are also required to submit a dissertation of original work as a requirement for partial fulfillment and present the same during his viva-voce examination. A two year (four semester) M.Sc (Energy Science) course which has also been recently initiated at the Madurai Kamraj University.

SCHOOL OF ENERGY STUDIES, FACULTY OF ENGINEERING SCIENCES, DEVI AHILYA VISHWAVIDALAYA, INDORE (M.P)

The School of Energy Studies in the Faculty of Engineering Sciences at the Devi Ahilya Vishwavidyalaya, Indore (Madhya Pradesh) offers Ph.D., M.Tech. and Post Graduate Diploma Courses. The course structure of the M.Tech. course includes nine core and four elective theory courses, seven core practical courses and a compulsory major project. Some of the theory core courses are (i) Conventional Energy Systems (ii) Non-Conventional Energy systems (iii) Direct Energy Conversion and (v) Computer Energy Management Systems.

The Postgraduate Diploma course is primarily a correspondence programme for candidates sponsored by energy related industries and organisations. The post graduate diploma is being offered with three specializations - Renewable Energy, Energy Conservation and Energy Planning. There is a provision for specializing in more than one area by taking more courses. The course structure of the programme consists of six core courses which are common to all three specializations, and five elective courses corresponding to the specialization and a project work in the area/areas of specialization.

Recently, a detailed questionnaire based survey of the existing university level teaching/training programmes in the area of new and renewable soruces of energy in Asia and Pacific was conducted (Garg and Kandpal, 1991). A breif summary of the relevant results for the Indian situation is presented in Table 1.

CONCLUDING REMARKS

University level teaching/training programmes in the area of renewable sources of energy in India have attained considerable maturity by now. However, several aspects of these programmes still merit further improvement for providing more efficient and effective renewable energy education to the students. The following recommendations are made for this purpose.

i) There is an urgent need of developing well structured curicula for teaching/training programmes in the area of renewable sources of energy. The curricula should, however, also integrate relevant aspects of conventional sources of energy, energy conservation, and ecological and environmental implications of energy production and use. The course curricula may be standardised at the national, regional or even global level to facilitate better technology transfer and exchange of know how.

ii) The renewable energy education programmes should offer a mix of academic as well as hands-on-skills training to the student. The later can be accomplished by conducting laboratory experiments,

2436

demonstrations, field visits and field installation of actual working system. The students may be encouraged to undertake hardware oriented projects as a partial requirement towards their degree/diploma.

iii) For the renewable energy education programmes to be successful at any level of education the preparation of suitable text books, laboratory manuals, teaching aids etc. is quite important. It is therefore recommended that sincere efforts be made in this direction as well. Modern techniques of communication and information processing may also be used for this purpose. Suitable case studies on relelvant aspects may also be prepared for class room discussion.

iv) Facilities and mechanism for providing in service training to existing teachers should be established in the country.

REFERENCES

Badran Adnan (1990). Towards the sustainable use of energy. Proc. 1st World Renewable Energy Congress, 23-28 September, 1990, Reading (U.K)

Berkovski Boris (1990). Renewable energies for sustainable development. Proc. 1st World Renewable Energy Congress, 23-28 September, 1990, Reading (U.K)

Garg H.P. and T.C. Kandpal (1991). Renewable energy education in Asia: Survey of existing graduate teaching/training programmes. Paper presented at the Int. Symposium on Renewable Energy Education, June 1991, Borlange (Sweden).

McVeigh J.C. (1982). Training in Solar Energy Curriculum development. Report submitted to UNESCO, Paris.

Shah Beena (1990). Energy Education. Northern Book Centre, New Delhi (India.)

Table 1. University Level Renewable Energy Teaching/Training Programmes in India

S.No.	University/Institute	Course Level	Offered Specialization	Duration	Maximum intake	Minimum Qualification Qualification for Eligibility	Renewable Resources covered
1.	Devi Ahilya Vishwavidyalaya, Indore	M.Tech	Energy Management	1.5 years	20	Bachelors degree in Engg., M.Sc (Physics)	Both Conventional and Non-Conventional Resources of Energy
		P.G. Diploma	Renewable Energy		20	-do-	-do-
2.	I.I.T. Bombay	M.Tech	Energy Systems Engineering	1.5 years	08	Bachelors degree in Engineering	Mainly Solar and Biomass Energy
3.	I.I.T. Delhi	M.Tech (Morning)	Energy Studies	1.5 years	30	Bachelors degree in Engineering	Primarily Renewable Resources of Energy
		M.Tech (Evening)	-do-	2.5 years	15	-do-	-do-
		P.G. Certificate	Renewable Energy Systems	8 months	08	Bachelors degree in Sciences/Engg.	-do-
4.	I.I.T. Kharagpur	B.Tech	Energy Engg.	4 years	–	XII level Physical Sciences	All Aspects of Energy Production and Utilisation
5.	Madurai Kamraj University	M.Phil	Energy and Environment	2 years	15	M.Sc with 60% marks	Mainly Solar and Biomass Energy
		M.Sc.	Energy Science	2 years	15	B.Sc. in any branch of physical sciences with 60% marks	Both Renewable and Non Renewable Sources of Energy
6.	University of Roorkee	P.G. Degree Diploma	Alternate Hydro Energy Systems	1.5 years	10	Bachelors degree in Engineering	Solar Thermal, Hydro, Biomass & Alternate fuels for Engines.

SOLAR HYDROGEN - WHY, POTENTIAL, WHEN ?

J. GRETZ

CEC, Joint Research Centre,
Ispra Site (VA), Italy

ABSTRACT

Hydrogen is a carbonfree fuel which oxidizes to water as combustion product. The generated water becomes, together with renewable primary energy for splitting it, a source of clean and abundant energy in a carbon-free, natural cycle.

Hydrogen is a fuel which can be transported over long distances and stored so that solar energy can be transported from energy rich countries over long distances in ships to Europe, stored and used in gaseous or liquid form in industry, households, power stations, motor cars and aviation propulsion and steel fabrication.

Solar energy as primary energy is discussed, a special form of it, the cheapest and by now largely available hydropower, is stressed.

Techniques of hydrogen production, vectorization and end use is discussed as well as safety aspects, costs and strategy for its implementation.

The Euro-Québec Hydro-Hydrogen Pilot Project is shortly described including its latest realisation of seven demonstration projects on the use of hydrogen in the fields of public transport, aviation propulsion, steel fabrication and storage.

KEYWORDS

Solar energy; hydrogen; economics of solar hydrogen.

WHY HYDROGEN?

Hydrogen is a carbon-free fuel which oxidizes to water as combustion product. The generated water becomes, together with renewable primary energy for splitting it, a source of clean and abundant energy in a carbon-free, natural cycle.

2439

The hydrogen technology is ecologically benign, entailing little or no local pollution.

Hydrogen can be transported and stored in gaseous and liquid form and used as fuel in industry, households, power stations, motor car and aviation.

The intermittence and the geographical distribution of solar energy require means of storing and transporting it to the user's place. An almost ideal means of doing this is to split water with this primary energy source in order to obtain hydrogen.

The characteristics of hydrogen assign to it a role which is complementary to that of electricity. But unlike electricity, hydrogen is a fuel which can, with by now established techniques, be transported over long distances and stored so that solar energy can be transported from energy rich countries over long distances in ships to Europe, stored underground or in containers and used in the different forms.

Hydrogen can thus open up clean and abundant renewable energy sources distributed all over the World and give countries of the Third World the opportunity to export energy to the developed countries. It leads by that virtue to more independence from the actual fossil fuel exporting countries and contributed thus to the economic cooperation of nations and the "North-South Compensation".

The transport of hydrogen entails no tanker oil spills (1989: Approx. 3 Mio To oil went into the oceans from tanker wrecks (13), tanker cleaning, pipeline and drilling platform leaks).

Hydrogen from hydropower is relatively cheap and in some cases already competitive (steel fabrication with Brazilian hydro/hydrogen).

Hydrogen allows smooths transition from the existing energy systems to clean and renewable energy systems since hydrogen energy systems need virtually no basic engineering breakthrough; all components of such systems exist, they need continuous, gradual development; large pilot/demonstration projects can today be built and run.

PRIMARY ENERGY

Hydrogen being a secondary energy - an energy vector - it has to be generated with primary energy i.e. direct or indirect solar energy such as hydropower and wind energy if the overall system is to be clean and renewable.

The most convenient process for the generation of hydrogen is electrolysis which requires electricity as primary energy.

Direct Solar Energy Conversion

The most developed technologies for helioelectricity generation are photovoltaic, thermal and wind conversion. All these technologies are today sufficiently developed and field-tested so that the erection and operation of pilot/demonstration plants in the 1-100 MW range can be undertaken.

A 100 MW el photovoltaic power plant, situated in an equatorial country with global yearly insolation of say 2300 kWh/m^2, y and 7.5% global efficiency of conversion from solar radiation into hydrogen would have a collector surface of about 1 km^2, covering some 3 km^2 land surface and would produce yearly $0.5 \cdot 10^9$ m^3 of hydrogen. New technologies like multi-band solar cells, capable of utilizing photons in a number of adjacent frequency bands and bringing the theoretical efficiencies up to 25% or 45% would reduce costs, material and land requirements proportionally.

Indirect Solar Energy Conversion Hydropower

Natural hydropower and hydropower of already existing dams are amongst the cleanest and most promising renewable energy sources. 23% of the solar energy falling on the earth operates the hydrological cycle, i.e. evaporation of water, precipitation and storage in water and ice.

Some characteristics of hydropower are given heareafter:
- 23% of the solar energy falling on the earth, i.e. $\sim 350 \cdot 10^6$ TWh/y, operates the hydrological cycle, i.e. evaporation of water, precipitation and storage in water and ice;
- the World's topological and technically exploitable estimated hydroenergy potential is $\sim 20 \cdot 10^3$ TWh/y, i.e. 0.0057% of the hydrological cycle energy;
- the yearly average utilization factor of hydroenergy is $\sim 65\%$ (in 1986, Worldwide installed 380 GW generated 2200 TWh electricity);
- today's hydropower electricity generation of about 2200 TWh/y represents $\sim 21\%$ of the World's electricity generation;
- the table gives the World's estimated hydropower potential and its approximate distribution:
  . Asia (without UDSSR) : $5.3 \cdot 10^3$ TWh/y
  . South America   : $3.8 \cdot 10^3$ TWh/y
  . Africa   : $3.1 \cdot 10^3$ TWh/y
  . North America   : $3.1 \cdot 10^3$ TWh/y
  . Ex UDSSR   : $2.2 \cdot 10^3$ TWh/y
  . Greenland   : $1.5 \cdot 10^3$ TWh/y
  . Europe   : $1.4 \cdot 10^3$ TWh/y
(for comparison, Europe's (EC) electricity consumption (1984) was $1.3 \cdot 10^3$ TWh/y).

ENVIRONMENTAL ASPECTS

The combustion of hydrogen does not produce CO_2, CO, SO_2 VOC and particles, but entails emission of water vapour and NOx.

Water vapour emissions from airplanes may be harmful since they generate - depending on the cruising altitude and latitude - ice clouds with ensuing greenhouse effects and ozone depletion. The problem is of great importance and actually under investigation.

The formation of NOx is a function of flame temperature and - duration. Considering the wide flammability range of hydrogen its combustion can be influenced by the design of the engine so that the NOx emission can be reduced.

The worldwide water evaporation from the oceans and rivers is $\sim 5 \cdot 10^{14}$ m^3 per year. If mankind's today's total energy consumption of sustained 11 TW would be effectuated by hydrogen, the ensuing yearly water evaporation would be some $2.5 \cdot 10^{10}$ m^3 i.e. about 1/20000 of the natural evaporation. Once hydrogen will be massively used, local considerations are obligatory like it is the case for today's wet cooling towers.

2441

END USE

Hydrogen is a carbonfree fuel with excellent combustion properties. This makes it a clean, universal fuel for use in industry, households, power stations, road vehicles and aviation.

The specific properties of hydrogen induce new end use techniques, like:

Fuel Cells

The key element of an efficient electricity generation will be the fuel cell for which hydrogen is the best fuel. Being devices which transform chemical energy directly into electrical energy its generation is not subjected to Carnot cycle limitations, is done silently without moving parts and with high efficiencies. Efficiencies of future fuel cells like molten carbonate cells operating at 600-800°C, with total energy efficiencies of 60-80% with waste heat at 600°C, make the fuel cell concept a potential candidate for use in power stations, road and space vehicles. Progress in catalyst and electrode technology and cost reductions have to be made, however, before fuel cells can competively be considered for general use.

Hydrogen is an excellent fuel for fuel cells, its more or less cold combustion will reduce the NOx emissions even down to zero.

Catalytic Burners

Hydrogen, having lower ignition energy than for instance natural gas, burns smokeless when in contact with a suitable catalytic surface at low temperatures between 20 and 400°C. This would be a preferred technology for generating low temperature heat in residential space heaters, cooking devices and industrial dryers and heaters.

Catalytic combustion has the advantage of high safety, very high efficiencies of up to 99% and negligible emissions of nitrogen oxides.

Super/Hypersonic Aviation

Jet engines at super/hypersonic speeds are advantageously - if not necessarily - powered with liquid hydrogen (LH_2) for mainly two reasons, its almost 3 fold gravimetric heating value compared with kerosene and its capacity of cooling parts (turbine inlet rim, wing leading edges, passenger cabine) of the plane exposed to outside stagnation temperatures of 1300°C at Mach 6, cryogenic storage temperature being at -252°C. Moreover, cooling of the wing skin induces laminar flow and therewith considerably reduces drag by up to, theoretically, 30% ("laminar flow control").

Water vapour emissions, however, are likely to form more or less long lasting ice clouds which enhance the greenhouse effect. The problem is of extreme importance and is seriously to be investigated not only in view of hydrogen - the combustion of hydrogen sets 2.5 times more water vapour free than that of kerosene - but also concerning kerosene combustion in general.

The elimination of CO, CO_2, SO_2 and VOC by using hydrogen not only is beneficial for the atmosphere, but reduces considerably airport ground pollution. For example, the NOx emissions resulting from the daily average 1259 landing/take-off cycles at Los Angeles International Airport are equivalent to the operation of about 1 Mio passenger cars.

Road Vehicle Propulsion

In Germany, for example, from the overall emissions, those resulting from road traffic are: 52% NOx, 70% CO and 49% VOC. The main problem of hydrogen utilization in vehicles is storage due to the low volumetric energy content of hydrogen of about one third of that of gasoline. From the three different storage technologies i.e. gaseous hydrogen in pressure vessels, cryogenic liquid hydrogen and hydrogen chemically bonded in metal and liquid hydrides (methylcycloexane), liquid hydrogen seems to be the most suitable solution.

The low ignition energy of hydrogen and its wide ignition range of hydrogen/air mixtures make gas turbines as well as piston engines well adaptable for hydrogen combustion. The ignition within a wide range of non-stoechiometric air/fuel mixture permits combustion with high amount of excess air and thus full and part load operation with low nitrogen oxides emission.

Since 1937, more than 30 vehicles with hydride storage have been built and operated in Europe by Daimler Benz, (more than 600.000 km), BMW and experimental trucks by Paul Scherrer Institute (Switzerland).

It should be emphasized that the building up the necessary infrastructure is at least as lengthy as the development of the technology itself.

Steel Fabrication

With ~ 2 kg CO_2 emissions per kg steel fabricated with carbon as reductant, the World's steel fabrication contributes 10-12% to the world's total anthropogenic CO_2 production and is therewith amongst the most intensive single CO_2 polluter.

Hydrogen is an excellent and clean reductant emitting water vapour instead of CO_2 and does not introduce extra impurities as coke does (sulphur in particular). If generated with relatively cheap and abundant hydropower, its application in steel fabrication is in some cases already competitive for instance Brazil which disposes of enormous resources of iron ore in the vicinity of large hydropower installations.

The penetration of hydrogen into the steel industry would be favorized by the fact that steel fabrication is rather centralized, the power per steel fabrication plant being in the order of GW compared with, for instance, 100 kW per motor car, which eases the logistics of intervention of hydrogen in steel fabrication.

Safety

Each energy carrier has its specific dangers and risks. Compared with natural gas and liquid fuel hydrogen scores less well since it:
- is flammable within a wide range of hydrogen - air mixtures of between 4 and 75 Vol. % vs. methane between 5 and 15%;
- has a low ignition energy of up to 15 times lower than methane;

2443

- is much more prone to leak; liquid hydrogen leakage rates are 50 times that of water;
- has high flame velocities of about 7 times that of methane and therewith reacts violently once it is inflamed.

Hydrogen is advantageous since:
- it has a high diffusivity of about 4 times larger air so that spills disperse rapidly to below flammability; there is higher probability of surviving an airplane crach - if one is not killed by the impact if the airplane - if it is fuelled with hydrogen rather than with kerosene;
- air flames are nearly invisible i.e. they release little heat by radiation implying both hazards in fire fighting because one does not feel the flame as well as advantages since fire fighters can work closer to the flame.

Experience from the use during decades of town gas containing 60 Vol. % hydrogen is available as well as from the operation of a 210 km long pipeline network operating with pure hydrogen at 25 bar in Germany since 1936, and by a 80 km hydrogen carrying pipeline in Houston, Texas. Considerable experience in liquid hydrogen storage and road/rail transportation has been gained in the course of space activities, in which no specific hydrogen fatality has been experienced.

Hydrogen is undoubtedly a hazardous material but there is general agreement that with proper use under adequate safety measures hydrogen will be no more hazardous than the fuels currently used.

On 18.12.1990 the BAM (German Federal Institut for Material Research) declared that liquid hydrogen would be not more dangerous than LNG and LPG and that it has no objection to the transport of liquid hydrogen in 2 G type ships (IMO rules).

COSTS

In the Euro-Québec Hydro-Hydrogen Pilot Project, for instance, the cost of liquid hydrogen, produced with hydropower at 2 $cents_{ECU}$/kWh, shipped to Europe and stored in the port of Hamburg are 14.8 $cents_{ECU}$/kWh at an annuity of 11.7%.

The specific product costs and their breakdown for the LH_2 vector are given in Figs. 1 and 2.

ECONOMICS

An adequate comparison of hydrogen costs with costs of the actual hydrocarbon energy has to take into account the external costs i.e. costs for pollution abatement/prevention and climate effects of fossil fuel burning. If the costs for pollution abatement originating from fossil energy use are evaluable only very approximatively and with great difficulties, the cost of the negative drawbacks on climate changes are clearly imponderable i.e. impossible to evaluate numerically.

The hydrogen cost of 15 $cents_{ECU}$/kWh are pictured in Fig. 3 together with the costs of taxed gasoline prices in Europe (average of the 12 EC countries, August 1990) of 8.5 $cents_{ECU}$/kWh which are made up of 3 $cents_{ECU}$/kWh for the crude oil itself, its transportation, refinement, manipulation and distribution and of 5.5 $cents_{ECU}$/kWh for taxes.

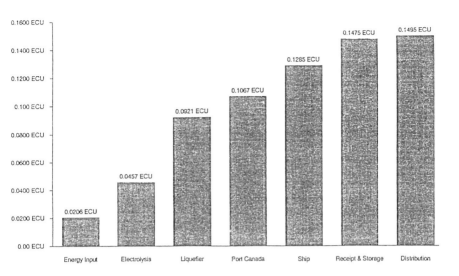

Fig. 1. Specific costs, ECU/kWh.

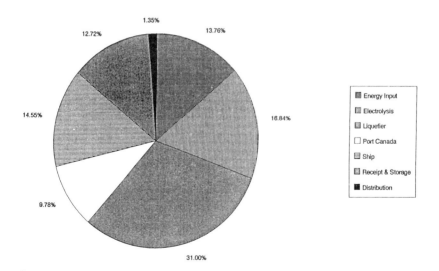

Fig. 2. Specific costs distribution.

With the above values and including the submerged costs of protection/repair of the damaged environment resulting from the use of fossil fuel (the value of 2.5 cents$_{ECU}$/kWh is the average from various literature sources), the "FUEL COST ICEBERG" would look as shown in Fig. 3.

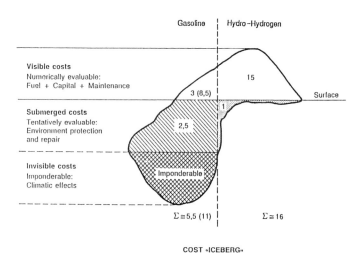

COST «ICEBERG»

(August 1990)

Fig. 3. Fuel cost comparison ($\mathcal{C}_{ECU}$/kWh; figures in
( ) are for taxed fuel).

With internalization of external costs, hydrogen energy would thus be about (1.45) 2.9 times higher than hydrocarbon energy costs, not taking into account the imponderable costs of climate effects.

WHEN?

Given the long lead times and the dynamics of the introduction of energy technologies, time is likely to become a precious raw material in view of the menacing deterioration of the environment.

Analysis of the molecular composition shows an increasing ratio of hydrogen to carbon molecule of the fossil fuels used worldwide since 1860, going from wood (H/C = 0.1) to coal (1), to oil (2) and finally - during the 90's - to methane (4). If that trend is to continue, extra hydrogen has to be provided from an external source i.e. water, see Fig. 4. This leads to the concept of water splitting using non-fossil fuels.

The analysis of the market penetration curve of new technologies, see Fig. 5, indicates that the conquest of the first 10% of the market is not very different from the one between 10 and 90%.

In the past it took about 100 years for a market penetration from 10 to 50%, which means that solar hydrogen should cover 10% of the energy market by the year 2000 if by the year 2100 it is assumed to cover 50% and if things would go at the same pace as in the past. Even if things move faster now solar hydrogen will not cover 10% of the market by the year 2000. Works on solar hydrogen should - by judiciously analyzing the penetration curve - have started before the beginning of the century in order to have 50% of mankind's energy consumption covered by solar hydrogen by the year 2100. The least to say is that it is not too early to start working on it now.

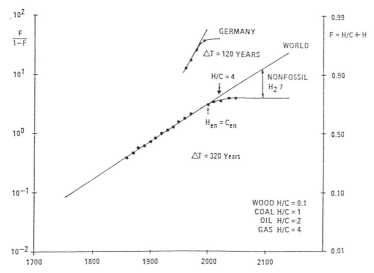

Fig. 4. Evolution of the hydrogen/carbon molecule ratio of the fossil fuels.

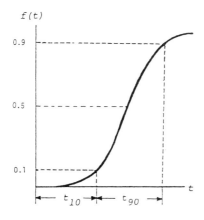

Fig. 5. Technology penetration.

STRATEGY FOR THE IMPLEMENTATION OF SOLAR HYDROGEN

The transition to the non-fossil energy era will take place the moment non-fossil energy becomes economically competitive, i.e. when the total costs including societal costs of fossil fuels, growing diseconomics of supply and environmental imperatives become greater than those for non-fossil energy.

Although it is not possible to predict neither the evolution of fossil fuel costs nor the moment the costs of fossil fuel pollution become internalized there are indications that the onset of a hydrogen technology within new clean and renewable energy supply systems will take place by 1995-2000, with pronounced deployment by 2020-2040.

A PRESENT REALIZATION OF THE TECHNOLOGY:
THE EURO-QUEBEC HYDRO-HYDROGEN PILOT PROJECT

The concept of a hydrogen-based, clean, renewable energy system, conceived by the Joint Research Centre Ispra of the Commission of the European Communities, is currently investigated by European and Canadian Industries, coordinated by the JRC-Ispra of the Commission of the European Communities and the Government of Québec.

The 100 MW pilot project is to demonstrate the provision of clean and renewable primary energy in the form of already available hydroelectricity from Québec converted via electrolysis into hydrogen and shipped to Europe, where it is stored and used in different ways: electricity/heat cogeneration, vehicle and aviation propulsion, steel fabrication and hydrogen enrichment of natural gas for use in industry and households, see Fig. 6.

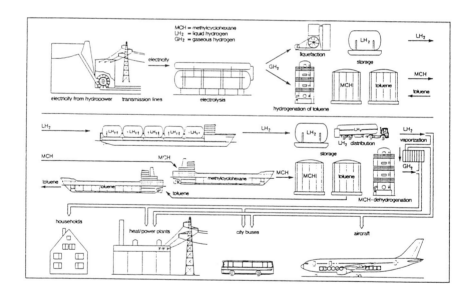

Fig. 6. The concept of the EQHHPP project.

Funds have been made available by the Commission of the European Communities in 1991 and by European industry as well as from the Government of Québec and from Canadian industry to undertake demonstration projects on the utilization of hydrogen in four fields where the use of hydrogen exhibits its attractiveness. In Europe, these projects are executed by industry on the basis of contracts following a tender action by the Commission with a cost sharing of at least 50% by the industrial partner:

- Vehicle propulsion: In Germany, for example, from the overall emissions those resulting from road traffic are: 52% NOx, 70% CO and 49% VOC.
Public transport is predestinated for the introduction of clean and therewith expensive fuel since fuel costs are only a small fraction of the overall operating costs of buses. A typical bus operation costbreakdown shows that fuel costs are only 7.3% of the total operation costs.
Four public transport buses of different concepts will be built and operated: internal combustion engine, fuel cell, Stirling engine.
- Aviation propulsion: The elimination of CO, CO_2, SO_2 and VOC by using hydrogen not only is beneficial for the atmosphere, but reduces considerably airport ground pollution. A problem to be solved, however, is the increased water vapour emission.
An Airbus combustion chamber designed for minimum NOx emission will be built and operated.
- Steel fabrication: With ~ 2 kg CO_2 emissions per kg steel fabricated with carbon as reductant, the world's steel fabrication contributes over 10% to the world's total anthropogenic CO_2 production. Hydrogen as excellent and clean reductant has a large potential to reduce CO_2 emissions.
A demonstration component including iron ore reduction with hydrogen by plasma arc process will be designed, built and tested.
- Advanced techniques of liquid hydrogen storage.
Large scale model containers will be built and tested, including accident simulation and rupture tests.

REFERENCE

Funk, J.E. and R. Reinström (1966). Energy requirements in the production of hydrogen of hydrogen from water. I&EC Process Design and Development, Vol. 5, No. 3, 1966.
Gretz, J. et al. (1992). Phase II and Phase III.0 of the 100 MW Euro-Québec Hydro-Hydrogen Pilot Project EQHHPP. Proceedings of the 9th World Hydrogen Energy Conference, Paris, June 1992.
Marchetti, C. (1985). When will hydrogen come? Int. J. Hydrogen Energy, Vol. 10, No. 4, 215-219, 1985.

OCEAN-WAVE ENERGY RESEARCH IN NORWAY

J. FALNES

Division of Physics, Norwegian Institute of Technology
University of Trondheim
N-7034 Trondheim, Norway

ABSTRACT

A historical outline is given of the research and development of ocean wave energy in Norway. Early theoretical achievements concern artificial wave focusing as well as optimum phase control of geometrically small wave energy converters ("point absorbers"). A short account is given of two full-scale demonstration prototypes, the Norwave "Tapchan" plant and the Kvaerner oscillating-water-column (OWC) plant. Some details are given on a phase-controlled wave-power buoy, as well as on a phase-controlled twin OWC wave-energy converter, which is a more advanced system than the type of single OWC built in full scale so far. A recent study of the Norwegian wave energy resource confirms previous studies for the northern part of the coast, while reduced updated numbers apply for the southern part. It is predicted that energy supply from waves will be commercial in developing countries in the 1990s and in industrialised countries, such as Norway, in the next century.

KEYWORDS

Phase control; point absorbers; oscillating water column; tapered channel; focusing.

INTRODUCTION

A substantial program on research and development of wave energy converters was launched in Norway in 1978, financially supported by the Ministry of Petroleum and Energy. This took place following a similar program start in the UK in the mid 1970s,

2451

after the oil crisis in 1973 and general public concern on environmental problems around 1970. However, some interested individuals at a research institute (Mehlum and Stamnes 1979) and at a university (Budal and Falnes 1975) started already in the early 1970s to make theoretical studies on conversion of the energy in ocean waves.

Based on these studies, various types of wave energy converters were proposed. Model experiments were carried out, and three different full-scale designs were assessed technically and economically (White paper, 1982). Moreover, the wave energy resource in Norwegian waters was estimated (Torsethaugen 1982).

In the early 1980s the oil price was declining and the general public was less concerned, than ten years earlier, about energy and environment problems. The governmental financing of wave power research was drastically reduced in the UK in 1982 and in Norway in 1983. However, since it was possible to find fifty percent financial support from private companies, the construction of two full-scale, shore-based demonstration prototypes was started on the Norwegian west coast, on the small island Toftestallen, 40 km NW of Bergen. The two prototypes were officially inaugurated in 1985 and connected to the electrial grid (Unknown, 1987). Both prototypes had a generator capacity of approximately 0.4 MW. One of the prototypes, a device of the OWC type, owned by the company Kværner Brug, was destroyed during a heavy storm after a few years of operation. After a reorganisation of the Kvaerner group of companies in 1989 it was decided to curtail Kvaerner's involvement in wave power.

Another company, Norwave, the owner of the other prototype, a horizontal tapered-channel device, is still engaged in the task of converting wave energy. Currently, private industry accounts for roughly two thirds of the budgets, while Government continues to contribute a "base load" of research funding. Agreements have been signed on developing wave-power plants in Indonesia and in Tasmania, Australia (Mehlum 1991).

Through NTNF (Royal Norwegian Council for Scientific and Industrial Research) and through NAVF (Norwegian Research Council for Science and Humanity) the Ministry of Petroleum and Energy has during recent years again increased the financial support for wave-power research aimed at developing more advanced wave power converters for the future. Support is also given to update and improve data on the natural wave energy resource.

Most wave-energy research groups around the world were started by researchers trained in civil engineering, mechanical engineering, naval achitecture or electrical engineering. In contrast, the early wave power research in Norway was initiated by phycisists who made theoretical studies for several years before support was made available for experimental work and substantial design studies. This fact had a particular influence on the types of wave-energy converters selected for extensive design studies and assessments in the Norwegian wave energy program.

A short review of this program will be given in the present paper, with emphasis on our own work. Finally, the future prospects for wave energy utilisation will be discussed.

WAVE ENERGY RESOURCE

Measurement of ocean waves was carried out for some years before 1980 at a few sites off the Norwegian coast. Further, using hindcasting models and meteorological data since 1955, a quantitative study of the Norwegian wave power resource was made some 10 years ago (Torsethaugen 1982). Among the research institutions taking part in early wave resource studies were NHL (Norwegian Hydrotechnical Laboratory), Det norske Veritas, NMI (Norwegian Meteorological Institute), IKU (Continental Shelf and Petroleum Technology Research Institute) and NTH (Norwegian Institute of Technology).

During the last decade hindcastig models have been improved. In addition, wave measurements have been extended both to new locations and to include directional wave measurements. Measurements taken in eight offshore locations, whose positions are indicated in table 1, have been used to calibrate the hindcast model. This model in combination with the wave measurements has improved our knowledge on the wave climate, such as information on wave direction, on duration of wave states, on seasonal variations, and on the variation of annual averages (Torsethaugen 1990). It should be noted that the wave energy resource off the Norwegian coast, as indicated by the numbers in the in the last line of table 1, agree fairly well with previous results (Torsethaugen 1982) for the northern part of the coast. However, south of 63^0 N the updated figures are smaller, by a factor of almost 2.

Table 1.   Norwegian wave energy resource (Torsethaugen 1990).
The first two lines indicate approximately the latitude and the longitude of offshore locations for wave measurements. The last line indicates the average power (in kW) transported by the waves through an envisaged vertical cylinder of unit diameter (1m).

59	62	64	65	66	70	72	72	(^{0}N)
5	5	9	8	9	20	19	31	(^{0}E)
22	31	33	39	41	24	33	20	(kW/m)

The wave state as well as other physical quantities of the oceans are measured not only in domestic waters, but also worldwide, by the Norwegian company OCEANOR (Barstow et al. 1991).

As the waves approach the coast and interact with the bottom topography they are refracted. The result is that waves are focused in some locations and defocused in other locations. The companies Norwave and OCEANOR, for instance, have developed computer programs which are useful when locating a wave power plant at a near-shore or on-shore site exposed to more wave energy than other sites. Because of the shielding effect of mainland and islands the direction of propagation of swells also plays an important role for the available wave energy at near-shore or on-shore sites. This effect may be of more significance than refraction (Løvhaugen 1991, private communication).

EARLY THEORETICAL ACHIEVEMENTS

In addition to natural focusing due to the bottom topography artificial focusing of waves may be utilised to concentrate wave energy from a certain coast length into a relatively narrow focal area. This idea led a group of physisicts in Oslo, working on optical and various other types of waves, into a research program to harness clean energy from the ocean waves (Mehlum and Stamnes 1979). Focusing of ocean waves may be achieved by using specially shaped, submerged bodies, which reduce the waves' phase speed, analogous to the reduction of the phase speed of a light wave in a glass lense. Later experiments in a big outdoor wave tank confirmed the theory of focusing (Mehlum 1982). The waves' variation in frequency, amplitude and angle of incidence can be accomodated. However, large variations require a more expensive lense system.

In Trondheim other physisists, with a background in electromagnetic waves and acoustical waves, considered a wave energy converter as analogous to a microphone or a receiving antenna. They pointed to the advantages of designing wave energy converter units of small geometrical extension. The term "point absorber" was introduced for a device whose horizontal extension is small compared to the wavelength (Budal and Falnes 1975). From this study (and independent simultaneous studies by D.V.Evans, University of Bristol, UK, and by C.C. Mei and J.N. Newman, M.I.T., Mass., USA) it is known that the maximum power that can be captured by an axisymmetric heaving buoy is equal to the incident wave power transport associated with a wave front of width one wavelength divided by 2π. Subsequently, the theory of an array of point absorbers was developed (Budal 1977, Falnes 1980), and it has been shown, also experimentally, that 50 percent of the incident wave energy can be absorbed under optimum conditions.

To obtain optimum conditions for maximum energy capture the absorbing system has to oscillate with a certain phase relative to the incident wave. For a single absorber this is achieved with a resonant oscillating system, which has a natural oscillating period equal to the wave period. However, since the wave period is a varying quantity and since real sea waves are not sinusoidal, a simple oscillating system with one natural period cannot, in general, accomodate the optimum condition. For this reason it was proposed by Budal (Budal and Falnes 1977, 1980), and independently by Salter (Salter et al. 1976), to use a control system to achieve the optimum oscillation condition.

Above we considered the theoretical maximum captured energy relative to the incident wave energy transport. Another theoretical upper limit, derived for a heaving point absorber (Budal and Falnes 1980) relates the maximum captured energy to the physical size of the device. It can be shown that the ratio of the power output P to the volume displacement change V (heave stroke multiplied by water plane area) is less than

$$(P/V)_{max} = (\pi/4)\rho gH/T$$

where ρ is the mass density of the sea water, g the acceleration due to gravity, T the wave period, and H the wave height ($H/2$ the wave amplitude). Thus, for typical wave parameters, $H = 2$ m and $T = 10$ s, we have $P/V < 1.6$ kW/m^3.

may be utilised to broaden the bandwidth and, hence, to improve the performance of the converter. A considerable theoretical and experimental program was carried out (Malmo 1984, Malmo and Reitan 1985). One of the results is that the optimum length of the "canal" is of the order of one quarter of the shortest wavelength for which the converter is designed. At first it was considered to construct an array of relatively large, bottom-standing structures in sea of depth 25 to 30 m. The envisaged power rating of each unit was 8 MW. (Ambli et al. 1982, White paper, 1982). Subsequent assessments indicated smaller units rated at 0.5 MW to be more economical.

Such a smaller unit, land-based on a steep rocky shore was constructed during 1984 and 1985 at Toftestallen. The lower part (up to 3.5 m above mean sea level) was built in concrete, while the upper part of the structure was built in steel. For financial reasons the "canal" in front of the OWC was made shorter than the optimum length, and it was shaped simply by rock blasting. The mouth of the OWC has a width of 10 m and a height of 3.5 m (ranging from level -7 m to level -3.5 m). The internal water surface of the OWC has an area of 50 m². Since the width is much smaller than wavelengths of interest, the OWC prototype may be classified as a point absorber.

After two years of testing the prototype it was reported that the basic principles developed theoretically and through laboratory testing were tenable (Unknown, 1987). In a heavy storm during the last week of 1988 the steel part of the plant was swept down, while the concrete part remains. The company Kværner Brug decided to rebuild the upper part in concrete. It also planned to build a wave-power plant in Tonga in the South Pacific. However, in 1990, when Kværner Brug had been fused into the company Kværner Eureka, it was decided to shelve Kvaerner's involvement in wave power.

PHASE-CONTROLLED POINT ABSORBERS

From the previous theoretical studies it became clear that it is advantageous to make the structure of a wave-energy converter small in comparison with the wavelength. However, since a small oscillating body has a narrow resonance bandwidth, this advantage cannot be realised with the real waves of the ocean unless the oscillatory motion can be optimised by using appropriate mechanical components, measuring gauges and an electronic computer. A method to obtain approximate optimum motion by a *mechanically* simple method was proposed by Budal (Falnes and Budal 1978). The idea is to latch a wave-absorbing oscillating buoy in fixed position during certain time intervals of each wave period. Various designs of heaving buoys were proposed to utilise this latching principle. We tested various buoy shapes. In some cases the conversion of the absorbed energy into useful form was achieved by means of a hydraulic piston pump and a hydraulic motor. However, most attention was paid to a buoy of spherical shape having a pneumatic power take-off system (Budal et al. 1982). During the years 1981 to 1983 a full-scale wave power buoy unit, 10 m in diameter and rated at 0.5 MW, was designed and assessed. Further, we tested a model in scale 1:10 in a wave channel and also in the sea near Trondheim.

Energy and labour are required to make any product. The energy invested in 1 ton of steel is approximately 11 MWh. It was estimated that our above-mentioned wave-power buoy

This relation is not widely known, and for most of the wave energy converter designs that have been assessed so far, the ratio P/V is less than the mentioned theoretical maximum by one order of magnitude or more. Instead most devices have been designed to capture a large fraction of the incident wave energy on a coast. However, this is not necessarily the best economic optimum, since the wave energy in the ocean is free.

THE NORWAVE TAPCHAN ENERGY CONVERTER

The above-mentioned research group which has studied focusing of ocean waves, proposed to convert the focused wave energy into useful energy by means of a horn-shaped or tapered horizontal channel which absorbs the incoming wave and converts most energy into potential energy in form of elevated sea water collected in a water reservoir. The waves enter the wide end of the tapered channel, and the wave height is magnified until the wave crests spill over the walls and into the water reservoir. Then the gradual narrowing of the channel causes a continuous sideways spill-off as the wave crests move along. In order to avoid reflection of wave energy the shape of the channel must be carefully designed. The term "Tapchan" has been adopted for this type of tapered-channel wave-energy converter. The envisaged height of the water reservoir is in the range 3 to 7 m above mean sea level (Mehlum 1985).

Funding has not been available for construction of a full-scale submerged wave-focusing lens. However, sufficient funding was available to construct one Tapchan demonstration unit in 1985, at Toftestallen. The water reservoir provides a head of approximately 3 m for a water turbine of the Kaplan type, which runs an electric generator connected to the Norwegian power grid. The rating of the machinery is 0.35 MW, and the area of the water reservoir is approx. 8500 m². The total length of the channel, including the wide wave-collecting part is approximately 170 m. The wide part is shaped simply by rock blasting. Half of the total length, the narrow part of the tapered channel, is made in concrete. The vertical concrete walls are 10 m high, of which 3 m is above the mean water level. The mean tidal range is 0.9 m.

Recent theoretical developments on the Tapchan converter indicate that it is possible to reduce the channel length considerably without significant reduction of the efficiency (Mehlum 1991, private communication). In order to demonstrate this experimentally, work has been started on reconstructing the Tapchan plant at Toftestallen.

THE KVAERNER MULTIRESONANT OWC

The first wave-energy converter unit considered by the company Kværner Brug, Oslo, was an offshore axisymmetric buoy, optimally operated by using high-pressure hydraulic machinery (Ambli et al. 1977). Around 1980 the company decided to develop a non-axisymmetric converter of the oscillating-water-column (OWC) type. A particular feature of Kværner's OWC is the pair of walls protruding from the sides of the mouth of the OWC. This has the advantage that in addition to the usual OWC resonance a "harbour" resonance (associated with the short "canal" between the OWC mouth and the open sea)

would have to operate for one to two years before it had recovered the energy associated with the steel and the other materials invested in it. This energy recovery time is much shorter than for other assessed wave energy converters. However, the labour invested in our wave power buoy would have to be relatively large. Since the phase-controlled power buoy contains some critical moving parts, it is also believed that relatively much labour is required for operation and maintenance, compared to the other assessed wave energy converters. It was judged that substantial development and testing of the critical mechanical components have to be carried out before reliability and lifetime reach a level required for a power plant. This development work was, unfortunately, discontinued in 1983, when governmental funding of this wave energy project was stopped.

With less external funding available for wave energy research, we have during recent years, as one of several projects, been applying the principle of phase-control to improve the OWC type of wave energy converter. Suppose that in an OWC structure, such as the one constructed by Kværner Brug at Toftestallen, a vertical separating wall has been erected. See fig. 1. The two adjacent OWCs may be controlled for maximum power conversion by opening and closing the two air valves V1 and V2 (fig.1) at optimum instants of every wave period. Valve V1 lets air in from the outer atmosphere, while valve V2 lets air out. Thus the valves also serve to rectify the air flow through the turbine, which means that a conventional air turbine of relatively high efficiency may be used.

We have made simulation studies (Falnes et al. 1989) and model experiments on such a twin OWC system. Let us now compare such a system with a corresponding single OWC

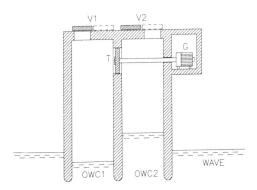

Fig. 1.   Schematical drawing of a twin-OWC wave energy converter. The two air chambers above the two OWCs communicate with the ambient atmosphere through the operable air valves V1 and V2 (shown in closed and open position, respectively). An air turbine T between the two chambers runs an electrical generator G.

system with a self-rectifying air turbine (Wells turbine) between the air chamber and the outer atmosphere. Let the geometry correspond to the Toftestallen prototype of Kværner Brug, except that the area of the mouth is increased from 35 m² to 50 m² (10 m width between levels -8.5 m and -3.5 m). At the resonance period 7 s the twin system cannot absorb more wave power than the single system at optimum loading. However, at longer wave periods, the twin system is the better. For instance, for wave periods 9.5 s and 11 s experiments on a model (in scale 1:10) has shown that the twin OWC system can convert 60 to 100 percent more wave energy than the single OWC system. This will more than outweigh the estimated 10 to 15 percent additional construction cost due to the separating wall and the two air valves of the twin OWC converter. An additional advantage of the twin OWC system is that the optimum turbine admittance is smaller (that is, the pneumatic power is converted by the turbine at a relatively higher pressure drop and a relatively smaller air flow). This means a turbine of smaller physical size, which probably has some economical advantage.

The model experiments have so far been carried out in regular (sinusoidal) waves, only. In order to experiment in more realistic, irregular waves, computer software is required to determine the optimum opening times of the valves. Work on this software development is in progress.

FUTURE PROSPECTS

A few successful, on-shore or near-shore, wave-energy converters, have been built. They are characterised by few moving components and by relatively large, fixed structures. It appears that such systems are going to be commercially competitive already in the 1990s in developing countries, where knowledge, skill an inexpensive labour are available, such as in China and India. Such converters will be of the greatest value for small island and other coastal communities, which at present have to rely on expensive diesel-generated electricity. Because of the expensive labour in industrialised countries, such large-structure wave-energy converters will hardly become commercial for general large-scale energy supply in those countries.

We will point out that the development of wave-power technology is still in its infancy. For instance, several of the previously mentioned theoretical achievements from the 1970s have not yet been brought into practice. A full-scale focusing plant has yet to be tested. In order to utilise the principle of optimally controlled point absorbers, it is necessary to develop new mechanical components, electronics and software. This, as well as other arguments which have been discussed in some detail elsewhere (Falnes and Løvseth 1991), suggests that there are good possibilities for developing more cost-effective wave energy converters, provided sufficient funding for the required research and development will be made available. Thus it seems to be possible that wave energy will provide a significant contribution to the general energy supply in the next century, in Norway, as well as in other industrialised coastal countries.

Unlike most other countries, Norway has its electricity supply almost exclusively from hydropower, which requires large water reservoirs to store water for the winter, when the precipitation in mountain areas is mostly in the form of snow. In contrast, wave energy matches better with consumers' need, since it is most abundant in the winter season. Thus it seems to be beneficial to supply the Norwegian electricity grid with additional energy from waves. In such a combined system the consumption of water from the reservoirs will have to be larger on days with calm sea. An advantage is that larger energy reserves are provided for the drier years.

REFERENCES

Ambli, N., Budal, K., Falnes, J. and Sørenssen, A. (1977). Wave power conversion by a row of optimally operated buoys. *10th World Energy Conference. Division 4. Unconventional Energy Resources. Studies of Development.* Paper 4.5-2. Istanbul, Turkey.

Ambli, N., Bønke, K., Malmo, O. and Reitan, A. (1982). The Kværner multiresonant OWC. *Proc. Second International Symposium on Wave Energy Utilization* (H. Berge, ed). pp.275-295. Tapir, Trondheim, Norway.

Barstow, S. F., Sørås, P.-E. and Selanger, K. (1991). Wave energy resource mapping for Asia and the Pacific. *Proc. OCEANS'91,* Honolulu, Hawaii, Vol. 1, pp. 385-390.

Budal, K. (1977). Theory for absorption of wave power by a system of interacting bodies. *Journal of Ship Research,* 21, 248-253.

Budal, K. and Falnes, J. (1975). A resonant point absorber of ocean-wave power. *Nature,* 256, 478-479. (With corrigendum in Vol.257, p.626).

Budal, K. and Falnes, J. (1977). Optimum operation of improved wave-power converter. *Marine Science Communications,* 3, 133-150.

Budal, K. and Falnes, J. (1980). Interacting point absorbers with controlled motion. In: *Power from Sea Waves* (B. Count, ed.), pp. 381-399. Academic Press, London.

Budal, K., Falnes, J., Iversen, L.C., Lillebekken, P.M., Oltedal, G., Hals, T., Onshus, T. and Høy, A.S. (1982). The Norwegian wave-power buoy project. *Proc. Second International Symposium on Wave Energy Utilization* (H. Berge, ed.), pp. 323-344. Tapir, Trondheim, Norway.

Falnes, J. (1980). Radiation impedance matrix and optimum power absorption for interacting oscillators in surface waves. *Applied Ocean Research,* 2, 75-80.

Falnes, and Budal, K. (1978). Wave-power conversion by point absorbers. *Norwegian Maritime Research,* Vol. 6, No. 4, pp. 2-11.

Falnes, J. and Løvseth, J. (1991). Ocean wave energy. *Energy Policy,* 19, 768-775.

Falnes, J., Oltedal, G., Budal, K. and Lillebekken, P.M. (1989). Simulation studies of a double oscillating water column. *Proc. Fourth International Workshop on Water Waves and Floating Bodies* (J. Grue, ed.), pp. 65-68. Department of Mathematics, University of Oslo, Norway, ISBN 82-553-0671-4.

Malmo, O. (1984) *A study of a multiresonant oscillating water column for wave-power absorption.* Thesis for the "dr.ing." degree. University of Trondheim, Norwegian Institute of Technology.

Malmo, O. and Reitan, A. (1985). Development of the Kvaerner Multiresonant OWC. *Hydrodynamics of Ocean Wave-Energy Utilization* (D.V. Evans and A.F. de O. Falcao, eds.), IUTAM Symposium, Lisbon, Portugal, 1985, pp. 57-67, Springer Verlag, Berlin.

Mehlum, E. (1982). Recent developments in the focusing of wave energy. *Proc. Second International Symposium on Wave Energy Utilization* (H. Berge, ed.), pp. 419-420. Tapir, Trondheim, Norway.

Mehlum, E. (1985). Tapchan. *Hydrodynamics of Ocean Wave-Energy Utilization* (D.V. Evans and A.F. de O. Falcao, eds.), IUTAM Symposium, Lisbon, Portugal, 1985, pp. 51-55. Springer Verlag, Berlin.

Mehlum, E. (1991). Commercial tapered channel wave power plants in Australia and Indonesia. Vol. 1, pp. 535-538. *Proc. OCEANS'91*, Honolulu, Hawaii.

Mehlum, E. and Stamnes, J. (1979). Power production based on focusing of ocean swells. *Proc. First Symposium on Wave Energy Utilization*, pp. 29-35. Chalmers University of Technology, Gothenburg, Sweden.

Salter, S.H., Jeffrey, D.C. and Taylor, J.R.M. (1976). The architecture of nodding duck wave power generators. *The Naval Architect*, Jan. 1976, pp. 21-24.

Torsethaugen, K. (1982). The Norwegian wave climate mapping programme. *Proc. Second International Symposium on Wave Energy Utilization* (H. Berge, ed). pp. 81-97. Tapir, Trondheim, Norway.

Torsethaugen, K. (1990). *Bølgedata for vurdering av bølgekraft.* Report no. STF60 A90120, Norwegian Hydrotechnical Laboratory, 1990, ISBN no. 82-595-6287-1

Unknown (1987). *Norwegian Wave Power Plants.* Report from MPE (The Ministry of Petroleum and Energy), Oslo, Norway.

White paper (1982). *Nye fornybare energikilder i Norge.* Stortingsmelding nr. 65 (1981-82). The Royal Ministry of Petroleum and Energy, Oslo, Norway.

RENEWABLE ENERGY PROGRAM OF THE NETHERLANDS

C.W.J. Van Koppen

Emeritus Professor Eindhoven University of Technology
Kosmoslaan 25, 5632 AT Eindhoven, Netherlands

ABSTRACT

In the course of time the renewable energy program of the Netherlands has developed from a single gouvernment action into a multiple action by central, regional and local authorities. These various actions are discussed for the fields of solar passive and active heating, photovoltaîcs, biomass and wind energy. History is indicated and the funding, short and long term objectives and general philosophy are outlined. The management and coördination of these actions is mostly taken care of or supported by the Netherlands Agency for Energy and the Environment, a semi-gouvernmental institute that manages the majority of the government programs in these areas. This organisational form proves to be effective and efficient.

KEYWORDS

Netherlands; R,D & D; renewable energy; program; passive solar; active solar; photovoltaïcs; biomass; wind energy.

INTRODUCTION

The Netherlands is a small country (land area 33920 km^2) and with 15.13 million inhabitants (ultimo 1991) it is rather densely populated (446/km^2). The country is the combined delta of the rivers Rhine, Maas and Schelde and consequently it is very flat. The climate is temperate because of the nearby North Sea and the prevailing westerly winds; average temperatures range from 2.0 $^{\circ}$C in January to 16.6 $^{\circ}$C in July. Solar radiation is among the lowest in the world, averaging 112 W/m^2 (980 kWh/m^2.yr) on a horizontal surface; only 35 % of this radiation is direct sunlight. Wind energy is rather strong in the Netherlands, in particular in coastal areas.

2461

The first in the Netherlands to draw attention to the potential of renewable energy was C. Daey Ouwens in 1972. A follow-up to this was given in 1973 by the author in the framework of an energy conservation study (Over, 1974). For reasons of historical depth we quote from this report: "A tentative evaluation shows that solar heating will only begin to make a significant contribution to national energy supply after the year 2000". The oil crisis in 1973 triggered a multitude of additional initiatives, almost all of them in the field of active solar heating. In 1976 the Dutch Section of the International Solar Energy Society was established and in 1978 the first formal National Solar Energy Program, mainly directed towards active solar heat was inaugurated. A formal program in the field of photovoltaïcs resulted around 1980 from an initiative of J.C. Francken, president of the Dutch Section of ISES and in the same period particular attention was drawn to passive solar by C. Zijderveld in the Municipality of Schiedam. The present programs have their roots in these initiatives.

The nature of the solar radiation in The Netherlands excludes such applications of solar energy as focussing collectors, power towers and large scale electricity generation. Further the potential of biomass is limited by lack of space and moderate temperatures, and the potential of hydropower by the flatness of the country. Lack of space also excludes an "all solar" energy supply for the country alone. In 1990 the primary energy consumption amounted to 2193 PJ (non-energy use of fossil fuels excluded), which corresponds to 2.05 W/m^2 horizontal. Solar supply of this energy would require a collecting area equal to a prohibitive 6% of the land area of the country, assuming a collecting efficiency of 33%, including (seasonal) storage losses. Official estimates state that in the year 2010 renewables might contribute about 150 PJ/yr to the national energy supply (in terms of displaced fossil fuel). This includes a provisional 80 PJ/yr from the combustion of waste and biomass, and 20 PJ/yr from energy conservation by recycling(!). It does not include the 100 PJ/yr contribution to space heating by "passive" insolation into houses. In the mid of the next century a total renewable contribution of 400 to 500 PJ/yr to a substantially unchanged final energy consumption is feasible.

PROGRAMS AT VARIOUS LEVELS

The national government funding for renewable energy R,D,D & I(nformation) programs is shown in table 1. The different kinds of renewables are discussed separately in the following sections. Here it is interesting to compare the Dutch efforts with those in some other countries. Restricting the comparison to the total R,D &D funding and taking the latter per capita it appears that the funding in The Netherlands is roughly equal to the funding in Japan, is double the funding in The United States and is half of the funding in Germany and Switzerland. However, the table does not include the Dfl. 3-4 million/yr government funding for wind energy R&D in separate programs,the Dfl.10 millions/yr that are made available for a 40% subsidy on new photovoltaïc and photothermal installations and the Dfl.35-40 million/yr investment subsidy for new windturbines (Dfl.300-600/m^2 swept rotor area). Adding these makes the government support for renewables per capita rank among the highest in the world.

All programs indicated in the table are managed by the Netherlands Agency for Energy and Environment (NOVEM), a semi-governmental organisation that also coördinates the programs,

2462

Table 1. The national government funding for renewable energy R,D,D & I in five subsequent years in The Netherlands (in Dfl.millions; 1 Dfl ≈ 0.45 US $)

Year	Solar Heat (act.+pass.)	Photo- voltaïcs	Wind- energy	Biomass	Geothermal energy	Other	All renewa- ble energy
1990	6.5	7.2	9.0	-	0.2	3.0	22.2
1991	6.0	10.7	9.0	preparatory	0.3	1.8	29.8
1992	6.0	9.5	9.1	work	2.3	0.2	27.1
1993	5.8	9.5	9.1	?	2 ?	0.2	26.6+?
1994	5.8	9.5	8.1	?	2 ?	?	25.6+?

as far as possible, with initiatives of local or regional governments or utilities. Most programs are handled by the Built Environment Division of NOVEM, but the Wind program is taken care of by the Energy Supply Division whereas Biomass is part of the responsability of the Industry and Environment Division (group: Energy from Waste). In this way each of the programs is brought as close as possible to the field of its eventual application.

Among the initiatives of local or regional governments or utilities the so called Environment Action Plan (MAP) of the Energy Distribution Sector (power stations not included) is the leading one. It aims at a reduction of the CO_2-emission by about 6% of the present total of 160 million tons/yr and a reduction of the emitted acidification by some 1.5% of the present precipitation of 16 billion acid-aequivalents per year, both in the year 2000. The efforts in the MAP are mainly concerned with all possible forms of energy conservation in the societal sectors that can be influenced by the Energy Distribution Sector, but in the period 1991-'95 also about Dfl.30 million/yr is made available to support the construction of windturbines and Dfl.1 million/yr to promote photovoltaïcs. Further, substantial help is planned for the development of geothermal energy (low temperature) and the rental of DHW-systems, but the financial consequences of these actions have not been clearly specified, as yet.

Execution of the MAP is mostly delegated to regional, sometimes local energy distributors and in many cases forms of collaboration with regional and local governments have been developed in this context. Also independent initiatives have been taken by the latter and some examples thereof will be given in the next paragraphs.

SOLAR HEAT

The main objective of the current photothermal program (1991 through '94) is a home market of 15,000 DHW systems per year after 1994, aiming for 300,000 systems installed in the year 2010. Quoting from (Havinga, 1992): Main areas of research are the development and testing of new cheaper materials and components and the development of new systems with lower production and installation costs. These systems should be appropriate for production at a large industrial scale and selling without subsidy.

An important difference compared with previous programs is the emphasis given in the program to creating the market boundary conditions for the large scale introduction of solar DHW systems. To structure the market introduction and information transfer activities in the coming years, a marketing communication strategy has been set up. Market studies showed that, within the Dutch population, several market segments can be distinguished with their own characteristics and attitude with respect to solar energy. The communication strategy aims to guide and schedule the market introduction activities and to make them as efficient as possible, taking into account all relevant boundary conditions. Important items will be knowledge transfer, information (local governments, housing companies, general public, etc.), and support activities for e.g. local activities of utilities and local governments with respect to introduction of solar DHW systems, especially by renting systems. Much attention is also given to installers of the systems, who are supported with training in sales and system installation, brochures, leaflets, meetings/workshops, etc.

By stimulating the solar market, sales will increase, which should result in series production of the solar DHW systems and subsequently lower prices. In term, this expected reduction in prices should make the present subsidy of 40% redundant. As a result of the market introduction activities, started in the beginning of 1991, the sales of solar DHW systems increased from 700 systems in 1990 to 2200 systems in 1991.

The utilisation of solar heat is preferably considered as part of energy conservation for the sake of the environment. A really fine example of this is the 100-houses project ECOLONIA in Alphen on Rhine. Here passive solar is one of the (many) means to reduce energy demand and active solar is applied to ten of the houses as a clean form of energy supply. Also a multitude of other ecological ideas have been incorporated in the layout and designs of ECOLONIA. The project has been funded by Novem, the Housing Fund of Dutch Municipalities, the occupants and some third parties. The results of the ongoing monotoring will become available in 1993.

Other subjects in the solar heat program are space heating and larger systems. In residential space heating the use of integrated heating equipment is increasing. Such appliances incorporate space heating, hot water preparation and storage, and ventilation in a single unit. Research on the application of solar collectors as (pre)heat sources to such units is part of the current program. Further field-experiments on larger, collective hot water systems for hospitals, homes for the aged, restaurants etc. are envisaged.

It needs little comment that all the work mentioned is supported by theoretical work on models and design methods, by (the development of) performance and certification tests, and by standardisation. International consultation and coöperation, mostly in the framework of EEC and IEA programs, plays an important role in these activities.

PHOTOVOLTAÏCS

It took more than ten years for the Dutch R&D program on photovoltaïcs to grow to maturity. Government funded research has been started in 1978 and up to 1990 this research was mainly concerned with (fundamental) research on cells, apart from the 50 kWp EC pilot project on the island of Terschelling. The funding was relatively low; Dfl. 2-3 millions. The current 1990-'94 program preludes on an strong increase of the practical use of photovoltaïc electricity in a not too distant future: it is not impossible that a break-even point for grid-coupled photovoltaïc electricity may be attained within two decades. The continual improvements of the price-performance ratio combined with the 40% government subsidy have already increased the selling rate of small stand-alone PV-applications considerably: it rose from 358 in the year 1988 via 1769 in 1989 and 2949 in 1990 to about 4000 in 1991. Parallel to this increase in number the totally installed power increased from 80 kWp to 700 kWp (ultimo 1990) and 1 MWp as of spring 1992. This process is expected to continue, maybe to the officially projected contribution of 2 PJ (displaced fuel) to the national energy supply in the year 2010. This contribution corresponds to an installed capacity of 230 MWp, which implies an average increase in newly installed PV of 40% per year during the next two decades.

Partially quoting from (Muradin 1991): The main part of the research budget (see table 1), roughly 65%, has been allocated to cell research, and the rest (35%) to system research, including technical demonstration projects. Cell research will examine three categories of solar cells: flat plate polycristalline silicon cells, amorphous silicon cells and cells based on other materials: III/V compounds (Galliumarsenide), copper indium diselenide and organic compounds.

The flat-plate polycristalline cells have priority, as a pioneer, preparing the way for other types of solar cells. The objective is a 16% encapsulated cell efficiency in 1993 whereby the costs for standard modules must be reduced from 10 to about 5 Dfl./Wp. In the years after 1994 a further reduction of the price below 2.5 Dfl./Wp will be necessary to achieve widespread grid-connected utilisation.

For the amorphous cells the objective is a 10% stabilized cell efficiency on a 10x10 cm^2 surface. Costs of standard modules must be reduced from 6 to 3 Dfl/Wp and eventually to half this amount. The Dutch aSi research participates in the EEC Joule program.

For III/V compounds the high efficiency is as yet offset by the high costs. Therefore emphasis in research will be on ways to reduce the costs, possibly by using alternate substrates such as germanium or, maybe, silicon. The long term economic viability of these compounds for flat-plate terrestrial applications should be known at the end of the current program.

The research on Cu-In-Se, CdTe, organic and electrochemical solar cells is in first instance restricted to assessment studies on their scientific, technical and economical prospects.

For all types of solar cells the environmental impact of manufacture, use and replacement is carefully established; this impact is one of the primary selection criteria for further research. Also an energy analysis is made in order to obtain more certainty on the energy payback time, because data in literature differ widely.

The objective of the system research is to build up know-how among the utilities and other involved parties, by means of technical demonstration projects. A series of primarily grid-connected projects is to be established. As The Netherlands is a highly built-up country photovoltaïc modules will be integrated in the built environment, e.g. on the roofs of buildings. The installed power of the projects will be scaled up from 3 kWp to 300 kWp (1990-'94). Feasibility studies concerning centralized PV systems are also conducted. The studies deal with both utility-scale systems, and medium-scale systems integrated in existing structures.

WIND ENERGY

It is estimated that in the 17th. century about 10,000 windmills with an average rated power of 25 kW were installed in The Netherlands. That total installed power of 250 MW compares in an interesting way with the 55 MW windturbine power installed as of ultimo 1991, and the official objectives of 1000 and 2000 MW rated windturbine power installed in respectively the years 2000 and 2010. In the last mentioned year wind energy should supply 7% of the electricity consumption, saving 35 PJ in primary energy. The eventual share of wind energy in electricity production might amount to 20%, requiring the maximum possible turbine capacity on land of 6000 MW.

The objectives of the research program T(oepassing=application)W(indenergie)I(n)N(ederland) are improved price/performance ratio, higher safety (fatigue studies) and reduced noise production of windturbines. The results of the research should support the many industrial developments that are currently going on in The Netherlands, and help in speeding up the realisation of new projects. The structure of the Dfl. 35-40 million/yr investment subsidy is such that large and "silent" turbines are promoted; the largest turbine presently being developed has a rated power of 1 MW.

Some years ago the production costs of wind-electricity amounted to Dfl. 0.20-0.25/kWh for the larger types of windturbines. For comparison: the production costs in a large conventional power station are roughly Dfl. 0.09/kWh, social and environmental costs not included. It is estimated that as a result of R&D and large scale production of turbines the costs of wind-electricity will eventually come down to Dfl. 0.10-0.12, and become fully competitive with conventional electricity production, all costs included.

It is still uncertain whether windturbines can profitably be located near the coast in the sea; investments there are substantially higher than for construction on land. However it seems that in a densely populated country like The Netherlands, locating of turbines in the sea is the only way in which the share of wind-electricity in total electricity production may be enhanced beyond the limit of 20% just indicated. Therefore that option is carefully studied, as well as the consequences that a larger share of grid-connected wind energy may have for the stability and performance of the grid. Offshore projects might be realised in 1994/95.

BIOMASS

R,D &D in the field of biomass is by nature highly diverse. It ranges from small wood-burning cook-stoves to MW-installations for electricity production and purely scientific research on photoinduced electron transfer in organic compounds. Yet all these activities have a considerable potential for energy conservation and/or sustainable energy supply. Even in The Netherlands, densely populated and with a temperate and cloudy climate, biomass residues might supply 2-3% (50 PJ) of the primary energy consumption and energy farming on the 5000 km^2 of surplus agricultural land might eventually add 6-7% (140PJ) to this. Moreover industrial products, like wood and paper, might first serve their primary purpose and thereafter be used as a bio-fuel. The energy potential of this "cascading" has not yet been established but is certainly significant.

Technologically the utilisation of biomass in the ways just indicated is relatively simple. Development of properly adapted equipment is required and new logistical problems will have to be solved but no fundamental problems are to be expected. Economically the prospects are such that niches may be found in the market. The current energy price is about Dfl.6/GJ for oil and gas and about Dfl.5/GJ for coal. Total production costs for Poplar and Miscanthus on surplus land have been calculated to amount to some Dfl.8/GJ, and costs for straw to about Dfl.3.5/GJ. From these prices it appears that both the expected higher price of oil and gas in the future and the including of environmental costs in the price of fossil fuel would make the just mentioned forms of bio-energy competitive. This is further illustrated by the costs of electricity production (via gasification) from biomass, which were calculated at Dfl. 0.10-0.14, as compared with Dfl. 0.08-0.10 in conventional power stations.

In a densely populated country the energy production density is also significant. For trees, windturbines and PV-cells this production density has been calculated as respectively 25, 160 and 560 MWh/ha/yr under Dutch conditions. Of course the respective investments also differ largely and the three production methods do neither exclude each other nor the alternate use of the land for recreation or light industry.

From table 1 it is clear that the national R,D&D program for biomass is in the final phase of formulation. It is part of the program "Energy from Waste and Biomass" and as such not always clearly distinct from other activities. The program is subdivided in two areas: thermochemical conversion and biochemical conversion. Subjects investigated are a.o.:

Thermochemical conversion:

- The gasification of biomass and the associated preparation of the material (drying, pelletising, chipping etc.).
- The co-firing of wood powder in coal fired power stations.
- The gasification of wood for the utilisation of the gas in a gasturbine.
- Pollutants emitted by wood burning stoves and open fires.
- Small and medium scale combustion of compacted straw, following a Danish concept.
- The energy potential of effluents containing organic substances from the foods and allied products industry.

Biochemical conversion:

- Production and use of landfillgas, as "follow up" to the extensive experience gained from earlier projects.
- Anaerobic fermentation of biomass, replacing composting and producing biogas as an energy carrier.
- Fermentation of organic mud, again in order to obtain a gaseous fuel.
- Potential of potato residues and of algae.
- Energy farming in its various forms; the main results of a recent study have been mentioned above and will probably lead to some new projects.

Information exchange in the framework of EEC and IEA programs is very well developed for most of the subjects mentioned.

GEOTHERMAL ENERGY

Volcanic activity in The Netherlands is virtually zero, and consequently geothermal energy is not a renewable energy source in the stricter meaning of the word. Earlier research has made clear that hot water masses deep below the surface might be used for many decades as a low temperature heat source for, say, space heating. However, scarcity of geological data makes the risk of drilling high and the present low energy price makes exploitation less attractive. Consequently activities in this direction are on a low level presently.

On the contrary substantial R&D is now focussed on the long term (seasonal) storage aspect as such. Here four related but different lines of research have developed, all concerning storage in aquifers:

- High temperature heat storage (40-90 °C), with attractive applications in co-generation, district heating and greenhouses.
- Low temperature heat storage (20-40 °C),with possible applications in the heating of office buildings and, again, greenhouses.
- Cold and heat storage (6-20 °C), serving air conditioning in offices, industrial workshops etc.
- Cold storage below ground water temperature (≈10 °C), for various cooling purposes.

As of ultimo 1990 no less than 21 projects with A(quifer) T(hermal) E(nergy) S(torage) had been realised in The Netherlands; the long term potential is estimated to amount to some 1000 projects.

The connection of ATES with renewables is that any heat source of adequate temperature may be combined with it. Therefore, also solar heat may be used, but no definite plans of that nature do exist at the moment.

The column "Other" in table 1 refers to some assessment studies concerning hydropower and tidal energy and an organisational shift with the geothermal program. No significant practical potential has as yet been established for the two first mentioned energy sources.

REFERENCES

Havinga, J (1992), Thermal solar energy research in The Netherlands, Sun at Work in Europe,7,1, pp 16-17.

Muradin-Szweykowska, M. and E.W. ter Horst (1991), The Netherlands Photovoltaïc National Program, Proceedings 10th European Photovoltaïc Solar Energy Conference, 8-12 April 1991, Lisbon, Portugal, pp 1390-93. Kluwer Academic Publishers, Dordrecht, The Netherlands.

Over,J.A. and A.C. Sjoerdsma (eds), (1974), Energy Conservation: Ways and Means, Future Shape of Technology Foundation (Stichting Toekomstbeeld der Techniek), The Hague, Report 19.

Sky brightness during twilight. Model and measurements

Rodolfo Guzzi

IMGA CNR, Via Emilia Est 770, Modena, Italy

1 Abstract

The twilight model presented is based on the radiative transfer equation solved for a curved medium. Results obtained for sky brightness at the zenith are in agreement with measurements. The sky brightness profile above the horizon is also derived using combined results of the model and measurements.
Key words: sky brightness, twilight model.

2 Introduction

Twilight is a natural phenomenon not only fascinating but also useful to investigate the structure of the atmosphere. The clear sky during twilight changes its brightness of several order of magnitude up to reach a background value indicated as zodiacal brightness. Depending on latitude and altitude and season twilight lasts from some hours to several days. Meinel and Meinel (1983) and Rozenberg (1966) have given several indications how to study the twilight. However only recently some models have been developed, solving the radiative transfer equation for curved media (Sobolev, 1975).

In this paper we use part of the work already done combining it with measurements that we have recently performed (October - November 1991).

3 Modelling the twilight

Consider the well known Radiative Transfer Equation (RTE):

$$\frac{dI}{ds} = \alpha \, (B - I) \tag{1}$$

where I is the light intensity, α is the absorption coefficient and B is the source function. Solutions for a plane parallel atmosphere have been developed by several authors; among the others we cite Guzzi (1990), and the selected papers by Kattawar (1991).

Solutions for spherical atmospheres were performed by Sobolev (1975). He proposed the following development, solving the RTE for a spherical atmosphere with anisotropic scattering.

If we specify (see Fig. 1) the position of the volume scattering of the atmosphere by the spherical coordinates r and ψ and the direction of the radiation, at that given volume characterized by the zenith angle θ and the corresponding azimuthal angle by φ, we can write that:

$$\frac{dI}{ds} = \frac{\partial I}{\partial r}\frac{\partial r}{\partial s} + \frac{\partial I}{\partial \psi}\frac{\partial \psi}{\partial s} + \frac{\partial I}{\partial \theta}\frac{\partial \theta}{\partial s} + \frac{\partial I}{\partial \varphi}\frac{\partial \varphi}{\partial s} \tag{2}$$

From the geometry of Fig. 1 we can also obtain that:

$$\begin{aligned}
\frac{dr}{ds} &= \cos\theta; & \frac{d\psi}{ds} &= \frac{\sin\theta\cos\varphi}{r} \\
\frac{d\theta}{ds} &= -\frac{\sin\theta}{r}; & \frac{d\varphi}{ds} &= -\frac{\cot\psi\sin\theta\sin\varphi}{r}
\end{aligned} \tag{3}$$

The source function $B(r,\psi,\varphi,\theta)$ in equation 1 can be expressed by:

$$B\left(r,\varphi,\theta,\psi\right) = \frac{\lambda}{4\pi}\int_0^{2\pi} d\varphi' \int_0^\pi I\left(r,\psi,\theta',\varphi'\right) x\left(\gamma'\right)\sin\theta' d\theta' + B_1\left(r,\varphi,\theta,\psi\right) \tag{4}$$

where $x(\gamma)$ is the phase function and the scattering angle γ is given by:

$$\cos\gamma = \cos\theta\cos\theta' + \sin\theta\sin\theta'\cos(\varphi - \varphi') \tag{5}$$

The first order scattering function B_1 is given by:

$$B_1\left(r,\psi,\theta,\varphi\right) = \frac{\lambda}{4}Sx\left(\gamma\right)\exp\left\{-T\right\} \tag{6}$$

where

$$\cos\gamma = -\cos\theta\cos\psi + \sin\psi\cos\varphi\sin\theta \tag{7}$$

T is the optical path from the sun to a given point of the atmosphere and is given by:

$$T\left(r,\psi\right) = \int_{r\cos\psi}^\infty \alpha\left(\sqrt{r^2\sin^2\psi + z'^2}\right) dz' \tag{8}$$

valid for $\psi \le \pi/2$ and $\psi > \frac{\pi}{2}$

The problem to determine the radiation field in a spherical atmosphere requires the solution of the system of equations (2) and (4) with the relation (3).

Due to the difficulty to obtain the exact solution of the equation we derive an approximate solution. We apply the directional averaging of the radiation intensity which corresponds to compute the simple scattering exactly and compute the higher

2470

order scattering by approximation. In such a case the phase function can be expanded in term of first two terms of the Legendre polynomials. So we have:

$$x(\gamma) = 1 + x_1 \cos\gamma \tag{9}$$

With this phase function the source function becomes:

$$
\begin{aligned}
B &= \lambda J + \lambda x_1 H \cos\theta + \lambda x_1 G \sin\theta \cos\varphi \\
&\quad + \frac{\lambda}{4}S(1 - x_1 \cos\theta \cos\psi + x_1 \sin\theta \sin\psi \cos\varphi)\exp\{-T\}
\end{aligned}
\tag{10}
$$

with the following notation:

$$J = \int I \frac{dw}{4\pi} \quad H = \int I \cos\theta \frac{dw}{4\pi} \quad G = \int I \sin\theta \ \cos\varphi \frac{dw}{4\pi} \tag{11}$$

Then the source function can be expressed in terms of J, H, G functions each of which depends on the variable r and ψ.

Multiplying equation (1) with the positions (2) and (3) by $\frac{dw}{4\pi}$ and integrating over all direction we obtain:

$$\frac{\partial H}{\partial r} + \frac{2}{r}H + \frac{1}{r}\frac{\partial G}{\partial\psi} + \frac{\cot\psi}{r}G = -\alpha(1-\lambda)J + \alpha\frac{\lambda}{4}S\exp\{-T\} \tag{12}$$

Multiplying the same equation first by $\cos\theta\left(\frac{dw}{4\pi}\right)$ and then by $\sin\theta\cos\varphi\left(\frac{dw}{4\pi}\right)$ we obtain:

$$(3-\lambda x_1)H = -\frac{1}{\alpha}\frac{\partial J}{\partial r} - \frac{\lambda}{4}Sx_1\cos\psi\exp\{-T\} \tag{13}$$

$$(3-\lambda x_1)G = -\frac{1}{\alpha r}\frac{\partial J}{\partial\psi} + \frac{\lambda}{4}Sx_1\sin\psi\exp\{-T\} \tag{14}$$

Substituting H and G of equation (13) (14) into equation (12) and considering α depending only on r and $(3-\lambda x_1) = cost$ we obtain a Laplace equation for J of the type:

$$\frac{\partial^2 J}{\partial r^2} + \frac{\partial^2 J}{\partial\psi^2} - \frac{\alpha'(r)}{\alpha(r)}\frac{\partial J}{\partial r} = \alpha^2(K^2 J - f) \tag{15}$$

where $K^2 = (3-\lambda x_1)(1-\lambda)$ and, with the aid of equation (8)

$$f = \frac{\lambda}{4}S[3 + (1-\lambda)x_1]\exp\{-T\} \tag{16}$$

The boundary condition at the upper boundary is:

$$J = 2H \tag{17}$$

and at the lower boundary is:

$$J + 2H = a[J - 2H + S\cos\psi\exp\{-T(R,\psi)\}] \tag{18}$$

where a is the albedo of planetary surface. When $\psi > \frac{\pi}{2}$, $T(R, \psi) = \infty$ and the exponent is equal zero.

Once J is obtained the quantity H and G are obtained by equations (13) and (14). Then when B is determined the intensity I is found by equation (1) and successive positions.

Consider now the case in which $\alpha(r) = \alpha(R)\exp(r - R)H$. Equation (15) takes form:

$$\frac{\partial^2 J}{\partial r^2} + \left(\frac{2}{r} + \frac{1}{H}\right)\frac{\partial J}{\partial r}\frac{1}{r^2 \sin^2 \psi}\frac{\partial}{\partial \psi}\left(\sin \psi \frac{\partial J}{\partial \psi}\right) = \alpha^2(k^2 J - f) \tag{19}$$

that it simplifies if we note that the scale height H is much less than the planet radius R. Then

$$\frac{\partial^2 J}{\partial r^2} + \frac{1}{H}\frac{\partial J}{\partial r} + \frac{1}{R^2}\frac{\partial^2 J}{\partial \psi^2} = \alpha(K^2 J - f) \tag{20}$$

that, introducing the optical depth $\tau = \alpha(r)H$ and the optical distance $t = \alpha(R)R\psi$ can be written as :

$$\frac{\partial^2 J}{\partial \tau^2} + \left(\frac{\tau_0}{\tau}\right)^2 \frac{\partial^2 J}{\partial t^2} = K^2 J - f \tag{21}$$

that applied to the entire atmosphere can be approximated by:

$$\frac{\partial^2 J}{\partial \tau^2} = k^2 J - f \tag{22}$$

Consider an isotropic atmosphere with single scattering albedo. The solution of equation (22) is:

$$J = C + D\tau - \int_0^\tau (\tau - \tau')d\tau' \quad \text{for } \lambda = 1 \tag{23}$$

Assuming the atmosphere has a finite optical thickness τ_0 and overlies a surface of albedo a, we can be obtained C and D with the aid of boundary conditions 17 and 18. Since in our case we have $x1 = 0$ and $3H = \partial J/\partial \tau$ we obtain:

$$C = 2D/3; \quad (4/3 + (1 - a)\tau_0)D = (1 - a)\int_0^\tau f(\tau)(\tau_0 - \tau)d\tau +$$
$$+ \frac{2}{3}(1 + a)\int_0^\tau f(\tau)d\tau + aS \cos \psi \exp(-T) \tag{24}$$

Since $I(\psi)$ is a function of zenith distance of the sun and is expressed by:

$$I(\psi) = \int_0^\tau J(\tau, \psi)\exp(-\tau_0 + \tau)d\tau + I1(\psi) \tag{25}$$

where $I1(\psi)$ is the zenith brightness produced by the scattering of first order, we obtain substituting equation (23) in equation (25) that:

$$I(\psi) = D(\tau_0 - \frac{1}{3} + \frac{1}{3}\exp(-\tau_0))$$

$$+ \int_0^\tau f(\tau)\left(1 + \tau - \tau_0 - \exp(-\tau_0 + \tau)\right)d\tau + I1(\psi) \tag{26}$$

substituting

$$f(\tau) = \frac{3}{4}S\exp(-\tau b(\psi)) \tag{27}$$

in equation (26) we obtain:

$$I(\psi) = D\left(\tau_0 - \frac{1}{3} + \frac{1}{3}\exp(-\tau_0)\right) + \frac{3}{4}\frac{S}{b(\psi)} \tag{28}$$

$$\left[\left(1 + \frac{1}{b(\psi)}\right)(1 - \exp(-b(\psi)\tau_0)) - \tau_0 - \frac{b(\psi)}{b(\psi) - 1}\left(\exp(-\tau_0) - \exp(-b(\psi)\tau_0)\right)\right]$$

$$+ I1(\psi)$$

with S is the extraterrestrial radiation and

$$I1(\psi) = \chi(\psi)\frac{S}{4}\frac{\exp(-\tau_0) - \exp(-b(\psi)\tau_0)}{b(\psi) - 1} \tag{29}$$

where $\chi(\psi)$ is the phase function related to ψ angle. The constant D is:

$$\left[\frac{4}{3} + (1-a)\tau_0\right]D = \frac{3S}{4b}\left[(1-a)\left(\tau_0 - \frac{1}{b(\psi)} + \frac{1}{b(\psi)}\exp(-b(\psi)\tau_0)\right)\right] + \tag{30}$$

$$\frac{2}{3}(1+a)(1 - \exp(-b(\psi)\tau_0)) + \frac{4}{3}a\exp(-b(\psi)\tau_0)b(\psi)\cos\psi$$

The preceding solution is based on the parameter $b(u, \psi) = b(\psi)$ which is linked to T by the relation $T = \tau b(u, \psi)$. Using the following relation $\tau = \alpha(r)H$ and $\tau/\tau_0 = u$ we find that:

for $H \ll r$, $b(u, \psi)$ can be expanded in series

$$b(u, \psi) = b_0(u, \psi) + b_1(u, \psi).... \tag{31}$$

where

$$b_0(u, \psi) = g(p)\sec\psi \tag{32}$$

and

$$b_1(u, \psi) = \left[1 + \left(\frac{1}{2p} - 1\right)g(p)\right]\frac{1 + 2\sin\psi}{2(1 + \sin\psi)^2}\cos\psi \tag{33}$$

2473

with

$$p = (1 - \sin\psi)(\frac{R}{H} - lnu) \tag{34}$$

and

$$g(p) = 2\sqrt{p}\exp(p)\int_{\sqrt{p}}^{\infty}\exp(-z^2)dz \tag{35}$$

In the case of $\psi \leq \frac{\pi}{2}$

$$b_0(u,\psi) = \sec\psi \tag{36}$$

For $\psi > \frac{\pi}{2}$ we obtain

$$b(u,\psi) = 2\frac{u_1}{u}b(u_1,\frac{\pi}{2}) - b(u,\pi-\psi) \tag{37}$$

where $u_1 = \frac{\tau_1}{\tau_0}$ and

$$ln\frac{u_1}{u} = (1 - \sin\psi)(\frac{R}{H} - lnu) \tag{38}$$

4 Results

Using algorithms presented with the proper ψ and the measured optical thickness we obtain the values plotted in Fig. 2a. Then a comparison between model and measurements can also be made. In fact we performed during October and November 1991 at M.te Cimone Observatory (near Modena) measurements of sky brightness above horizon during twilight periods. Model's results presented in Fig. 2a show a good agreement with measurements (see Fig. 2b); so we can conclude that it is possible to derive a realistic zenith sky brightness with the present model provided that optical thickness of the atmosphere is also measured. In order to obtain the sky brightness profile, in the western hemisphere, up to the horizon we can use our measurements. In Fig. 3 we show the sky brightness profiles above horizon during the twilight. Values are normalized to the zenith sky brightness.

From plots of Fig. 3 it is possible to obtain, graphically, the sky brightness at a certain angle above horizon and for a certain sun zenith angle during twilight.

Results show that the gradient of twilight brightness at zenith are quite equivalent to those one obtained by other authors (Meinel and Meinel, 1983). Results of the model are comparable with data obtained by measurements if additive data are also obtained, mainly phase function and optical thickness of the atmosphere. From the same model the variability of the radiation as a function of latitude can also be obtained on the basis of simple considerations.

Comparing the twilight brightness at zenith and that one at the horizon, we have that: i) at sun zenith angles near 90 degrees the sky brightness gradient between zenith and horizon is almost of an order of magnitude, while it increases remarkably at 98 degree. ii) at the horizon the red brightness is predominant respect to that one in the photometric and blue regions.

2474

5 Conclusion

A model of sky brightness at the zenith during twilight has been presented. Results are in agreement with measurements. From measurements of sky brightness above horizon a graphical form to obtain the sky brightness profile is shown. The present model indicates that it is possible to obtain sky brightness profile when the proper astronomical geometry and atmospheric status are known.

Acknoledgements

We thank Dr. Micaela Pàntano for her invaluable assistance in typing and editing the manuscript.

References

Guzzi, R., 1990: Radiative transfer equation in the atmosphere. It's application to the remote sensing passive measurements performed by airborne platforms. *Meteorology and Environmental Sciences*, Eds. R. Guzzi, A. Navarra, J. Shukla. World Scient.

Kattawar G.W. Editor, 1991. Selected papers on: Multiple scattering in plane parallel atmospheres and oceans: Methods. SPIE Milestones series, Vol. MS, 42.

Megrelishvili, T.G., I.G. Melk'nikova, G.V. Rozenberg and A.V Khovan-skiy, 1978: Mean twilight at the zenith and vertical profile of scattering coefficient according to observation at Abastumani Astrophysical Observatory (1942-1952). *Atmospheric and Oceanic Physics*, vol. 14, pp. 805.

Meinel, A. and M. Meinel, 1983: *Sunsets, Twilights and Evening Skies*. Cambridge University Press.

Rozenberg, G.V., 1966: *Twilight*. Plenum Press, New York.

Sobolev, V.V.,1975: *Light Scattering in Planetary Atmospheres*. Pergamon Press.

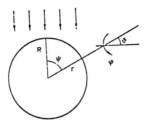

Fig.1 Definition of coordinates for a spherical atmosphere surrounding a planet of radius R. The direction of solar radiation is indicated by arrows.

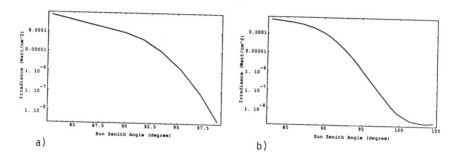

a)

b)

Fig.2a Computed sky brightness at the zenith during the twilight in the range 510-610 nm. The measured optical thickness is 0.25.

Fig.2b Sky brightness measurements at the zenith during twilight in the range 510-610 nm with the same optical thickness.

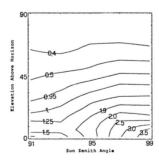

Fig.3 Contour plot of the sky brigthness for different sun zenith angles and elevation above horizon. Values are normalized at the zenith values and plotted as logarithm to base 10.

SOLAR INSTRUMENTATION USED IN METEOROLOGY

Inge Dirmhirn

Universität für Bodenkultur
Institut für Meteorologie und Physik
Türkenschanzstr. 18, 1180 Wien
AUSTRIA

Abstract

Solar Instrumentation relevant to solar conversion systems have not changed much throughout the last decades, they are readily available and, at least in part, used in meteorological services. Their accuracy is sufficient if properly maintained and if recalibrated regularly. Both requirements are not always fulfilled correctly. Hence available data have to be handled with care.

Solar radiation also has a naturally high variability. Hence, data over a limited period cannot be used for statements on the solar climate of a location. To determine the year to year and season to season variability of the solar climate requires careful handling of the equipment, so that changes in data are not attributed to the variability if actually caused by instrumental errors.

1. Introduction

Reasonable design of solar conversion installations requires the knowledge of the solar climate of the location on which the device shall be constructed. Mostly, some information is available through the national weather services that maintain solar radiation stations within their network. However, the permanent recordings are installed for diferent purposes, often for gaining knowledge of the energy exchange on the surface and are, hence, equipped with instruments to measure solar and scattered irradiance on a horizontal surface. In some major stations also scattered radiation on a horizontal surface, and in still fewer locations the direct solar radiation on a plane that follows the track of the sun are recorded. The heat loss of the natural surface by longwave terrestrial irradiance is measured only in still fewer cases and if so with low accuracy due to the lack of proper instrumentation. Independent recordings of the spectrally weighed radiation according to the sensitivity of silicon cells are not available in meteorological services.

Some of the required information can be deduced from the available data. The question that arises is what data exactly need to be known and to the degree of accuracy they need to be presented to be of any use in solar design. This then allows to view the available data in the light of the function of the instruments and the accuracy achievable under normal maintenance conditions.

2477

2. Major information for solar design

For most practical purposes the following information is n e e d e d   to provide an insight in the solar climate of a location and an adequate basis for solar designs:

a) Direct solar radiation on a plane perpendicular to the sun's rays (for solar tracking designs)

b) Global (solar and scattered) irradiance on a surface of a defined orientation (for stable conversion systems)

c) Terrestrial net radiation (for calculations of the heat loss from thermal designs)

d) Most of the above for a sensor with spectral sensitivity of a silicon cell (for direct conversion into electricity)

Instrumentation for these requirements is available. Also, data from recordings of the national weather services in many countries are   a v a i l a b l e   . The radiation components recorded by these services are the following:

A) Very few direct solar radiation

Global radiation and scattered radiation (with shading device or shadowband) from which the direct solar radiation can be deduced

B) Global irradiance on a horizontal surface

C) Several recordings of terrestrial net radiation

In using these data one needs to know how the data were taken, their reliability and accuracy. Hence, a general understanding of the function of the instruments as well as of their maintenance and recalibration pattern needs to be known.

Further, algorithms need to be known to convert the recorded data into the needed values (from a horizontal plane to a tilted etc.).

3. Solar instrumentation (their function, accuracy and maintenance)

3.1 Instruments to measure direct solar radiation on a plane perpendicular to the sun's rays (Pyrheliometers)

Pyrheliometers are instruments that carry a receiver on the bottom of a tube that allows the solar and circumsolar radiation from a total angle of 5 degrees to enter. The choice of the width of the angle was a compromise for mechanical adjustment of the opening of the tube towards the sun. Too small an angle may result in maladjustments and loosing the image of the sun. The compromise, however, leads to some possible errors in considering the measured solar energy at concentrating devices. The amount of the circumsolar radiation is dependent on the turbidity of the atmosphere and can vary considerably. Other designs with smaller apertures have been tried (and are used in some laboratories), they were, however, not successful in operational use.

The receiver of a pyrheliometer can be a cavity that is used as a calorimeter (to convert the absorbed solar radiation into heat that can be measured) or a simple plate with the same function. The former instruments can be brought to a precision of a few promille and are used as absolute instruments. The latter should be precise within 1 to 2%. Fig. 1 shows an example of each that are commonly in use today.

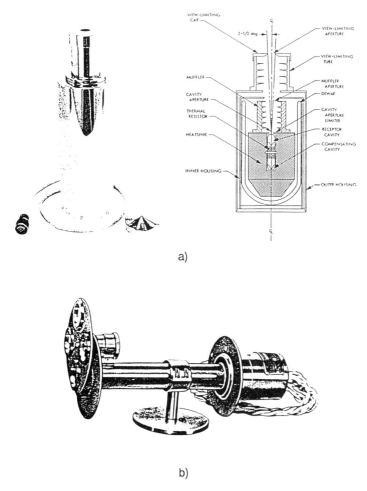

a)

b)

Fig. 1.   Pyrheliometers: a) Primary Absolute Cavity Radiometer (PACRAD) designed by . Kendall, b) Normal Incidence Pyrheliometer (NIP) from Eppley

The accuracy of both instruments seems to be sufficient for most solar design information. These instruments have also a history of longterm stability. In operational use only the second instrument type can be used at all weather conditions (entrance aperture covered by glass). They need a reliable solar tracking device. Additional observation of the proper tracking is recommended.

2479

3.2 Instruments to measure direct and scattered solar irradiance
on a plane surface (usually horizontal) (Pyranometer)

The receiver surface of pyranometers needs to be exposed to the whole hemisphere. Rays from either direction should be absorbed eaqually and due to the cosine of the angle they are impinging on the receiver. Conforming to the "cosine law" is an important requirement for these instrumants. Most of them meet the cosine law to a sun angle of approx. 15 to 20 degrees. Different claims of producing companies to meet the cosine law beyond these angles generally are not true. The cosine errors can be considered acceptable except at low sun angles. Because of the small energy contribution at these of the direct solar radiation (great path length through the atmosphere and small vertical component of the solar vector) the loss through nonideal cosine response of pyranometers is not of importance in daily totals of solar energy. However, it is if importance in calculating the direct solar energy from differences of recordings of global and scattered irradiance values, especially at seasons and in geographical latitudes when the solar angle remains below these low degrees.

Other designs, using a cavity (integrating sphere) to improve the cosine law have been tried but were not successful. The problem being that a knives edge at the entrance opening cannot be mechanically accomplished so that at low sun angles again errors of the cosine law occur.

The receiver surface of commercially available pyranometers is either painted black or is composed from black and white painted plates. The temperature difference between the black receiver plate and the instrument body, or between the black and white painted plates is measured thermoelectrically. The receiver surface is covered by one or two hemispherical glass domes. Both instrument types have their advantages and disadvantages. In both cases the precision can be brought to approx. 2% in operational use when careful maintained. In Fig. 2 one of each instrument types of common use is shown.

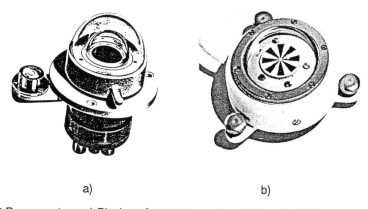

a)   b)

Fig. 2 Pyranometers: a) Black surface pyranometer (design by Moll
Gorczynski, manufactured by Eppley and Kipp) b) Black
and white pyranometer (designed in Vienna and manufactured
by Schenk), (Dirmhirn, I., 1964)

Though the precision of pyranometers might be sufficient for solar design needs (with the exception of low sun angles), the accuracy of the recorded data depends on the maintenance during operational use. Pyranometers require regular recalibrations if a high confidence in the data is needed. The paint on the receiver surface is exposed to high radiative power through long periods of time as well as to large temperature changes. In moist climates also the wettness combined with high radiation energy can cause changes in materials that are not known under such stringent conditions. Changes in one particular paint that was recommended and generally used on all radiation instruments years ago occurred after continuous use over approx. ten years. The material degradation was dramatic and so was the apparent climatic change.

With regular comparisons with a secondary standard that is calibrated in a radiation laboratory these degradations of the operational pyranometers can be detected and the instruments exchanged before irrepairable damage has been done to the data.

3.2.2 Measurement of the scattered (sky) radiation

Pyranometers are also used to measure scattered radiation from the whole sky using moving shading devices or shadow bands to obstruct the direct solar radiation. Though conceptionally simple maintenance here is of highest importance. Moving shadow devices can be easily displaced by wind and need to be frequently checked. Shadowbands introduce errors by itself (obstructing parts of the sky) and also need to be adjusted at least every other day. As obvious as this demand sounds as often it is violated. The errors are not easily discernable and have entered many publications of solra radiation data only to be discovered when the data are used and thoroghly checked. Unlike global radiation (for which restoration programs exist) erroneous scattered radiation data are unrestorable and are hopelessly lost if not carefully observed during the field phase. Fig. 3 shows a pyranometer equipped with a shadow band to record scattered radiation data.

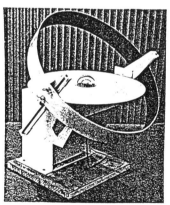

Fig. 3. Pyranometer with shadow band to record scattered
radiaiton (Robinson, N., 1966).

3.3 Instruments to measure terrestrial radiation (Pyrradiometers
and Pyrgeometers)

Unlike the two other categories of radiation instruments the precision of pyrradiometers and pyrgeometers are not of high degree. Pyrradiometers

measure the total irradiance (solar, scattered and atmospheric (terrestrial)) of a plane surface,
pyrgeometers only the longwave terrestrial. The receiver in both cases is a black plate similar to the one in pyranometers. The shielding, however, is a dome of poliethylene (because glass is opaque for the terrestrial waveband). Poliethylene, a soft variety of plastic of large molecular structure, shows only minor absorption bands in the total spectral range from 0.3 to 50 um. The net radiometer consists of two such instruments one exposed to the uper and one to the lower hemisphere. Several instruments of different design were tested in the fiftieth, however only the one with a shield against the weather (the poliethylene dome) can be used for recordings and has persisted till today. Even though and despite numerous efforts to improve the design the instrument is still unsatisfactory.

Up till a decade or so ago scientists have accepted the low precision of the pyrradiometer. The instrument was mostly used for the detection of the net radiation or radiation balance as an input into energy exchange measurements. The error was buried in one or all of the energy exchange components, flux into the air,
evapotranspiration and flux into the ground. Since the three energy exchange components can now be measured with more accuracy,
the deficiency in accuracy of the net radiometer becomes apparent and the demand for a better instrument becomes more urgent.

The pyrgeometer again uses the same principle of a black receiver whose temperature is measured against the temperature of the body of the instrument. Here, however, the shielding dome is made from germanium covered by an interference filter to act as a solar blind filter. This design certainly would have been a big step forward had not other errors reduced its precision. The solar blindness of the dome seems not to be complete, or at least other processes of heat transfer within the dome reduce the accuracy. Trials of compensation were not too successful up to today. Hence, the precision of the two instruments, the pyrradiometer and the pyrgeometer are in the range of approx. 5 to 10%. Fig. 4 shows one sample each of the pyrradiometer and the pyrgeometer.

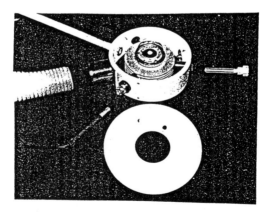

Fig. 4.a. Pyrradiometer design originally by Schulze (Coulson, K., 1975)

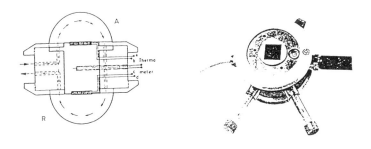

Fig. 4. b. Pyrradiometer manufactured by Schenk); the radiation of the upper and lower hemisphere can be measured independently and as difference (as net radiation = radiation balance).

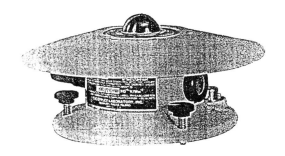

Fig. 4. c. Pyrgeometer (Eppley) with germanium dome and solar blind interference layer.

3.4 Silicon cells used as radiation sensors

Occasionally silicon cells are offered by some companies as "pyranometers". They do not fulfill the first requirement for a pyranometer, namely equal sensitivity over the total range of the solar spectrum and should not be called so. However, they provide direct information of the that part of the solar energy that can be converted into electricity.

Silicon cells must be supplied with a diffuser to follow the cosine law closely. One instrument tnat is used extensively is the Li-Cor "pyranometer". These instruments have the advantage to be small, well designed (with good cosine response), good temporal stability, and they are considerably less expensive than regular pyranometers. However, they should only be used for information of solar energy conversion into electricity. Comparisons with regular pyranometers show that for global irradiance a conversion factor can be established that is within 3 to 5%, for the direct solar radiation and particular for the scattered irradiance, however, the error can be much greater (at clear sky up to 30%). Such instruments, hence, should exclusively be used for information of solar energy conversion into electricity. Fig. 5 shows the silicon cell by Li-Cor primarily for measurements of solar energy conversion to electricity.

2483

Fig. 5   Well designed silicon cell (by Li-Cor) to measure
solar energy conversion to electricity

4. Calibration and Maintenance

Precision of an instrument alone does not garantee acceptable data. Proper maintenance and regular recalibration of the recording instruments are needed to achieve high accuracy of and confidence in the recorde4d data. I, personally, would settle for an instrument with lower precision (within limits, of course) than for a precise radiation instrument being left to its own.

4.1 Calibration

As soon as an instrument is installed in the field it is submitted to a score of influences that can alter the sensitive receivers. Most of the time this alteration is not a principal one and can be accepted if proper track of the change can be attained. This can only be done by calibrating or comparing the recording instrument with an secondary standard.

Two methods are presently being utilised:

the calibration of the sensor and
the comparison with a secondary standard in the field

4.1.1 Calibration of the sensor

In most of the services this method is presently being practised: The recording sensor is exchanged for an other and is recalled and calibrated in the laboratory.

The advantage of this method is that the calibration can be done under the surveillance of one and the same person and under controlled conditions. The disadvantage is the fact that this way only the sensor can be controlled. If there should be any deficiency in the recording system or in the connectors or even in the exposure of the sensor (is it exactly horizontally mounted?) the person in the laboratory would have no record of it and degradation of the data would go undetected.

4.1.2 Comparison of the whole system with a secondary standard

In this case a secondary standard complete with a recording system is sent to the station, mounted close to the permanently recording sensor (approx. 50 cm South or North in equal height) and left for data recording for a period not shorter than three days but not more than one week depending on weather conditions. The installation needs to be simple so that every (nontechnical) observer can easily do it. The parallel recordings can then be avaluated by one person in the central office and deficiencies can be noted and if necessary repaired. The time interval has to be at least one minute.

The secondary standard has to be compared with the primary standard after each return from the field.

It is astounding how much one can tell by comparing two recordings of daily courses of solar radiation. Maladjustments, close obstructions (by trees that have grown or buildings that have been erected), or general degradation of the sensor surface and systems resistances can be readily distinguished, though the two latter will need additional investigation to be separated.

If such comparisons can be done – depending on the importance of the station – in from 1/2 to 1 year intervals, there should be a good chance for excellent reliable data.

5. Time Interval

Meteorological services record daily totals of radiation parameters, some present also hourly values. These can be readily taken with proper caution from the published tables. Some services also provide the customer with datadiscs for easier processing. For some purposes in solar design shorter time periods will be needed. Then the data will have to be recorded by the interested person himself. Two more forms of caution shall be extended here:

5.1. Experiences of our own mislead to generalisation. Shortterm solar recordings must not be taken for a general solar climate of a place. Solar radiation undergoes strong variations that show in differences from year to year and between seasons of consecutive years, as could be shown for two stations in Austria (Dirmhirn et al. (1992). A general trend on either station could not be detected. But differences in the radiation climate of the two stations are obvious: so higher totals in Vienna during the summer and higher totals in Salzburg during the winter months. Variations of the total annual radiation sum of almost 20% can occur.

5.2 Some solar design will be satisfied with time intervals of the recorded data. If values in the minute interval are needed then data accumulation will be substantial. Such concise data cannot be taken for extended periods of time and will have to be evaluated shortly not to be overburdened by millions of useless data.

6. Precision and Accuracy of the instruments

The time interval also determines the precision of the instruments and the accuracy of the data. Hence, this chapter is placed at the end of this discussion on solar instrumentation.

An instrument that is exposed only for a short period of time under constant control will hold its precision of the lab. If operated over a considerable time changes in the precision can occur.

Data taken over short time intervals like minutes require an instrument with good temporal resolution.

The most important, however, for all radiation instruments is the maintenance during operation so that precision and accuracy of the same instrument can deviate considerably. Tab. 1 tries to give an insight into this for some common instruments under average conditions.

Tab. 1   Precision and accuracy of pyrheliometers, pyranometers and pyrradiometers under average maintenance conditions and one year continuous operation.

	prec. at good	acc. maintenance	acc. at no
pyrheliometer	1%	2%	2 – 4%
pyranometer	2%	2%	2 – 5%
pyrradiometer pyrgeometer	5 – 10%	5 – 10%	19 – 20%

7. C o n c l u s i o n s

Solar instrumentation used in meteorology today is generally sufficiently precise for considerations of solar energy conversion. Particularly the instruments used in meteorological services can be trusted per se.

The accuracy of recordings achieved with these instruments, however, depends on the maintenance and regular recalibration of the operationally used instrumentation. A simple checking of recorded data as described f.ex. by Dirmhirn et al., 1992, can give a hint of the reliability of the data and a means for rehabilitation of longterm recordings. This can be avoided if careful control of the instruments during operation is extended.

R e f e r e n c e s

Perrin de Brichambaut, Ch., 1963. Rayonnement Solaire et Echanges Radiatifs Naturels. Gauthier–Villars 300 pp.

Coulson, K., 1975. Solar and Terrestrial Radiation. Akademic Press 322 pp.

Dirmhirn, I., 1964. Das Strahlungsfeld im Lebensraum. Akad. Verl. Ges. 423 pp.

Dirmhirn, I.; Farkas, I.; Zeiner, B.. 30 Jahre Globalstrahlungsmessungen in Wie und Salzburg. submitted to Wetter und Leben.

Eppley, Data sheet for Pyrgeometer

Foitzik I. u. H. Hinzpeter, 1958. Sonnenstrahlung und Lufttrübung. Akad. Verlagsges. 309 pp.

Kendall, J. M., 1969. Primary Absolute Cavity Radiometer. Tech. Rep. 32–1396 NASA

Kleinschmidt, E., 1935. Handbuch der Meteorologischen Instrumente. Jul. Springer 733 pp.

MONITORING SOLAR INSOLATION AT GABORONE WITH A DEVICE WHICH INCORPORATES A HIGH PERFORMANCE SOLAR CELL AS A DETECTOR

By

Joe H. Prah
Physics Department
University of Botswana
Gaborone, Botswana

Abstract

The basis and structure of a low cost but accurate Solar Cell Characteristic Plotter (SCCP), is described. A modification of the SCCP to monitor solar irradiance continuously and reliably is given. Data obtained using the device during the period 1st July 1991 and 29 February 1992 are given. These results are then compared with those from two neighbouring stations within a radius of 10 km which employ conventional equipment. The linearity, between short circuit current and intensity will be demonstrated. The data was also used to determine the variation of fill factor of the solar cell with intensity. For the solar cell used, FF is related to the intensity E by:

$$FF = B \ln E + C.$$

B and C, the constants in this equation, were found to be:
$$B = 0.0109 \pm .0001; \quad C = 0.6958 \pm .0005$$

Introduction

Gaborone is the capital city of the Republic of Botswana. It shares borders with Namibia (N and SW), Zimbabwe (NE) and Republic of South Africa (S and SE). Gaborone is at latitude $24°$ 34'S and Longitude $25°$ 57'E. It is at an elevation of about 1000m above sea level. About two-thirds of the country is covered by the Kalahari desert but it is blessed with abundant sunshine throughout the year. Weather observations are carried out at nine synoptic stations dotted over the country. These observatories monitor solar irradiance mainly with the Campbell Stoke's Sunshine Recorder except Sebele 10 km north of the capital and Botswana Technological Centre (BTC) in the capital which employ Kipp and Zonen Pyranometers in addition.

Now solar energy is used extensively in the traditional way as one would expect. However, recently passive water heating in the cities and the generation of electricity using photovoltaics both in the cities and in communities remote from the National Grid Lines have been growing rapidly. One cannot therefore underscore the point that the rapid deployment of multiplicity of cheap reliable and easy to operate solar irradiance monitors would be an asset. Solar cell monitors are particulary pertinent because apart from their merit of being cheap and easy to operate, though

not as accurate as other instruments, they have the distinction
of producing solar irradiance data which have one-to-one
correspondence with the actual electric power that a photovoltaic
array or module produces.

Experimental Set Up

Generally a solar intensity measuring instrument which employs a
solar cell as the detector uses a milliammeter current
integrator. In the system under discussion, solar irradiance is
measured by determining the P.D. across a precision fractional
resistor when the short circuit current I_{sc} of an illuminated
solar cell flows through it. It was originally conceived as a
simple potentiometric method for plotting the I-V curve of a
solar cell (1). The scheme is shown in figure 1. When the
variable terminal, VT, of the helipot is shifted to position A,
the short circuit current corresponding to the intensity of the
illumination flows through R_1 but in position B when no current
flows through R_1 the digital voltmeter V_2 measures the open
circuit voltage Voc. As VT, is varied continuously from A to B,
the X-Y plotter automatically plots the I-V curve of a solar
cell. From the I-V curve, one obtains the cell parameters: I_0,
the reverse dark current, A_0 the diode quality factor the maximum
output current I_m the maximum output voltage V_m, the fill factor
FF and the efficiency EFF.

Figure 1: BLOCK DIAGRAM OF THE SOLAR CELL I-V PLOTTER	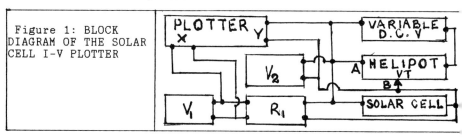

Now, the SCCP is readily modified into a device for plotting Isc
continuously for a solar cell under variable illumination, by
setting VT permanently to A and removing the D.C. biasing
voltage. The P.D. across R_1 can then be plotted by a chart
recorder or suitably recorded digitally or else fed directly into
a microprocessor-computer arrangement and analysed directly.

Some preliminary measurements embodying these ideas were first
announced in Accra[3] in August 1987. The first account of this
investigation, however, begun 1st July 1991, and was reported at
the ANSTI-UNESCO Conference in August 1991 at Gaborone[4] and
again at the American Solar Energy Society Conference at Cocoa
Beach Florida[5] in June 1992. That report covered the six month
period 1 July to 31 December 1991. This report spans the period
1 July to 29 February 1992.

Now, the P.D. across a Ø.5ØØ ohm precision resistor was plotted
with an Omniscribe Recorder-Houston Instruments. The value of
the resistor was checked with Wayne Kerr L-C-R Meter, Model 4225

which measures fractional resistance to an accuracy of 0.1%. Traces obtained gave visual information on how Isc varied with intensity during the day and with atmospheric conditions. Two Keithley Multirange Digital Multimeters, Model 197 were used to record the same P.D. and temperature separately. Temperatures were monitored with Copper-Constantan Thermocouple with an Omega International Electronic Ice Point. The Keithley 197 measures D.C. voltages in the range, microvolt to kilovolt with an accuracy of 0.2%. It also had a facility for storing data at various rates. A 2 x 2 cm^2 violet cell was used to monitor the radiations. Its parameters are shown in Table 1. The cell was mounted at 30° tilt angle and faced north. Such an orientation was chosen partly because, at Gaborone's latitude, it would give optimum response[7] and partly because most roofs in Gaborone are roughly so oriented.

TABLE 1: - CHARACTERISTICS OF VIOLET SOLAR CELL USED AS SOLAR RADIATION MONITOR

I_{sc} = 0.1611A;   T = 298K   ; A_o = 1.5063
I_m = 0.1475A; V_{oc} = 0.5981V ; V_m = 0.5025V
FF = 0.7692; E_{mo} = 10^{3}Wm^{-2} ; EFF = 18.53%
I_o = 3.1033 x 10^{-8}A

Theoretical Basis

For an ideal Solar Cell, Voc is related to Isc by the equation:

$$V_{oc} = (A_o kT/q)\ln((I_{sc}/I_o) + 1) \quad \ldots\ldots\ldots\ldots (1)$$

Where q = the electronic charge; k = Boltzmann's constant and A_o, the diode quality factor, is unity.

One defines a high performance solar cell (HPC) as one which has the following range of parameters

$$FF \geq .75; \quad EFF \geq 16\%; \quad R_s = 0, \quad R_{SH} = \infty$$

The cell used for the investigation has characteristics close to the ideal. Its parameters are shown in Table 1.

It can be shown (see Hovel[8]) that for such a cell, Vm is given in terms of Voc by

$$\exp(qV_m/kT)(1 + qV_m/A_o kT) = \exp(qV_{oc}/A_o kT) \quad \ldots\ldots (2)$$

Im is given in terms of Vm and V_{oc} by

$$I_m = (I_{sc} + I_o)(qV_m/kT)(qV_m/A_o kT) + 1)^{-1}\ldots\ldots\ldots (3)$$

These equations show that from a knowledge of I_{sc} pertaining to any time of the day, the corresponding V_{oc}, Im and Vm can be

computed. In order to find the intensity E associated with these
values, one uses the linear equation

$$E = (I_m . V_m . E_{m\emptyset})/(I_{m\emptyset} . V_{m\emptyset}) \dots\dots\dots\dots\dots\dots\dots (4)$$

Where $I_{m\emptyset}$ and $V_{m\emptyset}$ are the maximum output current and voltage
respectively, which correspond to a predetermined known intensity
$E_{m\emptyset}$. Linearity was checked in two ways. First, for any day, the
mean hourly intensities were plotted against the mean hourly
measured short circuit current. Such plots were found to be
exactly linear. Secondly using all the data obtained for the 8
months, temperatures which fell within the measured range 290-
310K were selected. Intensities corresponding to these selected
temperatures were plotted against their respective mean hourly
short-circuit currents. These plots were also found to be
remarkably linear.

Data and Data Analysis

Figures 2 and 3 illustrate typical traces obtained with the chart
recorder. Figure 2 appertains to a clear day and figure 3
illustrates a day which was partially cloudy.

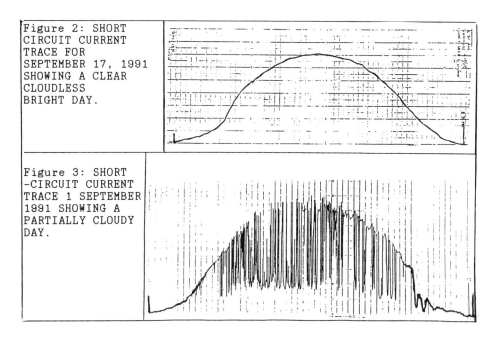

Figure 2: SHORT
CIRCUIT CURRENT
TRACE FOR
SEPTEMBER 17, 1991
SHOWING A CLEAR
CLOUDLESS
BRIGHT DAY.

Figure 3: SHORT
-CIRCUIT CURRENT
TRACE 1 SEPTEMBER
1991 SHOWING A
PARTIALLY CLOUDY
DAY.

At the end of each day, the data stored in the Keithley 197's
were retrieved and analysed to obtain hourly mean values of
temperature T_m, I_{sc}, V_{oc}, V_m, I_m, E and FF as illustrated in
Table 2. The total insolation I_D for the day was then obtained
by summing up the mean hourly intensities and multiplying by
3600s.

2490

TABLE 2: - DATA OBTAINED FOR AUGUST 14, 1991 WHICH IS TYPICAL OF DATA FOR CLOUDLESS DAYS.

Loc. Time hrs	07	08	09	10	11	12	13	14	15	16	17
I_{sc}/A	.0234	.0629	.1007	.1291	.1463	.1507	.1437	.1270	.0971	.0609	.0238
V_{oc}/V	.5193	.5572	.5753	.5848	.5896	.5907	.5889	.5842	.5739	.5560	.5199
V_m/V	.4246	.4592	.4751	.4844	.4888	.4898	.4882	.4839	.4745	.4589	.4251
I_m/A	.0215	.0581	.0932	.1196	.1357	.1398	.1332	.1177	.0898	.0562	.0218
E/Wm^{-2}	123	359	598	781	894	923	877	768	575	346	125
FF	.7499	.7606	.7658	.7675	.7687	.7689	.7686	.7669	.7649	.7617	.7500

Mean temperature of the day = 295.6K
Mean insolation of the day = 23.0 MJ m^{-2}
Comment: Clear bright and cloudless day.

At the end of the month the mean daily insolations were summed up and divided by the number of days in the month to obtain the monthly mean insolation I_m.
The mean monthly hourly intensities E_m for a particular month were also calculated by summing up the intensities obtained for a particular hour during the month and dividing by the number of days in the month. From these values another mean monthly insolation I_{mm} was obtained which served as a check on the value obtained for I_m.
To display the results to advantage histograms for E_m and I_m which are illustrated for February 1992 in figures 4 and 5 were drawn.

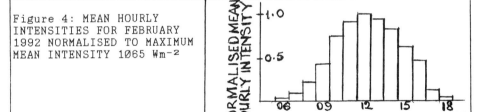

Figure 4: MEAN HOURLY INTENSITIES FOR FEBRUARY 1992 NORMALISED TO MAXIMUM MEAN INTENSITY 1065 Wm^{-2}

The data were also analysed to ascertain how FF varied with intensity. An inspection of the data acquired revealed that FF tended to saturate at intensities greater than 900 Wm^{-2}. Now such a behaviour indicates a logarithmic or exponential dependence, hence the data were investigated for the relation
$$FF = B \ln E + C,$$
where B and C are constants. Using least squares fit and three months data B and C were found to have the following values

$$B = 0.0109 \pm 0.0001$$
$$C = 0.6958 \pm 0.0005.$$

These values give excellent agreement with figures obtained for FF for other days.

Results, Discussion and Future Plans

Table 3 gives the mean monthly insolations obtained for the 8 months period under study, together with results from Sebele and BTC. The data from BTC also give information on insolation received and measured at a tilt angle of 30° to the horizontal. It is rather unfortunate that BTC had to fold up their data acquisition system in November 1991 because of technical difficulties.

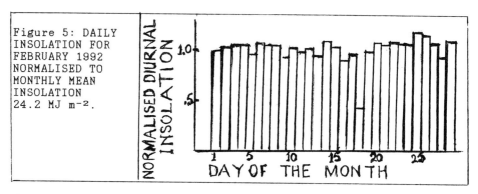

Figure 5: DAILY INSOLATION FOR FEBRUARY 1992 NORMALISED TO MONTHLY MEAN INSOLATION 24.2 MJ m⁻².

Table 3 - MEAN MONTHLY INSOLATIONS - MJm⁻²/DAY FROM THREE STATIONS FOR THE PERIOD JULY 1991 TO FEBRUARY 1992.

Month	Sebele Horizon-	BTC Horizon-	BTC 30° Tilt	HPC 30° Tilt	Sebele 10Yr Mean
July	17.0	15.3	21.5	18.5	15.5
Aug.	18.6	17.1	22.2	19.3	18.6
Sept.	17.9	16.0	19.4	19.7	20.7
Oct.	22.1	21.9	24.3	21.2	22.4
Nov.	22.0	21.1	19.8	20.9	24.2
		(½ month	(½ month)		
Dec.	22.4	-	-	19.7	25.4
Jan.	25.7	-	-	22.9	23.0
Feb.	25.5	-	-	24.2	22.7

One wishes to focus attention on a few points about these results.

The data from all three stations and the 10-year monthly average insolations are comparable.

As one would expect (7), the mean insolations measured at a tilt angle of 30° to the horizontal from the BTC data were always greater than the irradiance received with a horizontal collector. The HPC results show a similar trend for the three winter months, July, August and September. For the summer months, however, the trend is reversed and it is interesting to observe that the

2492

turning point occurs in September, and persists to the end of
February. Could there be another turning point in March?
One is not yet in a position to explain to what these deviations
are attributable. It could not be due wholly to reduction in the
performance of the solar cell arising from temperature increases.
Such an effect would account for only .05% decrease in
efficiency per kelvin[6]. The maximum differences in winter and
summer mean temperatures that were measured varied only between
10-15K.
One surmises that the deviations may be attributable mainly to
augmentation of ultraviolet and infra-red radiations in the
summer to levels higher perhaps than those present in the
simulator which was used for standardising the solar cell.
But one wishes to emphasize that this is only a guess until more
data have been accumulated; seasonal levels of infra-red and
ultra-violet radiations determined and the solar cell detector
standardised, in situ, using, for example, a normal incident
Pyrheliometer.

References

1.   PRAH, J.H., "Solar Cell Fubrication Studies Pertinent to
      Developing Countries".
      State University of New York, Albany Ph.D. Thesis 1983
      - pp 155-166.
2.   PRAH, J.H. and CORBETT, J.W., "A Simple Potentiometric
      Method for Plotting Solar Cell Characteristics" Proceedings,
      West Africa Science Conference, August 1987.
3.   PRAH, J.H. and MENSAH, E.S., "Preliminary Investigation of
      the Diurnal Variation of Solar Radiation Intensity Using a
      Standardised Solar Cell".
      Proceedings, Ghana Science Association Conference, August
      1988.
4.   "DIURNAL Solar Irradiance Measurements Employing
      Standardised Solar Cell" Proceeding, ANSTI (UNESCO) Physics
      Seminar; Gaborone, 5-7 August 1991.
5.   PRAH, J.H., "Solar Irradiance at Gaborone as Monitored with
      a device which Employs a High Performance Solar Cell as a
      Detector".
      Proceedings of American Solar Energy Society Conference
      Cocoa Beach, Florida, June 1992.
6.   HOVEL, J.J., "Semiconductors and Semimetals - Solar Cells",
      11.
7.   DUFFIE, J.A., and BECKMAN, W.A., "Solar Engineering of
      Thermal Processes".

Acknowledgements
My indebtedness to the University of Botswana and to the Physics
Department of this University is gratefully acknowledged.
I am particularly thankful to Professor K.R.S. Devan for his
encouragement and to Dr J. Andringa and Dr P.V.C. Luhanga for
their suggestions and advice.
My sincere appreciation go to Mr T. Tau who produced the diagrams
and to Ms Masego Phekonyane who typed the manuscript.

J.H.P.
28 March 1992

A MODEL TO ESTIMATE RADIATION ON HORIZONTAL AND INCLINED SURFACES
FOR MALAYSIA

A. ZAIN-AHMED

Department of Engineering Science, School of Engineering,
MARA Institute of Technology,40450 Shah Alam,Selangor,Malaysia.

M. M. SALLEH and M. N. DALIMIN

Department of Physics, Faculty of Physical & Applied Sciences
National University of Malaysia,43600 Bangi,Selangor,Malaysia.

ABSTRACT

A model to estimate the solar radiation for a hot, humid tropical climate is
presented. The model is used to compute the monthly average hourly, monthly
average daily and annual average daily values on inclined planes. The
horizontal hourly and daily values are generated by Angstrom-type equations
developed for this purpose. Subsequently the data are incorporated into the
model to generate the hourly and daily values for the inclined surfaces.
Some examples of the results are given.

KEYWORDS

Solar radiation; anisotropic model; diffuse radiation; global radiation;
horizontal and inclined planes.

INTRODUCTION

The knowledge of the monthly average hourly solar radiation on inclined
planes is important in the study and design of solar thermal and
photovoltaic systems. As an example, in the study of thermal comfort in
buildings and passive solar architecture, the radiation impinging on the
building envelope is required and this normally involves the utilisation of
radiation values on vertical and inclined surfaces. For less accurate
evaluations and other purposes, the monthly average daily or annual average
daily values are sufficient. In Malaysia, a hot humid country close to the
equator, the measurement of such data is made for horizontal surfaces in
many meteorological stations around the country, but radiation data on
inclined surfaces is not available. Hence, a simulation model is required to
produce such data and for the first time in Malaysia, that model has been
created. This paper briefly describes the model that has been designed to
suit the typical Malaysian climate.

The Malaysian climate

Malaysia recieves a relatively high solar radiation all the year round, with an average exposure of six sunshine hours. Although the mean annual daily temperatures do not vary more than $3^{\circ}C$, but the daily temperatures fluctuate between $21\ C^{\circ}$ and $35\ C^{\circ}$. This is due to heavy rains and clouds rather than seasonal changes. Relative air humidity is between 70-90%. The wind over the country is generally light and variable except during the monsoon periods when the wind speeds can reach between 10 to 20 knots. Peninsular Malaysia and the states of Sabah and Sarawak in Borneo are surrounded by sea and therefore the coastal and inland regions recieve maximum and minimum rainfalls during different periods.

EXPERIMENTAL DATA

Table 1 shows the types of data and the periods of measurements that were obtained for the development of the simulation model.

Table 1. Experimental radiation data for towns in
Peninsular Malaysia.

Location	Latitude	Height(m) above m.s.l	Radiation Data (Period) Sunshine hours	Radiation
Bayan Lepas	$05^{\circ}18'$	2.8	1968-1989	1975-1989
Kota Baru	$06^{\circ}10'$	4.6	1968-1989	1975-1989
Johor Baru	$01^{\circ}38'$	37.8	1974-1989	1985-1989
Kuala Lumpur	$03^{\circ}07'$	16.5	1975-1989	1968-1989
Petaling Jaya	$03^{\circ}06'$	45.7	1969-1989	1977-1989
Petaling Jaya			(Diffuse)	1989-1990

In Malaysia, the global radiation is measured in 22 meteorological stations, but the beam and total diffuse radiation are only measured at the Petaling Jaya meteorological station. This diffuse radiation data was used to develop the diffuse parameters for the model.

DEVELOPMENT OF THE MODEL

Estimating the solar radiation on a horizontal surface

Global solar radiation. The estimation of the global monthly average daily (MAD) radiation can easily be done by using the Angstrom-type equations by Chuah and Lee (1984) in the form of

$$\frac{\bar{H}}{\bar{H}_o} = a + b\ \frac{s}{S} \tag{1}$$

where a and b are constants particular to the chosen locality, s is the local sunshine hours and S is the corrected astronomical day found from standard equations. H is the theoretical MAD extraterrestrial radiation.

Diffuse solar radiation. Many regression models have been developed to estimate diffuse radiation but most of them are site and season-dependent (Liu and Jordan, 1960; Page, 1961 and Iqbal, 1978). However a recent study (Zain-Ahmed et al.,1990) showed that the Angstrom-type model produced by the Page (1961) was by far the most accurate when applied for the whole of Peninsular Malaysia. However, for the location of Petaling Jaya, the most accurate correlation is

$$\frac{\bar{H}_d}{\bar{H}} = 1.007 - 1.119 \frac{\bar{H}}{\bar{H}_o} \tag{2}$$

The linear regression analysis for the monthly average hourly (MAH) values produced the following correlation

$$\frac{\bar{I}_d}{\bar{H}_d} = 0.0001 \quad 0.850 \frac{\bar{I}}{\bar{H}} \tag{3}$$

Equation (3) can be applied for the locations of Petaling Jaya and Kuala Lumpur. For other locations, the MAH diffuse can be generated by utilising the equations developed by Whillier (1956) and Liu and Jordan (1960). Similarly, the MAH global radiation can be generated by using the Collares, Pereira and Rabl's (1979) equation.

Radiation on Inclined planes at arbitrary orientation

We can now estimate the radiation on a plane inclined at any angle β and surface orientation γ. First of all the configuration factor rb was calculated by solar geometry to project the radiation on horizontal to inclined surfaces. Secondly, the MAH radiation on an inclined plane at arbitrary orientation can then be estimated by applying the anisotropic model for the sky diffuse component

$$\bar{I}_{\beta\gamma}=(\bar{I} - \bar{I}_d)\bar{r}_b+\bar{I}_d\left[\left(\frac{\bar{I} - \bar{I}_d}{\bar{I}_o}\right)\bar{r}_b + \frac{1}{2} (1 + \cos\beta)\left[1 - \left(\frac{\bar{I} - \bar{I}_d}{\bar{I}_o}\right)\right]\right] + \frac{1}{2}\rho\bar{I}(1 - \cos\beta) \tag{4}$$

The albedo ρ in the above equation refers to the ratio of reflected radiation to the incident radiation on the ground or surrounding surfaces. The value of ρ can be determined from standard albedo tables. Finally, the MAD radiation on inclined planes at arbitrary orientation is calculated by integrating Equation (4) from apparent sunrise till apparent sunset relative to the surface and the annual average daily can be obtained by averaging the MAD values.

RESULTS

This model has been successfully applied to simulate radiation values for vertical and inclined surfaces in 8 different orientations for 5 major towns in Peninsular Malaysia. Examples of the simulated results are shown in Fig.1, Fig.2 and Fig.3 for the city of Kuala Lumpur. There is no limit to

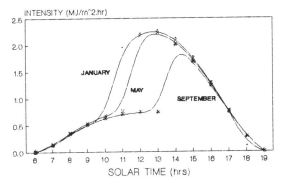

Fig 1. The monthly average hourly solar radiation on a southwest plane inclined at 13 degrees in Kuala Lumpur.

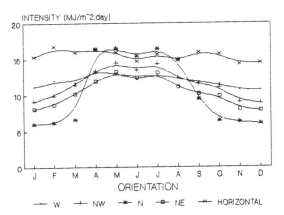

Fig. 2. The comparison of the monthly average daily solar radiation on a plane inclined at 13 degrees in 4 orientations with the horizontal plane in Kuala Lumpur.

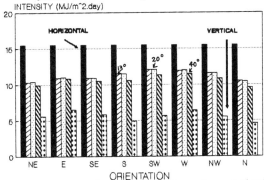

Fig. 3. The annual average daily solar radiation on inclined planes in 8 different orientations in Kuala Lumpur.

2497

the application of the model except for the availability of measured radiation data such as sunshine hours and MAD global radiation.

CONCLUSIONS

The calculations to estimate the MAH, MAD and the annual average daily radiation on vertical surfaces in 8 different orientations have been successfully carried out for 5 major towns in Peninsular Malaysia and the radiation on inclined planes was tested for the city of Kuala Lumpur. The generation of radiation data reduced the consumption of man-hours and computational time. There is no limit to the possible application of this model. This model may be improved further when a longer term set of horizontal radiation data is collected, especially of the diffuse type. Hence a more accurate Angstrom-type of correlation can be developed for the clearness index. Future work would include the radiation estimations for locations in East Malaysia.

ACKNOWLEDGEMENT

The authors wish to thank the National University of Malaysia for having provided the facilities, the Malaysian Meteorological Department for the data and MARA Institute of Technology for sponsoring the project.

REFERENCES

Chuah D. G. S. and Lee, S. L. (1984). In : Solar Radiation in Malaysia. A study on the availability and distribution of solar energy in Malaysia. Oxford Univ. Press, Singapore.

Collares-Pereira, M. and Rabl, A. (1979). The average distribution of solar radiation. Correlations between diffuse and hemispherical and between daily and hourly insolation values. Solar Energy, 22, 155-164.

Iqbal, M. (1978). Estimation of the monthly average diffuse solar component of total insolation on a horizontal surface. Solar Energy, 20, 101-105.

Klein, S. A. (1977). Calculation of monthly average insolation on tilted surfaces. Solar Energy, 19, 325-329.

Klein, S.A. (1978). Erratum. Solar Energy, 20, 441.

Liu , B. H. and Jordan, R. C. (1960). The interrelationship and characteristic distribution of direct, diffuse and total solar radiation. Solar Energy, 4, 1-19.

Page, J. K. (1961). The estimation of monthly mean values of daily total short wave radiation on vertical and inclined surfaces from sunshine records for latitudes 40°N - 40° S. Proc. UN Conf. on New Sources of Energy, Paper No. 5/98.

Whillier, A. (1956). The determination of hourly values of total solar radiation from daily summaries. Archiv fuer Meteorologie, Geophysic und Bioclimatologie Series B, 7, 197-244.

Zain-Ahmed, A. (1991). Sinaran suria pada sampul bangunan di Semenanjung Malaysia. MSc. thesis. Universiti Kebangsaan Malaysia.

Zain-Ahmed, A., M. M. Salleh and M. N. Dalimin. To be published. Computing the monthly average daily global and diffuse radiation from sunshine hours and clearness index for Petaling Jaya, West Malaysia.

Zain-Ahmed, A. To be published. Modelling the sky diffuse radiation for a hot, tropical climate.

THE SPECTRUM OF THE SOLAR GLOBAL RADIATION ON HORIZONTAL SURFACES IN POLLUTED AND PURE AIR ZONES OF GREAT CAIRO

M. A. MOSALAM SHALTOUT, A. H. HASSAN, AND A. E. GHETTAS

National Research Institute of Astronomy and Geophysics, Helwan - Cairo, Egypt

ABSTRACT

Measurements for the spectrum of the global solar radiation on horizontal surfaces in polluted (Helwan) and pure (Shebein El Kanater) air zones of great Cairo are carried out during clear sky (cloudless) days of March 1992. There were made with a spectrometer in which a polycrystalline and amorphous silicon solar cells mounted as detectors behind a rotating circular variable contains 32 metallic interference glass filters. It is noticed, the air pollutants affecting the spectrum intensity, mostly at the noon time.

KEYWORDS

Spectrum, solar radiation, silicon solar cells, interference glass filters, air pollutants.

INTRODUCTION

It is recognized the need for spectral solar radiation data that the photovoltaic community can use to design and predict the performance of single and multi - band gap PV devices throughout the day, and seasonally, for various climates. The electric current produced by a PV device can be predicted by multiplying spectral solar radiation by the spectral response of each layer in the PV device and integrating over the response region. A spectral data set that includes a range of outdoor conditions can be used to test PV device sensitivity to natural variations and determine if efficiency is compromised by these variations (Riordan et al , 1989).

The spectral distribution of solar radiation is known to vary considerably due to the influence of the changing air mass, the latitude of the measurements site and the local climatic and environmental conditions. Furthermore, each particular type of solar cell is characterized by a different spectral response curve, depending primarily on the value of the energy bandgap for the semiconductor used (Wilson and Hennies, 1989; Myers, 1989).

Many sites around Cairo area are in need to electrical power supply, where it in the dep desert, and far from the national electric net. In recent years, some of these sites used photovoltaic systems as electrical power generation, where Egypt have high solar energy potential. Great Cairo

2499

is high air polluted area, where are 1. 137. 555 cars (from the statistics of Cairo Traffic Authority, at July 1991). These cars work by the internal combustion engines, and produce aboat 60% of the total air pollutants in Cairo atmosphere. The industrial activities in Cairo North (Shoubra El Kheima), and in Cairo South (Helwan) caused 40% of the total air pollutants at the great Cairo.

The objective of this work is to measure the spectrum of the global solar radiation on horizontal surfaces in polluted and pure air zones of great Cairo. These were made with a spectrometer, in which a polycrystalline and amorphous Si solar cells mounted behind a rotating circular variable filter acts as the detectors.

MEASUREMENTS

Helwan (Lat. 29°52'N, Long. 31°20'E) is the castle of the heavy and cement industry in Egypt. The annual average of the total suspended particles is 946 µgm / m^3 for 1990, and the precipitant cement dust is 478.000 kg / square mile / month. It is air is very polluted. In contrast, Shebin El Kanater (Lat. 30°20'N, long. 31° 23'E) is an agricultural area in the far north of Cairo. It is air is clean (Mosalam Shaltout, 1988).

We selected Helwan as polluted zone, and Shebin El Kom as clean zone for our measurements in the great Cairo.

The used filters are 32 metallic interference glass filters from Carl Zeiss Jena (Germany), the band width of each filter is 25 nm. They cover the spectral range (350 - 1100 nm).

The detectors are : (1) polycrystalline silicon solar cell, from Telefunken System Technik (Germany), and (2) amorphous silicon solar cell, from Chronar (UK).

The performance of the detector cells was measured for 32 wavelengths in the spectral range (350 - 1100 nm), the sampling wavelengths were spaced at equal internals of 25 nm.

High precision pyranometer from Eppley (USA) was used for the simultaneous pyranometer measurements in horizontal mode, and in the outdoor conditions as the detectors. The absolute calibration factor was obtained by determining the ratio of the integrated spectrum, to the product of the simultaneous pyranometer measurements.

The measurements are carried out in clear sky day (cloudless) at 24 March 1992 at different air masses in Helwan, and at 23 March 1992 at for Shebien El Kanater.

RESULTS

Figures (1 and 2) show the spectral distribution of the global solar radiation on horizontal surface for Helwan and Shebin El Kanater as detected by Polycrystalline silicon solar cell, at air masses m = 1.14 and m = 1.6.

Figures (3 and 4) show the spectral distribution of the global solar radiation on horizontal surface for Helwan and Shebin El Kanater as detected by amorphous silicon solar cell, at air mass m = 1.3 and m = 1.6.

2500

We can notice the following :

1. For the polycrystalline silicon solar cells, there is a clear decrease in the intensity of the spectrum due to the air pollutants at Helwan. The decrease is $\cong$ 30% and cover all the spectral range at air mass m = 1.14. But, a decrease of $\cong$ 7% cover the infra red region only at m = 1.6.

2. For the amorphous silicon solar cells, there is a decrease in the intensity of the spectrum due to the air pollutants at Helwan. The decrease is $\cong$ 10% and cover the peak, and the blue wing of the peak of maximum response at m = 1.3. But, it is not valuable at m = 1.6, where its value is $\cong$ 3% .

3. The decrease in the intensity of the spectrum occur at low air masses, near to the noon time, for the both types of the silicon solar cells. The noon time is the optimum time to utilize the solar energy, and to convert solar radiation to electricity by PV systems.

CONCLUSIONS

We conclude that the air pollutants affect the spectral range (350 - 1100 nm) for the polycrystalline silicon solar cell, by decreasing the intensity of the spectrum by about 30% at the noon time. The effect is limited for the amorphous silicon solar cell by about 10 % at the peak, and the blue wing of the peak of maximum response at the noon time. This recommend the amorphous silicon solar cells for PV application at the polluted Cairo area.

REFERENCES

Mosalam Shaltout, M. A. (1988). Solar radiation and air pollution at Cairo. Third Arab International Solar Energy Conference, 21 - 24 February 1988, Baghdad - Iraq, Proceedings, 1 - 53 to 1 - 58, Editor : N. I. Al - Hamdani and et al.

Myers, D. R. (1989). Estimates of uncertainty for measured spectra in the SERI spectral solar radiation data base, Solar Energy, 43 , 347 - 353

Riordan, C., D. Myers, M. Rymes, R. Hulstrom, W. Marion, C. Jennings, and C. Whitaker. (1989). Spectral Solar radiation data base at SERI. Solar Energy, 42 , 67 - 79.

Wilson, H. R., and M. Hennies. (1989). Energetic relevance of solar spectral variation on solar cell short circuit current. Solar Energy, 42 , 273 - 279.

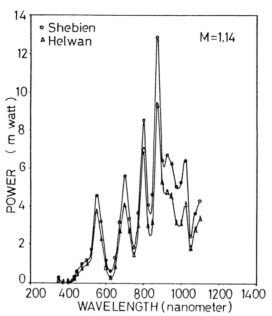

Fig. 1. Solar spectrum detected by polycrystalline silicon solar cell at Helwan and Shebien El - Kanater at air mass m = 1.14.

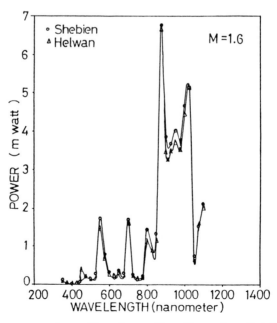

Fig. 2. Solar Spectrum detected by polycrystalline silicon solar cell at Helwan and Shebien El - Kanater at air mass m = 1.6.

2502

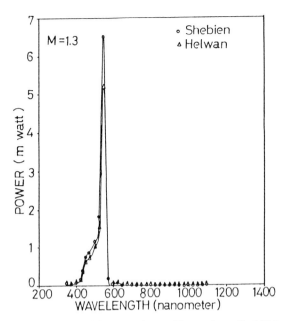

Fig. 3. Solar spectrum detected by amorphous silicon solar cell at Helwan and Shebien El Kanater at air mass m = 1.3.

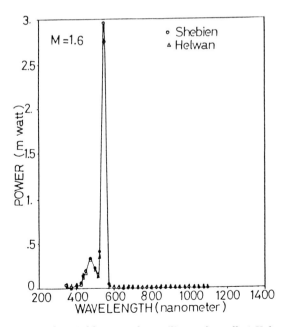

Fig. 4. Solar spectrum detected by amorphous silicon solar cell at Helwan and Shebien El Kanater at air mass m = 1.6.

2503

AIR POLLUTION MODIFICATION DUE TO ENERGY SUPPLY DIVERSIFICATION FOR MEXICO IN THE YEAR 2000

Manuel Martínez

Laboratorio de Energía Solar, Universidad Nacional Autónoma de México, Apartado Postal No. 34, 62580 Temixco, Morelos, México

ABSTRACT

Possible amounts of air pollutants due to the net internal energy supply are studied for Mexico in the year 2000. Two scenarios are considered: historical trend and a possible energy diversification scheme. For the year 2000, the considered viable participation of nuclear, solar and wind energies amounts, in percent, to: for NIS, 2.3; for ED, 5.3, and for TD, 1.4. Also, the reduction in air pollutants, in kg per kJ/person, could be: for electric demand, 9 percent in SOx, 13 in NOx, 13 in CO, 19 in HC and 6 in PARTS, and for thermal demand, 6 percent in SOx, 5 in NOx, 5 in CO, 5 in HC and 0.6 in PARTS.

KEYWORDS

Energy pollution; year 2000; Mexico; air emissions.

INTRODUCTION

The key role of energy in causing environmental problems has been confirmed by several systematic studies on risks related to its production, transportation, storage and transformation.

The primary energies significantly consumed now in Mexico are: coal, oil, natural gas, hydroenergy, geoenergy, bagasse and firewood. The incorporation of nuclear, solar and wind energies have been considered for the year 2000.

The most important local air pollutants due to the energy usage were considered: sulfur oxides (SOx), nitrogen oxides (NOx), carbon monoxide (CO), gaseous hydrocarbons (HC) and particulates (PARTS). The amounts emitted depend on the specific technological process, the fuel composition, the fuel heat content and the efficiency of the system used, as reported by UNEP (1985).

Considering that these emissions are mainly due to the transformation of primary energies into electric or thermal secondary energies, the national energy balances have been re-written. For each primary energy, the flow from the net internal supply (NIS) to electric demand (ED), thermal demand (TD) and operation and transformation (OT) has been established.

Due to the large variation in the reported figures, a selection of average values of the heat contents for the energies supplied and of the systems efficiency in Mexico has been chosen. Also, the amounts of pollutants emitted to the air, for each energy source and use were considered according to Martínez (1992).

The purpose of this paper is to study the possible medium-term variations in the amount of several air pollutants due to the internal supply of energy in Mexico. The evolution of each air pollutant in a more diversified energy supply programme is evaluated.

ENERGY SCENARIOS FOR THE YEAR 2000

For the year 2000 there are two scenarios: the Mexican energy development varies according to an historic trend or to a specific energy diversification scheme proposed by Martinez and Best (1991). In both cases, the energy flow is: NIS, 7300 PJ/year; ED, 2272; TD, 3173, and OT, 1855. The national energy balance for the year 2000, according to historic trend is shown in Table I. The national energy balance for the year 2000, according to the proposed energy diversification scheme is shown in Table II.

Table I. MEXICAN ENERGY BALANCE, YEAR 2000 (PJ/YEAR).
HISTORICAL TREND SCENARIO

	NIS	ED	TD	OT
COAL	309	242	57	10
HYDROENERGY	268	268	0	0
GEOENERGY	492	492	0	0
BAGASSE	103	0	94	9
FIREWOOD	371	0	371	0
OIL	3786	1089	1600	1097
GAS	1971	181	1051	739
TOTAL	7300	2272	3173	1855

Table II. MEXICAN ENERGY BALANCE, YEAR 2000 (PJ/YEAR).
ENERGY DIVERSIFICATION SCENARIO

	NIS	ED	TD	OT
COAL	306	239	57	10
HYDROENERGY	598	598	0	0
GEOENERGY	373	287	86	0
BAGASSE	103	0	94	9
FIREWOOD	371	0	371	0
OIL	3539	884	1496	1159
GAS	1842	143	1023	676
NUCLEAR	60	60	0	0
SOLAR	106	60	46	0
WIND	2	2	0	0
TOTAL	7300	2273	3173	1854

2505

For the year 2000, the considered viable participation of
nuclear, solar and wind energies amounts to: for NIS, 2.3
percent; for ED, 5.3, and for TD, 1.4.

RESULTS

The evolution, from 1965 to 1988, of these air pollutants is
reported in kg of pollutant per kJ of energy supplied per
capita. The emissions due to electric demand, for the historic
trend and diversification scheme, are shown in Figs. 1 and 2,
respectively. The emissions due to thermal demand, for the
historic trend and diversification scheme, are shown in Figs. 3
and 4, respectively. In order to satisfy the electric demand,
sulfur and nitrogen oxides have increased very fast, due to a
larger use of oil and coal. In order to satisfy the thermal
demand, nitrogen and sulfur oxides have increased fast, gaseous
hydrocarbons and carbon monoxide have been kept constant, while
the participation of particulates has decreased; which is a
consequence of a larger consumption of oil and natural gas, a
constant participation of coal and firewood, and a lesser use
of bagasse.

Fig.1.AIR POLLUTION EVOLUTION DUE TO ED
HISTORICAL TREND SCENARIO

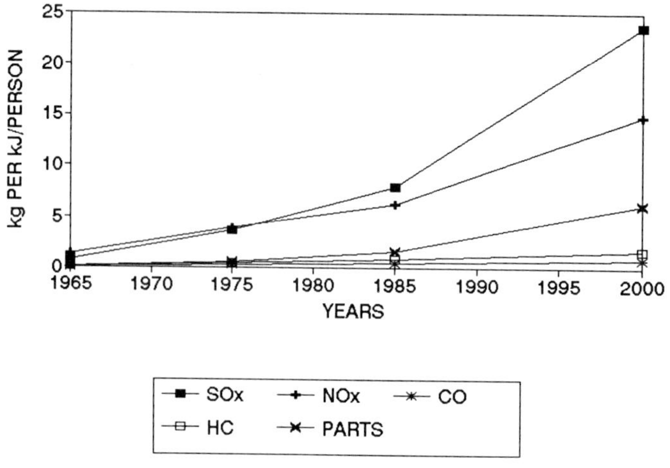

Fig.2.AIR POLLUTION EVOLUTION DUE TO ED
ENERGY DIVERSIFICATION SCENARIO

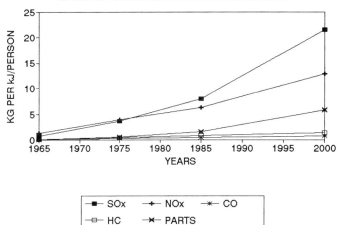

Fig.3.AIR POLLUTION EVOLUTION DUE TO TD
HISTORICAL TREND SCENARIO

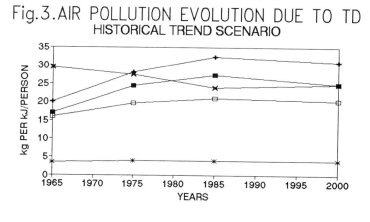

Fig.4.AIR POLLUTION EVOLUTION DUE TO TD
ENERGY DIVERSIFICATION SCENARIO

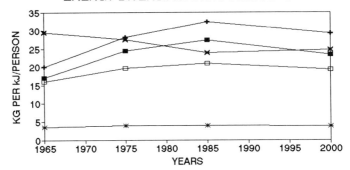

CONCLUSIONS

For the year 2000, the considered viable participation of nuclear, solar and wind energies amounts to: for NIS, 2.3 percent; for ED, 5.3, and for TD, 1.4. Also, the reduction in the amount of air pollutants, in kg per kJ/person, due to this diversification scheme could be: in relation to ED, 9 percent in SOx, 13 in NOx, 13 in CO, 19 in HC and 6 in PARTS, and in relation to TD, 6 percent in SOx, 5 in NOx, 5 in CO, 5 in HC and 0.6 in PARTS.

The feasibility to decrease the emissions of air pollutants related to energy supply has been shown for an energy diversification scheme as the one proposed in this paper.

REFERENCES

1. UNEP (1985) _Energy Report Series_, The environmental impacts of production and use of energy, part IV, Nairobi.
2. Martínez, M. (1992) Air pollutants due to energy supply in Mexico _Renewable Energy J_ to be published.
3. Martínez, M. and Best, R. (1991) Developments in Geothermal Energy in Mexico-Part Thirty-Two. Supply and Demand Perspectives for the Year 2000 _Heat Recovery Systems & CHP_ _11_ (1) 91-98.

DIRECTLY COUPLED TURBINE-INDUCTION GENERATOR SYSTEMS FOR LOW-COST MICRO-HYDRO POWER

N.P.A. Smith, A.A. Williams, A.B. Harvey*, M. Waltham*, A-M Nakarmi**

Department of Electrical Engineering, Nottingham Polytechnic,
Nottingham NG1 4BU, U.K.

*Intermediate Technology Development Group,
Myson House, Railway Terrace, Rugby CV21 3HT, U.K.

**Kathmandu Metal Industries,
Naghal Quadon, Cha 3-812, Block No. 12/514, Kathmandu-3, Nepal.

ABSTRACT

Electricity is increasingly in demand by rural communities in developing countries. Often this demand can be met by stand-alone micro-hydro schemes. Currently it is conventional for the turbines and generators to be coupled through pulley-and-belt drives, though this approach has a number of disadvantages. This paper describes how, by using multi-jet pelton turbines and pumps as turbines, direct drive generating systems can be used for a wide range of micro-hydro sites. It also describes how by using induction generators this range can be increased and the reliability of the system improved. The practical application of these technologies is discussed, with reference to installations in Nepal, Pakistan, Sri Lanka and the U.K.

KEY WORDS

Rural electrification; micro-hydro; induction generator; pelton turbine; pump-as-turbine.

INTRODUCTION

On large hydro schemes, where the generators and turbines are custom-built, the generators are always coupled directly to the turbine. On micro-hydro schemes (capacities of less than 100 kW) standard generators and to an extent standard turbines are used. The generator usually requires an operating speed that is higher than the best efficiency speed of the turbine and as a result a speed increasing drive is used. The most common type of coupling is by belt, though occasionally chains or gears are used. The disadvantages of such drives are lower system efficiency, reduced turbine and generator bearing lifetimes, increased cost, additional maintenance requirements and more complex installation. They do have the advantage that one or more mechanical loads can be belt driven from the turbine. There are nevertheless an increasing number of micro-hydro schemes where electricity generation is the sole requirement and for these a direct drive approach is the preferred option.

Pelton turbines are an obvious candidate for direct coupling because, since their width is usually quite small, the runner can often be mounted on the generator shaft. A less obvious example is a pump-motor unit, with the motor operated as a generator as explained in the next section. Directly-coupled pump units, sometimes known as 'mono-blocks' have the pump impeller mounted on an extended motor shaft.

INDUCTION MOTORS AS GENERATORS

An induction motor can be operated as a stand-alone generator by connecting sufficient capacitors across the stator windings to compensate for the lagging power factor of the machine. The main advantages of induction generators are low cost, robustness and availability. An added advantage, particularly for direct drive applications, is the availability of slow speed machines of 1000 rpm and, from some manufacturers, 750 rpm.

Whilst synchronous generators can be purchased ready for use, the induction machine will not work without capacitors of the correct value being fitted. A sufficiently accurate calculation of the capacitance required can be made from the power rating of the machine and an approximate value of its power factor. If the power factor is not quoted on the name plate then data from another machine of the same rating and speed can be used. The capacitance required is approximately equal to that required to fully power factor correct the machine when operated as a motor. Fine tuning can be achieved on site by making small changes to the amount of capacitance connected or, more easily, by allowing the generator to run a little above or below its normal operating speed. This is explained in the next section.

In Nepal more than sixty induction generators have been installed on micro-hydro schemes in the past six years and they have proven to be considerably more reliable than synchronous generators. The main reasons for this are the robustness and simplicity of the generators and the fact that they are inherently overload safe, since when overloaded the voltage falls limiting the winding currents to safe levels. Most of these generators are only used for village lighting, though recently schemes with a variety of loads, including low-wattage cooking and agroprocessing equipment, have been installed. For the variable load schemes an electronic controller is fitted in order to regulate the voltage and frequency of the supply (Smith *et al.*, 1990). This unit, known as an induction generator controller (IGC) has been installed on ten schemes in Nepal. Two Nepalese engineers were trained in the design, construction and installation of the controller. They have built and installed all of the controllers that are now in use. Due to poor facilities, the quality of these units is slightly lower than that of imported units. However, this is more than compensated for by reduced costs and ease of repair.

MATCHING THE GENERATOR TO THE TURBINE

A major advantage of a belt driven system is that the sizes of the pulleys can be chosen such that the turbine operates near its best efficiency point when the generator is running at its design speed. This allows a standard range of sizes of turbine to be used, thereby reducing costs. With directly coupled units, unless the site happens to be suitable for a standard turbine, either a custom-built turbine must be used or a standard turbine operated at below its best efficiency in order to drive the generator at its rated speed. However, this assumes that the generator must be driven at the speed required to produce its rated frequency. This is rarely, if ever, the case. Allowing the generator to be run within a speed range enables direct coupling to a standard turbine operating at, or near to, its best efficiency point. The limitations on the operating speed of the induction generator depend on the effects that this has on the generator and on the loads.

In the case of the generator, the absolute limits are set at the lower end by the maximum allowable winding temperature and at the upper end by the limits of stable generation. Both of these limits are due to the variation in generator magnetising current with frequency. This is illustrated in figure 1. Since the magnetising current passes through the stator windings of the machine, if the magnetising current is too high the windings will overheat and burn out. Modern induction machines are highly saturated and at rated voltage and frequency can have a magnetising current that is as high as 80% of the full load current rating of the machine. Operation below rated frequency and at rated voltage must be avoided both to protect against overheating and to reduce losses. Increasing the operating frequency reduces the magnetising current and improves the efficiency. However, there is a limit to how high the frequency can be increased since for stable operation the generator must operate with a degree of saturation and above a certain frequency the machine will not be saturated at rated voltage. The maximum frequency is typically at or above 1.5 times the rated frequency. Hence, induction generators can normally be run at between about 48 and 75 Hz for a machine rated at 50 Hz.

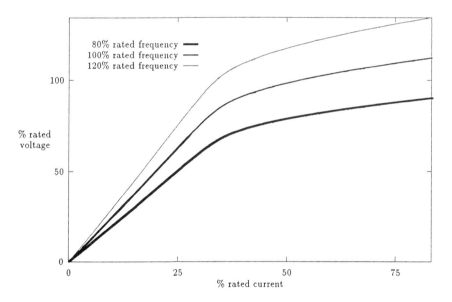

Fig. 1. Terminal voltage against no load current for an induction machine.

Most types of synchronous generator can be operated over a wide frequency range. However, they must have specially strengthened rotors to withstand the overspeeding that will occur if the load on the generator is removed. If the generator is operated at 50% above its rated speed under normal load conditions then under runaway conditions the speed may be up to 200% above its normal speed. No special protection is required in the case of induction machines with 4 or more poles as the rotors have cast aluminium bars instead of windings. 2 pole generators at 200%

overspeed will run at 9000 rpm which could cause damage to the bearings through overheating if the overspeed is allowed to persist. However, use of 2-pole generators is very unusual for micro-hydro schemes. Whereas small synchronous generators with more than four poles are not usually manufactured, induction machines of six and sometimes eight poles are mass produced. Hence, with induction generators and allowing operation at up to 50% above rated speed the speed range 750 rpm to 2250 rpm is covered or 1000 rpm to 2250 rpm where 8 pole generators are unavailable.

The limitations imposed by the loads depend on the type of loads connected. In the case of purely resistive loads, such as standard light bulbs and water or space heaters, there is no limit on the frequency. With loads that contain transformers, such as radios and televisions, operation at rated voltage and below rated frequency causes overheating due to increased saturation. Overfrequency is quite acceptable and may actually prolong the life of the transformer. Commutator type motors, commonly found in hand-held tools, are not significantly affected by the supply frequency. As explained in the previous paragraph, induction machines benefit from being run at higher frequency as their magnetising current is reduced. However, if loads, such as fans or pumps, are connected whose power requirements increase with speed the current drawn by the motor will increase and may more than offset the benefit of reduced magnetising current. If such loads are to be used on the system the maximum frequency should be limited to about 10% above the rated frequency of the machines. For the majority of micro-hydro schemes, which are used just for lighting and sometimes water heating or cooking, there is little or no constraint on the generated frequency imposed by the consumers.

PELTON WHEEL TURBINES

Nowadays, more micro-hydro turbines are made in Nepal than probably any other country, excluding China. The industry in Nepal developed some twenty years ago to fabricate steel water turbines to power agricultural processing machinery. The majority of the machines installed are of the cross-flow (Mitchell-Banki) design which are well suited to medium flows and medium heads. Today there are around 600 cross-flow turbines in Nepal's middle hills, between the high Himalayas and the flat Terai area.

More recently, there has been a growing interest in Pelton wheel turbines in Nepal. As low specific speed turbines, Pelton wheels are suitable for sites with high heads and low flows, although the use of two or more jets enables these machines to be installed on sites that would previously have been equipped with cross-flow turbines. The move towards Pelton turbines has been prompted by various factors. Firstly, Pelton wheels exhibit much higher efficiencies and greater reliability than cross-flows, and are therefore very attractive for remote areas. A second factor is that the large number of cross-flow turbines already installed has saturated the market for turbines in Nepal's middle hills, forcing manufacturers to work in more remote and mountainous regions. Finally, the development of a simple technique for manufacturing Pelton wheel runners has enabled local manufacturers to produce Pelton turbines at less than the cost of equivalent cross-flow machines (Waltham, 1991).

An important contribution to the promotion of Pelton wheels was the introduction of 'peltric' sets by a small Nepali company called Kathmandu Metal Industries. These very small machines can produce between 1 kW and 5 kW of power and have the Pelton runner hung directly on the shaft of an induction generator. The runner is enclosed in a lightweight fabricated casing and provided with a manifold of two or three jets.

Assuming a runner velocity of half the jet velocity and a maximum jet diameter of 11% of the pitch circle diameter, standard pelton turbine theory can be used to show that the net head and maximum flow per jet can be determined from the following formulae:

$$H = \left(\frac{\pi \, N \, D}{30}\right)^2 \frac{1}{2 \, g} \; ; \quad Q_{\text{per jet}} = \frac{\pi}{4} \, \sqrt{2 \, g \, H} \, \left(0.11 \, D\right)^2$$

where N is the generator r.p.m., D is the pitch circle diameter in metres, H is the head in metres and Q is the flow rate in cubic metres per second.

Originally, peltric sets were designed to be retro-fitted to the hundreds of existing sprinkler irrigation schemes in Nepal, mostly to supply electricity for evening lighting. Typically, the schemes use 75 mm pipe to convey about 5 l/s of water over heads of around 35 m. Sufficient electricity can be produced for a small hamlet and, because the civil works are already in place and the peltric sets are cheap, at a cost of less than £500 per installed kilowatt. More than forty schemes have been installed within one year.

Fig. 2. 4 kW Sri Lankan peltric set before installation

The concept of a Pelton wheel runner directly coupled to the shaft of an induction generator has been adopted by other Nepali maunfacturers as an extremely reliable and cost effective type of micro-hydro-electric plant. Using Pelton runners of pitch circle diameters between 100 mm and 200 mm with one, two or three jets, it is possible to drive induction generators of up to 20 kW output and to accomodate a wide range of head and flow conditions. The success of this technology in Nepal has encouraged Sri Lankan engineers to manufacture peltric sets. A three jet 4 kW peltric set built and installed in Sri Lanka is shown in figs. 2 and 3.

Fig. 3. 4 kW Sri Lankan peltric set installed

PUMPS AS TURBINES

Reverse running pumps have been used as turbines for micro-hydro schemes in various parts of the world (Williams et al., 1989). Most of these have used a belt-driven synchronous generator. Using a pump as turbine has certain advantages over a conventional turbine because pumps are produced in large quantities in standard sizes. They can often be purchased off-the-shelf at lower cost than an equivalent turbine, and spare parts are widely available. There are additional advantages in employing a combined motor-pump unit as a direct-coupled turbine and induction generator. All of the advantages of a direct drive mentioned in the introduction apply to these 'mono-block' units, which are the most common type of pump for outputs of 10 kW and below. A wide choice of units is available, covering different heads and flows.

Experience from Northern India and Pakistan suggests that with belt-drive pump-as-turbine installations, there are often problems with the pump bearings. The use of 'mono-block' pumps reduces bearing wear, but regular maintenance of the pump seals is still required to ensure trouble-free running. The motors in these pump units are standard squirel-cage induction machines, with a shaft extension on which is mounted the pump impeller. Motor speeds are usually either 2-pole or 4-pole, but in some cases 6-pole or 8-pole motors can be ordered.

If the turbine performance of a pump is not known, and cannot easily be obtained from tests, then an estimate of the turbine operating point for maximum efficiency can be obtained by using a simple calculation based on the pump performance (Sharma, 1985). If the pump is rated to produce a flow of Q_p at a head of H_p, motor speed N_m, and pump maximum efficiency of e_p, then the turbine operating point will be given approximately by:

$$Q_t = \frac{N_{gen} \; Q_p}{N_m \; e_p \; 0.8} \quad ; \qquad H_t = \frac{N_{gen} \; H_p}{N_m \; e_p \; 1.2}$$

where N_{gen} is the generator speed. The vlues predicted by this method have been compared for 35 pumps with published turbine test data. Of these, 20% of the calculated turbine operating points lay outside acceptable limits relative to the actual performance. As shown earlier, the generator speed can be varied over a wide range, which enables a certain amount of inaccuracy in the prediction of the turbine performance to be compensated for. The generator speed can be altered on site during commissioning by taking a range of capacitor sizes to site and changing the capacitance to obtain the best performance.

Fig. 4. Pakistani pump operating as a turbine

The efficiency of a pump operating as a turbine is usually within +2% and -5% of the pump efficiency. The range of efficiencies is large, decreasing with head and increasing with power output. Typical turbine efficiencies vary from 45% for a 1 kW unit running from 50 m head to 85% for a 9 kW unit running from 13 m head. The range of heads and flow-rates for pumps-as-turbines overlaps with pelton and crossflow turbines. There is not much advantage to be gained from using a pump instead of a pelton turbine, but for slightly higher flow rates and lower heads, where the only other option would be a crossflow turbine, a pump-as-turbine has several advantages. Crossflow turbines tend to be relatively expensive for low power outputs, because of the high manufacturing costs for small units. In this part of the range diagram, pumps are widely available at lower cost than an equivalent crossflow turbine, and operate at a higher speed, thus enabling the use of a direct drive. A medium-head pump-as-turbine is also likely to be be more efficient than an equivalent crossflow unit.

At a demonstration site on a remote farm in the north of England, three different types of pump unit have been installed. The generators, each integral with its pump, vary in size from 0.8 kW to 4 kW. One of the pump units being monitored has been imported from a small manufacturer in Pakistan. This pump is shown in fig. 4. Experience over the first 12 months of the scheme's operation suggests that there are no major problems in running these pump units at the higher heads and flows required for turbine operation.

FUTURE DEVELOPMENTS

The technology of induction generators, induction generator controllers, peltric sets and pumps as turbines is being transferred to other countries, through training courses and demonstration installations. There are plans to develop a directly-coupled propellor turbine-induction generator unit for low-head micro-hydro sites.

REFERENCES

Sharma, K.R. (1985). Small Hydro Electric Projects - Use of Centrifugal Pumps as Turbines. Kirloskar Electric Co., Bangalore, India.

Smith, N.P.A., Williams, A.A., Brown, A., Mathema, S. and Nakarmi, A.-M. (1990). In: *Proceedings of the First World Renewable Energy Congress* (A.A.M Sayigh, ed.), Vol. 5, pp. 2904-2908. Pergamon Press, Oxford.

Waltham, M. (1991). The Manufacture of Pelton Wheel Runners. Intermediate Technology Development Group, Rugby, U.K.

Williams, A.A., Smith, N.P.A., Mathema, S. (1989). In: *Science Technology and Development.* Vol 7, 2, pp. 98-112. Frank Cass, London.

Hydropower in China's Energy System

Dr. Kurt Wiesegart

"Water as the basis (for power generation), steam as a supplement" (Shui zhu, huo fu) was in 1958 the pithy slogan for China's announced medium and long-term development strategy on the energy sector. The country with the world's largest hydro-energetic resources (out of 680 GW of hydropower reserves about 370 GW are considered technically usable for power generation) once planned to prioritize the utilization of these enormous, inexhaustible reserves and to use hydrocarbons only as a complementary resource for power generation.

Whether it was ecological prudence or (more probably) long-term supply considerations that led Chinese planning strategists to formulate this objective, the fact is that not only has the goal not been achieved, but during the following two decades the trend has even been quite the opposite. While the share of hydro-electric installations in the country's total power plant capacities was as high as 50 % back in 1957, it had sunk to about 30 % by the late 70s, with the proportion of primary electricity from hydropower dropping from 25 % to 17 - 18 %[1] during the same period. And the thermal power plants erected since then under a priority program, most of them coal-fired, are a source of massive environmental pollution.

Although China, with a capacity of about 36,000 MW installed in hydro-electric power plants, is already one of the world's largest users of this energy source (behind the United States, the CIS countries, and Canada), only just under 10 % of the country's total technically utilizable hydropower reserves have so far actually been employed for power generation.
The objective formulated back in 1958 was abandoned almost before it had been announced. The main reasons behind this U-turn were as follows:
-   the rapidly growing demand for electricity had to be met
-   high growth rates in coal production, and successes in building up an oil industry
-   the (unfavourable) regional distribution of hydropower resources
-   repeated changes in development policy concepts

---

1   Wiesegart, Kurt: Die Energiewirtschaft der VR China, published by Weltarchiv GmbH, Hamburg, 1987, pp. 91 ff.

Growing demand for electricity

Economic development in China during the 50s and 60s was characterized by concentration on expanding heavy industry, which was regarded as the motor of economic progress: the iron and steel sector, the mechanical engineering industry, and the power industry. During the 60s and 70s, the chemical and petrochemical industries, likewise energy-intensive operations, were also prioritized.

During the period starting in the early 50s and ending in the late 70s, a cutoff date which marked the beginning of a rethink in economic policy, industry's gross production value showed an average annual growth of over 11 %, while the increase in power generation and power consumption over the same period was 14 - 15 % per annum. [2]

Although investments in expanding power generation capacities were extremely high compared to other countries, expansion of power plant capacities fell increasingly behind the demand for electricity. [3] Since the late 70s, China's eastern, comparatively industrialized coastal belt has been "living" with a supply/demand shortfall totalling about 70 billion kWh, or 25 %, which means that 25 % of the industrial production capacities cannot be utilized due to the lack of electricity. The resulting production losses are estimated at 400 billion Yuan (1988).

Fossil fuels

Consequently the Chinese leadership has since the late 50s not been able to rely on a "cushion" of existing reserve capacities which would enable them to conceive and implement a long-term national policy for energy (electricity) supply. Instead, in view of the increasing imbalance manifested by ongoing development, they had no alternative but to utilize the resources available (capital, labour, and energy) in a manner enabling an expanded range of power generation capacities to be made available as quickly as possible. And the convenient solution was to build thermal power plants based on the enormous coal deposits (with their comparatively favourable regional distribution), and on oil production, which up to the late 70s was prospering rapidly.

Coal production, for example, thanks to massive employment of labour (during the period of the Great Leap Forward) was tripled from 131 million tons to 397 million tons between 1957 and 1960, while in the same period power generation from thermal power plants was increased from 14.5 billion kWh (1957) to 52 billion kWh (1960), oil production showed average annual growth rates of approximately 25 % between 1958 and 1978. [4]

Compared to hydro-electric power plants, thermal power plants had the advantages of shorter construction times and lower capital investment costs. Whereas investment costs in thermal power pants in China averaged 660 Yuan

---

2   Wiesegart: Die Energiewirtschaft ..., pp. 89 ff.

3   Wiesegart: Die Energiewirtschaft ..., pp. 57 ff., pp. 203 ff

4   Guojia Tongjiju; Zhongguo Tongji Nianjian 1989, Beijing 1989; own calculations

per installed kilowatt output, the figure for hydro-electric power plants was almost double, at 1170 Yuan (period under review: 1953 - 1980).[5]

Potential and distribution of hydropower resources

Another impediment to prioritized utilization of hydropower resources is the regional distribution of the hydropower potential involved. Out of the 370 GW of technically utilizable resources, 70 % are concentrated in the south-west of the country, just under 10 % in central China, and 12.5 % in the north-west - far away from the country's industrial regions. The hydrocarbon deposits, conversely, exhibit a comparatively favourable distribution, when measured against the regional distribution of electricity demand.[6]

Lasting utilization of the gigantic hydropower potential in the south-west would have entailed, in addition to the relatively high capital investment costs for the power plants, the building of transmission lines (another capital-intensive project) over distances of 1500 to 1800 kilometres; prohibitive considerations in view of the lack of capital resources and technical know-how available. Up to the early 80s, establishment of a power transmission grid remained one of the Cinderellas in state investment policy.[7]

After the political and economic links with the Soviet Union had been broken off in the late 50s, there was also a shortage of qualified Chinese technicians and engineers, and thus of the requisite know-how for building large-scale hydro-electric power plants. Until the middle and late 70s, domestic machinery manufacturers were not able to provide turbine and generator units with the large capacities needed.[8]

Nonetheless, those hydropower resources which could be accessed with the resources available were indeed utilized - currently far more than 100,000 miniature hydro-electric power plants are installed throughout the country, offering a total capacity of 13,000 MW (cf. Table I).

5   Li Rui: Guanyu jiakuai fazhan Woguo shuidian jianshe de jijian yijian, in: Shuili Fadian, Beijing, No. 1, 1980

6   Li Wenyan: Woguo kuangchan ziyuan yu dili weizhi de diqu chadao - gongye buju ruogan tiaojian de jingji dili fenxi, in: Dili Yanjiu, Beijing, No. 1, 1982

7   Wiesegart: Die Energiewirtschaft ..., p. 98

8   Qian Jing, Dai Dingzhong: Woguo heliu nisha wenti jiqi yanjiu jinzhan, in: Shuili Fadian, Beijing, No. 2, 1980

These power plants (together with a similarly large number of small thermal power plants) were of considerable significance for building up China's small industries in the countryside - a program which China has been progressing in parallel with the establishment of large-scale industrial conglomerates throughout the various phases of development.

Tidal energy

In addition to the utilization of "conventional" hydropower, the utilization of regenerative energy sources has already achieved respectable dimensions (cf. Table I). In the context of this paper's subject-matter, the tidal power plants deserve particular attention.

Table I

INSTALLED SMALL-SCALE RENEWABLE ENERGY CAPACITY IN PR CHINA (End of 1990)	
Solar water heaters	1.8 million m2
Solar PV cells	1.5 MW
Solar cookers (parabolic dish)	120,000 units
Small wind generators (total capacity: 13 MW)	110,000 units
Wind farms (total capacity: 4.5 MW)	6 units
Wind pumps (total capacity: 2.1 MW)	1,600 units
Domestic biogas digesters	4.7 million units
Biogas generators	25 MW
Geothermal power stations	25 MW
Tidal generation	8.3 MW
Mini hydro generation	13,200 MW

Hu Chengchun: The Development of New and Renewable Sources and Improvement of Environment in China, in: Environmentally Sound Coal Technologies, Conference Paper pp. 221-223, International Conference, Beijing December 1991.

The total capacity of tidal power plants in China comes to 8.3 MW, not much compared with the world's largest tidal power plant at the Rance estuary (France, 240 MW), but remarkable enough when one considers that this energy source was opened up back in the 70s using the technological/scientific basis of a developing country.

The first, (and so far the largest) tidal power plant, Jiangxia (in Zhejiang Province), was begun in 1973, and completed in 1980. Its capacity is 3 MW.[9]

---

9   He Yougen: Jiangxia chaoxi shiyan jianzhan yihao ji fadian yunxing de tedian ji dan ji yunxing shi de shuiku tiaodu, in: Xin Nengyuan, Chongqing, No. 3, 1984

According to Chinese authors, three sites are particularly suitable for building large-scale tidal power plants:[10]
- the mouth of the Qiantang river at Hangzhou, utilizable output: 4500 MW
- the Changjiang (Yangtse) estuary: 800 MW
- the Queqing bay in Zhejiang: 500 MW.

Table II

REGIONAL DISTRIBUTION OF THE TIDAL ENERGY POTENTIAL IN CHINA		
	Utilizable power generation potential in tidal power plants (billion kWh/year)	Share of regions in total potential (%)
China overall	275.16	100.0
Zhejiang	114.6	41.6
Fujian	108.1	39.4
Shandong	16.5	6.0
Guangdong	13.3	4.8
Liaoning	11.3	4.3
others 1	10.9	3.8

1   including Taiwan. The statistical share of Taiwan is 0.2 %.
    Zhu Chengzhang: Wo guo de chaoxi dongli ziyuan, in: Xin Nengyuan, Chongqing, No. 3, 1981

According to calculations prepared by Chinese proponents of tidal power utilization, though the capital investment costs for tidal power plants are three to four times as high as in thermal power plants per kilowatt output, tidal power plants are most definitely competitive if one takes into account the costs for building production capacities for hydrocarbons, for transport capacities, the environmental costs, etc., plus the significantly lower operating costs involved.[11]

China's total technically utilizable tidal potential, distributed along its 14.000 kilometres of coastline, is estimated as 110 GW.[12] The distribution of this potential is shown in Table II. So far, however, no concrete plans for the erection of additional tidal power plants are known.

---

10   Wiesegart, Kurt: Gezeitenkraftwerke in der VR CHina, in: wasser, energie, luft - eau, énergy, air, 1984, Issue 7/8, CH Baden

11   Li Zhenrong: Jiangxia chaoxi shiyan dianzhan sheji yunxing chubu songjie, in: Shuili Fadian, Beijing, No. 4, 1984

12   Xu Shoubu: Nengyuan jishu jingjixue, Changsha 1981, p. 31

Administration - long-term perspectives lacking in planning and plan implementation

Besides the factors mentioned above which have hindered a prioritized utilization of this hydropower potential, the administrative system is a further contributory factor. The political upheavals and internecine struggles which took place between the late 50s and the late 70s (Great Leap Forward / consolidation phase / Cultural Revolution / overthrow of the "Band of Four" / beginning of the reform policy stressing decentralization and rapprochement towards the West) were not exactly conducive to the implementation of long-term energy policy concepts.

Inadequate coordination between decision-making bodies, and a competitive relationship between the institutions involved, frequently delayed project implementation; almost no cost-benefit analyses were carried out for power plant construction before the early 80s.[13] To enable the ministries for coal, for oil and natural gas, for water supply and electricity, and for nuclear energy, to cooperate effectively and to reduce the problems resulting from competitive behaviour patterns, a formally superior "Energy Commission", for example, was set up in the early 80s; since it proved a failure, it was dissolved after two years and finally, in the mid-80s, the ministries mentioned were amalgamated into a single body, the Ministry of Energy. There has since been a certain continuity of development in the energy sector.

China's present energy supply situation is characterized by the above-mentioned shortfall of approximately 25 % between energy supply and demand in the industrialized coastal region. This energy gap goes hand-in-hand with a simultaneous energy wastage of major proportions. Since for more than three decades China has had a centrally planned economy, with resources (labour, capital, including energy sources) being allocated on an administrative basis, there were no (economic) incentives to save energy - political appeals were just as ineffective as in other administratively managed economic systems.[14]

The future of hydropower in China

Under the projects planned by the Chinese leadership, another 40 GW of capacity is to be installed in hydro-electric power plants, plus 4 to 6 MW in pump storage plants, by the turn of the century, out of a total of 130 MW power plant capacity planned.[15] Thus by the end of the decade, with around 80 GW installed output, China is set to have the world's largest utilized hydro-electric power plant capacities, whose share in the total of power plants built would then be 37 - 40 %.

---

13   Wiesegart: Die Energiewirtschaft ..., pp. 115 ff., pp. 165 ff.

14   Wiesegart, Kurt: "Utilization of Energy and Potential of Energy Saving in PR China", paper presented at the UN Conference Series "Environmentally Sound Coal Technologies" in Beijing (China) and Madras (India) 1991/92

15   Zheng Bokun: Woguo de dianli nengyuan fazhan ji ji jiegou bianhua qushi, in: Zhongguo Nengyuan, Beijing, No. 4, 1990

The majority of the new hydro-electric power plants will be constructed in the upper and middle reaches of the Yangtse, and in the catchment area of the

Hongshui river in the south-west of China, and on the Yellow River in the north-west of the country. The electricity will be transferred through extra-high-voltage lines (500 kV) into the demand regions in the east of the country. It is also planned to locate more of the energy-intensive industries near the large hydro-electric power plants.

The Sanxia power plant (Three Gorges)

Since the mid-80s, discussions on building the Sanxia power plant (Three Gorges) have seen a vigorous revival. The decision to implement this project was taken in early 1992 after several years of exhaustive feasibility studies. On completion, the Sanxia power plant, with 17.68 GW, will be about 70 % larger than the world's present largest hydro-electric power plant, Itaipu. The first serious considerations given to utilizing the immense hydro-potential of the Changjiang (Yangtse) go back as far as Sun Yatsen (1919). Geological studies were carried out in the 30s and 40s, and in the 50s as well, without any actual measures for implementation being taken.

The hydro-electric power plant will be designed to generate approx. 84 billion GWh of "clean" power per year - that is about one-fifth of the entire amount of electricity currently being produced - and will "replace" the consumption of 40 to 50 million tons of coal. Besides its use as a supplier of electricity, the power plant's proponents point to spin-off benefits for flood protection and improved navigability of this dangerous section of the river.

The project is, however, by no means uncontroversial - not feast due to the ecological effects involved. Critics of the project object that more than one million people from 145 towns and 4500 communities will have to be resettled, that about 30,000 hectares of agricultural land and a road network of 1000 kilometres will be lost beneath the waters of the 600 km long lake behind the dam. There are also fears of unforeseeable ecological consequences (climatic alterations included) which a dammed-up lake of these dimensions may cause, and of the potential dangers presented by the enormous quantities of water (40 billion tons) stored behind the 180-metre-high dam in the event of war; and last but not least, the gigantic costs of this mammoth project. While Chinese authorities five years ago were still estimating the costs involved at 10 billion Yuan (3.2 billion US dollars at the then rate of exchange), official sources today are already speaking of sums in the order of magnitude of 57 billion Yuan (in early 1992: 10 billion US dollars).[16] This would be just under half the capital investment expenditure which during the period from 1953 to 1980 had been provided from the national budget for building up all the energy sectors put together.[17]

---

16 If we remember that the actual costs for the Gezhouba hydro-electric power plant (currently under construction) are already five times as high as the original estimates, it can be assumed that the provisional cost prognoses for Sanxia will likewise have to be corrected upwards. And measured against international yardsticks (approx. 1300 US dollars per kW of installed output), it can be anticipated that the costs actually incurred will be at least substantially higher than official forecasts predict.

17 Wiesegart: Die Energiewirtschaft ..., p. 307

The determinant reasons behind the swing towards hydropower, perceptible since the early 80s, may have been both ecological ones, and the foreseeable limits of fossil fuels.

China's environment is being polluted each year with about 15 million tons of SO_2 by the combustion of 1.1 million tons of coal alone; 85 % (1988: 516 million t) of China's entire CO2 emissions result from the burning of coal.[18]

Utilization of nuclear energy as an alternative has so far been only hesitantly pursued by the government. The first nuclear power plant, with an output of 300 MW, to the south of Shanghai, is scheduled to go on line during the course of this year; the second one, at the border to Hong Kong (900 MW), is due for completion in 1993/94. By the end of the century, a total of 4 to 5 GW of installed nuclear power plant output are planned - only half of the capacities envisaged as the plan target in the early 80s.

The primary factor behind this renewed prioritization of hydropower will essentially be the realization that the supply of fossil fuels is a limited one. Expressed in absolute figures, China is the world's third-richest country in terms of energy resources (behind the United States and the CIS countries): the geological coal deposits are estimated at 5000 billion tons, with over 700 billion tons regarded as proven; the substantiated oil reserves are quantified at 2 billion tons. With 500 to 600 tons of fossil fuel reserves for each member of the population, however, the reserves are only about half of those in the CIS countries, and a tenth of those applying in the United States. When we take into account the actual technical/economic rate of fuel production, the *per-capita* reserves come to only 250 - 300 tons. Since the present *per-capita* energy consumption in China is hardly 10 % of the equivalent figure in the industrialized countries of the West, increasing consumption can be anticipated. The Chinese leadership thus has no alternative but to devote more resources to utilizing regenerative energies in particular, in order to assure the country's supply of electric power in the medium and long term.

---

18   Cf. Zhao Dianwu, Wu Baozhong, Zhuan De'an: Control Technology and Policy of Energy-Related Environmental
      Pollution in China, Conference Paper for "Environmentally Sound Coal Technologies, Beijing, December 1991

THE UTILIZATION OF THE GENETIC ALGORITHM FOR THE OPTIMAL DESIGN OF A PNEUMATIC HYDROPOWER DEVICE.

I. C. Parmee* and G. N. Bullock**

*Plymouth Engineering Design Centre, University of Plymouth, Devon, UK.
**Dept. of Civil and Structural Engineering, University of Plymouth, Devon, UK.

ABSTRACT

The paper outlines the operation of a novel,run-of-river, low head Pneumatic Hydropower Device and the development of a hydraulically efficient system configuration. The application of evolutionary design techniques incorporating the Genetic Algorithm to the device is described and the generic potential of these techniques within the renewable energy field is discussed. Other relevant, energy- related projects currently under study at the Plymouth Engineering Design Centre are described.

KEYWORDS

Hydropower; low-head; pneumatic; evolution; genetic algorithm.

THE DEVICE

The Pneumatic Hydropower Device (Bullock and Parmee,1989; Parmee,1990) utilizes a high speed, small diameter, self-rectifying air turbine to capture the energy potential of a differential head of water. The turbine is housed in a structure which converts this potential into air power. In its simplest form the device achieves this in the manner shown in Fig. 1. The turbine is situated within an opening in the roof of a chamber which is, itself, an integral part of a river barrage across which the differential head is generated. Upstream and downstream gates alternately open and close creating an oscillating water column within the chamber which drives air to and from atmosphere through the self-rectifying turbine. In this manner the high costs normally associated with the manufacture and installation of a large diameter water turbine are avoided.

It is necessary to ensure that the overall hydraulic efficiency of the system is maximised in order that viable energy capture is achieved from the low-head flow. Small-scale laboratory testing has

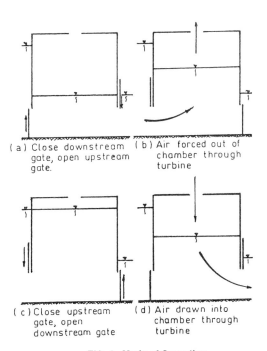

(a) Close downstream gate, open upstream gate.

(b) Air forced out of chamber through turbine

(c) Close upstream gate, open downstream gate

(d) Air drawn into chamber through turbine

FIG. 1. Mode of Operation

2525

indicated that the well-shaped water passages of the configuration shown in Fig. 2 present the least resistance to the flow and therefore minimise energy losses. As can be seen, the turbine is now situated in a duct which links two chambers in a closed, dynamic system. A particularly novel aspect of this configuration has been the development of the flow-actuated butterfly gate (Parmee and Bullock,1991) the operation of which relies solely upon the hydrostatic and hydrodynamic characteristics of the flow. Parasitic power requirements are therefore minimal.

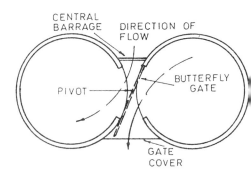

PARAMETRIC TESTING

The physical modelling was complemented by the concurrent development of a comprehensive mathematical model of the system which, when suitably coded, provided a flexible computer simulation (PWETWO) that could be utilized as a design tool. A series of parametric studies utilizing this model were embarked upon to investigate the parameter relationships of the system and to gain a better understanding of the system's operating characteristics. It rapidly became apparent that the interactive nature of the main parameters presented a complex, multi-dimensional search space within which only the broadest physical constraints could be identified by simple trial and error techniques. It was also apparent that there existed a large number of local optima for each parameter set that was investigated. This suggested that traditional linear optimization techniques could only be relied upon to place the investigator in the "better" design regions as opposed to converging upon a global optimum solution.

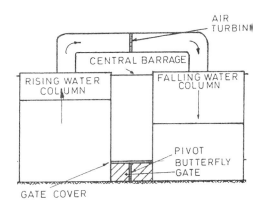

Fig. 2. Twin Chamber Configuration (Plan and Upstream Elevation)

Typical parameters thus investigated included the upper and lower bounds of the water column stroke and the damping characteristics of the turbine. Another more complex design problem that required attention was the definition of an optimal shape for the twin chambers of a linked system. There must exist a common chamber shape that will allow a compromise between the air capacity of the system, the water column velocities and the air flow rate through the turbine and result in an optimal average power output. This shape must provide an optimal solution where water columns are rising and falling in each chamber and must therefore involve the pressures, flow rates and compressibility characteristics of the closed system.

Finding this optimal shape by simple parametric study presented an impossible task. The effects of all but the simplest variations in plan cross-sectional area of the chambers upon each of the interdependent variables already studied was difficult to ascertain. The application of traditional optimization techniques would inevitably result in premature convergence upon a local optima.

In order to tackle this problem with any degree of success it was therefore necessary to adopt a methodology that utilised the random, non-linear elements of the parametric studies whilst ensuring that the search effort was concentrated within the "better" areas of the design space. In this manner, an acceptable number of chamber shapes could be investigated, local optima could be avoided and a global solution would be found. Study of the field of such adaptive search techniques revealed the existence of Evolution Strategy (Rechenburg, 1984) and the Genetic Algorithm (Holland, 1975). The Genetic Algorithm (GA) seemed to present the powerful processing capabilities required to achieve the objective.

THE GENETIC ALGORITHM

The GA offers a robust non-linear search technique that is particularly suited to design problems involving large numbers of variable, interactive parameters. Unlike other more traditional optimization techniques the GA will avoid local optima and rapidly identify the global solution. The algorithm achieves this by the random exchange of information between increasingly fit parameter combinations and the introduction of a probability of independent random change.

The designer must first select those parameters that appear to affect the overall design of an element / system and determine the characteritic that he wishes to optimize. At this stage little, if any, knowledge of the interactive relationships of the chosen parameters is required. A population of parameter combinations is randomly generated and each combination is processed by a computer model of the system under design. The relative fitness of each combination is determined from the model output and, based upon this relative fitness, combinations are either discarded or reproduced. Those parameter combinations that are considered to be particularly fit may be reproduced more than once.

The reproduced combinations then represent the second generation. A Crossover Operator exchanges randomly selected information between the parameter combinations and a Mutation Operator may then randomly change parameter values depending upon a preset mutation probability. Other operators can also be introduced if required. Having applied the various operators to the parameter combinations of the second generation they are further processed by the computer model and the cycle is repeated until a desired level of convergence is achieved. It is the Fitness Proportionate Reproduction and Crossover which ensure that each successive generation consists mainly of increasingly fit chromosomes whilst Random Mutation ensures that wide areas of the design space are constantly being sampled thereby preventing the GA from converging on local optima.

The analogy to the natural evolution process is apparent. The fitness of a particular offspring of any species is determined by its ability to survive in that environment to which it is subjected. If the design parameters are coded in a binary form then the parameter combinations may be regarded as chromosone strings, the individual binary digits as genes and the values of each digit as alleles. The manipulation of these strings by the Crossover and Random Mutation Operators is again analogous to the natural process.

METHODOLOGY

The GA was first applied to the less complex relationships previously investigated during the parametric studies. The results were very encouraging with significant increases in average power output being achieved within twenty to thirty generations. This familiarization with the technique provided the confidence to apply the GA to the more challenging problem of finding the optimal chamber shape. A square chamber shape was chosen via PWETWO in order to simplify the hydraulics of the system for this attempt. Upstream and downstream water surface elevations were set at 5.0m and 2.0m respectively in order to provide a head differential across the chambers of 3.0m.

The chamber roof elevation was set to 5.6m and nodes were introduced at 200mm centres between the roof and the downstream water elevation. In this manner 18 positions were established where the plan cross-sectional area could be varied. An upper and lower bound of 28.0m^2 and 12.0m^2 respectively was applied to these plan cross-sectional areas. The resolution of any variation was set at 0.25m^2. This is adequately illustrated by the first generation, randomly generated chamber profile of Fig. 3. The controlling GA ensured that the plan cross-sectional areas of each chamber varied in an identical manner.

The 0.25m^2 resolution of the area variation creates 64 possible plan cross-sectional areas between the upper and lower bounds of each of the eighteen nodes. This results in a design space consisting of 3.3×10^{32} possible area combinations each representing a different chamber shape. Each plan cross-sectional area is represented as a six-bit binary form to produce 108 bit chromosone strings. The population size was set at 20 chromosones and a random mutation probability of 0.01 was introduced.

The evolution of the chamber shape can be seen in Fig. 3. Two hundred generations were required before the process converged upon an average power output of 90.45kW. As the shape of the chamber containing the volume of air remaining above the maximum water level has no significance with regards to the mathematical model it can be idealized in the manner shown without significantly affecting the predicted power output.

The performance of the optimum chamber shape was then compared with that of a uniform, cylindrical chamber system of the same chamber capacity and with identical turbine and gating characteristics. This comparison revealed that the evolutionary process had optimized the chamber shape in a manner in which the resulting air flow/pressure differential regime was better suited to the operating characteristics of the air turbine. The chamber shape achieves this by accelerating the rise and fall of the water columns in those regions of their stroke within which the head differentials driving the flow are diminished. Water column velocities are also significantly increased in the better regions of the stroke. During any one cycle the turbine of the optimized system was operating at efficiencies greater than 70% for 80% of the cycle whereas the uniforn chamber turbine could only maintain similar efficiencies for 65% of its cycle time. This ensured an improved performance over the complete cycle and an increase in average power output of 7.7%. A better relationship between the incompressible water flow and the compressible air flow within the chambers may be a contributory factor to this improved overall performance.

OTHER ENERGY-RELATED APPLICATIONS

The Plymouth Engineering Design Centre (PEDC) is one of the six Engineering Design Centres established by the SERC at academic institutions. The main objectives of the PEDC is to carry out fundamental research into the application of the Genetic Algorithm and related adaptive search techniques to engineering design. As has been previously explained a primary use of these adaptive search techniques is in the optimization of engineering system design. The Genetic Algorithm can offer a means of improving the efficiency of existing systems thereby reducing overall cost and improving performance. Improved efficiency will lead to savings in energy usage and raw materials and in some cases to an attenuation of the pollutive aspects of the system under design.

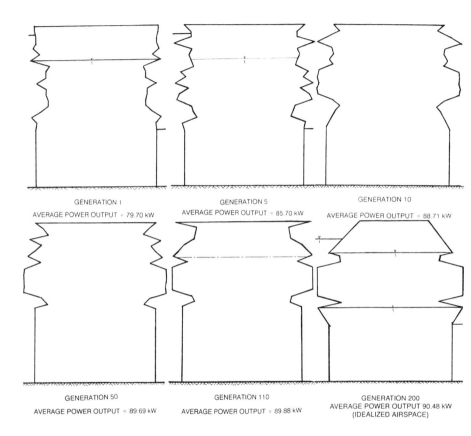

GENERATION 1
AVERAGE POWER OUTPUT = 79.70 kW

GENERATION 5
AVERAGE POWER OUTPUT = 85.70 kW

GENERATION 10
AVERAGE POWER OUTPUT = 88.71 kW

GENERATION 50
AVERAGE POWER OUTPUT = 89.69 kW

GENERATION 110
AVERAGE POWER OUTPUT = 89.88 kW

GENERATION 200
AVERAGE POWER OUTPUT 90.48 kW
(IDEALIZED AIRSPACE)

Fig. 3. The Evolutionary Development of the Chamber Shape

The PEDC is currently working in collaboration with Rolls Royce investigating the application of the GA to the optimization of various aspects of gas turbine engine design. General Electric are already some way along this optimization path using hybrid techniques incorporating the GA. They are claiming an increase of between one and two percent in engine efficiency. The significance of such an improvement in terms of savings in fuel oil and in the reduction in the discharge of pollutive exhaust gases illustrates the potential energy and environmental benefits of system optimization.

Any improvement in gas turbine efficiency would benefit those renewable energy technologies that generate combustible gases such as methane and hydrogen Such technologies would also benefit from the application of adaptive search techniques to dual-fuel, reciprocating engine design. Similar applications to the steam turbine generating systems of Solar Thermal plant and to the binary systems utilized by the goethermal industry could also improve the economic characteristics of the technologies.

The same philosophy applies throughout the renewables field. Improvements in component design and in the operational characteristics of the various technologies could provide the small increases in efficiency which would contribute directly to the ability to generate an acceptable economic return. It is probable that the non-linear approach of the GA will result in further improvement in those cases where traditional

2528

mathematical techniques have already been applied especially where particularly complex multi–dimensional search spaces are involved.

Another PEDC project involves the optimal design of concrete arch dams. The objective function in this case is minimum volume whilst retaining acceptable construction characteristics. The achievement of this objective will result in significant cost savings which may contribute directly to the economic viability of a hydropower project. Related to this but taking account of the overall hydrology, the application of these techniques to river basin planning and system operation could result in the optimal use of a specific water resource for large and small–scale cascade hydropower schemes and associated irrigation networks.

The development of the use of adaptive search techniques has been hindered by the computational resource required to allow their powerful processing capabilities to be used to the full. However, the increase in availability of computational power at an acceptable cost has been phenomenal in recent years and there is little sign that this progress is slowing. Use of the GA on desktop machines is already a reality. Further hardware development will result in more complex problems being attempted and the possibility of the algorithm being applied to several aspects of a system's design both at a macro and micro level.

Bearing this in mind perhaps we should look closely at those renewable technologies that are technically proven but still cannot penetrate the market through being considered to possess marginal economic potential. In some cases slight improvements in system efficiency would be all that is required to remove this economic ambivalence and to ensure that the technology achieves the initial market penetration required to become established. This efficiency increase may be achieved by identifying those areas of system design which will benefit from the application of the correct adaptive search technique.

Suitable applications could vary from the operational optimisation of a large scale tidal power scheme to the optimal design and operation of those manufacturing processes required for the production of photovoltaic cell materials. Other potential areas of use currently being explored by the PEDC include the energy efficiency / best practice fields and energy efficient building design.

SUMMARY

The Pneumatic Hydropower work has shown that the powerful and robust processing capabilities of the Genetic Algorithm can provide an optimal design even in those multi–dimensional cases where traditional techniques could only provide sub–optimal alternatives. The economic sensitivity of the renewable energy technologies to system efficiency suggests that that they could benefit greatly from the application of such adaptive search techniques. The optimization of not only power output but overall system design and even manufacturing processes could result in significant improvements in economic potential.

ACKNOWLEDGEMENTS

Initially, the work described in the paper was supported by the Department of Civil and Structural Engineering of Polytechnic South West, Devon, UK. It was a continuation of a study funded by the UK Energy Technology Support Unit, Harwell. Further work utilizing PWETWO is now being carried out at the Plymouth Engineering Design Centre which is funded by the UK Science and Engineering Research Centre. The Authors wish to thank these organizations for their support.

REFERENCES

Bullock, G. N. and Parmee, I. C. (1989). The Performance and Economics of a Pneumatic Water Engine. ETSU Report No. SSH 4043.
Holland, J. H. (1975). Adaptation in Natural and Artificial Systems. Ann Arbor, The University of Michigan Press 0–472–08640–7.
Parmee, I. C. (1990). Pneumatic Hydropower Devices. PhD Thesis, Polytechnic South West, Devon, UK.
Parmee, I. C. and Bullock, G. N. (1990). The Development of a Flow–actuated Butterfly Gate for a Pneumatic, Low–head Hydropower System. Proceedings of the ISES Solar World Congress, Denver, Colorado.
Rechenberg, I. (1984) The Evolution Strategy – A Mathematical Model of Darwinian Evolution. In: Synergetics – From Microscopic to Macroscopic Order. (E. Freland, Ed.) Springer Series in Synergetics, Vol. 22, Springer, Berlin – Heidelberg; 122–132.

FIELD EXPERIMENTS OF A WAVE POWER CONVERTER WITH CAISSON BREAKWATER

Hiroaki Nakada[1], Hideaki Ohneda[2], Shigeo Takahashi[3],
Masazumi Shikamori[4], Tadashige Nakazono,

Coastal Development Institute of Technology, Japan

Sumitomo-hanzoumon Build.3-16, hayabusa-cho
Chiyoda-ku, Tokyo, Japan 102

ABSTRACT

The Ministry of Transport and the Coastal Deveropment Institute of
Technology in Japan are jointly developing a fixed-OWC wave power
converter. This paper describes the results of the field experiments of
the converter conducted from 1987 to 1991.
A prototype devise was installed in a breakwater of Port Sakata and the
system started electric power generation in the Winter of 1989.
Various measurements were made in the experiments; incident wave power,
power conversions by the system, the wave forces on the caisson, etc.
The experiments were conducted very sucsessfully and the design method of
the system was confirmed by the experiments.

Introduction

As global environmental issues become increasingly, important the
importance of developing natural energy resources also increase. Under
such circumstances, the development and utilization of clean as well as
unlimited ocean energy is desirable.
Basic studies on a wave power extracting caisson breakwater, which absorbs
the wave's energy into the caisson and converts it into electric power,
have been conducted since 1982 by the Harbor Research Institute, Ministry
of Transport, Japan.
In order to put the wave energy utilization system to practical use
through the joint efforts of government, universities and industry, the
Ministry of Transport and the Coastal Development Institute of Technology

---

[1]The First District Port Construction Bureau, Ministry of Transport,
Japan
[2]The First District Port Construction Bureau, Ministry of Transport,
Japan
[3]Port and Harbour Research Institute Ministry of Transport, Japan
[4]Coastal Development Institute of Technology, Japan

began field experiments at Sakata Port, Yamagata Prefecture, and installed a wave power extracting caisson breakwater and generating sets there to conduct various measurements, analyses and experiments to utilize generated electric power.

Outline of the breakwater to generate electric power from wave power

As shown in Fig. 1, the wave power generation breakwater is a caisson breakwater with a sloped wall in its upper portion. At the front this is a hollow box called an "air chamber", while the rear is the basic part of the caisson breakwater. The air chamber is the primary wave energy converter of the oscillating water column (OWC) type.

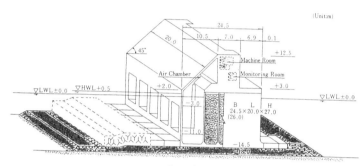

Fig.1 The shape of Caisson Breakwater

The turbines and generator are installed at the center of the machine room which is installed at the upper part of the caisson. The turbines are rotated by the air flow, while the generator is rotated by the turbine, and ultimately coverts the wave energy into electricity. Two Wells Turbines, providing one-way rotation, are installed sandwiching the generator along the same axis. In the machine room, three types of control devices are installed in addition to the above equipment. These control devices include an air flow regulation valve, a pressure release valve and an emergency valve. Figure 2 shows diagram of the machine room.

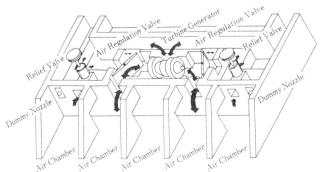

Fig.2 The Conception of the Machine Room

Summary of field experiments

Figure 3 gives a plan view of SAKATA Port, where verification experiments were carried out using a caisson of the Second North breakwater at SAKATA Port to measure the stability of marine weather and the breakwater, the stability of the component members, and the power generation efficiency. Collected data was converted into optical signals at an offshore observation station before they were sent by an opto-electric power combination cable to an observation station on land, where data is processed by a personal computer for real-time analysis.
Field experiments were carried out from 1987 to 1991 to study and verify methods of design, construction and utilization of electric power to create a caisson that could use wave power to generate electric power.

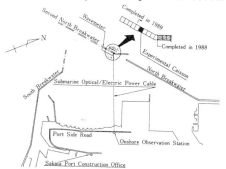

Fig.3 Location of Field Experiments Site (SAKATA Port)

The task of verification included (1) ensuring an excellent wave dissipation capability and stability against the force of waves, (2) the development of methods of designing a proper air chamber, turbine and generator as major components of the wave energy converter, and (3) the development of a method to utilize electric power generated by waves. As the result of the verification experiments conducted so far, the safety and appropriateness of methods of designing the wave resistance of the breakwater have almost been verified.
Power generating operations have also been completely ensured and the performance of the energy converter has been confirmed as conforming to the initial design. In the studies onthe utilization of electric power generation by wave energy, assuming thepotential use of electric power generation in practical applications, various devices were actually operated, and satisfactory results were obtained.

Power operation

The wave power generation breakwater has been in full-scale power generating operation since December 1989 and has obtained results close to those initially anticipated by the design. Figure 4 shows the changes in mean electric power output per hour during the winter in January 1991 when the wave power was at its highest level. The mean electric power output of the operations during this month was 13.25KW. However, power operations were carried out 6 days a week for 8 hours during the daytime only.

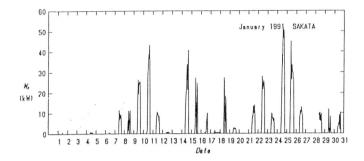

Fig.4 Variation of Electric Power

Figure 5 shows the typical conditions of power generation before and after 14:00 on November 8, 1990. This figure provides the water level in the air chamber η, pressure in the air chamber P_a, opening of the air flow regulation valve V_B, turbine pressure difference P_d, turbine revolutions N_T, and time-based changes of the electric power output W_g for 20 minutes. The water-level fluctuation was 2 to 4 meters and the electric power output fluctuation was 20 to 60KW.
The significant wave height was 3.0 meters, the significant wave period was 7.9 seconds, the incident wave power was 539KW and the mean electric power generation was 36.4KW.
In this case, regarding the energy conversion efficiency, the air chamber efficiency was 0.59, the turbine efficiency was 0.38, the generator efficiency was 0.91, and the product of these efficiencies was 0.20.

Utilization of wave power generation technology

Electric power generated by waves produces energy that depends substantially on time-based or seasonal changes. If an inexpensive electric power storage device with a large capacity is developed, it will be possible to supply electric power in a more consistent manner. In these experiments, no special devices were used to smooth the fluctuating electric power; however, methods of electric power utilization depending

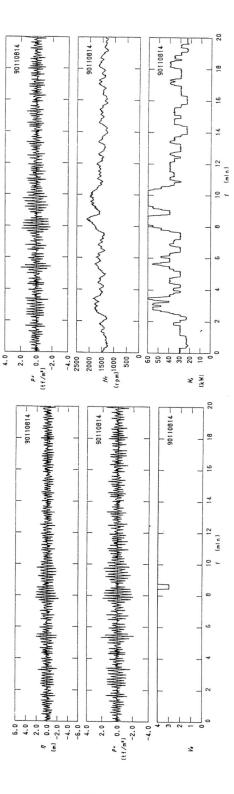

Fig.5 Example of Profile of Time Series Date

2534

on the fluctuation were studied. Specifically, demonstration equipment
intended for the illumination of electric lights, road heating to melt
snow, pumping for sea water exchange, etc. were used.

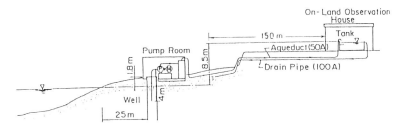

Fig.6 Water Pumping System

Figure 6 shows a pumping system which pumps up sea water from the beach to
the onshore observation station. Water is pumped although the frequency
fluctuates. This system can be applied to improve the quality of water in
a closed harbor. The system can be used for subsand filter construction
to prevent erosion of the seashore.
The road heating system is used to melt snow through the thermal energy
obtained from the heating wire embedded in the road, and is suitable for
use in areas such as Sakata City, which is located adjacent to the Sea of
Japan, which has much snow and high waves in the winter.
According to the wave date collected during the field verification
experiments the cost of wave power generation in the coastal waters of
Japan was estimated to be ¥13 to ¥36/KWh.

Acknowledgement

Field experiments were satisfactorily completed just as initially
designed, and electric power generation technology by the wave power
generation breakwater was demonstrated. As a result, the functions of
breakwaters that protect harbors can be improved, and it will soon be
possible to effectively utilize and practically use the clean and
unlimited energy of the ocean.

REFEERENNCES

Takahashi,s.: A study on design of a wave power extracting caisson
  breakwater. Report of Wave Energy Lab., Port and Harbour Reseach
  Institute, 1988.337p.
Goda, Y. et.al: Field Verification Experiment of a Wave Power
  Extracting Caisson Breakwater ,-Design and Construction of the System
  and Plan for Its Test Operation-, International Conference on Ocean
  Energy Recovery, Hawaii 1989 11.
Ojima,R. et.al.: Theory and experiments on extractable wave power by
  oscillating water-column type breakwater caisson. Coastal Engineering
  in Japan, Vol.27,1984,pp.315-326.

TIDAL ENERGY POTENTIAL ALONG INDIAN COAST

K.V.S.R. Prasad and V. Ranga Rao

Department of Meteorology & Oceanography
Andhra University, Visakhapatnam 530 003   INDIA

ABSTRACT

India has a long coast line of about 6,100 km. The rhythmic rise and fall of sea levels in the form of tides constitutes a source of energy which is continuously being replenished. Studies on energy related to tides in India are very limited. Studies on these aspects help to identify country's alterante energy resources. In the present paper an attempt is made to study the tidal energy potential available along Indian coast. For this, the tide data at a number of places along the Indian coast were collected and analysed. The tidal ranges during spring tide time and neap tide time have been discussed. Three promising sites have been identified for large scale power generation- Gulf of Cambay, Gulf of Kutch and the region of Ganga delta.

KEY WORDS

Tidal power; Gulf of Cambay; Rann of Kutch.

INTRODUCTION

Tidal power is produced in slugs, in two periods of a few hours duration each lunar day which is slightly longer than a solar day. Thus the time periods during which tidal power is available shift slowly in relation to the solar day. The quantity of power that becomes available during each tidal generation period varies, moreover overtime, according to the tidal range. The absorption of tidal energy in power systems that are characterised by variation in load demand derived from daily, weekly and seasonal cycles based on the solar system, poses certain problems, as a result of this irregular and variable supply. Systems in the maritimes experience peak loads during the late afternoon and early evening hours of each day. This peak loads are highest during week days in the winter and lowest during week end days in the summer. The lowest load is experienced during summer nights. Whereas a supply of tidal power during a winter peak load will be very welcome, the same supply during a summer night could be a little value or might be unusable in the maritime market.

The tidal energy is derived by using the difference between the load and high tides. In specific form, the method of harnessing this form of energy is through filling up the estuary basin with water during high tides and using the stored water for generating electricity during low tides. For continuous generation of electricity both high and low tides may be used by installing special turbines.

Some hundred sites suitable for the construction of tidal power plants exist in the world, some of which could be coupled for still greater efficiency and productivity. India has long coast line which can offer many sites for harnessing of this form of energy. An expert panel of the National Committee on Science and Technology has examined the tidal data and prosepctive sites for tidal power generation in India (1974). Sastry (1974) while reviewing the prospects of harnessing tidal power on Indian coast has stated that the configuration of the Indian shoreline does not encourage the production of tidal power but a careful examination of the shoreline near the Gulf of Cambay suggests the presence of a potential site for tidal plant. He has further suggested that the Dhadar estuary can be used for the construction of an artificial plant for creating storage reservoir.

The first study conducted in the Gulf of Cambay aimed at installing a 30 MW facility that would provide some 75 GW-hr annually. Both the Gulf of Cambay and the Rann of Kutch was stated as suitable for large scales while small plants could be built in the Sunderbans area of Gangas delta (Subramanyam, 1978).

The object of this paper is to assess the possibilities of locating tidal power generation schemes on the Indian shoreline.

METHODOLOGY

The selection of suitable site remains the primordial step in developing tidal energy utilization. Until recently, only locations with high tidal ranges could be considered, but technological advances may well reduce that factors impact.

Bernshtein (1939) and Mosonyi (1963) has calculated the potential energy of tidal power sites. Tidal energy is primary concern when considering a site. The incremental amount of energy (E) in KW-hr/year obtain from the tidal flow of water is given by

$$E = K. 10^6 AR^2 \qquad (1)$$

Where the basin area (A) in square kilometers, the average tidal range (R) in meters and K the constant depending on the type of scheme of a single basin with a double tide. We find that the actual developable energy E is given by

$$E = 0.67 \times 10^6 AR^2 \text{ KW-hr/yr} \qquad (2)$$

Accuracy of this calculation depends on the basins sufficient depth and a tide range sufficiently high, and provided the dimension of the basin in the direction of the travel of the tide's wave is equal or large than the tide wave's length.

DISCUSSION

Although India has a long coast line, the tidal variations are large only few locations, that is Gulf of Kutch and the Gulf of Cambay in west coast and the delta of Ganga and Godavari estuary in east coast of India (Fig. 1). To study the tidal variations of the above places, tidal data have been obtained from Survey of India, Dehradun, the area of the basins have been obtained from past studies (Subramanyam, 1978). To compute the tidal energy potential, the Gulf of Cambay and the Gulf of Kutch divided into C_1, C_2 and K_1, K_2, K_3 respectively. The Sunderbans and Godavari estuary has been divided into S_1, S_2, S_3 and G_1, G_2 basins (Fig. 1). For all the above stations the tidal

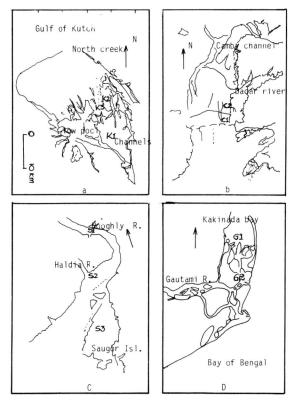

Fig. 1. Map showing the locations of
a) Gulf of Kutch, b) Gulf of Camby
c) Sunderbans and d) Godavari estuary

2538

energy potential have been computed by using the equation (2) and presented in Table.

Table . Suggested areas of tidal power plants
in India

| Place | Tidal Range (m) | | | Area (km) | Tidal power 10^6 KW-hr/yr |
	Spring tide	Neap tide	Mean		
Gulf of Camby					
C_1	10.81	4.33	7.57	1972	75713.5
C_2			7.57	1751	67228.4
Gulf of Kutch					
K_1			5.03	639	10832.0
K_2	6.88	3.18	5.03	538	9119.1
K_3			5.03	278	4712.1
Sunderbans					
S_1			3.54	0.76	6.38
S_2	4.31	1.75	3.54	0.83	6.97
S_2			3.54	13.57	113.90
Godavari estuary					
G_1			1.11	11.2	9.24
G_2	1.56	0.65	1.11	14.4	11.06

The maximum tidal range in the Gulf of Camby is quite large 10.81 m which qualifies this estuary for further considerations for location of a tidal power scheme. There are two possible sites in the Gulf of Camby C_1 and C_2 . The probable potential of the scheme has been assessed to be about 75713.5 x 10^6 KW-hr/yr and 67228.4 x 10^6 KW-hr/y respectively. The silt index of the Gulf of Camby is about 5000 ppm which is considered very high.

Though in the case of Gulf of Kutch, the maximum range of tides is smaller than that in the Gulf of Camby (6.88 m maximum spring tidal range agianst 10.81 m). The silt charge in the Gulf of Kutch is much smaller (1000 ppm). The estimate of probable potential power in this estuary is about 10832 x 10^6 KW-hr/yr.

In the case of West Bengal, Sunderban Islands the tidal range is maximum of about 4.31 m. The estimated tidal energy in this estuary is about 113.9 x 10^6 KW-hr/yr. The Sunderbans area is criss-crossed by a number of creeks and many of them have tidal conditions at both the ends. These characteristics often possibilities for tidal power development and schemes including those which involve filling and/or emptying at both ends can be designed.

CONCLUSIONS

The studies carried out so far indicate that there is large tidal power potential available in the west coast of India in Gujarat and there are possibilities of small scale developments in the Sunderbans area. Tidal energy development scheme will essentially involve (1) construction of a long embankment across the estuary so as to create a large reservoir on the landward side which may be filled up during flood (rising tide) period through a system of large sized sluices (2) Turbogenerators capable of efficient generation at low head consequently handling large flows, so that energy is produced during falling tide (3) Aerial and hydrographical survey of the estuaries to be undertaken (4) Investigation be undertaken by setting up complete tidal models of the estuaries so as to determine the best locations of establishing tidal power plants.

REFERENCES

Bernshtein, L.B. (1974). Russian tidal power station is precast offsite, floated into place. Civ. Eng., 44, 46-49.
Mosonyi, E. (1963). Utilizable power in seas and oceans. Water power Development I. Moscow State Publishing House.
Report of the Fuel Policy Committee, India (1974).
Sastry, J.S. (1974). Harnessing tidal power. The Tribune, Punjab, India.
Subramanyam, K.S. (1978). Tidal power in India. Water power and Dam construction, 6, 42-44.

"TOWARDS A PROTOTYPE FLOATING CIRCULAR CLAM WAVE
ENERGY CONVERTER

L J Duckers, F P Lockett, B W Loughridge, A M Peatfield, M J West,
P R S White, Coventry University, Coventry, CV1 5FB, UK.

ABSTRACT

The clam wave energy converter concept has been developed over a number of
years by Coventry University. It is intended to be a simple, reliable and
cost effective system which will be moored in moderately deep water a few
kilometers from land.

This paper will describe the operation of the system and will go on to show
how the current state of development: model scale trials, mathematical
modelling and component fabrication and testing; can be taken forward to
a full scale sea going prototype with a projected mean electrical output
of 1MW.

Successful completion of the prototype tests will lead to deployment of
numbers of units off the west coast of the UK.   These could make
significant contribution to satisfying the UK electrical demand, and the
projected costs are in the region of 5p/kWh, which should be seen as
attractive from an environmentally benign, sustainable energy source.

KEYWORDS

Wave Energy; Circular Clam; Design Optimisation; Prototype.

INTRODUCTION

The Coventry University Energy Systems Group have worked on the development
of wave energy devices since 1975, initially on the Salter Duck, then on
the invention and development of the STRAIGHT CLAM which, in 1982, was
considered by the Department of Energy assessment programme to be the most
promising device for further development.

However the costs predicted by the final report, 7-10 p/kWh at 1982 prices,
were considered to be too high to warrant further government support and
wave energy in general was put into the "longshot" category for future
development.
Although general funding for wave energy research was phased out at this
time, the Coventry Energy Systems Group carried out further development of
the CLAM, supported as previously by RMC and with some funding from the
Department of Energy. This work lead to the evolution, initial design, and
performance model testing of the CIRCULAR CLAM. This circular design was
then assessed by the Department of Energy in 1986 and even though it was
an initial design with no optimisation of performance or structural design,
the predicted costs were down to 6-8 p/kWh at 1986 prices.   These were
still regarded as too high and so, with no further funding, the development
of the circular CLAM came to a near halt.

As part of its ongoing programme of re-assessment, the UK Department of
Energy have recently undertaken a comprehensive review of the future
prospects for wave energy, with past and newly funded performance and cost

optimisation work on the circular CLAM forming a major input. The results of a modest programme to improve the cost effectiveness of the CLAM suggest costs of the order of 4-6p/kWh. There has been strong agreement between the group and the Department of Energy's Review and it is likely that the CLAM will be classified as 'promising' but uncertain!

THE CLAM

The CLAM converter currently envisaged is a rigid floating torus 60m or more in diameter and 8m high consisting essentially of 12 interconnected air cells with rectangular membranes on their outer faces separating the sea from the closed air system. With ballast, the structure would weigh about 5000 tonnes and be moored a few kilometers off shore in water 40m to 60m deep. The air system is inflated to about 15 kPa with the membranes approximately 3/4 submerged.

Differential wave action around the torus then causes air to be pumped back and forth between the cells. This flow is resisted by Wells turbines, one located between each pair of neighbouring cells in air duct running around the torus. The turbines are coupled directly to generators and the power is fed to shore via a subsea flexible cable. With compliant mooring and low freeboard the device experiences moderate mooring and structural forces, even in storm conditions. The CLAM is thus highly modular in its construction with simple multiple components which, though novel in their application, require only conventional technology for their series production in factory conditions.

THE MEMBRANE

The use of flexible rubber membranes as the interface between sea and air is clearly the innovative feature of the CLAM which calls for major development work. It is the key moving part of the system which affects the conversion of wave to pneumatic power yet at the same time offers the potential of low manufacturing cost. The membrane has to be strong enough to survive the forces acting upon it yet must be able to flex and accommodate shape changes as its air cell inflates and deflates in each wave cycle.

To achieve these anisotropic properties, the CLAM membrane will be constructed with two plies of reinforcing cords laid at around 15 degrees to the vertical with a 2mm covering of blended rubber to make it airtight and resist wear and abrasion. In ten years of operation the membrane is expected to undergo 30M a significant flexings, about the same as the number of revolutions of a truck tyre carcase during its lifetime. A ten year life for the CLAM membrane is therefore not impossible given sufficient development in the research phase and on the seagoing prototype. In any case, the membranes represent a few percent of capital cost, so regular replacement should not affect energy cost unduly.

It is also proposed that any cell with a damaged membrane can be isolated from the rest of the air system, by closure of an emergency valve, to limit water ingress and to allow the rest of the system to perform much as it did before.

For many years the ESG has worked closely with Avon Rubber plc. to develop the material and one-piece fabrication technology. The research staff at Avon are confident of producing a durable CLAM membrane at acceptable cost. The following table shows calculated details of its operational duty.

Length	m	16	
Height	m	8	
Thickness	mm	6	
Strength	kN/m	40	(Vertically)
Elasticity	%	0	(Vertically)
Strength	kN/m	20	(Horizontally)
Elasticity	%	10	(Horizontally)

This specification is well within the performance characteristics for aramid, polyester, and nylon reinforcing cords. Experience of fabrication, installation, operation and maintanence of a similar membrane used with a much more arduous duty cycle in a hydro electric scheme has proved extremely valuable. Detailed design for the Clam membrane is the next step in the development programme, and it is proposed that this is undertaken by constructing a full scale air chamber, complete with membrane, and testing this by attaching it to a harbour wall located in the North Sea. A second full scale air chamber would be mounted on the interior side of the same harbour wall, to act as an air exchange for the first cell. Thus confidence in the design and durability of the membrane will be established.

THE WELLS TURBINE

The Wells Turbine is an ideal prime mover for a wave energy converter which uses air as the working fluid because it can operate in reversing air flows, and the linear relationship between flow rate and induced pressure drop provides optimum damping to achieve maximum energy extraction from the waves. Theoretical considerations and laboratory and other experimental work are reported in (5). A 343kW prototype Wells Turbine was fabricated and tested in a hydro electric scheme. Further full scale testing in a wide range of steady flow conditions is still required to accurately establish the full scale characteristics of the Wells, and hence provide design information to yield the optimum performance in a sea going clam scheme.
Some economic methods of Wells turbine manufacture have already been developed, and although further refinements will be undertaken the team feel that the turbines do not represent a major cost centre of the overall scheme.

Peak conversion efficiencies of about 70% air to shaft power are common for wells turbines. Research is being conducted to increase this figure, by considerations of blade profiles and of guide vanes amongst others. Although this peak value is lower than that for a conventional turbine, the overall cycle performance is significantly better.

STRUCTURAL DESIGN

The 1986 design of the circular clam was based upon a modular space frame design produced by Howard Doris Ltd, formely of London. The design was reviewed by Rendel Palmer and Tritton Ltd (RPT) in October 1986 and accepted as a possible route forward.

As a result of work carried out from January to July 1991 by the Energy Sytems Group working with RPT, two new designs have been identified which substantially move forward the structural efficiency of the circular clam. One of the designs is in steel; the other is a steel reinforced concrete design with a similar section has described below. These two designs have been costed by RPT and compared directly with the earlier 1986 solution. Costs at second quarter 1991 rates for the concrete design are given below

1991 Concrete Hull design

The hull consists of twelve modules separated by watertight bulkheads. Each module is divided into six watertight cells by horizontal mid-height floor and two intermediate bulkheads. Each module supports one cell connected to a 1.5m common manifold duct. The central upper section in each module contains the turbine unit

concrete	1900 cubic metres
Reinforcement	290-310 tonnes
Prestressing	50 tonnes
Steel ducts	32 tonnes

10 off production run, 1992 Budget estimate, £1.18-1.80M

The 1991 versions of the circular Clam are designed to take full account of Lloyd's 'Rules and Offshore Guidance Notes' applicable to wave energy structures.

MATHEMATICAL MODELLING FROM WAVE TO WIRE.

Complementing the experimental and component development work, mathematical investigations have focused on three areas of interest. These are a linear hydrodynamic model of the device in three dimensions, the computer simulation of the air turbine and generator power take-off system and the analysis of the geometry of the flexible membrane.

During the years of evolution of the CLAM concept, a comprehensive linear mathematical model of its hydrodynamic operation has been developed. The model is based on linearized wave diffraction theory in which the CLAM is viewed as a freely floating body moving in heave, surge, pitch, sway and roll together with a further 12 modes to represent the surging motions of the flexible membranes which form the outer wetted surfaces of the CLAM air cells. Wave diffraction and radiation forces are calculated numerically using the well-known source distribution technique. Air flow within the device is modelled using a linearised gas law applied to the separate air cells connected by linear dampers to represent the Wells turbines. Frequency domain analysis then determines the system response to arbitrary excitation.

The model is able to predict well CLAM performance observed in the scale model tests conducted on Loch Ness in 1984. Absorption efficiency, that is the ratio of air power captured to wave power incident on a diameter, was calculated from the system response to incident plane waves of periods from 3 to 18 seconds. Mean efficiency in seas with a Peirson-Moskowitz spectral spread were then deduced. The results show good agreement with experimental values bearing in mind that the spectral spread of the seas typical of Loch Ness is narrower than the P-M spread of the mathematical results.

The future objective of the work is to quantify the performance changes and productivity gains to be expected from the variation of a wide range of design parameters, well beyond the scope of a practicable experimental programme. For example, structure diameter and height, the number of CLAM cells per structure and details of the air system configuration and turbine damping rate can be optimised in this way.

The simulation work takes data on air flow rate either observed in scale model tests or generated synthetically from the linear model, together with information on full-scale turbine and generator performance and calculates the response of the power take-off system in the time domain. In particular mean efficiency of turbine and generator and the variation of electrical output are determined in order to assess the impact of integrating the power output into isolated grid systems.

The membrane shape analysis has been successfully completed in 2-dimensions and the difficult move to three is contemplated. The objective is to provide detailed information on stress patterns to aid the design process. Effective tailoring of the membrane for its application is essential for its survival, on which the cost-effectiveness of the CLAM depends.

FUTURE DEVELOPMENT

If the viability of floating wave energy devices, in particular one as promising as the circular CLAM, are to be rigorously evaluated, then a coherent, properly funded programme of component development, detailed prototype design, construction and testing needs to be set in motion in the near future.

Coventry Energy Systems Group is proposing a development timetable which recognises that further design and testing of the bag and turbo-generator

Table

Outline Research, Development and Deployment Programme
for the Circular CLAM.

	State of Development	1st Stage 2½ Years 1993-1996 £3-4M	2nd Stage 3 Years 1996-1999 £7-10m	3rd Stage 2000
Membrane	Some full scale fabrication and testing. Some mathematical modelling.	Full scale cell to be tested on harbour wall.		
Wells Turbine	Various wells turbines tested.	Steady flow full scale test follwed by flow cycle testings.	Full scale sea going protoype construction installation and testing.	Deploy arrays of CLAM units in the North Atlantic.
Structural Design	Full scale outline design completed.	Detailed prototype design.		
Mooring Details	Outline mooring scheme designed.	Detailed scheme design.		
Electrical Generation	Outline scheme designed.	Detailed electrical, transmission and integration design.		
System Optimisation	Mathematical model.	Further optimisation using test results.		

needs to be carried out in parallel with the general design and performance optimisation and come together with the detailed, fully costed prototype design at the end of a two and a half year design phase.

The estimated cost for this phase is £3-4M and would bring the development to the stage where accurate future energy cost could be confidently predicted and decision made about proceeding to the three year programme of construction, installation and testing of the prototype at a further cost of approximately £7M.

The table sets out the outline programme which we feel will lead to a successful CLAM scheme and to the future deployment of arrays of CLAMS in the North Atlantic.

CONCLUSION

The CLAM is a concept with many merits. It is based on a floating structure which can be deployed in deep water where power densities are high, it can accept energy from all directions, performs well, has simple components and a modular design to take advantage of volume production.

The economics, performance and integrity of the CLAM will only be finally proved by the deployment of a sea going protype, with sufficient funds and commitment this could be acheived by 1999.

The CLAM then offers a substantial opportunity for the UK to derive a significant contribution to its electricity needs for an economic, environmentally benign resource. This resource could be exploited in a programme manner over a number of years by deploying clam units, at the rate of some tens or hundreds per year in the North Atlantic.

REFERENCES

1.   Duckers L J, Lockett F P, Loughridge B W, Peatfield A M, West M J, and White P R S, Membrane and Turbine Developments for the Circular Clam. Third Symposium on Ocean Wave Utilization, Tokyo (1991).

2.   Hogben N and Standing R G. Wave Loads on Large Bodies. International Symposium on the Dynamics of Marine Vehicles and Structures in Waves, IMechE London, (1974) paper 26, 258-277.

3.   Newman J N, The Motions of a Floating Slender Torus. J Fluid Mech. (1977) Vol 83 part 4, 721-735.

4.   Lockett, Duckers, Loughridge, Peatfield, West and White. The CLAM Wave Energy Device Component Development and Hydrodynamic Modelling. The Solar Energy Society Conference C57 "Wave Energy Devices". Coventry November (1989). ISBN 0-904963-56-X.

5.   White P R S W. A Phenomenological Design Tool for Wells Turbines. Meeting at Institution of Mechanical Engineers "Wave Energy" November (1991).

6.   Duckers, Lockett, Loughbridge, Peatfield, West, White; Offshore Wave Energy Devices; World Renewable Energy Congress, Reading (1990).

7.   Lockett, Duckers, Loughbridge, Peatfield, West, White; The CLAM Wave Energy Converter; ISOPE91, Soc Offshore, Polar Engineers, Edinburgh, (1991).

8.   The Development of the Circular SEA Clam. 1st July 1985 to March 1986. D of E Contract no E/5A/CON/1676/1381 (1986).

SOLAR ENERGY APPLICATIONS IN BOTSWANA

PUSHPENDRA K. JAIN

Department of Physics, University of Botswana,
Private Bag 0022, Gaborone, Botswana

ABSTRACT

Various applications of solar thermal and photovoltaic devices in Botswana which include water heating, water desalination, solar passive buildings and photovoltaics for lighting, water pumping, refrigeration, communication and fence electrification are reviewed.

KEYWORDS

Solar energy; water heaters; passive solar buildings; solar desalination ; PV-lighting; PV-water pumping; PV-refrigeration; PV-communication; Botswana.

INTRODUCTION

The Republic of Botswana in Southern Africa lies between the latitude 17° S and 27° S and longitude 20° E and 30° E. Gaborone is the capital of the country, (Fig. 1). The country covers an area of 581,730 km² and has an estimated population of 1.35 million of which about 74% lives in rural areas. The country is an importer of refined petroleum products, has no potential for hydro-power generation and has vast reserves of coal. The thinly populated and remotely separated rural areas together with abundant sunshine and favourable economic conditions offer a vast potential for the application of solar energy devices in Botswana. The rate at which their applications have increased during the last decade and the role they are playing in development in the country are both quite significant and noticeable. According to the Botswana Energy Master Plan (Energy Unit, 1987), in 1985 solar energy contributed 16 TJ to 25,125 TJ of energy consumed. In the year 2010 solar energy is estimated to provide 51 TJ to a total energy demand of 45,455 TJ.

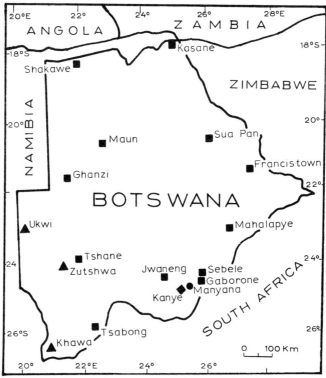

Fig. 1: Republic of Botswana with some locations of
interest with activities in solar energy.

SOLAR RADIATION CHARACTERISTICS AND MEASUREMENT

Botswana receives over 3200 hours of sunshine per year. Completely cloudy days are few
and far between. Average daily irradiation on a horizontal surface is 21 MJ m^{-2}.The daily
sunshine duration is measured at twenty-three locations. Sebele, 10 km north of
Gaborone is monitored by the Department of Agricultural Research. The remaining
twelve synoptic stations (two in Gaborone) (Fig. 1), and ten agrometeorological or
climatological stations are monitored by the Department of Meteorological Services
(DMS). Jwaneng and Suapan were commissioned in 1988 and 1991 respectively. Data
from Sebele and other synoptic stations are available for over ten years. Intensity of solar
radiation is measured at two locations, Sebele and Botswana Technology Centre (BTC),
Gaborone. At Sebele daily global irradiation is measured on a horizontal surface whereas
BTC which started measurements in November 1988 measures the daily global
irradiation on a horizontal and a tilted surface at 30° to the north. For other locations
the daily global irradiation has been estimated by Bhalotra (1987) from sunshine
duration using the Angstrom formula. Table 1 on the next page gives annual averages
of the characteristics of solar radiation for ten locations in Botswana. The averages
reported by Andringa (1987, 1989) are based on ten years data from 1975 to 1984
whereas Bhalotra (1987) has used data for different durations for different locations.

2548

Table 1. Annual averages of the characteristics of solar radiation
for ten locations in Botswana.

LOCATION	Sunny days per year[A1].	Frequency of 1 to 4 successive days with irrad. below 10 MJ m^{-2} per day[A2].				Daily Sunshine duration, (hours per day[B]).	Total hours of sunshine per year[A1].	Cal. global irrad. on a horizontal surface, MJ m^{-2} per day[B].
		1	2	3	4			
Gaborone	304	5	2	0	0	9.0	3230	20.8*
Francistown	282	7	1	0	1	8.7	3070	21.3
Ghanzi	316	3	1	0	0	9.2	3390	21.8
Kasane	-	-	-	-	-	8.2	-	20.8
Mahalapye	296	5	2	0	0	8.5	3120	20.7
Maun	306	2	1	0	0	8.9	3320	21.7
Sebele	314	6	1	0	0	-	3320	19.6*
Shakawe	299	2	0	0	0	8.6	3130	21.3
Tsabong	330	5	1	0	0	9.7	3510	22.0
Tshane	324	4	1	0	0	9.4	3460	21.8

A1 - Andringa (1987); A2 - Andringa (1989); B - Bhalotra (1987); * measured values.

SOLAR THERMAL APPLICATIONS

Thermal applications of solar energy in Botswana include water heaters, passive solar buildings and solar stills.

Water Heaters

Water heating contributes to the largest use of solar energy in Botswana. Residential sector accounts for 90% of the heaters in use with 10% going to other sectors. Both direct and indirect heat transfer types of collectors are common. Mostly the systems are imported except two companies which manufacture them within the country. Poor installation, inadequate maintenance and use of inappropriate plumbing materials had resulted in the failure of some systems. The problems faced are calcification and blockage of cooling channels, damage to collectors due to freezing of water or due to over heating and leakage due to galvanic action and pitting corrosion. With the formulation of a code of practice for solar water heaters in Botswana (Energy Unit, 1990) the problems have mostly been resolved.

Passive Solar Buildings

Under the climatic conditions in Botswana, passive solar buildings offer an economic alternative to the use of expensive means for heating and cooling. The BTC headquarters with a covered area of 400 m^2, a newly added workshop with floor area of 160 m^2, a guest house and some staff residences are some of the few passive solar buildings in the Southern African and SADCC (Southern African Development Coordination Conference) regions. A larger BTC headquarters with a covered area of 2000 m^2 and more staff residences are expected to be completed by 1992-93.

Solar Desalination

In the Kgalagadi District in the south-west the ground water is highly saline. The Rural Industries Innovation Centre (RIIC), Kanye has been working on solar desalination of water since 1977 as a means to provide potable water to the remote settlements in Kgalagadi. Among the various techniques developed and tested, Mexican and brick-type solar stills are found to be best suited to the needs of Botswana. In the first phase, small scale desalination farms to provide drinking water to settlements of up to 300 people were set up in three locations at Ukwi, Zutshwa and Khawa, (Fig. 1). A total of 304 Mexican stills and 10 brick stills with base areas of 1.6 m^2 and 9 m^2 each respectively were installed (Yates et al., 1990). The still farm from Ukwi has recently been removed after potable water was discovered there. The next phase is to set up medium and large scale desalination farms for settlements of up to 1000 people.

PHOTOVOLTAIC APPLICATIONS

Photovoltaic conversion of solar energy (PV) in Botswana is used for lighting, water pumping, refrigeration, communication and for fence electrification by some wildlife game reserves to control the movement of wild animals.

PV -Lighting

PV-lighting was first installed in nineteen primary schools in Kgatleng District in 1986. Later, Botswana National Library Service (BNLS) also installed it at village reading rooms and community halls in nine locations. A few rural clinics use PV-systems for emergency and spot lighting and to recharge torches for nurses at night duties. About 700 houses in rural areas also have PV-lighting. All the lighting systems are 12 V-dc and include batteries with charge controllers. All school and BNLS systems are functioning well. The main problem with PV-lighting has been the insects that are attracted to light. Use of fly screens is not very successful as people often remove them.

PV - Water pumping and Refrigeration

PV power has been used successfully for water pumping in some places in Botswana. The maintenance problems faced are due to breakdown of motor and controller units. It shows favourable economics as compared to diesel pumps, is free from the need to maintain a regular supply of fuel and is simpler to size because water can be pumped during sunshine hours. The use of solar refrigerators in rural clinics has not been successful and it is only used by rural veterinary clinics.

PV - Communication

Botswana Telecommunication Corporation uses 12 V and 48 V PV-arrays for over 80 rural pay-phones and about 70 microwave repeater stations with plans for further expansion of the network. Botswana Police force operates microwave communication system for data and facsimile transmission, to link the nation wide police stations to the headquarters in Gaborone and to provide cross links between major police stations. Thirteen of the repeater sites operate from 24 volt PV-panels at 50 to 170 watts of power. Botswana Railways uses PV-power for hot-box detectors to detect the overheating of bearings as the trains pass over the device and for radiocommunication and signalling.

SOLAR ENERGY RESEARCH AND DEVELOPMENT

The R & D work in solar energy in Botswana is mainly testing and evaluation of devices and their adaptation to local requirements. In addition to work discussed above, RIIC, Kanye has developed low-cost water heaters and a thermal water pump which uses air and water on a diurnal cycle with solar heat and nocturnal cooling. BTC has developed a battery charge controller. The Energy Unit is monitoring a pilot project for the demonstration and assessment of various renewable technologies at Manyana, (Fig. 1) and it plans to set up a Renewable Energy Technology Centre. Work in energy at the University of Botswana is mainly theoretical and of academic interest.

CONCLUSIONS

A low population density, remotely isolated rural areas, scarcity of fuelwood, abundant sunshine, simplicity of the routine maintenance of solar devices and a flourishing economy strongly favour the use of solar energy in Botswana. Solar technology has a well established base in the country and its potential has been demonstrated with a wide range of applications in the last decade. Tapping of solar energy for rural development is expected to emerge as an important source of renewable energy in the country.

REFERENCES

Andringa, J. (1987). Pattern of global irradiation in Botswana. In: *Advances in Solar Energy Technology* (W. H. Bloss and F. Pfisterer, ed.), Vol. 4, pp. 3887-3891. Pergamon Press, Oxford.
Andringa, J. (1989). Some characteristics of the pattern of solar radiation in Botswana. *RERIC Intl. Energy J.*, 11, 69-78.
Bhalotra, Y. P. R. (1987). *Climate of Botswana, Part II - Elements of climate : 2. Sunshine and solar radiation & evaporation.* Dept. of Meteorological Services, Gaborone.
Energy Unit (1987). *Botswana Energy Master Plan - Final Report.* Ministry of Mineral Resources and Water Affairs, Gaborone.
Energy Unit (1990). *Code of practice for domestic solar water heating in Botswana.* Ministry of Mineral Resources and Water Affairs, Gaborone.
Yates, R., T. Wato and J.T. Tlhage (1990). *Solar powered desalination - A case study from Botswana.* International Development Research Centre, Canada.

SOLAR ENERGY IN BOTSWANA-
THE IMPORTANCE OF CORRECT PRACTICES IN THE ADOPTION OF NEW TECHNOLOGIES

R. BURTON,

Botswana Technology
Centre
P/Bag 0082
Gaborone
Botswana

S. CARLSSON & B.MOGOTSI

Energy Unit
Ministry of Mineral Resources
and Water Affairs
P/Bag 0018
Gaborone, Botswana

ABSTRACT

The paper discusses the problems which can arise with the introduction of solar technologies, in what climatically is an ideal situation, if attention is not given to correct design, equipment selection and installation practices. Some solar energy technologies performed so poorly in Botswana that their use was virtually abandoned. The role of industry, government and other organisations in rectifying this situation is described.

KEYWORDS

Solar energy, water heaters, photovoltaics, codes of practice.

BOTSWANA

In terms of climate and population density Botswana is ideally suited to the application of solar energy. The country is situated on the high velt of southern Africa at an altitude of 1000m straddling the tropic of capricorn. Average annual rainfall varies from 550mm in the east to less than 200mm in the south-west, so that there is a low incidence of cloudy days. Solar radiation data is only available for Gaborone, and then only for the past eight years, but sunshine hours data is available for most major centres. Estimated average total daily insolations over the country range from a minimum of $14MJm^{-2}$ to a maximum of $28MJm^{-2}$ in the south (depending on the season) and $17MJm^{-2}$ to $23MJm^{-2}$ in the north (Andringa 1989). Successive days with total irradiation below $15MJm^{-2}$ are rare, making for ideal conditions for the application of solar energy technologies. Further incentives to use solar energy are the low population density (the 1.25 million people of Botswana occupy a land area of $582,000km^2$, which makes the distribution of conventional energy expensive) and the relative wealth of the country (due to income from diamonds) means that there are few constraints to the importation of solar energy equipment.

EXPERIENCE WITH SOLAR TECHNOLOGIES IN BOTSWANA

Given the situation described above, it is hardly surprising that solar energy was perceived to have great potential in the country and that its use has been actively promoted. In excess of 6000 solar water heaters and 0.5MW of PV equipment are currently in service, but this has not been achieved without problems, in some cases serious enough to cast doubt over the future of some solar technologies.

Solar water heaters

Both the government and the private sector have installed domestic solar water heaters and two companies manufacture them in Botswana. Unfortunately most of the water used in Botswana is highly aggressive borehole water and this caused the majority of solar water heaters installed to fail, either due to corrosion or calcination, within two to three years. The Botswana Housing Corporation placed a moratorium on the use of solar water heaters in its housing projects, confidence in the technology was at a very low level.
Eventually a working group, consisting of the Energy Unit of the Ministry of Mineral Resources and Water Affairs, installers/manufacturers and users was formed to look into the problem. Their efforts led to the drawing up of a code of practice for solar water heater installations (MMRWA 1990), adherence to this code is now mandatory for all companies tendering for government business and is required by virtually all private organisations which have solar water heaters installed. It is hoped that this will result in viable systems being installed in the future.

Photovoltaic lighting systems

PV lighting is used both by the government and by private householders. The most common application of this technology by the government is for village reading rooms, which are primary school classrooms equipped with lights to allow both school children and adults to read and hold meetings at night. There are over sixty such facilities in Botswana at present and ten new ones are added each year. Private PV lighting systems number in the hundreds, most being located in urban self-help housing areas (which are not provided with grid electricity).
Prior to the National Library Service reading room programme a number of school rooms were fitted with PV lighting under an aid funded renewable energy project (Zietlow et al 1985). Experience with these systems indicated that a number of technical shortcomings existed, mainly in relation to the battery and charge controller. It was decided therefore to draw up a code of practice for such installations (MLGL 1990). When the reading room programme commenced this code was already in existence and all installations were carried out according to the standards set out in the code. Follow up inspections of the first group of 20 of these systems after four years of use indicated that there had been only two failures in service (BTC 1990).
Unfortunately this good experience was not repeated in the private market for PV lighting systems, which expanded rapidly in the late 1980's. A considerable number of consumers were provided with PV systems which were of sub-standard design and workmanship (BTC 1991). In some instances the installers involved were unaware of the poor quality of the systems which they were installing, having taken up PV work with a background of conventional electrical wiring. Without the necessary knowledge of the special requirements for PV installations, they were unable to design and install good quality systems. The Botswana Technology Centre, together with installers and manufacturers of PV equipment,

therefore decided to use the code of practice for reading rooms as the basis for a code for domestic lighting systems. This code (BTC 1991) was published late last year and will, it is hoped, lead to an improvement in the standard of PV lighting systems in the country. Unlike the solar water heater code (which is in effect enforced by the government tender board) the lighting code will be enforced by certification of installers, following inspection of their work by Botswana Technology Centre staff. In effect this scheme will constitute voluntary self-regulation by the solar industry itself, since certification will not be compulsory.

PV water pumping

As with PV lighting some of the earliest experience in Botswana with PV water pumping was gained by non-government aid funded organisations (Hodkin *et al* 1988). The technology is particularly attractive in Botswana since large areas of the country have only intermittent surface water, leaving many communities dependent on diesel powered borehole pumps.
Because of high costs the first PV pumps installed in the country were relatively small (400 to 500W) and comparatively low lift (less than 30m). Submersible 12V D.C. pumps were used with battery storage, but these proved to be unreliable due to battery failure and pump corrosion. Later versions used constant voltage trackers to eliminate the batteries, but pump corrosion remained a serious problem. Most of these systems were removed within a year or two and none remain in service today.
More reliable were A.C. submersible and progressive cavity pumps, both of which avoided the use of batteries and cathodic corrosion of the submerged components. Several of these kinds of pumps, dating back to the early/mid 1980's, are still operational, but the majority are no longer in service due to lack of systematic maintenance. This has come about because some of the agencies involved in installing these pumps were involved in programmes which were of fixed duration and they failed to ensure that the systems which they had installed were handed over to the relevant local authorities when their projects came to an end. The result was that no one took responsibility for their maintenance once the projects which installed them were wound up. In most cases the local authorities did not in any case have the requisite technical capacity to carry out maintenance A few of these early pumps, installed by the Botswana Technology Centre, have been kept running because the Centre has a permanent presence in the country. However it has not proved possible to hand over the maintenance of these systems totally to the respective local authorities.
All villages in Botswana are provided with water supply systems by the central government's Department of Water Affairs, smaller communities are catered for by District Councils.The definition of village is somewhat involved, but in effect it excludes communities which are small enough to be served by the relatively low output of PV pumps. District Councils can obtain technical assistance from Water Affairs to install PV pumps, but Water Affairs has no mandate to maintain them, unless they serve recognised villages. Since District Councils do not, generally speaking, have the necessary expertise to maintain PV pumps, those pumps which have been installed are often out of commission.
Most of the PV pumps installed of recent years, and those planned for the immediate future, serve recognised villages, in parallel with diesel pumping equipment. Typically a borehole equipped with a diesel pump will, after a few years, fail to provide sufficient yield to cope with the increasing demand for water. So additional boreholes will be sunk a few kilometers from the village. These boreholes can be equipped with PV pumps and, since the communities concerned are large enough for Water Affairs to cater for, they can be effectively maintained. Systems of this kind permit relatively small output PV pumps to be used to supply villages and so overcome the institutional barriers to their use.
It is unfortunate that institutional arrangements between government agencies inhibit the use of PV water pumping technology by small, remote communities which should logically be the main beneficiaries of such systems. The situation does however illustrate how technological advances can be at odds with existing structures set up to cater for earlier technologies.

PV in telecommunications

PV is widely used in Botswana for public telecommunications, police communications and by the railways. It was an unfortunate coincidence that these systems first began to be used towards the end of an extended period of drought and that, in the absence of long-term insolation data, the solar energy resource was therefore overestimated. As a result, when the drought eased, some of these early systems proved to have been underdesigned and some doubts were raised as to the viability of PV for telecommunications. Fortunately the modular nature of the systems allowed them to be upgraded to match the actual insolation regime. Of recent years both PV water pumping and telecommunications installations have become subject to vandalism (mainly by children with catapults) and theft. This has led to consideration of a code of practice for the protection of large arrays.

PV vaccine storage

One would expect that ,with its favourable solar regime and scattered population, Botswana would make extensive use of PV vaccine storage. In fact this is not the case. As with PV lighting for reading rooms at schools, and PV water pumping, PV vaccine storage was first introduced to Botswana by a non-government aid project. A combination of failure to consult with the relevant people and make effective arrangements for maintenance of the equipment at the end of the pilot project, resulted in the technology being perceived, by the end-users, as both inconvenient to use and unreliable (McGowan 1985). The standard vaccine storage refrigerator in Botswana is therefore an LPG absorption unit (except ironically for the storage of animal vaccines, where the Department of Veterinary Services does use solar).

ACTIONS TAKEN TO FACILITATE THE INTRODUCTION OF SOLAR ENERGY TECHNOLOGIES

It will be evident from the above that, despite highly favourable climatic, geographic and economic conditions, the introduction of solar technology into Botswana has not been without problems. Both the Energy Unit of the Ministry of Mineral Resources and Water Affairs and the Botswana Technology Centre have been involved in attempting to overcome at least some of these problems by making solar energy technology more readily available to potential users and at reducing the risk for those who choose to invest in it.

Codes of practice

As noted above codes of practice have been formulated for the installation of solar water heaters and PV lighting systems. These codes are the result of cooperation between industry, government and private individuals and their formulation has served as a learning process for all involved.

Working groups

In the area of PV the Technology Centre acts as the convenor of the PV Working Group, which acts as a forum for government and private sector interests (including some from other countries in the region) to discuss any matters relating to the use of PV technology. This includes encouraging the local technical college to mount courses in PV installation and maintenance; private companies have indicated a willingness to contribute equipment to be used as teaching aids. Standards are also being established for appropriate lighting levels for reading rooms and for private houses. Several PV related products are currently under development at the Technology Centre as a result of ideas arising from the PV Working Group, these products will be manufactured in Botswana.

Information

A public awareness radio programme is broadcast monthly to, amongst other things, inform the general public concerning the potential of solar technologies to improve their well-being. Seminars and workshops have been run to inform decision makers of the advantages of solar energy, and of the pitfalls to avoid when purchasing the various technologies.

CONCLUSIONS

When introduced, solar energy technologies often cut across established skill demarcations and institutional arrangements, unless concerted efforts are made to overcome the problems which arise from this situation the process of introduction can be seriously set back. Educators, private companies and government must all cooperate to establish the relevant codes, training programmes and information systems for solar technologies to achieve wide acceptance, even in countries with ideal solar conditions. Reliable solar radiation data must also be collected to allow designers to size systems accurately.

REFERENCES

Andringa, J. (1989). Some Characteristics of the Pattern of Solar Radiation in Botswana. RERIC International Energy Journal Vol 11 # 2.

Code of Practice for Domestic Solar Water Heating in Botswana (1990). Energy Unit - Ministry of Mineral Resources and Water Affairs. Gaborone.

Zietlow, C.P. and J.Oki. (1985). Botswana Renewable Energy Technology Project Final Report. BRET Gaborone.

Specifications for Photovoltaic (12V DC) Energy Installations for Local Authorities Architectural and Buildings Unit, Ministry of Local Government and Lands. Gaborone 1990.

Report on Planned Maintenance of PV Lighting Systems at Kgatleng Schools. Systems and Testing Unit, Botswana Technology Centre. 1990.

A Report on Survey of Private PV Systems in Gaborone. Botswana Technology Centre. 1991

A Code of Practice for Photovoltaic Installations. Botswana Technology Centre. 1991.

Hodkin, J., R. McGowan and R. White (1988). Small - scale Water Pumping in Botswana. Volume IV: Solar Pumps. Associates in Rural Development Inc. Burling ton USA.

McGowan, R. Draft Report of Photovoltaic Electrification of Health Clinics and Village Schools in Botswana. (1985). BRET Gaborone.

Technical Potentials of Renewable Energies at the Example of a Bundesland in the Federal Republic of Germany

Andreas Wiese, Martin Kaltschmitt
Institute for Energy Economics and the Rational Use of Energy, University of Stuttgart
Pfaffenwaldring 31, Tel.: (international)49/711/78061-0
7000 Stuttgart 80, Federal Republic of Germany

The objective of this paper is to analyze the technical potentials of renewable energies. Therefore, potentials for the different renewable energy sources are determined for every community in a federal state in Germany. The methodical approach for the investigation of the technical potentials is presented first. Then the technical energy potentials for every community in the state are calculated. The results are analyzed taking local differences into account. Next, the energy potentials per inhabitant for different districts are estimated and compared. Finally, the share of the currently given energy demand potentially available for renewable energies is calculated.

Keywords: Potentials of renewable energies, wind energy, photovoltaic electricity production, solarthermal heat production, straw, residual wood, biogas, energy plants.

1 Introduction

Not only because of the danger of a possible global climate change the increased usage of renewable energy sources has been strongly suggested in the energy economics discussion. Therefore, it is the aim of this investigation to analyze the technical potentials of energy production from renewable sources, using Baden-Württemberg, a state in the south west of the Federal Republic of Germany, as example. Here we distinguish between the possibilities of producing electricity from wind power and solar energy, the production of heat using agricultural and forestry waste products (biogas, straw and wood residuals) and the biomass from planting energy crops.

2 Renewable Energies for the Electricity Production

In this paper, only two of the possibilities to produce electricity from renewable sources are considered: the photovoltaic conversion of solar energy into electrical energy and the usage of energy in moving air masses with the help of wind-energy generators.

The capacity for **photovoltaic electricity production** is limited by the amount of area available to install photovoltaic modules. Consequently the technical area potential of this option is defined by the available installation area. In principle, only the usable roof areas in residential and non-residential buildings are available for photovoltaic generators with low capacity as principally is farmland for larger plants. To determine the technical roof potentials the total roof area has to be estimated from the amount of residential and non-residential buildings in different building categories. By subtracting the roof area which is not available for solar technology (i.e. chimneys and sky-lights) the total area for possibly installing modules can be calculated. Only those areas at present used for agriculture are of interest for larger plants because residential areas have already been considered with the roof potential and woodland areas are not acceptable for installing solar farms. By taking into consideration certain restrictions such as fruit plantation or the northern slope of hills the likely collector area potential can be determined on the base of an optimal positioning of the modules. Considering the dependency on the locality and optimal installation of the modules, a specific yearly mean energy yield can be estimated for photovoltaic modules on roofs and in free areas. By this means, the corresponding energy potential can be calculated.

To determine the technical potential of the possible **wind energy electricity production** it is necessary to consider restrictions such as the fact that wind-energy generators cannot be installed in residential or transport areas. Considering the mean wind speeds the available wind energy supply in each area can be calculated. The potential area for the installation of converters is then deduced in different wind classes (3 to 4 m/s, 4 to 5 m/s and 5 to 6 m/s) for each community. Using average currently commercially available wind converters and taking an average area usage for each wind converter into consideration to reduce the shadow effect, it is possible to calculate a wind-dependent average yearly energy yield. The energy potentials for Baden-Württemberg shown in the following table were estimated in this way.

windpotential 5 to 6 m/s	300 GWh/a
windpotential 4 to 5 m/s	1 640 GWh/a
windpotential 3 to 4 m/s	1 510 GWh/a
solartechnical roof area potential	7 270 GWh/a
solartechnical free area potential	29 400 GWh/a
total	40 120 GWh/a

Consequently, about 40 000 GWh/a electrical energy could be gained if a full usage of the potential would be assumed. The largest part is the free area solar potential (29 400 GWh/a). With approximately 3 450 GWh/a the wind potential is almost half of the solar roof area potential (7 270 GWh/a). The potentials can also be deduced for different communities. Figure 1 (the left side) shows the energy potential per inhabitant of the options discussed for seven communities and the state average. The range is from 17,0 MWh/Inh/a (rural areas, district 1 to 5) to 0,6 MWh/Inh/a (city area, district 6 and 7) with an average of 4,2 MWh/Inh/a. The high specific potential per inhabitant is found in rural areas where the technical wind potential is larger and the availability of free areas for photovoltaic plants is above average. The highest specific potentials for both options are in areas with low population and consequently areas with generally lower than average demand for electrical energy. The wind potential is influenced more by local conditions than the solar potential. This is due to the fact that the supply of solar radiation in a given area fluctuates less than the energy supply from moving air.

3   Renewable Energies for Heat Production

Next, the following options to produce low-temperature heat from renewable energy sources are discussed: solarthermal heat production, the usage of residuals from agricultural and forestry products and the usage of biomass produced by energy crops.

As far as **solarthermal heat production** is concerned, only the solar heating of water is analyzed. Roofs of buildings are possible installation areas. The restrictions for the use of solarthermal energy are the amount of heated water required and the average amount of energy that can be covered by solar energy in an average year. Assuming a typical number of people for different residential buildings, and considering the total roof area which is not suitable for solar collectors, assuming further an average end energy requirement to heat water and an average share of 60 % of this requirement that can be covered by solar energy, it is possible to calculate the energy potential for private households. Also the share of solar energy available for the industrial consumption of heated water has to be considered. Out of this, the technical potential of solarthermal water-heating can be estimated for each community.

Considering the usage of **agricultural crop waste**, the main sources of energy are straw and other crop rests. Based on the statistically documanted size of agricultural land and the specific grain production it is possible to calculate the amount of straw with a grain to straw ratio for each type of grain for each community. Allowing for ecological and other restrictions the total amount of usable straw can be computed which can be removed from the nutritive substance- and humus-circulation. The technical usable energy potential is now derived from the hear value of straw.

The energy potential from **residuals from the forestry industry** is estimated using the registered woodland areas in each community. Taking into account the different growing areas and tree sorts and the long-term average wood yield it is possible to work out how much residual wood is available. Allowing for restrictions such as protected areas and national parks, a usable energy potential can be calculated which would not cause substantial damage to the woodlands resources. Assuming an average energy content of wood with the mean rest moisture of about 15 % it is possible to determine the total technically usable energy potential.

Using the animal registration statistics for each community it is possible to calculate how much **biogas production** is possible from excrements. Allowing for the not available part, an average daily amount of organic mass can be determined. Subtracting the energy required for the biochemical decomposure, the biogas potential can be estimated. If further restrictive parameters such as a minimum required quantity of organic mass for an efficient management of a biogas plant are incorporated, then the technical biogas potential can be ascertained.

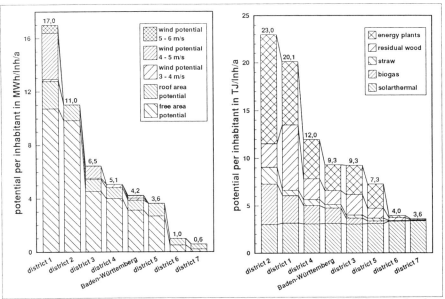

Figure 1: Inhabitant specific potentials for electricity (left side) and heat (right side) production from renewable energies in different districts in Baden-Württemberg

Of the various potentials to use **energy plants** only the usage of whole grain plants and grass or reeds are examined here. On the basis of agricultural areas which are theoretically not required for the production of food, a total theoretical biomass for energy production can be calculated assuming an average mix of different possible energy plants. Allowing for rest moisture in grain plants and grass or reeds the technical energy potential of this option can be determined.

Given by this procedure the renewable energy options for heat production can be summerized in the following table for Baden-Württemberg. The technical potential for energy plants is half from grain and half from grass or reed cultivation.

solarthermal	29 500 TJ/a
biogas	13 690 TJ/a
residual straw	3 610 TJ/a
residual wood	13 745 TJ/a
energy plants	19 840 TJ/a
total	80 385 TJ/a

The table shows that the solarthermal potential makes the largest contribution (29 500 TJ/a) of all renewable energy options. Straw (3 610 GWh/a) contributes the lowest energy potential. The technical potential of all options together is approximately 80,4 PJ/a. Figure 1 (right hand side) shows the potentials for different communities in relation to the number of inhabitants. The deviation in certain communities from the state average of 9,3 TJ/Inh/a is relatively high. The highest specific potential in a rural area is 23,0 TJ/Inh/a, while in a city a potential of only 3,6 TJ/Inh/a was determined. Because the solarthermal energy potential was calculated with respect to personal requirement it is more or less independent of the community and is always about 3 TJ/Inh/a. The reason that energy plant potentials in rural areas are above average lies in the large size of available agricultural areas. There, the biogas potential is also high due to the intensive animal farming. The cities, on the other hand, have only solarthermal potential at their disposal.

4   Comparison of the Total Potential

The energy potentials discussed above could substitute the currently used energy sources and therefore compared with the energy system at the present time. However, by a full usage of the options described some would exclude one another. For example, roof areas can be used either for the installation of solarthermal collectors or photovoltaic modules. If part of the agricultural land is used for cultivating energy plants, the technical potential of the residual straw is reduced. Figure 2 shows the technical potentials and the final energy requirement for 1989 allowing for the excluding factors discussed above. Furthermore, as far as the wind and photovoltaic electricity production is concerned, net losses of about 4 % have to be taken into account.

Figure 2 shows clearly that the main contribution of the options analysed is to the electrical energy production, and here the photovoltaic electricity production contributes the largest part (121,0 PJ/a). Wind energy, on the other hand, contributes only a small amount (11,8 PJ/a). The total contribution to the final energy requirement for heating from solarthermal, biogas, straw, residual wood and energy plants is only a maximum of 8,1 %. The largest part is solar energy with 29,5 PJ/a, the smallest part is straw with 3,1 PJ/a.

5   Conclusions

From the technical potentials of renewable energy sources analysed for Baden-Württemberg, a state in the south west of the Federal Republic of Germany, the following conclusions can be drawn:
- Renewable energy sources can only substitute a part of the total energy requirement in the state because of the low technical potentials.
- The highest potential is provided by photovoltaic electricity production. This is partly given on roofs which are not used at the present. Therefore, it could provide an environmentally neutral and socially acceptable alternative and could be incorporated into the existing energy system with less technical problems. This fails due to the too high specific costs of electricity production of this option.
- Of the renewable energy options for heat supply only the solarthermal heat production is characterised by a high potential. Here again, the high specific costs mean that this option is not acceptable to provide warm water or support heating requirements.

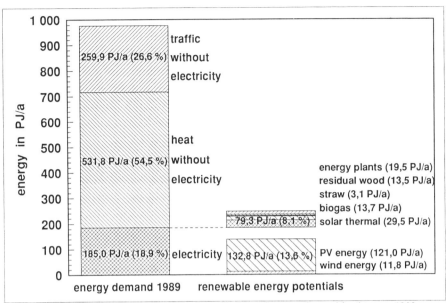

Figure 2: Comparison of the technical potentials with the end energy demand for the year 1989

- Only those options with a relatively low potential remain. For the production of electricity this is wind energy which is characterised by relatively low specific cost and which is already promoted by the government. However, only a small contribution to the electrical energy demand is possible. Furthermore, the energetical usage from organic residuals from agriculture and forestry is characterised by low specific costs but have low potentials too. With favourable conditions this option is economically viable today in rural areas; another advantage is an indirect reduction in CO_2 emmissions.

Because of these considerations the renewable energy sources for electricity and heat production can only provide about one fifth of the energy requirement for Baden-Württemberg in the short and middle term. Here, the quota of the technical potential to produce electricity of 63 % of total requirement is far higher than the potential to produce heat of only 37 % of requirement. Although the technical potentials are low, it must still be a primary aim of energy politics to promote these energy options by providing favourable conditions. This is not only a contribution to the environment but also helps to save fossile energy resources.

6   Literature

/1/ German Bundestag (Ed.) (1990). Protecting the Earth; A status report with recommendation for a new energy policy, volume 3; Economica Verlag and Verlag C. F. Müller, Bonn and Karlsruhe.

/2/ Wiese, A.; Kaltschmitt, M. (1992). Potentiale und Kosten regenerativer Energieträger in Baden-Württemberg; research report, Institute for Energy Economics and the Rational Use of Energy, Stuttgart.

/3/ Kaltschmitt, M. (1991). Möglichkeiten und Grenzen einer Stromerzeugung aus Windkraft und Solarstrahlung am Beispiel Baden-Württembergs; research report, Institute for Energy Economics and the Rational Use of Energy, Stuttgart.

SOLAR ENERGY IN CYPRUS: FACTS AND PROSPECTS

I. M. MICHAELIDES
Department of Mechanical Engineering, Higher Technical Institute,
P.O.Box 2423, Nicosia, Cyprus.

ABSTRACT

This paper reviews the application of solar energy technology in Cyprus and presents an energy analysis with emphasis on the contribution of solar energy to the energy consumption in the island. The almost full reliance of Cyprus on imported oil to meet its energy demand, together with the abundance of solar radiation and a good technological base, created favourable conditions for the exploitation and development of solar energy in the island.

Cyprus began manufacturing solar water heaters in the early sixties and today it produces more than 30,000 m^2 of solar collectors yearly. It is estimated that more than 130,000 solar water heaters are in operation providing the equivalent of 9% of the total electricity consumption in the country; this corresponds to, approximately, 4% of the national energy consumption. However, the use of solar energy for space heating and cooling provides a further challenge, because it does not appear to be economic under the climatic conditions and system design practices currently prevailing in Cyprus.

KEYWORDS

Solar energy, solar collector, thermosyphon, solar water heater, payback period, solar heating, solar fraction, simulation.

1. INTRODUCTION

Cyprus is the third largest island in the Mediterranean and is situated at 33°E of Greenwich and 35°N of the equator. It has an area of 9,251 km^2 and a population of 698,600 (end of 1989). Its main sources of income are agriculture, industry and tourism.

The climatic conditions are predominantly very sunny with a daily average solar radiation of about 5.4 kWh/m^2 on a horizontal surface. In the lowlands the daily sunshine duration varies from about 5.5 hours in winter to about 12 hours in summer. On the high mountains, the

2562

cloudiest winter months have an average of nearly 4 hours of bright sunshine per day and in July the figure reaches 12 hours.

Mean daily global solar radiation varies from about 2.3 kWh/m² in the cloudiest months of the year, December and January, to about 7.2 kWh/m² in July (Meteorological Service, 1985). The amount of global solar radiation received on a horizontal surface with average weather conditions is 1725 kWh/m² per year (Hadjioannou, 1987). Of this amount, 69% reaches the surface as direct solar radiation (1188 kWh/m²) and 31% as diffuse radiation (537 kWh/m²). For the year as a whole, the amount of global solar radiation received by a surface inclined at 50° to the horizontal and facing south is 1817 kWh/m² or 105% of that on a horizontal surface. However, the most important fact is that in the winter months, when the energy demand for heating is at its highest, this percentage is considerably higher.

2. THE ENERGY CONSUMPTION PROFILE

The annual energy consumption in Cyprus for 1989 was about 1.077 million tons of oil equivalent (Toe). This corresponds to a per capita annual energy consumption of about 1.91 Toe. The forms of energy used in Cyprus and their percentage contribution to the final energy bill for 1989 are shown in Fig. 1. The contribution of domestic resources to meeting the energy needs of the country is estimated to about 6%–7%, mainly from solar energy and wood. The rest is met by imported oil and coal.

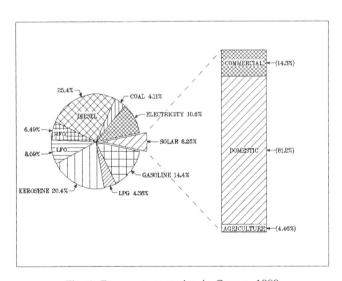

Fig. 1. Energy consumption in Cyprus, 1989.

The almost full reliance of Cyprus on imported oil in meeting its energy demand, together with the plentiful availability of solar radiation and a good technological base, created favourable conditions for the exploitation and development of renewable energy sources. The first application that has been historically developed in Cyprus, regarding renewable energies, is the use of windmills for water pumping for irrigation purposes. In the early 1930's, hundreds of windmills were set up in the south–east coastal areas to irrigate small plots of vegetables. This application grew quickly but declined in the 1960's, while diesel pumps proliferated due to the very low price of oil at the time.

A second mass–extended utilisation of a renewable source of energy appeared in the early sixties; as solar water heaters were developed on a large scale in Israel, Cypriot manufacturers based their designs on Israeli products and quickly created a major national industry. Cyprus produces more than 30,000 m² of collectors per year.

3. SOLAR WATER HEATING

Cyprus began to manufacture Solar Water Heaters (SWH) in the early sixties. Cypriot industry of solar water heaters quickly expanded to reach an annual production of about 30,000 m² of collectors by more than 20 manufacturers. Fig. 2 shows the evolution of the production of solar water heaters in Cyprus for the years 1978–89 as compared to electric water heaters (Department of Statistics and Research, 1989).

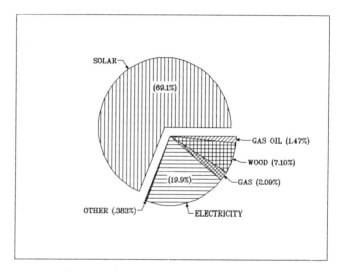

Fig. 2. Production of water heaters in Cyprus.

A typical solar water heater in Cyprus consists of two flat–plate solar collectors having an absorber area of 3 m², a storage tank of 180 litres equipped with an auxiliary electric immersion heater of 3 kW, and a cold water feed tank on top.

There is an increased interest in employing solar water heating in dwellings. According to Construction and Housing Statistics (Department of Statistics and Research, 1987), 87.3% of new dwellings built in 1987 have been equipped with solar water heating as compared to 69.7% in 1982. An important market potential exists for the development of solar water heaters in the residential and commercial sectors. There is much to be done in the tourism sector and in collectively owned buildings. It can be assumed that the number of collectively owned buildings using solar water heating, currently at around 15%, would gradually increase up to 40% due to the development of new techniques for using solar energy in collective systems. In the tourism industry, it is estimated that about 55% of the existing hotels are equipped with solar-assisted water heating systems and the contribution of solar energy to the final energy consumption in the hotels industry is about 2%.

The Cyprus Organisation for Standards and Control of Quality (CYS), has formulated two standard specifications for solar water heating systems. These were adopted by nearly all major Cypriot manufacturers to specify the required design characteristics and the manufacturing tolerances for their products.

The first standard is the CYS 119 :1980 (Cyprus Organisation of Standards, 1980), which describes the method of testing the performance of flat plate solar collectors. For this purpose, a testing facility has been set up for the outdoor testing of collectors. The second standard is the CYS 100:1984 Standard Specification for Solar Water Heaters (Cyprus Organisation of Standards, 1984). It deals with solar water heaters intended for domestic use and specifies the requirements for materials, construction, and marking of solar water heaters using water as the heat transfer medium. The standardisation work continues at a more rapid pace aiming at harmonizing the initial standards with the European requirements and directives in the field of solar energy. For this purpose the Technical Committee for solar water heaters is currently preparing the issue of a series of new standards to cover the durability tests and the short and longterm test methods for solar water heating systems, based on the recommendations of a Specialist Group of European solar energy experts (Aranovitch *et al*, 1989). In order to provide the test facility which is required by the new standards, the Cyprus Government has set up a modern and fully equipped testing centre for monitoring and analyzing the performance of solar collectors and solar water heating systems.

3.1. Economic aspects of Solar Water Heating.

It is estimated that the number of Solar Water Heaters (SWH) installed in Cyprus exceed 130,000 units. This means one solar water heater for 5 people in the island. The estimated collector area installed in Cyprus, including central systems in hotels and hotel apartments, is about 400,000 m^2 out of which 360,000 m^2 in houses, 25,000 m^2 in flats and the rest in hotels, hospitals and clinics.

Simulation studies conducted at the Higher Technical Institute (HTI), showed that the annual solar fraction of a typical SWH varies from 64% to 89% depending on the hot water consumption pattern (Michaelides *et al*, 1991). The monthly and annual solar fraction of a SWH for a high and a low hot water consumption pattern is shown in fig. 3. In terms of energy, this means an average annual energy contribution of about 1350 kWh per SWH. Thus, the annual energy savings provided by the SWHs installed in the island are about 160 GWh, based on a total number of 130,000 units and assuming that the rate of utilisation[1] of the present stock of SWHs is 0.9. This represents savings of about 9% of the national electricity consumption and crresponds to about 4% of national energy consumption.

With regards to the cost effectiveness of SWHs, simulation studies showed that this depends primarily on the competing source of energy. It has been found that SWHs are very

---

[1]*This coefficient takes into account that part of the present SWH stock is not optimally used: SWH under repairs, SWH oversized compared to consumer needs, houses and flats unoccupied, etc.*

2565

competitive compared to the electric heating systems. Pay-back periods as low as 3 years have been observed for individual houses and 5 years for apartment buildings and hotel apartments. However, as long as the competitor is an oil-fired boiler system the payback period ranges from 5 to 10 years.

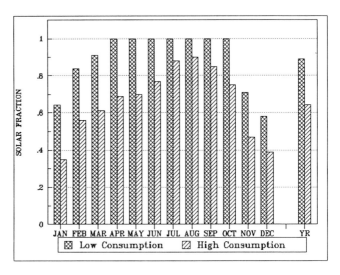

Fig. 3. Predicted Solar Fraction of a Thermosyphon Solar Water Heater.

4. Solar Space Heating

The climatic conditions in winter are such that space heating is needed for a period of 4 to 5 months and the average number of degree-days (base 18 °C) for winter is estimated to 950 degree-days. For space heating, LPG and kerosene are the most widely used forms of energy, followed by oil and electricity.

Solar energy has not been put into use for the heating of buildings except from very few cases where solar active systems have been combined with oil-fired central heating systems and floor heating in residential dwellings. However, performance data for these installations is not available and therefore no comprehensive conclusions as to their energy and economic performance can be drawn.

Computer simulations carried out at the HTI, showed that the payback period for residential solar heating systems using diesel oil as auxiliary, is about 10 to 12 years depending on the economic parameters used. Based on the findings of the above simulation studies and on the results obtained from an experimental installation which was monitored at the HTI (Michaelides, 1985), it can be concluded that the use of solar energy for space heating does not appear to be an attractive proposition under the conditions prevailing in Cyprus, due mainly to the very high investment costs, the short heating periods which lead to high depreciation costs and long payback periods and the poor thermal insulation of buildings.

However, the use of solar energy as a means of providing space heating for residential and commercial buildings must be promoted by the government. This can be achieved by providing the necessary funding and manpower which is required for the construction, monitoring and analysis of performance data of trial installations. In parallel, research and development work can also be carried out through modelling and simulation of solar systems aimed at developing flexible design tools which can be validated against actual performance data. This will enable the thorough investigation of the capabilities and the performance characteristics of solar assisted heating systems at the design stage. Consequently, it will be possible to conduct parametric studies to determine the optimum system size for a particular application and to develop the necessary control strategies for optimum operation.

REFERENCES

Aranovitch, E., D. Gilliaert, W.B. Gillet, and J.E. Bates (1989). *Recommendations for Performance and Durability Tests of Solar Collectors and Water Heating Systems.* EUR 11606 EN, Commission of the European Communities, Brussels.

Cyprus Organisation for Standards and Control of Quality, (1980). *CYS 119:1980 Method of Testing the Performance of Flat-plate Solar Collectors.* Ministry of Commerce and Industry, Nicosia.

Cyprus Organisation for Standards and Control of Quality, (1984). *CYS 100:1984 Specification for Solar Water Heaters.* Ministry of Commerce and Industry, Nicosia.

Department of Statistics and Research, (1982). *Census of Housing 1982.* Ministry of Finance, Nicosia.

Department of Statistics and Research, (1987). *Construction and Housing Statistics for 1987.* Ministry of Finance, Nicosia.

Department of Statistics and Research, (1989). *Industrial Statistics 1989.* Ministry of Finance, Nicosia.

Hadjioannou, L. (1985). *Three Years of Operation of the Radiation Centre in Nicosia – Cyprus.* Meteorological Service, Nicosia.

Meteorological Service (1985). *Solar Radiation and Sunshine Duration in Cyprus.* Ministry of Agriculture and Natural Resources, Nicosia.

Michaelides, I.M. (1985). *Performance of the HTI Experimental Solar Heating System, Technical Report.* Higher Technical Institute, Nicosia.

Michaelides, I.M., W.C. Lee, D.R. Wilson and P.P. Votsis (1991). Computer Simulation of the Performance of a Thermosyphon Solar Water Heater. *Applied Energy,* accepted for publication, April 1991.

SUSTAINABLE DEVELOPMENT IN INDONESIA :
A RENEWABLE ENERGY PERSPECTIVE

Rahmat Rozali[1] & Spencer Albright[2]

1) Dept. of Physics Engineering
Bandung Institute of Technology
Ganesa 10, Bandung, 40133, Indonesia.
2) Texas Christian University
Fort Worth, Texas, 76129, USA.

ABSTRACT

The concept 'Sustainable Development' has become popular in recent years. Different from the previous development concepts, this concept has begun to be discussed along with a greater emphasis on ecological concerns.

In Indonesia, fossil fuel plays a big role in increasing the economic growth. But the problem is the reserve of this fuel supply is very limited, and its use has environmental impact.

Seeing these facts, the role of renewable energy in Indonesia becomes important. Renewable energy sources such as geothermal, hydro-power, solar energy, biomass and tidal, which haven't been optimally used, are expected to assist in the development of Indonesia in the future.

This paper describes sustainable development with a perspective on renewable energy in Indonesia. Aspects which will be emphasized are geography, geology, energy demand, and renewable energy resources. The following is a discussion of these issues.

KEYWORDS

Sustainable development, renewable energy, energy demand, resources.

INTRODUCTION

In the beginning, development was only identified with economic growth. Its quantitative measurement was GNP or GDP. This development concept has been refined and finally a new concept has emerged called 'Sustainable Development'. This concept was begun to be discussed along with development's environmental impact.

The phenomen of greenhouse effect, acid rain, and pollution are the global issues which are widely discussed today. These are hot topics because of the rapid increased use of fossil fuel.

The available renewable energy resources in Indonesia include geothermal, hydro-power, solar energy, biomass, tidal, ect. Each of these resources offer great potential, but as yet they are not being used optimally. Some effort on increasing the renewable energy sources will be discussed later.

GEOGRAPHICALLY AND GEOLOGICALLY

Indonesia is a geographically complex archipelago consisting of approximately 17,500 islands spreads out over a distance af about 3,200 miles along the equator. It is the fourth most populous nation in the world, with over 180 million people.

Geologically, Indonesia is richly endowed with a variety of mineral and energy resources. Non-renewable energy resources include are oil, natural gas and coal. Renewable energy resources are solar, geothermal, hydro-power, wind, tidal, biomass, etc.

ENERGY DEMAND

From 1980 until 1990, Indonesian energy consumption increased on average of 10 %. Compared to the world's average (1.8 %), the Indonesian consumption is rate considerably higher. Most of it (65.5 %) is provided by oil.

At this time Indonesia is proceeding with an economic restructurization towards industry. With 180 million people and its rapid economic growth each year (over 6 %), it is estimated that energy needs will be increased.

For instance, lets examine the electric energy needs for industry and housing in the city region. For the 6th national five-year development plan (1993 - 1998), an additional capacity of 13,432 MW or 141 % compared to today capacity (9,515 MW) is required.

2569

In rural areas at the end of 1991, the villages with electricity were only 22,288 or 36 % of the total number of villages in Indonesia. In addition most of Indonesians live in the villages.

RENEWABLE ENERGY RESOURCES

Geothermal and hydro-power

Geothermal and hydro-power have large potential in Indonesia. Geothermal resources could generate electricity up to 16,000 MW. The geothermal sources currently used amount to about 142 MW (less than 1 %).

Indonesia also has hydro-power potential, equivalent to 75,000 MW. Indonesia has many rivers and half of them are capable of generating electricity over 100 MW. Until now only 2,889 MW or 3,8 % of the potential is being used.

Solar energy

Since Indonesia lies in a tropical climate, it has certain advantages in receiving solar radiation. For instance, it receives Average daily radiation of 6 hours. The solar radiation energy application such as photovoltaic system, is very potential for rural electrification, water pumping, medical care, etc.

Other resources

Other renewable energy resources such as wind, biomass, tidal, etc also very potential. Until today they are still being researched.

DISCUSSION

Since the country has a great number of people and it is experiencing rapid economic growth, the role of energy has become an important factor. For domestic needs, the energy is used primarily for industry, transportation and household needs.

In the rural sector, where energy demand hasn't been sufficiently fulfilled, especially electric power, it is still a problem. The low financial resources of the rural community to pay for electricity, distribution systems for electricity, and the very remote and widely dispersed location of the villages are the major issues that need to be given special attention.

If electrified, it is expected that economic activity will increase and this will obviously change the prosperity of that community.

By this time the role of renewable energy in Indonesia hasn't yet reached its optimal point because of some handicaps. For insteance, the application of the geothermal and hydro-power require a considerably high initial investment for the power station construction and also for the distribution network. In addition the hydro-power and geothermal energy sources are only available in certain locations.

Regarding the PV electrification for rural areas, the low financial resources of the community has become the dominant factor. However PV electrification has great potential, especially in rural communities which are very remote and widely dispersed.

To exceed the need for energy in the future, a few steps should be promoted :

- The role of renewable energy systems must be heightened, so that over time, it can subtitute for fossil fuel as much as possible.

- All energy, both renewable and non-renewable, should be used more efficiently.

- Because of the financial situation in the rural community, a bigger role of the private sector and government is needed for implementing the application of PV electrification in rural areas.

- Universities must provide leadership in the development and adoption of technology for renewable energy systems.

- Regional and international interaction must be heightened. Information exchange, training programmes, low interest grants, and production facilities are necessary in aplication of the renewable energy systems.

CONCLUSION

Along with increased economic activity in Indonesia, the need for energy also will increase. The optimal use of renewable energy sources is one of answers for Indonesian development in the future.

REFERENCES

1. Dunn, P.D., (1990), Renewable Energy and Developing Countries, Energy and Environment, vol 1, pp. 28 - 38, Pergamon Press.
2. Salim, Emil, (1992), Towards a sustainable future, Development 1990:2, pp. 61 - 63, Christengraf.
3. Speth, J., (1990), Environmental Security for the 1990s, Development 1990:3/4, pp. 9 - 15, Christengraf.
4. Andoyo, A., (1989), Implementation Strategies for Photovoltaik Rural for Electrification, Jakarta.

THE RENEWABLE ENERGY
IN TUNISIA

GUELLOUZ Khereddine

AGENCE POUR LA MAITRISE DE L'ENERGIE
3, Rue 8000 MONTPLAISIR TUNIS – 1002
TUNISIE

ABSTRACT

    TUNISIA is situated at the south Coast of mediteraneen sea between the 30th and 37 th parallel
and has a high time duration of sunlight and for some regions a good speed of wind. Tunisia is also an
agricultural country. (Annex 1).
For these reasons, renewable energies could have a large range of utilisations there.
The A.M.E was founded to put on and to manage the government policy in the Renewable Energies Field.
This report will summarise the Tunisian Program as to the use of the différent Renewable Energy Sources,
the results and the future actions to be taken.

TUNISIA CHARACTERISTICS

    TUNISIA is characterised by a long coast (1300 km) where 70% of the population is
agglomerated in towns and villages. In the Northen and central Inland areas, where agriculture is
prevailing, the population is disseminated ; Towns and Villages are less important.

In the South, witch has a subsaharian climate, the population is agglomerated around the OASES.

The sun is shining all over the year on Tunisia, there are a few days which are cloudy but there are some
differences between winter and summer and between Nord and South :

	North	South
Sunshine duration/year	2800 hours	3200 hours
Medium Sun energy radiation		
Winter	2,3 kwh/m²/day	4,1 kwh/m²/day
Summer	5,8 kwh/m²/day	7,5 kwh/m²/day

The wind potential is situated mainly on the North Coast, the medium speed is about 6–7 m/s
Forests are in the North west on the Atlas mountains which do not exceed 900 m in altitude ; and cover
about 6% of the country surface. Wood is used almost exclusively for cooking , by rural population.
Animal husbandry and extensive agriculture are the main activities in the N.W. and west central areas.

RENEWABLE ENERGY POLICY IN TUNISIA

TUNISIA is an oil producing country and exports part of its production, but the growth of its
local energy consumption limits this part very seriously .

Therefore, in the eighties, the Tunisian government founded the A..M.E. to manage its policy in the field of
Energy Conservation and Renewable Energy.

In this field three aspects are considered :

  - The first one is to decrease oil and gas consumption
  - The second is to give some comfort and services to rural areas where population is
    disseminated and where are missing electricity and roads..., and to reduce the pressure
    on towns by limiting rural depopulation.
  - The third aspect concerns the best way of using agricultural wastes in energy production
    mainly for rural consumption.

The National policy consists in some key steps to agree Renewable Energy programs and to provide
facilities and tax exemptions.

The Key steps are

  - the feasability study which includes the calculation of energy potential, the market and social
    and economic surveys when necessary.
  - the demonstration program and the results analysis
  - Technology selection and local industry promotion when demonstration results prove to be
    positive.

The facilities and exemption consist :

  - in generally financing the feasability study
  - in financing the demonstration program partially or exclusively
  - in exempting Renewable Energy goods and equipments from duty and some local Taxes

THE ACHIEVED PROGRAMS

A lot of projects are achieved for Research, for demonstration or in industry, using different
sources of Renewable Energy (photovoltaïc, wind, Biomass...)

FOR RESEARCH

1- Low temperature thermosolar plant of 10 kw;
This plant is running at normal level. But this technology is not efficient enough and should be
discarded. Although it has a didactic role.
2- water desalination with photovoltaïc and wind energy
Osmose inverse and electrodialyse systems were tested using the electricity from a photovoltaïc
filed (4 kw) and from two wind generators (4,5 and 1kw).
the technical results are positive but the cost of the water production is prohibitive and this
experience cannot be generalized.

In the other hand small green house desalinators, were assembled and could have a good use in rural areas where water is brackich.
FOR DEMONSTRATION

1) PHOTOVOLTAÏC SYSTEMS

– Household photovoltaïc electrification

. HAMMAM–BIADHA village

a photovoltaïc plant of 29 kw supply the Hammam–Biadha village (43 household and utilities). This plant is running normally with more than 95% of disponibility since 1984. Energy storage and installation cost are the weak point of this system.
The connection to the grid which is now near the village would be made.

. Rural dispersed houses and schools :

For the dispersed rural houses, autonomous systems are tested for covering the minimal needs of the rural household in two areas, 90 in le KEF and 45 in KAIROUN.
AHUNDRUND and FIFTY FIVE (155) schools were also electrified with small photovoltaïc systems.

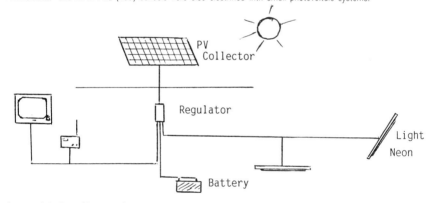

Compared to the grid connection these systems cost fairly less and solve most of household problems, knowing naturally that the grid service is much better.

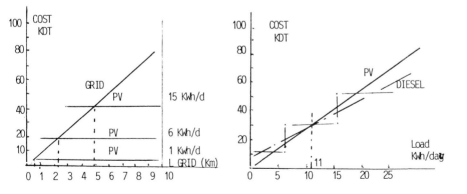

Photovoltaïc water pumping :

Only two units where installed for irrigation, the results are mitigated because of the water needs exceeds pumping capacity.
Another program is now, underway it concern the installation of fifteen (15) photovoltaïc pumps on surface wells for drinking water destined to rural population.

2) WIND SYSTEMS

Water pumping

Widely diffused, then left off when Diesel Engine became popular, there's now new attempt to repromote this technology in the agricultural sector, but The cost of equipment slows this action down.

Wind generator

In this field, two experiments where made the first one consisted in connecting two units (12,5 and 10 kw) to the grid. Because of the limited number of the units and of the problems occasioned by some electronic and mechanic parts, the results of the experience are not very significant.
The second one is the electrification of Jabouza village with two wind generators (2 X 10) which supply in electricity seventy two (72) house holds.
Some mechanic and electronic problems appeared, but in general, the energy produced by the two engines covered about 90%of the population needs, the cost and the energy storage are the weak points of this system

3) BIOMASS

In this field only two subjects are of some interest to date ; the Biogaz production and the combustion amelioration of rural stoves using wood

. Biogaz

In the North—west about thirty thousand (30 000) little cattle breeders have more than four animals.

The general layout of rural homes including cowsheds and the household needs in energy for lighting and cooking compared to the biogaz production from four animal dung showed that it is possible to promote Biogaz production in this region.

Forty Biogaz digestors were constructed for demonstration
The chineese model was chosen and adapted to local conditions. They have 16 m³ and 20 m³ of capacity and produce between 2 m³ of gaz in winter and 2,7 m³ in summer

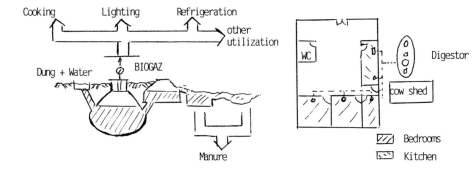

Encouraging results and a very good acceptance by population are noted. The generalization of the systems can be made in puttingon an adequate financing scheme.

.Wood stoves

Wood is the main combustible utilized in rural areas for cooking (mainly the bread} : 80 % of the consumption.)
The goal of this project was to reduce wood consumption by improving stoves efficiency and then to protect forest cover.
Laboratory tests were made on different models, they showed that some models can save 50 % to 60% of wood consumption, without a great price increase, while preserving taditional cooking methods.
If the prototype is generalized a hundred thousand tons of wood could be saved.

4) GEOTHERMICS

There are in the south of Tunisia a lot of hot water sources which are utilized for irrigation   and require cooling equipments.

Demonstration was made sucessfully for heating greenhouses with geothermal water and now a great program with hundreds of hectars is launched in the south.

5) SOLAR WATER HEATING AND MECHANICAL WIND PUMPS

It's noteworthy that these equipments are being commercialized since 1985. They are manufactured   in Tunisia for both local use and export.

PROGRAMMATION PLANNED :

The AME activity will finalise some of the demonstration projects witch are mature for application and start others for demonstration, in particular :

. Extenseve use of small photovoltaïc systems in rural areas for houses, schools and utilities

. Installation of a wind farm (5IMW), the feasibility study shows that it is economically
  acceptable

. Scheduling plantation of quick-growing trees which will provide wood for energy production.

. generalization of biogaz digestors
. research on other Renewable energies like solar ponds, bioclimatic construction

. feasibility study on high temperature solar plant
. solar drying of land products
. Low and high temperature geothermic potential study in the north of the country and
  possibilities of applications.

. Energie conversion of urban wastes.

AIR POLLUTION MODIFICATION DUE TO ENERGY SUPPLY
DIVERSIFICATION FOR MEXICO IN THE YEAR 2000

Manuel Martínez

Laboratorio de Energía Solar, Universidad Nacional Autónoma de
México, Apartado Postal No. 34, 62580 Temixco, Morelos, México

ABSTRACT

Possible amounts of air pollutants due to the net internal
energy supply are studied for Mexico in the year 2000. Two
scenarios are considered: historical trend and a possible
energy diversification scheme. For the year 2000, the
considered viable participation of nuclear, solar and wind
energies amounts, in percent, to: for NIS, 2.3; for ED, 5.3,
and for TD, 1.4. Also, the reduction in air pollutants, in kg
per kJ/person, could be: for electric demand, 9 percent in SOx,
13 in NOx, 13 in CO, 19 in HC and 6 in PARTS, and for thermal
demand, 6 percent in SOx, 5 in NOx, 5 in CO, 5 in HC and 0.6 in
PARTS.

KEYWORDS

Energy pollution; year 2000; Mexico; air emissions.

INTRODUCTION

The key role of energy in causing environmental problems has
been confirmed by several systematic studies on risks related
to its production, transportation, storage and transformation.

The primary energies significantly consumed now in Mexico are:
coal, oil, natural gas, hydroenergy, geoenergy, bagasse and
firewood. The incorporation of nuclear, solar and wind energies
have been considered for the year 2000.

The most important local air pollutants due to the energy usage
were considered: sulfur oxides (SOx), nitrogen oxides (NOx),
carbon monoxide (CO), gaseous hydrocarbons (HC) and
particulates (PARTS). The amounts emitted depend on the
specific technological process, the fuel composition, the fuel
heat content and the efficiency of the system used, as reported
by UNEP (1985).

2578

Considering that these emissions are mainly due to the transformation of primary energies into electric or thermal secondary energies, the national energy balances have been re-written. For each primary energy, the flow from the net internal supply (NIS) to electric demand (ED), thermal demand (TD) and operation and transformation (OT) has been established.

Due to the large variation in the reported figures, a selection of average values of the heat contents for the energies supplied and of the systems efficiency in Mexico has been chosen. Also, the amounts of pollutants emitted to the air, for each energy source and use were considered according to Martínez (1992).

The purpose of this paper is to study the possible medium-term variations in the amount of several air pollutants due to the internal supply of energy in Mexico. The evolution of each air pollutant in a more diversified energy supply programme is evaluated.

ENERGY SCENARIOS FOR THE YEAR 2000

For the year 2000 there are two scenarios: the Mexican energy development varies according to an historic trend or to a specific energy diversification scheme proposed by Martinez and Best (1991). In both cases, the energy flow is: NIS, 7300 PJ/year; ED, 2272; TD, 3173, and OT, 1855. The national energy balance for the year 2000, according to historic trend is shown in Table I. The national energy balance for the year 2000, according to the proposed energy diversification scheme is shown in Table II.

Table I. MEXICAN ENERGY BALANCE, YEAR 2000 (PJ/YEAR).
HISTORICAL TREND SCENARIO

	NIS	ED	TD	OT
COAL	309	242	57	10
HYDROENERGY	268	268	0	0
GEOENERGY	492	492	0	0
BAGASSE	103	0	94	9
FIREWOOD	371	0	371	0
OIL	3786	1089	1600	1097
GAS	1971	181	1051	739
TOTAL	7300	2272	3173	1855

Table II. MEXICAN ENERGY BALANCE, YEAR 2000 (PJ/YEAR).
ENERGY DIVERSIFICATION SCENARIO

	NIS	ED	TD	OT
COAL	306	239	57	10
HYDROENERGY	598	598	0	0
GEOENERGY	373	287	86	0
BAGASSE	103	0	94	9
FIREWOOD	371	0	371	0
OIL	3539	884	1496	1159
GAS	1842	143	1023	676
NUCLEAR	60	60	0	0
SOLAR	106	60	46	0
WIND	2	2	0	0
TOTAL	7300	2273	3173	1854

For the year 2000, the considered viable participation of nuclear, solar and wind energies amounts to: for NIS, 2.3 percent; for ED, 5.3, and for TD, 1.4.

RESULTS

The evolution, from 1965 to 1988, of these air pollutants is reported in kg of pollutant per kJ of energy supplied per capita. The emissions due to electric demand, for the historic trend and diversification scheme, are shown in Figs. 1 and 2, respectively. The emissions due to thermal demand, for the historic trend and diversification scheme, are shown in Figs. 3 and 4, respectively. In order to satisfy the electric demand, sulfur and nitrogen oxides have increased very fast, due to a larger use of oil and coal. In order to satisfy the thermal demand, nitrogen and sulfur oxides have increased fast, gaseous hydrocarbons and carbon monoxide have been kept constant, while the participation of particulates has decreased; which is a consequence of a larger consumption of oil and natural gas, a constant participation of coal and firewood, and a lesser use of bagasse.

Fig.1.AIR POLLUTION EVOLUTION DUE TO ED
HISTORICAL TREND SCENARIO

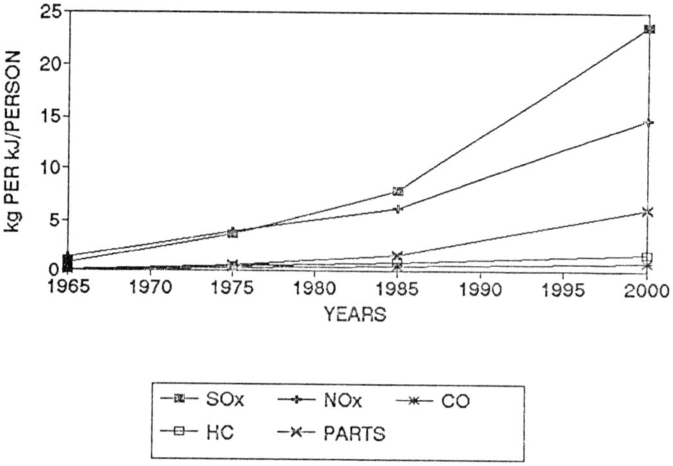

—▣— SOx   —+— NOx   —✻— CO
—▱— HC   —✕— PARTS

Fig.2.AIR POLLUTION EVOLUTION DUE TO ED
ENERGY DIVERSIFICATION SCENARIO

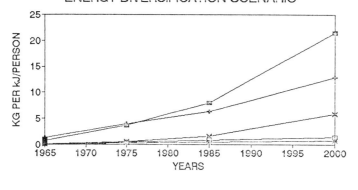

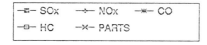

Fig.3.AIR POLLUTION EVOLUTION DUE TO TD
HISTORICAL TREND SCENARIO

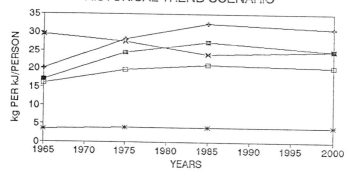

Fig.4.AIR POLLUTION EVOLUTION DUE TO TD
ENERGY DIVERSIFICATION SCENARIO

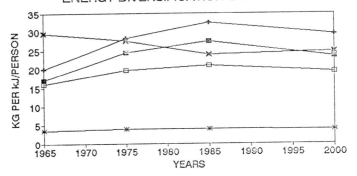

CONCLUSIONS

For the year 2000, the considered viable participation of nuclear, solar and wind energies amounts to: for NIS, 2.3 percent; for ED, 5.3, and for TD, 1.4. Also, the reduction in the amount of air pollutants, in kg per kJ/person, due to this diversification scheme could be: in relation to ED, 9 percent in SOx, 13 in NOx, 13 in CO, 19 in HC and 6 in PARTS, and in relation to TD, 6 percent in SOx, 5 in NOx, 5 in CO, 5 in HC and 0.6 in PARTS.

The feasibility to decrease the emissions of air pollutants related to energy supply has been shown for an energy diversification scheme as the one proposed in this paper.

REFERENCES

1. UNEP (1985) _Energy Report Series_, The environmental impacts of production and use of energy, part IV, Nairobi.
2. Martínez, M. (1992) Air pollutants due to energy supply in Mexico _Renewable Energy J_ to be published.
3. Martínez, M. and Best, R. (1991) Developments in Geothermal Energy in Mexico-Part Thirty-Two. Supply and Demand Perspectives for the Year 2000 _Heat Recovery Systems & CHP_ __11__ (1) 91-98.

HYDROGEN PRODUCTION BY STEAM GASIFICATION OF COAL
IN MOVING BED USING NUCLEAR HEAT

S. EL ISSAMI   A. BELGHIT

Laboratoire de Mécanique des Fluides et d'Energétique
Faculté des Sciences Sémlalia. Département de Physique
Boulevard Prince Moulay Abdellah BP S : 15
MARRAKECH   MORROCCO

ABSTRACT

The purpose of this paper is to develop a theoretical model of a chemical
moving bed reactor for gasifying coal with steam. The heat used to drive
the endothermic reaction is provided by a high temperature nuclear reactor.
The model describes the complex physical and chemical processes taking
place in the multiphase moving bed using mass and energy balances and gives
information about rates of chemical reaction and physical transport
processes. The dependence of the reactor performance on control parameters:
gas flow rate and inlet gas temperature is also studied.

KEYWORDS

Heat and mass transfer, porous media, gasification, energy demand,
hydrogen, fission reactors.

INTRODUCTION

Gasification of coal is a well-known technique which has been performed
word-wide on an industrial scale and in many forms. In all conventional
plants, coal serves as raw material for gas and as source of reaction heat,
that leads to a high consumption of coal. In this meaning, the use of heat
requirement of endothermic process provided by a high temperature nuclear
reactor is more advantageous compared with conventional process. Indeed,
above the manifold uses of steam gasification of coal using nuclear heat,
as shown in fig. 1 (JÜNTGEN et al. 1975), this process permits the
saving of coal reserves, the production of small amount of carbon dioxide
in the gasification plant and a lower costs of gas production (VAN HEEK.,
1977). The heat is discharged from the nuclear reactor by a helium circuit,
and then transferred from it into the steam gasifier. The coal is gasified
according to the reaction : $C + H_2O \longrightarrow CO + H_2$ $\Delta H = 131,4 \ kj/mol$.

Our hope is to develop a theoretical model of moving bed which permits the
determination of the temperature profiles for gas and solid, the
concentration profile in gas and the radiative flux density asafunction of
control parameters.

In the gasifier, the rate of gasification is determined by chemical, mass transport or mixed control. The heat transfer is governed by the three exchange modes : conduction, convection and radiation. In this work we present the functioning of the moving bed reactor with mass transport control.

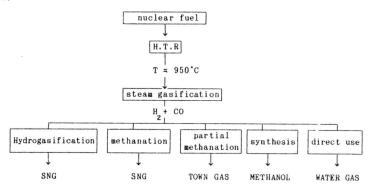

Fig. 1. Manifold uses of steam gasification
of coal using nuclear heat.

HEAT AND MASS TRANSFER EQUATIONS

Attempt to simplify the problem, we consider : the steady state, the section of a gasifier is large enough that its side may be considered as adiabatic, the flow is one dimensional, the physical properties are the same for all points located in the same section and the bed is assumed to be a gray optically thick medium characterized by extinction coefficient K and emissivity ε_p. Heat and mass transfer equations can be written as follows (EL ISSAMI., 1991) :

* For the gas

To study the mass transfer in the gas, we defined the advancement degree of the reaction X ,which represents the ratio of H_2O moles consumed at a distance x into the gasifier to the number of moles entering at x=0, by :
$X = (C_o-C)/C_o$.

mass

$$\frac{\partial X}{\partial t} + V \frac{\partial X}{\partial x} = \frac{\partial}{\partial x} (D \frac{\partial X}{\partial x}) - \frac{6 (1-\varepsilon) K_g}{\varepsilon d (1+X_1)}(X-CS) \qquad (1)$$

energy

$$\rho C_p\frac{\partial \theta}{\partial t} + \rho V C_p\frac{\partial \theta}{\partial x} = \frac{\partial}{\partial x} (\lambda(\theta) \frac{\partial \theta}{\partial x}) + \frac{6(1-\varepsilon)}{\varepsilon d} h (T-\theta) \qquad (2)$$

* For the solid

mass

$$\frac{\rho_c}{M_c} \frac{\partial r_c}{\partial t} - \frac{\rho_c}{M_c} V_s \frac{\partial r_c}{\partial x} = \frac{K_g}{(1+X_1)} (C-C_s) \qquad (3)$$

continuity

$$\rho_s \frac{\partial Vs}{\partial x} = \frac{-6}{d} \frac{K_g M_{H2O}}{(1+X_1)} (C-C_s) \qquad (4)$$

energy

$$\rho_s C_{ps} \frac{\partial T}{\partial t} + \rho_s C_{ps} V_s \frac{\partial T}{\partial x} = \frac{\partial}{\partial x} (\lambda^* \frac{\partial T}{\partial x} + \frac{16\ \sigma\ T^3}{3\ K} \frac{\partial T}{\partial x})$$
$$- \frac{6}{d} h(T-\theta) - \frac{6}{d} \Delta H \frac{K_g}{(1+X_1)} (C-C_s)$$

(5)

In this equation, the radiative transfer contribution is introduced by the ROSSELAND approximation (ROSSELAND., 1936).

These equations are fitted with the following boundary and initial conditions.

$$\text{at } t=0 \begin{cases} X=0 \\ \theta=\theta_o \\ T=Tamb \\ V=0\ ;\ r_c=r(0) \end{cases} x=0 \begin{cases} X=0 \\ \theta=\theta_o \\ T=Tamb \\ V_s=0 \end{cases} x=L \begin{cases} \frac{\partial X}{\partial x} = \frac{\partial T}{\partial x} = \frac{\partial \theta}{\partial x} = 0 \\ P=Patm\ ;\ r_c=r(0) \end{cases}$$

The set of equations (1), (2) et (5) are solved using the finite difference method proposed by PATANKAR (PATANKAR., 1980). The solution of a steady-state equations is obtained through the use of the discretisation equations for a corresponding unsteady situation that can be recognized as simply a particular kind of under-relaxation procedure. The integration of equations leads to the algebraic equations with a tridiagonal matrix which is resolved by direct method "Thomas's method ". The resolution of equation (3) is obtained using the finite difference method with Crank-Nicolson scheme and equation (4) is integrated using the trapezoidal rule.

BEHAVIOUR OF THE MODEL

Most of the results are obtained using these parameters : $\theta_o = 1400°K$; $V_o = 0.06 m/s$; $F_{H2O}^o = 30\ \%$; $F_{He}^o = 70\ \%$; $\varepsilon=0.45$; $d_L = 3mm$; $K=2000m^{-1}$. The model predicts the temperature distribution and the gas composition in the gasifier.

Fig. 2a shows the temperature distribution for the gas and the solid. Two temperature zones are presented : the first corresponds to solid heating by the use of the high helium temperature and the second to an equilibrium thermal zone between solid and gas. Fig. 2b presents the molar fraction of H_2, H_2O and H_e in the moving bed. It is noted a high consumption of H_2O at the region neighboring the bed surface.

REACTOR PERFORMANCE

The primary use of a simulation model is to allow computer experiments to be carried out in order to study the effect of changes of operating condition on reactor performance. The control parameters chosen are : gas flow rate, inlet gas temperature and effective conductivity of the bed. The knowledge of this performance necessitated to define the specific productivity of the gasifier presenting the amount of treated coal :
$$P_s = (1-\varepsilon)\ \rho_s\ V_s\ .$$

The increase of the inlet gas temperature leads to an increase of the advancement degree of the reaction and therefore to a large consumption of coal as shown in fig. 3a.

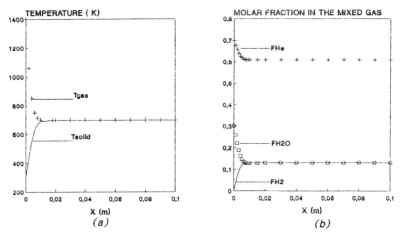

Fig. 2. Temperature distribution and molar fractions
along the gasifier.

An increase of the inlet gas temperature leads to a high temperature
gradient in the zone neighboring the bed surface and therefore to an
important radiative flux density in this zone ; see fig. 3b.

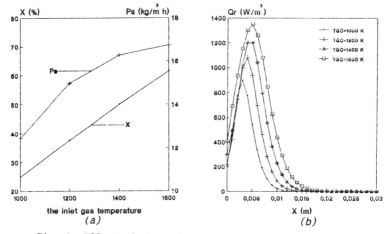

Fig. 3. Effect of the inlet gas temperature.

Fig. 4a shows that an increase in the flow rate induced an increase in the
temperatures and in the transformation rate. Fig. 4b presents the evolution
of the radiative flux density in the bed as a function of the inlet gas
velocity. It is noted an increase of the radiative flux density with the
flow rate.

CONCLUSION

A theoretical model of a chemical moving bed reactor for gasifying carbon
with steam using the heat from the H.T.R is presented. It permits the
determination of gas and solid temperature distribution and gas composition
along the gasifier . It includes the effect of gas flow rate and inlet gas

temperature on its functioning and performance.

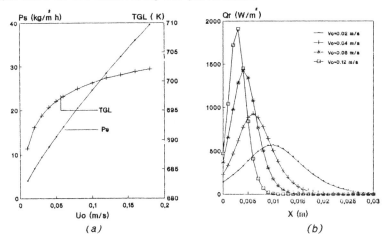

(a)   (b)

Fig. 4. Effect on the inlet gas velocity.

NOMENCLATURE

C : H_2O molar concentration
Cp: heat capacity
D : diffusion coefficient of
    H_2O in He
d : particle diameter (2 r)
Kg: mass transfer coefficient
K : extinction coefficient
h : particle-fluide heat
    transfer coefficient.
L : lenght of the moving bed.
M : molar density
P : pressure inside the reactor
Qr: radiative flux density.
X_1: H_2O molar fraction at the
    solid surface

X : degree of advancement of reaction
x : axial coordinate (positive in the
    flow direction).
ΔH: molar enthalpy of reaction.
T : solid temperature .
t : time
V : fluid velocity
Vs: solid velocity
ε : bed void fraction
θ : gas temperature
ρ : density
σ : stefan-Boltzmann constant
λ_*: thermal gas conductivity
λ^*: effective bed conductivity

subscripts : o : inlet ;   L : outlet ;   c : coal ;   s : solid surface

REFERENCES

EL ISSAMI, S. and A. BELGHIT,(1991). Etude des transferts de chaleur et de
masse dans un milieu poreux réactif soumis à une source de chaleur
nucléaire. Troisième Colloque Maghrébin sur les Modèles Numériques de
l'Ingénieur. 26-29 Novembre. Tunis.
JÜNTGEN,H. and K.H. VAN HEEK.,(1975). gasification of coal with steam
using heat from HTRS. Nuclear Engineering and Design,34, pp : 59-63.
PATANKAR SUHAS V.,(1980). Numerical heat transfer and fluid flow. Mc.
GRAW-HILL, LONDON.
ROSSELAND.,(1936).Theoretical astrophyxies. OXFORD UNIVERSITY PRESS.
CLARENDON (ENGLAND).
VAN HEEK.,K.H.,(1977). Hydrogen from coal gasification. ISPRA COURSES. The
Hydrogen-Energy concept. May 9-13 H/77 No : 5 .

EVOLUTION OF HYDROGEN FROM WATER
OVER ZEOLITES

MAGDALENA MOMIRLAN

Institute of Physical Chemistry, Romanian
Academy, Spl.Independentei 202, Bucharest
77208 ROMANIA

ABSTRACT

A catalytic process for hydrogen generation from water by the
decomposition of water in presence of usual zeolite catalysts
impregnated with non-noble metals of variable valences and
activated in vacuum was developed. For optimizing the technological flow, continuous decomposition of water in reaction-
regeneration catalyst cycles in moderate vacuum at temperatures
up to 500°C was effected.

KEYWORDS

Hydrogen; water; zeolite.

INTRODUCTION

Different methods of producing of hydrogen from water are well
known   (Bicelli,1986; Bockris et al.,1985; Warner and Berry,
1986; Simarro et al., Getoff,1984). The present paper searches
a less energointensive process for hydrogen generation from
water (Dorrance,1981). The process advanced by us eliminates
some disadvantages by effecting continuous water decomposition
in reaction-regeneration catalyst cycles in moderate vacuum up

to 10^{-2} mm Hg and temperature to 500°C.

EXPERIMENTAL DETERMINATIONS

Experiments were effected in a quartz reactor connected to a vacuum line. The experimental installation with the components is shown in Fig.1. The continuous catalytic process takes place as follows: first stage: activation of catalyst at temperature up to 500°C and a vacuum of 10^{-2} mm Hg; second stage: introduction of water in the reactor, over the activated, heated catalyst and collection of the evolved hydrogen at several temperatures: 200°C, 300°C, 400°C, 450°C, 500°C. For the complete evacuation of hydrogen from the reactor the system was purged with an inert gas (nitrogen or argon); the collected gases were analyzed periodically by gas chromatography. In the installation represented in Fig. 2 the activation of zeolites was achieved by UV irradiation.

The two methods of zeolite activation (vacuum outgasing and activation with UV radiations) were efficient.

RESULTS

Regarding the evolution of hydrogen by decomposition of water in presence of thermally activated zeolites, using the installation shown in Fig.1, the amount of hydrogen was approximately 27.96 ml to 1 g PbO from the modified zeolite in a thermocycle. The reaction cycles refer to the fact that after the effecting one experiment of successive generation of hydrogen, the catalyst has undergone a second process of activation, followed by reaction. After several activations, the catalyst does not present any efficiency decline. The experiments show that the reaction cycle with modified zeolite is impeded by the reduction of the oxidized form of Pb from zeolite. The oxidation of the reduced Pb and the generation of hydrogen takes place much easier. The present procedure presents the following advantages:

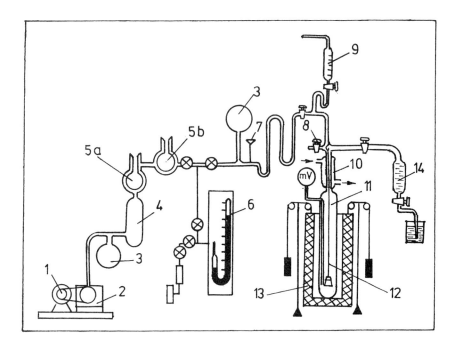

Fig. 1. Experimental device for hydrogen
evolution by thermocatalytic decomposition of water over zeolites
thermally activated in vacuum.

1,2-preliminary vacuum pump; 3-preliminary vacuum flask; 4-di-
ffusion pump; 5a, 5b-nitrogen traps; 6-Hg manometer; 7-vacuum
measuring lamps; 8-vacuum valves; 9-burette; 10-cooling water
tube; 11-quartz reactor; 12-thread for hanging the sample (zeolite); 13-furnace; 14-hydrogen collecting phial.

- catalyst activation, that is its reducing to an lower va-
lence, can be achieved without consumption of any reagent;
- catalyst activation can be achieved at a lower vacuum (10^{-2}
mm Hg, against 10^{-5} mm Hg mentioned in literature, without
superatmospheric pressure of steam);
- working temperature in all phases does not exceed 500°C;
- the catalyst does not contain noble, expensive metals;
- the process runs continuously.

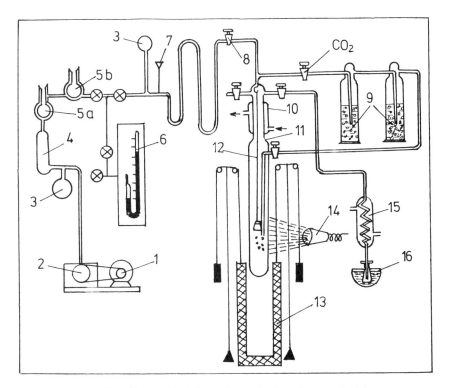

Fig. 2. Installation for study of evolution
of chemically bonded hydrogen (methanol), passing CO_2 over the hydrogen evolved following water decomposition.

1,2-preliminary vacuum pump; 3-preliminary vacuum flask; 4-di-
ffusion pump; 5a,5b-nitrogen traps; 6-Hg manometer; 7-vacuum
measuring lamps; 8-vacuum tap; 9-washing vessels; 10-cooling
water sheath; 11-quartz tube; 12-thread for hanging of zeolite;
13-furnace; 14-UV irradiation lamp; 15-cooling tube; 16-ice
vessel.

CONCLUSIONS

The experimental conditions for evolution of hydrogen by thermocatalytic decomposition of water with Pb-modified zeolite as

well as with other zeolites impregnated with non-noble metals as catalysts were established. The decomposition of water takes place in reaction-regeneration catalyst cycles. Both before and after each reaction cycle, the catalyst was thermally activated in vacuum at a pressure of 10^{-2} mm Hg and a temperature of 500°C.

Acknowledgment

The author thanks to researchers A.Wohl, Gr.Pop, G.Musca, I.Iosif, T.Danciu and to prof.dr. I.V.Nicolescu for helpful discussions.

REFERENCES

Bicelli, L.P. (1986). Hydrogen: a clean energy source. Int. J. Hydrogen Energy, 11, 555-562.

Bockris, J.O´M., B.Dandapani, D.Cocke and J.Ghoroghchian (1985). On the splitting of water. Int. J. Hydrogen Energy, 10, 179-201.

Dorrance, W.H. (1981). Method for producing oxygen and hydrogen from water. Patent USA 4, 278-650 Int.Cl.Col B 1/08, 13/06.

Getoff, N. (1984). Basic problems of photochemical and photo-electrochemical hydrogen production from water. Int. J. Hydrogen Energy, 9, 997-1004.

Simarro, R., S. Cervera-March and S. Esplugas (1984). Hydrogen photoproduction in a continuous flow system with u.v.-light and aqueous suspensions of $RuO_x/Pt/TiO_2$. Int. J. Hydrogen Energy, 10, 221-226.

Warner, J.W. and R.S. Berry (1986). Hydrogen separation and the direct high-temperature splitting of water. Int. J. Hydrogen Energy, 11, 91-100.

AN INVESTIGATION ABOUT THE KOH CONCENTRATION EFFECT ON HYDROGEN PURITY

SOFRATA, H., AL-SAEDI, Y
Energy Research Institute, Solar Programs,
King Abdulaziz City for Science & Technology (KACST)
P.O. Box: 6086, Riyadh 11442, Kingdom of Saudi Arabia

ABSTRACT

In the Solar Hydrogen production, purity of hydrogen is one of the main objectives. Among other factors KOH concentration has a significant effect, specially in small current intensities.

An experimental investigation reveals that a critical current intensity ranges form 1000 ampere/m^2 to 2500 ampere/m^2 exists. Varying the KOH concentration between 50 gm/lit. to 300 gm/lit. the critical current intensity oscillates within the above mentioned range.

Such finding is very important for Solar Hydrogen production due to the fact that KOH concentration affects electrode activation and its working life.

INTRODUCTION

One of the most promising method in Solar Hydrogen production is electrolytic splitting of water to H_2 and O_2 gases [1-24]. Using activated electrodes improves the system performance especial under solar intermittent operation. Since KOH concentration plays also a main role in lifetime of electrodes; an in depth investigation has been carried out. Such investigation may help in the evaluation process of KOH concentration parameter.

EXPERIMENTAL SETUP

A laboratory size solar powered hydrogen generator has been considered [25]. The system has been operating utilizing an activated electrodes. Special very thin KREBSKOSMO P.S. diaphragm have been utilized. Using activated cathode, a maximum hydrogen flow of 18 normalized liter/hour and oxygen flow of 9 normalized liter/hour have been collected. At a current of 40.5 ampere 2.5 volts. The hydrogen purity has been measured as 99.67 vol.%.

2593

The system has been powered by 6 photovoltaic modules each of 50 watts peak. These photovoltaic panels are originally designed for a nominal output of 12 volts. All panels have been adapted to produce a nominal 3 volts output.

The test rig Fig. (1) consists of three main components, namely: the photovoltaic panels set, the storage batteries, controlling switches, the electrolyzer and measuring instruments.

The photovoltaic panels set consists of nine locally manufactured specially designed modules. Each module consists of parallel connected four sets of series connected nine cells rows. The nominal peak power of each module is 50 watts, 4 volts and 30 amperes at the peak point. The peak power is 50 watts at Vp of 4.31 volts and Ip of 11.75 amperes. The open circuit volt is 5.38 volt and Ip of 70 amperes.

In order to minimize the power losses, blocking diodes have been connected only when the system is running in battery mode; otherwise, the system is connected directly to the electrolyzer.

Two control methods have been introduced to give an easy, simple and cheap, full control on the current flow; namely:

1.   On/Off switch control for each individual solar panel and an overall Master Switch for emergency cases.

2.   The orientation to the sun may be also used as a controlling parameter. Putting the solar modules out of sun direction, the flowing current may be controlled.

EXPERIMENTAL DOMAIN

Experimental tests have been carried out for KOH concentration ranging from 50 gm/liter to 300 gm/liter. The current intensity has been varied between 0-4000 amperes/m^2.

The hydrogen purity against current intensity for KOH 50, 100, 125, 150, 200, 250 and 300 gm/liter is displayed in figures 1-7.

An in-depth investigation shows that the purity increases for the same current intensity for KOH 50, 125 and 150 gm/liter as shown in Fig. 8.

Increasing the concentration 150 and 200 gm/liter, the purity decrease for the same current intensity as may be shown in Fig. 9.

Again the purity increases by in creasing the KOH concentration from 200 to 250 gm/liter as seen in fig 10. Further increase of KOH 250-300 gm/liter the purity decreased (See Fig. 11). This observed oscillation of purity is likely to occur around a critical current intensity ranging from 1000-2500 amperes /m^2 for activated electrodes.

CONCLUSION

In this experimental investigation it has been observed that a criti-cal current intensity of 1000-2500 ampere/m^2 for activated elect-rodes exists. Such critical current intensity acts as pivote where the hydrogen purity varies for different KOH concentrations varying between 50-300 gm/lit. Decreasing the KOH concentration will contribute to electrodes life time.

ACKNOWLEDGEMENT

The authors would like to thank Solar Programs, King Abdulaziz City for Science and Technology (KACST)for funding this research. Also, thanks are due to Mr. Kamal Abdulzaher, Thermal Department Laboratory Supervisor, KACST, for his efforts in doing the experi-mental part of this work.

REFERENCE

1.   Zahed, A.H., Bashir, M.D., Ali, T.Y., Najjar, Y.S.H., "A Perspective of Solar Hydrogen and its Utilization in Saudi Arabia", Int. Journal of Hydrogen, Vol. 16, No. 4, pp 277-281, 1991.

2.   Nitsch, J., and Klaib, H., "Solar Hydorgen and its Importance and Limits", ISES, Solar World Congress, Hamburge, F.R.G., (13-18 September, 1987).

3.   Champagne, R.D., "Economics and Markets for Hydrogen", 6th World Hydrogen Energy Conference, Vienna, Austria, (20-24 July 1986).

4.   Technical Bullitin, Costs of Hydrogen in Plants. The Chemical Engineers, London (November 1989).

5.   Zahid, A.H., Bashir, M.D., Ali, T.Y., "Presentation of an Overview of the Important Aspects of Solar Hydrogen Technology and its Utilization", Report prepared for the Joint German-Saudi Arabian Research Program on Solar Hydrogen Energy (HYSOLAR) (June 1988).

6.   Najjar, Y.S.H., Zahed, A.H., Bashir, M.D., and Alp, T.Y., "Feasibility of Hydrogen Utilization in Gas Turbines, Energy and Environment.

7.   Williams, L.O., "Hydrogen Power - Introduction to Hydrogen Energy and its Application", Pergamon Press, Oxford (1980).

8.   Zahed, A.H., Bashir, M.D., and Alp, T.Y., Najjar, Y.S.H., "Solar Hydrogen Production and its use in Saudi Arabia", Report prepared for the Joint German-Saudi Arabian Research Program on Solar Hydrogen (HYSOLAR) (December 1989).

9.   Varey, P., "Summary of World Energy Conference, Montreal, Canada", held on 8-12 September 1989. The Chemical Engineer, London (November 1989).

10.   Serfass, J.A., Nahmias, D. and Appleby, A.J., "A Practical Hydrogen Development Strategy", Int. Journal of Hydrogen Energy, Vol. 16, No. 8, pp 551-556, 1991.

11.   Cervera-March, S. and Smotkin, E.S., " A Photoelectrode Array System for Hydrogen Production from Solar Water Splitting", Int. Journal of Hydrogen Energy, Vol. 16, No. 4, pp 243-247, 1991.

12.   Bowen, C.T., Davis, H.J., Henshaw, B.F., Lachance, R., Le Roy, R.L. and Renaud, R., Int. Journal of Hydrogen Energy 9, 59 (1984).

13.   Divisek, J., Malinowski, P., Mergel, J., Schmitz, H., Int. Journal of Hydrogen Energy 10, 383, (1985).

14.   Justi, E.W., Wasserstoff, Energie fuer alte Zeiten, Chapter 8, Udo Pfriemer Verlag, Muenchen, (1985).

15.   Gericher, H., in Schiavello, M., (ed) Photoelectrochemis-try, Photocatalysis and Photoreactor, Fundamentals and Developments, NATO ASI Ser. 146, 39 (1985).

16.   Heller, A. in Graetzel, M., (ed), in Energy Resources through Photochemisty and Catalysis, p 385 (1983).

17.   Kharkats, Yu. I. and Pleskov, Yu. V., " A Plant for Solar Energy Conversion and Storage, 'Solar Array + Electrolyser + Storage Battery', Computation of the Non-Steady-State Operating Conditions and Design Optimization", Int. Journal of Hydrogen Energy, Vol. 16, No. 10, pp 653-660, 1991.

18.   Divisek, J., Mergel, J., Schmitz, H., " Improvement of water Electrolysis in Alkaline Media at Intermediate Temperatures", Int. Journal of Hydrogen Energy, Vol. 7, No. 9, pp 695-701, (1981).

19.   Carpetis, C., "An Assessment of Electrolytic Hydrogen Production by Means of Photovoltaic Energy Conversion", Int. Journal of Hydrogen Energy, Vol. 9, No. 12, pp 989-991, (1984).

20.   Winter, C.J., Nitch, J., "Hydrogen Energy - a Sustainable Development - Towards a World Energy Supply System for Future Decades", Int. Journal of Hydrogen Energy, Vol. 14, No. 11, pp 785-796, (1989).

21.   Dutta, S., Block, D., Port, "Economic Assessment of Advanced Electrolytic Hydrogen Production", Int. Journal of Hydrogen Energy, Vol. 7, No. 9, pp 695-701, (1981).

22.   Winter, C.J., Nitch, J., "Wasserstoff Als Energietrager", Springer Verlag, Berlin, W. Germany.

23.   Sidz, J., "Modem Concepts of Electrochemistry", No.18, pp 304, 1986.

24.   Krebskomo, "Private Communication", Berlin, W. Germany

25.   Sofrata, H., Al-Saedi, Y., Aba-Oud, H., "A Laboratory Size Solar Hydrogen Generator", Proceeding of Project Hydrogen 92, 1992, USA.

Fig.1. TEST RIG FOR LABORATORY ELECTROLYZER

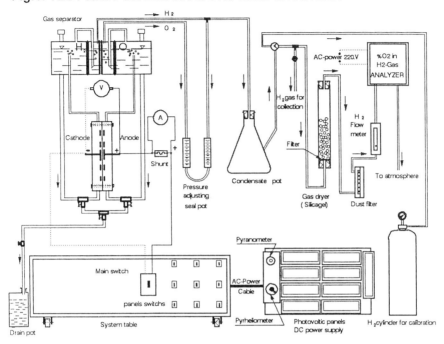

Fig.2. HYDROGEN PURITY AGAINST CURRENT INTENSITY
THROUGH ELECTROLYZER FOR KOH = 50 gm/lit

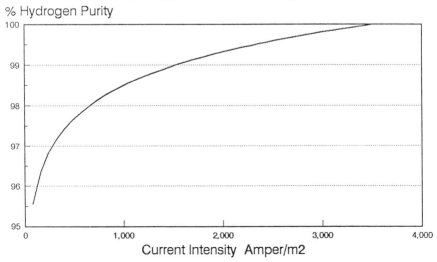

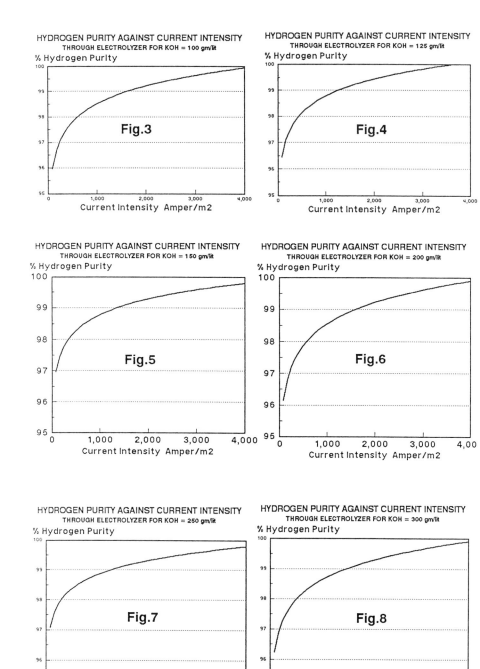

HYDROGEN PURITY AGAINST CURRENT INTENSITY
THROUGH ELECTROLYZER FOR KOH = 100 gm/lit
% Hydrogen Purity
Fig.3
Current Intensity Amper/m2

HYDROGEN PURITY AGAINST CURRENT INTENSITY
THROUGH ELECTROLYZER FOR KOH = 125 gm/lit
% Hydrogen Purity
Fig.4
Current Intensity Amper/m2

HYDROGEN PURITY AGAINST CURRENT INTENSITY
THROUGH ELECTROLYZER FOR KOH = 150 gm/lit
% Hydrogen Purity
Fig.5
Current Intensity Amper/m2

HYDROGEN PURITY AGAINST CURRENT INTENSITY
THROUGH ELECTROLYZER FOR KOH = 200 gm/lit
% Hydrogen Purity
Fig.6
Current Intensity Amper/m2

HYDROGEN PURITY AGAINST CURRENT INTENSITY
THROUGH ELECTROLYZER FOR KOH = 250 gm/lit
% Hydrogen Purity
Fig.7
Current Intensity Amper/m2

HYDROGEN PURITY AGAINST CURRENT INTENSITY
THROUGH ELECTROLYZER FOR KOH = 300 gm/lit
% Hydrogen Purity
Fig.8
Current Intensity Amper/m2

2598

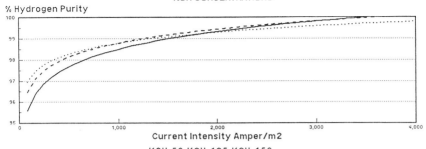

Fig.9. HYDROGEN PURITY AGAINST CURRENT INTENSITY
THROUGH ELECTROLYZER FOR DIFFERENT
KOH CONCENTRATIONS

% Hydrogen Purity

Current Intensity Amper/m2

KOH=50 KOH=125 KOH=150

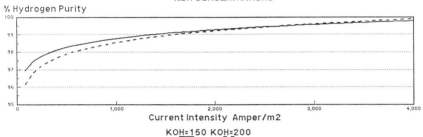

Fig.10. HYDROGEN PURITY AGAINST CURRENT INTENSITY
THROUGH ELECTROLYZER FOR DIFFERENT
KOH CONCENTRATIONS

% Hydrogen Purity

Current Intensity Amper/m2

KOH=150 KOH=200

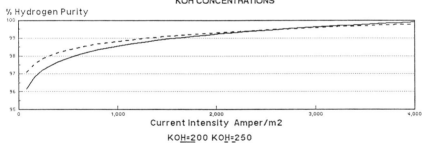

Fig.11. HYDROGEN PURITY AGAINST CURRENT INTENSITY
THROUGH ELECTROLYZER FOR DIFFERENT
KOH CONCENTRATIONS

% Hydrogen Purity

Current Intensity Amper/m2

KOH=200 KOH=250

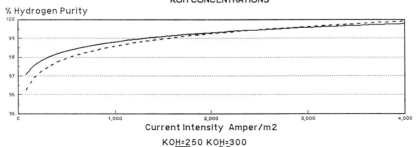

Fig.12. HYDROGEN PURITY AGAINST CURRENT INTENSITY
THROUGH ELECTROLYZER FOR DIFFERENT
KOH CONCENTRATIONS

% Hydrogen Purity

Current Intensity Amper/m2

KOH=250 KOH=300

2599

Accumulation of Hydrogen in Metal Hydrides and Optimization of Hydrogen Storage Tanks in Respect to Pressure Drop and Mechanical Tensions within Granular Bulks

I. Haas

Technische Universität Berlin , Thermodynamik und Reaktionstechnik / FG Reaktionstechnik , RDH9 , Fasanenstraße 89 , D-1000 Berlin 10 , Germany

Abstract - Measurements by other investigators and the research presented here have shown that the amount of hydrogen, which can be absorbed by metal hydrides under *equilibrium conditions*, strongly depends on the *mechanical pressure tensions* exerted on the *metal hydride* by the walls of the storage container. As the ab- and desorption of hydrogen in metal hydrides generally is controlled by the heat exchange with the environment, the fastest *hydriding rates* can be achieved by compacting the granular metal hydride under *high mechanical pressure*. Within *highly compacted metal hydride bulks*, though, the *pressure drop* of the ab- or desorbing hydrogen on its way to or from the metal hydride surface becomes the controlling process of the hydrogenation. By means of measuring the *equilibrium data* and the *stationary pressure drop* of hydrogen in highly compacted granular metal hydride bulks with a voidage of less than 25% an optimal value for the storage capacity (equilibrium data) and the convective hydrogen transport (pressure drop) within the bulk can be estimated. The estimated values can be compared with the results of heat transfer measurements thus indicating the future method for *construction of metal hydride storage containers* and the connected systems.

Nomenclature

Symbol	Definition	Name; Units
a	$\lambda /(\rho\, C_V)$	temperature conductivity; $m^2\, s^{-1}$
c	n/V	concentration; $mol\, m^{-3}$
C	$\dfrac{1}{T'}\int\limits_0^{T'}(\dfrac{dh}{dT})_p\,dT$	mean specific heat capacity; $Ws\, kg^{-1}\, K^{-1}$
d	(SI)	diameter of particle; m
$\underline{F}$	m $\underline{a}$	force; N
m	(SI)	mass; kg
M	m/n	molecular weight; $g\, mol^{-1}$
n	(SI)	number of moles; mol
p	$|\underline{E}_{gas}|/A$	gas pressure; $N\, m^{-2}$, bar
R	(p V)/(n T)	ideal gas constant; $J\, mol^{-1}\, K^{-1}$
Re	eq.(13)	Reynolds number
T	(SI)	temperature; K, °C
$\bar{w}$	$\dot{V}/A$	mean velocity of empty reactor
ZY	---	number of active cycles
Z(p,T)	(p V)/(n R T)	real gas factor
χ_H	eq.(1)	degree of sorption
$\Delta...$	---	difference of ...
ε	$(V_S-V_M)/V_S$	voidage
η	$\underline{\underline{\tau}}=-\underline{\underline{\eta}}\cdot\!\cdot\mathrm{grad}\,\underline{w}$	dynamic viscosity; $N\, s\, m^{-1}$

λ	$\varphi_F=-\lambda\,\mathrm{grad}(T)$	heat conductivity; $W\, m^{-1}\, K^{-1}$
μ_i	$(\dfrac{\delta G}{\delta N_i})_{T,p,N_{j\neq i}}$	chemical potential; $W\, s\, mol^{-1}$
ρ	m/V	density; $kg\, m^{-3}$
σ	$|\underline{E}_{mech}|/A$	tensile stress; $N\, m^{-2}$, bar ($\sigma>0$ means pressure tension)
ψ	eq.(11)	friction factor

Indices

abs	absorbed
des	desorbed
e	end, termination, final condition
H	hydrogen
M	metal alloy, metal
mech	mechanical
MH	metal hydride
MH,2	metal dihydride
P	pipe, tube (V_P); particle, grain (d_P)
PS	pressure force sensor
R	reactor
S	storage tank (V_S); specimen, bulk, fixed bed (A_S, $\epsilon(V_S)$)
sol	soluted
tot	total
0	initial condition(σ_0,ϵ_0); reference values (p_0,T_0; fig.8: $\lambda_{eff,0},\chi_{H,0},\Delta p_0$);

1. INTRODUCTION

In the literature can be found most different equilibrium values for metal hydrides. For a less complicated proceeding the expression **absorbed** hydrogen shall conclude both ad- and absorbed hydrogen in the chapters following. The concentration-pressure-isotherms (CPI) of $FeTiH_x$ calculated by Wicke [1] and Pedersen [2] (ref. Fig.1) indicate a relative deviation of more than 100%. The concentration is determined as degree of sorption, which is defined by:

$$\chi_H = \frac{m_{H,abs}}{m_M} \quad (1).$$

The alloy, which was investigated by the author, also shows a similar behaviour (ref. fig.2). The specifications of the alloy DB5800^R are listed in table 1.

A merely different preparation of the sample could be excluded with sufficient security studying the authors´ description about the preparation of the test samples. By studying literature it seemed most likely, that an other property of the metal hydrides was the reason for the difference of the CPI´s. Especially the swelling of the metal hydrides during uptake of hydrogen [3-6] and external mechanical stress (pressure-tension) on the alloys during hydrogenation [1, 7-12] were found out to influence the equilibrium values of the hydrides.

Flanagan [7] introduced the influence of mechanical stress on the equilibrium data by adding a stress-dependent term to the common equation for the chemical potential of hydrogen μ_H during its transition of gaseous hydrogen into its metal soluted form:

$$\Delta\mu_H(\sigma>0) = \Delta\mu_H(\sigma=0) + \sigma\frac{V_H M_H}{m_M \chi_H}$$

$$= \Delta\mu_H^0(\sigma=0) + RT\ \ln(\frac{\chi_H}{\chi_{H,max}-\chi_H})$$

$$+ \sigma\frac{V_H M_H}{m_M \chi_H} \quad (2),$$

For verification of the influence of the mechanical stress on the equilibrium data of the metal hydride and in order to make mechanical stress and pressure drop within the bulk measureable a reactor was constructed.

2. APPARATUS

Measurements of the absorbed amount of hydrogen $\Delta n_{H,abs}$ were taken by means of pressure and temperature sensors, being part of an isochoric apparatus of known volume (fig.3). The readings were evaluated assuming that there is no leakage in the measurement equipment. Thus all the hydrogen vanishing out of the gas volume shall be absorbed by the specimen.

The reactor (fig.4) is equipped with a sensor for detection of pressure forces at its bottom face. The cylindrical metal hydride bed is radially enclosed by an extremely thick tubular wall in order to keep the volume of the bed constant in radial direction, even during the episodes of large mechanical pressure-tension in the specimen under hydrogen absorption. In axial direction the swelling of the bed is recorded by the pressure-force sensor, only at a minimum of axial bed-dilatation.

3. SPECIFICATIONS OF THE METAL HYDRIDE

For experiments the low-temparature metal hydride code DB5800^R was used. The alloy can be characterized by its superior ability to be activated and by a high uptake of hydrogen ($\chi_H>1\%$) at room temperature and behaves similar to $LaNiH_x$. The specifications of the metal hydride are listed in table 1.

4. MEASUREMENT PROCEDURES

The granulated metal alloy with a coarse initial grain size $d_{p,0} \approx 3mm$ was loosely poured into the ractor or was inserted pre-pressed without foreign matter (aluminium, copper, etc.). In case of a pre-pressed bulk, it was further compressed by the pre-stressing screw of the reactor. The alloy was cyclically exposed to hydrogen and evacuated at alternating low and high temperatures ($T_{low} \approx 5°C$; $T_{high} \approx 80°C$). The activation process was interrupted at the commencement of detectable absorption of hydrogen by the metal hydride and was followed by alternating absorption and desorption cycles until a constant degree of sorption χ_H was achieved. Below, the proceeding during absorption and desorption is explained. The measuring volume for both, absorption and desorption, consists of V_S, V_R and V_{P1}.

Absorption

- The amount of gaseous hydrogen at the beginning of the absorption $n_{abs,0}$ within the volumes V_S, V_R and V_{P1} is calculated by measuring the temperatures T_i and p_i ($i=P1,R,S$) and using the real gas equation (3). The hydrogen pressure in V_R has to be lower than the pressure in V_S and V_{P1}.

- Absorption is started by opening cock CR.

- After obtaining absorption equilibrium (i.e. T_i, p_i, σ=const.) the values for p_i, T_i and σ are registered and the amount of gaseous hydrogen at the end of the absorption $n_{abs,e}$ is calculated.

Desorption

- The proceeding for desorption is the same as for absorption. The only difference is,

that the hydrogen pressure in V_R has to be higher than the pressure in V_S and V_{P1}.

Pressure-Drop Measurements

Stationary pressure drop measurements were taken by seepage of hydrogen from the gas cylinder through V_{P1}, V_R, V_{P2}, the volumetric flow meter and the bypass BP of the vacuum pump VP. As the equilization time of the mechanical pressure-tension σ was about two times as long as the settling time of a stationary pressure drop, the stationarity of the mechanical pressure-tension was taken as criterion for taking the readings. The pressure drop of the aparatus without bulk was determined and subtracted from the readings of the measurements with the bulk in-situ. The pressure drop was determined for pure hydrogen and nitrogen (both $V_i/V_{tot} \geq 99,999\%$ with i=H,N) in order to compare the pressure drop of the loaded bulk (i.e. dilatated particles) with the unloaded bulk.

Different values for the mechanical pressure-tension at the end of absorption ($\sigma_{abs,e}$), desorption ($\sigma_{des,e}$) and pressure drop measurement could be obtained for the same metal temperature T_M and hydrogen pressure p_H by changing the initial compaction of the bed before inserting it into the sample holder and by changing the pre-tension of the pre-stressing screw (compare fig.4).

5. EVALUATION OF THE EXPERIMENTS

In order to determine the degree of sorption χ_H, the molar amount of hydrogen n_H in the gas volume of the apparatus was calculated at the beginning and at the end of the experiment using the real gas equation (3).

$$p \, V = n \, R \, T \, Z(p,T) \qquad (3),$$

whereby:

$$Z(p,T) = K_1 + K_2(T-T_0) + [K_3 - K_4(T-T_0)](p/p_0)$$

2603

and

$K_1 = 0,9979$, $K_2 = 1,628*10^{-5} K^{-1}$,
$K_3 = 6,6944*10^{-4}$, $K_4 = 1,7069*10^{-6} K^{-1}$,
$T_0 = 273,15 K$, $p_0 = 1$ bar

The difference of the molar amount of hydrogen in the gas volume $\Delta n_{H,gas}$ has been absorbed by the metal alloy.

$$\Delta n_{H,abs} = -\Delta n_{H,gas} \qquad (4).$$

The mass of the absorbed hydrogen in the metal hydride $m_{H,abs}$ is proportional to the difference of the molar amount of hydrogen $\Delta n_{H,abs}$:

$$m_{H,abs} = \Delta n_{H,abs} M_H \qquad (5).$$

The degree of sorption χ_H is determined by inserting equation (5) into equation (1).

For determination of the mechanical pressure-tensions the simplyfying assumption was made, that the pressure tension in the fixed bed shall behave isotropic, i.e. $\sigma \neq \sigma(x,y,z)$, and that the mechanical tension σ can be calculated from the measured force F_{PS} and from the cross section $A_{S,C}$ of the fixed bed:

$$\sigma = \frac{F_{PS}}{A_{S,C}} \qquad (6).$$

The stationary pressure drop was evaluated by using the Ergun equation (ref. Ergun [13]):

$$\frac{\Delta p}{\Delta z} = 150 \frac{(1-\varepsilon)^2}{\varepsilon^3} \frac{\eta_g}{d_p^2} \bar{w} + 1.75 \frac{(1-\varepsilon)}{\varepsilon^3} \frac{\rho_g}{d_p} \bar{w}^2 \quad (7)$$

The gas data η_g and ρ_g were calculated using the temperature T_R and mean reactor pressure

$$\bar{p}_R = \frac{(p_1 + p_2)}{2} \qquad (8).$$

The initial voidage ε_0 (i.e. with metal unloaded) was determined by measuring the outer diameter d_{bulk} and height h_{bulk} of the bulk already inserted into the reactor and by determination of the bulk´s weight using the equation:

$$\varepsilon_0 = \frac{V_S - V_M}{V_S} \qquad (9)$$

whereby:

$$V_S = \pi \left(\frac{d_{bulk}}{2}\right)^2 h_{bulk} \text{ and } V_M = \frac{m_M}{\rho_M}$$

The Ergun equation (7) can also be written as an implicit equation for the voidage ε:

$$\varepsilon^3 + K_1 \varepsilon^2 + K_2 \varepsilon + K_3 = 0 \qquad (10)$$

with:

$K_1 = -(B/A)$ $A = (\Delta p/\Delta h_{bulk})$
$K_2 = (2B+C)/A$ $B = 150 (\eta \bar{w}) / d_p^2$
$K_3 = -(B+C)/A$ $C = 1,75 (\rho_{gas} \bar{w}) / d_p$

The voidage ε can be determined from eq.(10) by using the cardan formula for cubic equations (ref. Bronstein [14], p.183 ff.), if the mean diameter of the bulk´s granular particles $\bar{d}_p$ is known. The mean particle diameter can be determined from a particle analysis after taking out the bulk from the reactor or during pressure drop measurements by a method of successive approximation of equation (10). If the bulk is fully activated, the particles stop to disintegratet to smaller size. Thus the mean particle diameter from external particle analysis has to equal to the diameter, which is determined from eq.(10).

The value of the initial voidage ε_0 determined from the bulk´s external dimensions and weight has to be equal to the voidage, which was determined from pressure drop measurements with nitrogen.

The difference between the voidage of the unloaded bulk ε_0 and the voidage of the bulk, which had absorbed hydrogen ε, could only be determined by pressure drop measurements.

6. MEASURED EQUILIBRIUM DATA

With a number of test beds the results displayed below could be reproduced within the employed aparatus. The same behaviour of the degree of sorption χ_H versus mechanical pressure tension σ could also be verified with other reactors by variation of the bulk's geometry, though the tensile stress in the bulk of those reactors could not be measured.

The presentation of equilibrium data by means of **concentration-pressure-isotherms** (**CPI**) was expanded by introducing the additional parameter **mechanical tension**.

Due to the simple construction of the reactor it was not possible to change the pre-tension (pressure tension) of the test bed after the metal hydride had been activated. Hence, exactly equal tensions could repeatedly be measured only by accident. It was necessary, therefore, to form classes of most similar mechanical tensions in order to present the equilibrium data. Below, those curves of most similar mechanical tension shall be refered to as **isotenses**, notwithstanding the restrictions outlined above.

Caused mostly by the activation of the metal hydride the structure of the test bed is changed due to grain disintegration with an initial grain size $d_{P,0} \approx 3$mm and a final grain size $d_{P,e} = 1..100\mu$m, depending on the initial pressure tension σ_0 and the initial voidage ε_0 of the unloaded bulk. The grain disintegration causes also a reduction of the initial pre-tension of the unloaded bulk. For this particular reactor, the metal alloy

investigated and the pre-tensions exerted on the bulk, the equilibrium data turned out to become reproducable at cycle numbers ZY>20 (only the cycles after the first detectable hydrogen uptake were counted). The following figures are based on equilibrium data which were collected from different test samples of the alloy DB5800[R] for cycles numbering ZY>20.

Isothermal isotenses, i.e. curves of constant metal temperature T_M and constant mechanical tension σ, are shown in figure 5 (refer to the restrictions of the term **isotenses** as discussed above).

The relationship between the mechanical pressure-tension σ and the degree of sorption χ_H is clearly indicated: An increase in the external mechanical pressure-tension will cause a decrease in the degree of sorption, if all other conditions of experiment stay unchanged.

The same measurements can also be presented as **isobaric isotherms** (fig.6) demonstrating the dependency of the degree of sorption χ_H from the mechanical pressure-tension σ even more apparent. The data points scatter about the smooth equalization curve because the isobaric isotherms are not derived from equilibrium data of equal hydrogen pressure but instead from data of mostly equal hydrogen pressure.

7. MEASURED PRESSURE DROP

The most common kinds of displaying pressure drop measurements are:

- pressure drop (Δp or $K^*\Delta p$) over mean fluid velocity $\bar{w}$ of the empty reactor
- dimensionless friction factor ψ over Reynolds number Re
- dimensionless bulk friction factor ψ_ε over bulk Reynolds number Re_ε.

2605

The friction factor ψ and the bulk friction factor ψ_ε are defined as:

$$\psi := \frac{\overline{\Delta p}}{\Delta h_{bulk}} \frac{d_p}{\rho_{gas} \overline{w}^2} \tag{11}$$

$$\psi_\varepsilon := \frac{\overline{\Delta p}}{\Delta h_{bulk}} \frac{d_p}{\rho_{gas} \overline{w}^2} \frac{\varepsilon^3}{(1-\varepsilon)} \tag{12}$$

and the Reynolds number Re and the bulk Reynolds number Re_ε are determined by:

$$Re := \frac{d_p \overline{w}}{\nu} \tag{13}$$

$$Re_\varepsilon := \frac{d_p \overline{w}}{\nu} \frac{1}{(1-\varepsilon)} \tag{14}.$$

After activation the bulk consists of a mixture of different sized particles.

Therefore a mean particle diameter $\overline{d}_P$ is introduced instead of the diameter d_P and is calculated from a mixture of i classes of granules each with a mean class diameter $d_{P,i}$ and a volume of the particles within this class $V_{P,i}$ by:

$$\overline{d}_p := \frac{1}{\frac{V_{p,1}}{V_{p,ges}d_{p,1}} + \frac{V_{p,2}}{V_{p,ges}d_{p,2}} + .. + \frac{V_{p,i}}{V_{p,ges}d_{p,i}}} \tag{15}$$

If the pressure drop measurements are displayed, it can be noticed that pressure drop measurements with $\psi_\varepsilon/Re_\varepsilon$-axes nicely agree with the Ergun equation. Though, the plotting within ψ/Re-axes more distinctly shows the increase of the pressure drop because of absorption, which is accompanied by a volume-dilatation of the particles. Nevertheless, for linearization of measurements with different gases and

voidages the $\psi_\varepsilon/Re_\varepsilon$-diagram is much more convenient.

In figure 7 the pressure drop of nitrogen and hydrogen in a pre-pressed bulk with an initial voidage $\varepsilon_0 = 9,66\%$ is displayed.

8. INFLUENCE OF THE VOIDAGE ON THE SORPTION RATE

During ab- and desorption it can be distinguished between three rate-limiting steps, which influence the time for reaching equilibrium conditions:

- **transport** of hydrogen from the gas phase to the particle (pressure drop)
- **conversion** of hydrogen from gas phase to the ad- and absorbed phase (adsorption, dissociation, solution and reaction)
- **heat exchange** of the bulk with the thermostat bath (heat conduction)

In context with the rate limiting steps the **absoption equilibrium** of the metal hydride (degree of sorption) should be considered, which controls the total amount of hydrogen, that can be absorbed.

The conversion of hydrogen seems to be important for the investigated metal hydrid only before and during activation, i.e. as long as the oxide layers on the metal surface prevent the hydrogen from entering the alloy. Since after activation an influence of the conversion on the ab- and desorption was not detectable, it will not be considered in the discussion below.

For the fictive case, in which the absorption of the fully activated metal hydride is controlled only by one of the three rate-limiting steps, the dependence of the relative pressure drop ($\Delta p/\Delta p_0$), of the relative effective heat conduction ($\lambda_{eff}/\lambda_{eff,0}$) and of the relative degree of sorption ($\chi_H/\chi_{H,0}$) from the voidage are displayed in figure 8.

The quantities of figure 8 are defined as follows:

- $\Delta p := (p_1 - p_2)$
stationary pressure drop of the metal hydride bulk only (pressure drop of apparatus neglected)

- $\lambda_{eff} := \varepsilon_0 \, \lambda_H + (1-\varepsilon_0) \, \lambda_{MH}$
volumetric setup for the heat conduction, employing the data of table 1 as values for λ, with

$$\lambda_H = \lambda(\bar{p}_R, T_M)$$

- χ_H ref. to eq.(1)
degree of sorption, determined from equilibrium measurements, with

$$\chi_H = \chi_H(\bar{p}_R, T_M)$$

The quantities Δp_0, $\lambda_{eff,0}$ and $\chi_{H,0}$ are introduced as reference values in order to avoid a confusing number of units on the axes.

CONCLUSION

External mechanical pressure-tensions σ significantly influence the degree of sorption χ_H of hydrogen in metal hydrides. Mechanical pressure-tensions of $\sigma \approx 1000$bar reduce the uptake of hydrogen at equilibrium conditions at as much as ($\chi_{H,\sigma=1000bar}$ / $\chi_{H,\sigma=0bar}$) $\approx 50\%$ in comparison to a not pre-stressed bulk.

Pressure drop measurements show that the absorption of hydrogen is accompanied by a considerable volume-dilatation of the particles, thus causing an increase of the mechanical tensions within the bulk. Optimum values for the construction of metal hydride containers can be derived from measurements of the bulk's pressure drop, equilibrium data and heat conduction.

Any further investigation of equilibrium data of metal hydrides should account for the external mechanical tension exerted on the bulk by the reactor walls. It also can be expected, that a better correlation of the deviating equilibrium data of various investigators can be provided, if mechanical tension will enter consideration during research on metal hydrides.

REFERENCES

[1] Wicke, E.; Chemie-Ingenieur-Technik 54 No.1 (1982) p.41 ff.

[2] Pedersen, A.S.; metallurgi-afdelingen, DK-4000 Roskilde, Risoe Kem. Lab. IV, Koebenhavns Universitetet (1981)

[3] Brose, J.; Dissertation Universität Dortmund (W.-Germany) (1985)

[4] Matsumoto, T.; Journal of the Less-Common Metals 88 (1982) p.443 ff.

[5] Mayer, H.W.; Journal of the Less-Common Metals 88 (1982) L7 ff.

[6] Peisl, H.; in "Hydrogen in Metals I", ed. by G.Alefeld and J.Völkl, Progress in Applied Physics 28 Berlin (1978) p.53 ff.

[7] Flanagan, T.B.; in "Metal Hydrides", ed. by G. Bambakidis, New York (1981) p.361 ff.

[8] Magerl, A.; Physica Status Solidi / A (Appl. Res.) 36 (1976) p.161 ff.

[9] Riecke, E.; Zeitschrift für Metallkunde 75 (1984) p.76 ff.

[10] Schmidt-Ihn,E.; Daimler-Benz AG, D-7000 Stuttgart, Zentrale Forschung (1984) Jahresbericht EHC-44-015-D

[11] Schober, T.; Journal of the Less-Common Metals 89 (1983) p.63 ff.

[12] Wicke, E.; Zeitschrift für Physikalische Chemie Neue Folge 31 (1962) p.222 ff.

[13] Ergun, S.; Fluid Flow Through Packed Columns, Chemical Engineering Progress 48 (1952) S.89/94

[14] Bronstein, I.N., Semendjajew, K.A.; Taschenbuch der Mathematik (17.Auflage), Zürich, Frankfurt/Main, Thun (1977)

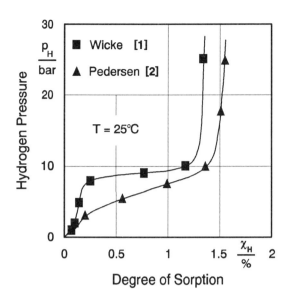

fig.1 Absorption isotherms for FeTiH$_x$ published by different authors; effects of mechanical tension not taken into account

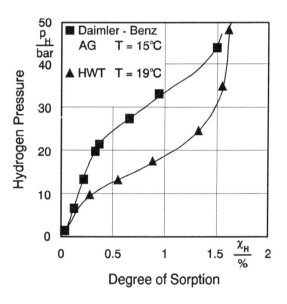

fig.2 Absorption isotherms of the metal alloy DB5800^R (ref. table 1), measured by different authors; effect of mechanical tension not taken into account

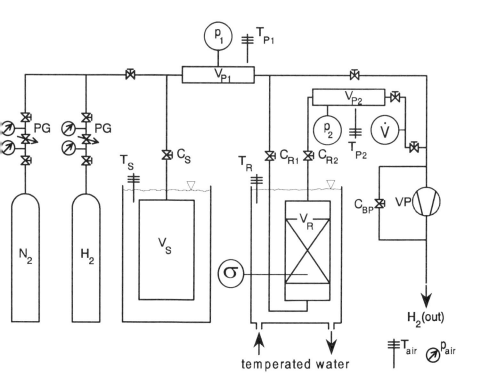

fig.3 **Isochoric apparatus** for measuring equilibrium data (C_i= cock i; H_2= gas cylinder hydrogen; N_2= gas cylinder nitrogen; p_i= pressure sensor i; PG= pressure gauge; T_i=thermometer i; TH= thermostat; V_i=volume i; $\dot{V}$= volumetric flow meter; VP= vacuum pump; σ= sensor for pressure-tension; *Subscripts*: BP= Bypass; P_i= pipe i; R= reactor; S= storage tank)

2609

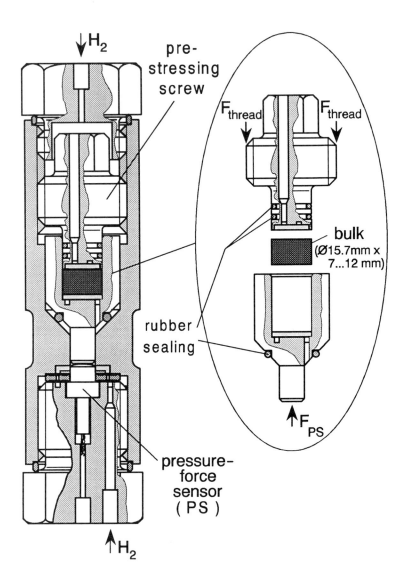

fig.4 **Reactor** for measurement of mechanical pressure-tension in the metal hydride bed during hydrogenation (▨ = copper seal ■ = metal hydride ▢ = metal filter ▨ = stainless steel)

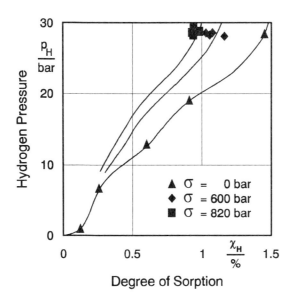

fig.5　Isothermal isotenses of the examined low-temperature metal hydride DB5800[R]

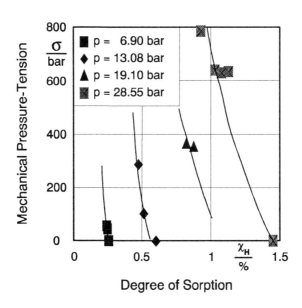

fig.6　Isobaric isotenses of the tested low-temperature metal hydride DB5800[R]

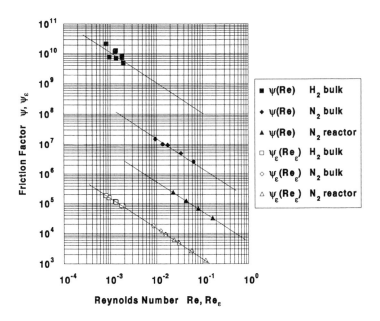

fig.7 **Pressure drop** of the empty reactor and of hydrogen and nitrogen within a pre-pressed bulk of DB5800^R plotted in both ψ/Re- and ψ_ε/Re$_\varepsilon$-axes

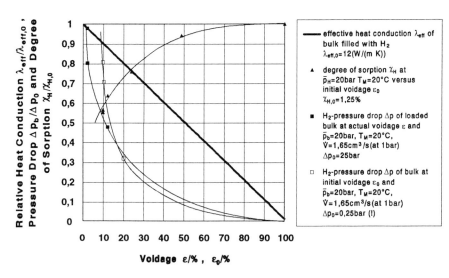

fig.8 Dependence of the relative **pressure drop** ($\Delta p/\Delta p_0$), of the relative **effective heat conduction** ($\lambda_{eff}/\lambda_{eff,0}$) and of the relative **degree of sorption** ($\chi_H/\chi_{H,0}$) from the **voidage** of the bulk

metal alloy: code DB5800[R]	$Ti_{0.98}Zr_{0.02}V_{0.43}Fe_{0.09}Cr_{0.05}Mn_{1.5}$		
metal density (unloaded)	ρ_M	6300	kg/m^3
metal hydride density (loaded*, $\sigma \approx 0bar$)	$\rho_{MH,max}$	5000	kg/m^3
mean molar mass of metal	M_M	0,1607	kg/mol
spec. heat capacity of metal (unloaded)	C_M	540	$J/(kg\ K)$
spec. heat capacity of metal hydride (loaded*)	$C_{MH,max}$	870	$J/(kg\ K)$
heat conduction of metal	λ_M	12	$W/(m\ K)$
temperature conduction of metal	a_M	$3,53*10^{-6}$	m^2/s
heat conduction of poured bed ($\varepsilon_{bulk} \approx 50\%$)	λ_{bulk}	0,5...0,8	$W/(m\ K)$
temperature conduction of poured bed ($p_H \approx 30bar; \varepsilon_{bulk} \approx 40\%$)	a_{bulk}	$0,29*10^{-6}$	m^2/s
heat conduction of hydrogen ($p_H = 20bar$, $T_M = 20°C$)	λ_H	$\approx 0,2$	$W/(m\ K)$
spec. surface of granular metal particles in original form (not yet activated)	$A_{M,0}/m_{M,0}$	$\approx 0,005$	m^2/g
spec. surface of fully activated metal particles after oxidation (unloaded)	$A_{M,a}/m_{M,a}$	≈ 5	m^2/g
enthalpy of formation of dihydride (exothermal reaction)	h_{MH2}	≈ -25	kJ/mol
maximum degree of sorption (loaded*, $\sigma \approx 0bar$)	$\chi_{H,max}$	1,7	$\%$
total concentration of hydrogen in metal (loaded*)	$c_{H,M,max}$	≈ 94	kg/m^3
[R] = pat. pend. by Daimler-Benz AG, FRG			
* = at: $p_H = 30bar$, $T_M = 20°C$, $\varepsilon_S \approx 45\%$			

tab.1 Specifications of the examined low-temperature metal hydride code DB5800[R]

THERMOELECTRIC GENERATION, A POTENTIAL SOURCE OF LARGE SCALE
ELECTRICAL POWER

D.M. ROWE

School of Electrical, Electronics and Systems Engineering
University of Wales College of Cardiff
Cardiff, CF1 3YH.

ABSTRACT

Thermoelectric generators have no moving parts, are silent, reliable and able to operate unattended in hostile environments. Generally they are employed in specialised applications where combinations of their desirable properties outweigh their relatively low conversion efficiency and high initial cost. Using a variety of high temperature heat sources they have been used to provide electricity over a power range which extends from the millimicrowatts produced in miniature integrated type thermoelectric circuits to approaching a megawatt in the nuclear reactor powered thermoelectric generator employed in the NASA SP-100 space project. A recent Watt Committee on Energy Report identified thermoelectric generation as a possible method for converting large amounts of low temperature waste heat into electrical power. This paper reviews the present status of "low temperature" thermoelectric generating technology and discusses its potential as a large scale source of electrical power.

The basic principles of thermoelectric generation and the factors which determine the conversion efficiency of this process are outlined. Materials suitable for use in the fabrication of thermoelectric modules intended for operating over a temperature range from ambient up to around 420K are identified and an estimate made of the conversion efficiency of thermoelectric devices fabricated from the best available materials. A number of sources of low temperature "waste heat" suitable for thermoelectric conversion are discussed. The design of a large scale "low temperature" thermoelectric generator is outlined and suggestions made for improving the performance of the thermoelectric modules. Finally the economics of this method of generating electrical power are examined. It is concluded that provided the constructional engineering problems associated with building a multimegawatt thermoelectric system can be overcome, electricity can be generated at a price which is competitive with current conventional power stations.

INTRODUCTION

Most people are familiar with the use of metal/metal alloy thermocouples to indicate temperature, but are probably less familiar with the use of modern semiconductor thermocouples to convert heat into electrical energy.

The principles of thermoelectric conversion are well understood and there are a number of modern texts on the subject (Rowe and Bhandari 1983, Goldsmid 1988). The Seebeck effect is the phenomenon underlying the conversion of heat energy into electrical power and is explained with reference to Figure 1. Here, a single thermocouple circuit is formed from

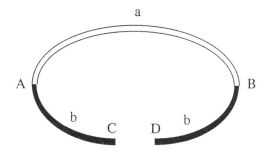

Fig. 1. Basic thermocouple.

two dissimilar conductors a and b. When the junctions A and B are maintained at a temperature difference ΔT an open circuit potential difference V is developed between C and D. The differential Seebeck coefficient $\alpha_{ab} = V/\Delta T$. If a load is connected across C and D, the heat energy supplied at the hot junction drives a current through the circuit and delivers power to the load.

A complementary effect, the Peltier effect, is the phenomenon utilised in thermoelectric refrigeration. If an emf is applied across C and D which drives a current I around the circuit there will be a rate of heating q at one junction between a and b and cooling −q at the other depending on the direction of current flow. The differential Peltier coefficient $\pi_{ab} = q/I$.

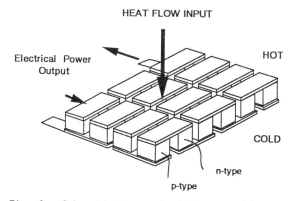

Fig. 2. Schematic thermoelectric generating module.

In a practical thermoelectric convertor (refrigerating and generating) a large number of thermocouples are connected electrically in series to form a module as shown schematically in Figure 2. Thermoelectric generators find use in a number of specialised military, medical and space applications. Heat input is obtained from a variety of sources; fossil fuel, isotopic, waste heat and nuclear reactors. Electrical power output extends over 16 orders of magnitude, from the millimicrowatts produced in miniature integrated type thermoelectric circuits (Rowe et al 1989) to approaching a megawatt in the nuclear reactor powered thermoelectric generators employed in the NASA SP-100 space project (Mondt 1990).

Thermoelectric refrigeration has attracted more general commercial interest and this technology has found considerable use in the electronics industry in applications where components require a stable, cool thermal environment. In order to meet an expanding market a wide variety of Peltier modules are readily available off the shelf. In contrast modules for thermoelectric generation are usually designed and assembled to meet the requirements of a specific application and are an integral part of a generating system. Consequently generating modules are of very limited availability and those which are, are designed for high temperature operation. Fortunately, Peltier modules can operate reliably in the generating mode at temperatures up to about 412K and are fabricated from semiconductor materials which are the best available for operation over this low temperature range.

In this paper the present status of 'low temperature' thermoelectric generating technology is reviewed and its potential as a large scale source of electrical power discussed.

GENERAL PRINCIPLES

The thermoelectric conversion efficiency ϕ is given by the ratio of the electric energy supplied to the load to the heat energy absorbed at the hot junction. It can be shown that the maximum conversion efficiency can be expressed as

$$\phi_{max} = \eta\gamma \text{ where } \eta = T_{hot}-T_{cold}/T_{hot}$$

$$\gamma = \frac{(1+Z_c \, T)^{\frac{1}{2}} - 1}{(1 + Z_cT)^{\frac{1}{2}} + T_{cold}/T_{hot}} \qquad \text{with } T = \frac{T_{hot} - T_{cold}}{2}$$

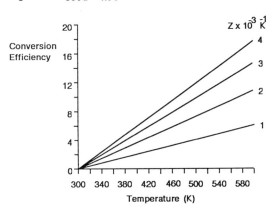

Fig. 3. Thermoelectric conversion efficiency as a function of hot junction temperature

The first term η in the efficiency equation is the thermodynamic efficiency for an ideal reversible engine (Carnot efficiency) with the second term γ incorporating the irreversible heat losses. Evidently, the conversion efficiency depends upon the temperature difference over which the device is operated, its average temperature of operation, and the "goodness factor" or so called figure of merit "Z_c" of the thermocouple material $Z = \alpha^2 \sigma / \lambda$ where α is the Seebeck coefficient, σ the electrical conductivity and λ is the thermal conductivity. Evidently an increase in conversion efficiency can be achieved by increasing the figure of merit and operating at higher temperatures. In Figure 3 is plotted the thermoelectric conversion efficiency as a function of hot junction temperature for different values of the figure of merit. A material with a figure of merit of $3 \times 10^{-3} K^{-1}$ when operating with a hot junction of 373K and a cold junction at ambient would have an efficiency of around 4%.

THERMOCOUPLE MATERIAL

Solid state theory provides an insight into materials likely to be good thermoelectric candidates. The parameters which occur in the figure of merit are dependent on carrier concentration. Metals have a high electrical conductivity, but the Seebeck Coefficient is too low.

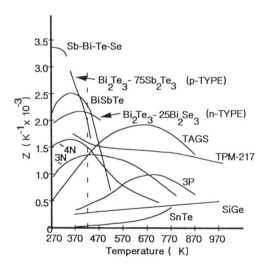

Fig. 4.   Figure of merit of thermoelectric semiconductors.

Insulators can have high Seebeck coefficients but are electrically resistive. A compromise is reached in heavily doped semiconductors (carrier concentrations $10^{19} cm^{-3}$–$10^{20} cm^{-3}$). In Figure 4 is displayed the figure of merit of established thermoelectric semiconductors. The materials conveniently fall into three groups, depending on their temperature range of operation. Over the temperature range of interest, viz from room temperature to about 423K alloy/compounds based on the bismuth chalcogenides exhibit the highest figures of merit.

$Bi_2Te_3 - 75Sb_2Te_3$ is the best p-type material and $Bi_2Te_3 - 25Bi_2Se_3$ the best n-type with an average figure of merit over the temperature range of interest in excess of $2.5 \times 10^{-3}K^{-1}$. These materials are capable of converting at maximum about 20% of the available heat energy, (Carnot fraction) into electrical energy, when operating at a hot junction temperature of 373K and the cold junction at 293K, the Carnot efficiency is 27%; this gives a conversion efficiency in excess of 5%.

SOURCES OF LOW TEMPERATURE HEAT

Enormous quantities of low temperature heat are available on Earth from a variety of sources which are not confined to industrial nations. This heat which is currently wasted or under utilised, can be thermoelectrically converted into electrical power.

Geothermal energy

This is the Earth's natural heat, which is derived mainly from the radioactive decay of long lived isotopes of uranium, thorium and potassium. Geothermal heat sources vary considerably in quality and accessibility and are generally classified into those with temperatures ranging from 423K to 473K (high enthalpy) and those sources at a temperature of less than 423K (low enthalpy). Generally high enthalphy geothermal energy sources, in which naturally occurring steam or hot water are available have been used directly with conventional steam turbines since the turn of the century. Steam turbines typically operate at an inlet temperature of about 453K and an outlet temperature of around 323K and appreciable amounts of electrical power have been generated by this method (Di Pippo, 1984). Techniques have also been developed to extract heat from hot dry rocks (HDR system), where there is insufficient natural fluid available to transfer the heat to the surface. This concept has been put into practice notably by the Los Alamos National laboratory in the USA and by the Camborne School of Mines, Cornwall (Downing and Grey, 1986). In the UK a study has also been undertaken to assess marine based geothermal operations and in particular to convert redundant oil platforms into geothermal power stations (Turner, 1989).

Ocean thermal energy

Temperature gradients of 20-30K over a depth of a few hundred metres exist in a number of tropical and subtropical regions of the oceans (Brandstetter, 1983). Over a twenty year period attempts have been made in the United States (Lewis et al, 1987), Japan (Tokyo, 1981) and the United Kingdom, (Laughton, 1990) to develop Ocean Thermal Energy Conversion (OTEC) systems which utilise the differences in temperature to drive a low temperature heat engine.

Solar ponds

These have been employed in many parts of the world as energy storing devices and as a means of distilling saline water (Garge, 1985). A modification of the solar pond is the so-called gradient stabilised solar pond. In this modification a non-convecting density gradient is established in the intermediate levels of the pond which effectively suppresses heat transfer to the upper regions and consequently reduces radiative heat losses. Solar energy is collected with an efficiency of up to 30% and heats up the pond to temperatures as high as 363K. In locations where the average insolation is $200W/m^2$ about 50 Watts of thermal energy/m^2 can be extracted from the pond.

Waste heat

It has been estimated that 30% of the total energy consumed in the United States industrial sector is wasted. Modern electricity generating installations operate at best with an efficiency of around 40%. A considerable proportion of the waste energy is used in heating the cooling water, the temperature of which in some installations approaches 308K. Higher temperature sources of heat are associated with other industrial operations such as steel making. The amount of waste heat from a typical steel making plant is 100MW, with large volumes of water associated with blast furnace operation at temperatures in excess of 370K).

LOW TEMPERATURE THERMOELECTRIC GENERATOR

Of the sources of heat identified above ocean thermal gradients exhibit the smallest difference in operating temperature and consequently the lowest Carnot efficiency. Nevertheless OTEC systems have been developed which employ heat engines using low boiling point organic liquids such as ammonia or freon and driven by the 20 – 25K temperature difference between the cold deep water and the warm surface water encountered in a number of tropical seas. An alternative generating system that has been considered for OTEC is one based upon thermoelectrics. Considerable progress has been made in developing this technology and a laboratory system has been constructed and operated for several years which generates 50 watts (e) when the temperature difference between the hot inlet and cold outlet is 25K (Matsuura, 1982). This technology is being extended to operate at higher efficiencies at the considerably greater temperature differences associated

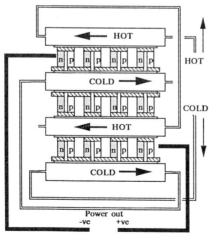

Fig. 5. Schematic low temperature generator (Matsuura *et al*, 1991).

with water from redundant North Sea oil wells and blast furnace operations (Matsuura *et al*, 1991). A schematic of a low temperature thermoelectric generating system is shown in Figure 5. Basically it consists of a number of parallel plate heat exchanges with the thermoelectric generator

2619

sandwiched between the hot and cold water flow channels. Assuming a temperature difference of 85K and a thermoelectric conversion efficiency of 3% a hot water throughput of 10^5 litres a minute, typical for a North Sea oil platform, would provide several megawatts of electrical power.

MODULE IMPROVEMENTS

In large scale generation the thermoelectric modules are likely to be the major capital cost item. Currently attempts are being made to improve the competitiveness of a thermoelectric conversion system by (1) increasing the figure of merit of the thermocouple material, (2) reducing the material's cost and (3) improving the power output per module surface area.

Increasing the figure of merit.

Attempts to increase the figure of merit have concentrated on reducing the thermal conductivity. It has been established that the thermal conductivity of high temperature thermoelectric semiconductor alloys can be reduced by employing very small grain size compacted materials: the small grains increase the phonon-boundary scattering. This has been demonstrated in silicon germanium alloys (Rowe *et al*, 1981) and in materials based on lead telluride (Rowe and Clee, 1987). It is uncertain that this technique can reduce the thermal conductivity of low temperature thermoelectric materials such as those based on bismuth telluride as the thermal conductivity is already very low. However, there is a suggestion that this phenomenom has been observed in thin films of bismuth telluride, (Boikov, 1978). Bulk material presents a problem as bismuth telluride type materials, unlike silicon-germanium and lead telluride, exhibit anisotropic transport properties. Consequently, Z is maximised along a particular direction. Isostatically pressed and sintered material exhibits complete anisotropy and consequently a degraded figure of merit. A method has been developed for preparing compacted material with a preferred grain orientation (Situmorang, 1988) but to date its application has been restricted to powders with grains too large to exhibit phonon-grain boundary scattering.

Material Production

Thermoelectric semiconductor materials are expensive partially due to the cost of the constituent elements and partially due to processing/ manufacturing methods employed. Conventionally thermoelectric elements are made from semiconductor material which been has 'grown from the melt' or cast, ground and compacted by metallurgical methods. Two methods which appear to offer savings in energy and cost are the so called PIES method and electrolytic deposition from aqueous solutions.

In the PIES method the constituents of the alloys are subjected to a high energy ball milling procedure. Alloying occurs near room temperature through solid state diffusion (Schwarz and Koch, 1986). This technique offers a cost saving alternative to the conventional methods of alloy preparation. Materials based on silicon-germanium alloys have been prepared by this method (Cook *et al*, 1990). Bismuth telluride type materials have also been prepared and initial results are encouraging (Ohta *et al*, 1990).

Electrodeposition readily lends itself to the economic production of 'area films' of desired thickness and compostion. The application of electrodeposition techniques in the production of thin film semiconductor alloys for photovoltaic materials has been extensively studied. This preparation technique has been extended to other applications such as high temperature superconductors and magnetic material. Recently the possibility of preparing thermoelectric materials using this method has been reported, (Muraki and Rowe, 1990, 1991). The technique has an additional attraction in that the carrier concentration can be controlled by altering the stoichiometry of the deposited alloy. This method of preparation is being extended to materials based upon the bismuth chalcogenides.

OPTIMISATION OF GEOMETRY

As indicated previously commercially available modules have been designed for use in the Peltier mode, as refrigerating devices. However it has been reported that these devices will operate at the highest efficiencies that thermoelectrics can currently attain at temperatures up to 500K. This is because the thermoelements are fabricated from materials which possess their highest figures of merit over this temperature range of operation (Burke and Buist, 1984). However, whether the thermoelectric convertor is used as a generator or a refrigerator, the geometry of commercially available modules has been optimised for the device to operate at maximum efficiency or to attain the greatest temperature difference.

In thermoelectric generation using 'waste heat' conversion efficiency is not a main consideration. An increase of up to 75% in power output per unit area can be obtained by optimising the thermocouple geometry. Unexpectedly this increase in power output is not accompanied by a drastic reduction in conversion efficiency and is typically less than 10%, (Min and Rowe, 1991).

ECONOMICS

In general the cost of generating electrical power depends upon a number of factors and can be simplistically expressed as (Benson and Jayadev, 1980).

$$\text{Cost } (\$/\text{Wh}) = \frac{\text{Cost of Fuel}}{\text{Conversion efficiency}} + $$

$$\frac{\text{Capital investment}}{\text{Hrs per yr of operation}} \times \text{FCR} + \text{O and M}$$

It is apparent from the first term that when the cost of fuel is negligible or very low, as is the situation being considered in this paper, conversion efficiency is not a major consideration. Consequently there is no cost advantage in attempting to improve the figure of merit of the thermoelectric materials used in fabricating the thermoelectric modules; although improvements in efficiency will be a factor in determining the final size of the generating system.

The third term includes fixed charge rate (FCR) which incorporates the cost of realising the initial capital investment and the charges such as tax and operational and maintenance costs (O and M) expressed as a

fraction of the investment capital per year. An estimate of this cost
contribution can be obtained by assuming a 10% interest rate and 5% for
other charges; the fixed charge rate over a 20 year operating life is
16.7% per year.

It is evident that in this proposal, capital investment and the operating
lifetime of the conversion system are the dominant factors. Ongoing
device lifetime studies confirm that availablle Peltier modules can
function as reliable thermoelectric generators. Consequently a reduction
in device costs is a main objective. Thermoelectric Peltier modules
(coolers) are available which when operated in the generating mode,
produce about 1 W(e) when a temperature difference of about 80K is
maintained across the device. The module costs at present around $20. If
ordered in sufficient numbers the cost may be reduced to about $10. The
calculated cost of generating electrical power using available
thermoelectric modules is displayed in Figure 6. Over a 20 year operating

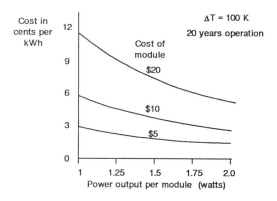

Fig. 6. Cost of generating electrical power.

period the cost of generating power from a single module is less than 6
cents per kWh. A module with optimised geometry for power generation (Min
and Rowe, 1991) generates 1.75 W(e) and reduces the cost to 3.5 cents per
kWh. The modules will be the most costly item in any large scale
generating system. Consequently doubling the cost per unit when extending
to large scale generation would overestimate the cost. Nevertheless, the
total thermoelectric generating cost of around 7 cents per kWh still
compares very favourably with the price of between 5 and 7 cents paid for
electricity generated by conventional methods, (Crook, 1992).

CONCLUSION

Thermoelectric generation does offer a potential for converting low
temperture (waste heat) into electricity and a large amount of suitable
waste heat is available worldwide from a variety of sources. This
technology has been successfully demonstrated on a laboratory scale. A
simplistic economic analysis indicates that provided the engineering
constructional problems of building a large-scale thermoelectric
generating system can be overcome, electrical power can be generated at a
cost which is competitive with existing utilities.

REFERENCES

Benson, D.K., and Jayadev, T.S., (1980), Thermoelectric energy conversion Proc. Third Int. Thermoelectric Conference, Arlington, Texas, (K. R. Rao, Ed.), 27-56.

Boikov, Yu, A., Gol'tsman, M.B., and Kutasov, V.A., (1978), Sov. Phys. Solid State, 20, 757.

Brandsetter, A., (1974), The hot deeps of the Red Sea as a potential heat source for thermoelectric power generation, Proc. 9th IECEC., San Francisco, Cal., 277-80.

Burke, E.J., and Buist, R.J., 1984, Thermoelectric coolers as power genertors, Proc. Fifth Int. Thermoelectric Conference, Arlington, Texas, (K.R. Rao, Ed.), 91-94.

Cook, B.A., Beaudry, B.J., Harringa, J.L., and Barnett, W.J., (1990), Mechanical alloying as an alternative method of producing n-type $Si_{80}Ge_{20}$ thermoelectric materials, Proc IX Int. Conf. on Thermoelectrics, JPL Labs., USA, (C. Vining Ed), 234-241.

Crook, J., (1992), Some aspects of low grade heat, to be published in Power Generation Technology, Sterling Publications Int. Ltd., London.

Di Pippo, R., (1984), Worldwide Geothermal Power Development Bulletin, Geothermal Research Council, 13, 4.

Downing, D and Gray, A., (1986), British Geological Survey, ISBN, 0 11 8843664.

Garg, H.P., (1985), Solar ponds as an energy storage device, Proc. Workshop on the Physics of Nonconventional Energy Sources and Materials Science for Energy, Trieste, Italy, H4SMR/156-12.

Goldsmid, H.J., (1986), Electronic refrigeration, Pion Ltd., London.

Lennard, D.E., (1990), Ocean thermal energy conversion in Renewable Energy Sources, Watt Committee on Energy Report, (M. A. Laughton, Ed.), 22, 75-86.

Lewis, L.F., Trimble, L., and Bowers, J., (1987), Open-cycle OTEC seawater experiments in Hawaii, Proc. Oceans 87 - The ocean an international workplace - 12th Annual Conf., IEEE Marine Technology Soc. 2, 397-402.

Matsuura, K., Honda, T. and Kinoshita, H., (1982), Tech. Reports of Osaka University, 33, 59-62.

Matsuura, K., Rowe, D.M., Tsuyoshi, A.S., and Min, G., (1991), Large scale generation of low grade heat, Proc. X International Thermoelectric Conference, University of Wales, Cardiff, (D. M. Rowe, Ed.), Babrow Press, ISBN 095129286, 0 0 233-241.

Muraki, M., and Rowe, D.M., (1990), On the possibility of preparing thermoelectric semiconductor films from aqueous solutions, Proc. IX Int. Conf. on Thermoelectrics, JPL Labs., USA, (C. Vining Ed.), 62-75.

Muraki, M., and Rowe, D.M., (1991), Structure and thermoelectric properties of thin film lead telluride prepared by electrolytic deposition, Proc X Int. Thermoelectric Conf., University of Wales, Cardiff, (D.M. Rowe, Ed.), Babrow Press ISBN, 095129286 0 0, 174-177.

Ohta, T., Sugimoto, K., Tokiai, T., Nosaka, M. and Kajikawa, T., (1990), Solid solution formation processes on $(Bi,Sb)_2$ $(Te,Se)_3$ based N-type thermoelectric material by PIES method, Proc. IX Int. Conf. on Thermoelectrics, JPL Labs. USA., 16-26.

Rowe, D. M., and Clee, M., (1990), Preparation of high-density small grain size compacts of lead tin telluride, Journ. of Mat. Sci., Materials in Electronics, 1, 129-132.

Rowe, D. M., Shukla, V., and Savvides, N., (1981), Phonon scattering at boundaries in heavily doped fine grained silicon-germanium alloys, Nature, 290, 5806, 765-766.

Rowe, D. M. and Bhandari, C. M, (1983), Modern Thermoelectrics, Holt Saunders, London.

Schwarz, R.B., and Koch, C.C., (1986), Formation of amorphous alloys by the mechanical alloying of crystalline powders of pure metals and powders of intermetalics, Applied Physics Letters, 49.3, 146-148.

Situmorang, M., (1988), Sintered bismuth telluride alloys for thermoelectric generators, PhD thesis, University of NSW, Australia.

Tokyo, (1985), Japan's sunshine project summary of comprehensive research. Sunshine project promotions headquarters, Agency for Industrial Science and Technology, Ministry of International Trade and Industry, Tokyo.

Turner, M. J., (1989), Preliminary assessment of offshore platform conversion to geothermal power station, United Kingdom Department of Energy Report, ETSU G 145.

ELECTROLYSIS OF WATER USING PHOTOVOLTAIC SYSTEMS

C.ALKAN, M.ŞEKERCİ, Ş.KUNÇ

Department of Chemistry, Faculty of Arts and Sciences, Fırat
University, Elazığ, 2300 TURKIYE

ABSTRACT

It is believed that future fuel will be hydrogen within next century. A study
has been conducted to produce the hydrogen gas using an electrochemical cell
and solar energy in Elazığ meteorological condition. The variables of electrochemical cell were optimized at laboratory conditions using D.C.power supply. Later, hydrogen production using photovoltaic cell was carried out at
the optimized conditions.

V-I curve and efficiency of photovoltaic cell were obtained by using with and
without fresnel lens. The efficiency of photovoltaic cell was found to be 5%.
It seems to be more economical when the fresnel lens was used.

KEYWORDS

Hydrogen production; electrolysis; photovoltaic cell; solar energy; fresnel
lens.

INTRODUCTION

The fossil fuels are the main energy sources of the industrial world. Since
the resources of these are limited, the alternative energy resources must be
urgently discovered. After the energy crisis in 1973, many scientists started
intensive researches on the production of hydrogen. Because of its physical
and chemical properties, hydrogen is considered as an ideal energy source in
the future. It can be produced from water which has a large reservoir on the
earth.It can be stored and its combustion product is water which has no pollution problem. A lot of hydrogen production methods have been developed.
These can be grouped in four headings;

1- Thermochemical process
2- Hydrogen-hybrid process
3- Electrochemical process
4- Photoproduction-biological process

Non of them are economical by now and a lot of researches have been carried
out on these topics. Getoff (1988) claims that, the most promising technique
seems to be the photoelectrochemical and the hybrid process.

EXPERIMENTAL

The fresnel lens was obtained from Oreshaza Glass Plant-Checoslovakia with a size of 37.2 x 35.7 cm. Its physical properties are given by Maly et al (1987).

Metallic electrodes (zinc, lead, brass, aluminium, copper, silver, iron and nickel) with a size of 1x1 cm were prepared from their pure metal sheets. As a membrane, synthetic fibers, glass woöls and asbestos were used. All chemical used in the electrolytes were analytical grade.

In the experiment, the amorphous silicon solar cell produced by Aromco with a size of 7.5 x 5 cm were used. The electrochemical cell were produced in our laboratory from glass tubes. A rough drawings is shown in Fig.1. As a DC source, Raysel LPS 30-10 model was used. The solar radiation measurements were carried out with Kipp and Zonen CM 11 model pyronometer.

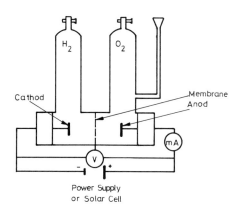

Fig. 1. The electrochemical cell set up

RESULTS AND DISCUSSIONS

Generally, zinc, lead, brass, aluminium, copper and silver oxidized under the working conditions. Iron as cathod and nickel as anod were found to be more stable. Best current efficiency obtained with 15% NaOH electrolyte and synthetic fiber membrane. Theoretical hydrogen production efficiency of the cell is given as 0.66 by Louw (1978), however it was obtained between 0.70-0.80 in this work.

The measured voltage-current curves of the solar cell at different solar radiations are given in Fig. 2.a and Fig. 2.b without and with fresnel lens, respectively.

The efficiency of solar cell, without and with fresnel lens at different solar radiations, was given in Fig.3. The concentration of fresnel cell of day light at different hours is given in Fig.4. It changes from 1.9 to 3.8 depending upon the solar angles.

The transmittance of fresnel lens is too low (around 70%) and concentration ratio is not too high. Therefore the efficiency of solar cell with fresnel lens are found to be low at the same solar radiation values.

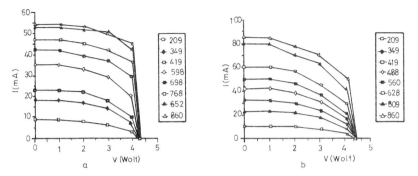

Fig. 2. Voltage-Current curves at different solar radiation.
a. Solar cell alone. b. Under the focus of fresnel lens.

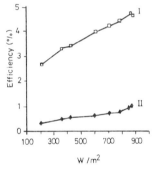

Fig. 3. Efficiency of solar cell without (I) and with fresnel lens (II)

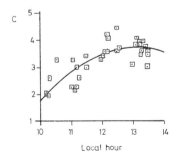

Fig. 4. The concentration ratio of fresnel lens at different solar hours.

2627

The total system efficiency at different solar radiations is given in Fig.5

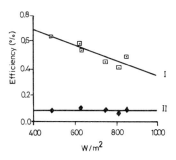

Fig. 5. Total hydrogen production efficiency at different
solar radiation. I. Solar cell alone. II. With
fresnel lens.

The economical analysis of system with and without fresnel lens are given in
Table 1.

Table 1. The capital cost analysis of the system.

	Without fresnel lens	With fresnel lens
Cost of solar cell ($/m^2)	830	830
Cost of fresnel lens ($/m^2)	–	20
Cost of energy produced ($/W)	139	112

REFERENCES

Getoff, N. (1988). Photoelectrochemical and photocatalytic methods of hydrogen production, A short review, Hydrogen Energy Progress VII. Proceedings
of the 7 th. World Hydrogen Energy Conference Moscow USSR, 25-29 September, pp 235-256.
Louw, N.J. (1979). Mobile Electrolytic Hydrogen Generators, Int. J. Hydrogen
Energy, 4,pp 187-192.
Maly, M. V. Sırka, F. Franc, B. Nabelek (1987). Lineer raster lens designed
for oblique incidence of rays, Proceedings of Applied Optics in Solar Energy II. Praha 7-9 July pp. 177-180.

RENEWABLE ENERGY, ECONOMIC DEVELOPMENT AND ECO-AGRICULTURE FOR A SUSTAINABLE FUTURE FOR RURAL AREAS OF THE WORLD

D.R. Neill CEO, HAWAII-CHINA-ASIA ENERGY & TECHNOLOGY CORP.
S.H. Xu, State Science & Technology Commission, P.R. China/Fellow East-West Center
Z.Q. Guo, Mechanical Engineering Dept., Beijing Union University

HCA-EAT, 98-859 Olena Street, Aiea, Hawaii, 96701, U.S.A.

ABSTRACT

About 70% of the world's population lives in rural areas and have very low incomes and standards of living. Their energy needs are mainly for cooking and heating. This paper review the FAO report on the environment and energy. It summarizes a planning paper: "Strategy for Renewable Energy and Eco-Agriculture for China." Finally, it provides some suggestions for an **ACTION PROGRAM**. The goal is to help advance a sustainable future for the world's rural areas. Renewable energy, eco-agriculture, sensitivity to the environment, and the need for economic development opportunities for rural areas form the basis a desirable future. They are keys to a sustainable future. Systems must be **technically viable, economically feasible, socially acceptable, environmentally desirable, and politically achievable.**

KEY WORDS

Sustainable Future; Rural; Renewable Energy; Economic Development

I. BACKGROUND

There are about three billion poor in the world. They have very low incomes and standard of living. Their primary energy use is for cooking and heating their homes, using primarily firewood or fuelwood. They earn their limited income from agriculture. They have few opportunities to improve their living conditions. There is a growing concern as the conditions of the rural poor worsen. Their primary energy supply is becoming less sustainable. The firewood supply is not being replenish fast enough to keep up with its use and the population growth. The new concern about the environment has a greater adverse effect upon their lives/future.

FAO FINDINGS: The United Nations Food And Agricultural Organization (FAO) had a conference in Beijing, China in 1989. The FAO Environment and Energy Paper 12 "A New Approach to Energy Planning for Sustainable Rural Development" published the results. The forward states: "The approach set out in this document stresses that energy planning, without effective mechanisms to convert its findings into implementable projects, becomes an academic exercise. National energy polices which do not converge with the interests of the smallest rural communities can only be theoretical. Mature rural technologies such as small diesel engines, solar dryers, wind generators, small hydropower schemes, biomass gasifiers or rural electrification programmes, can only be successfully implemented if they fulfill **identified and assessed energy requirements for rural and agricultural activities and if they are supported by technical financial and policy measures.**" pg.iii

This program will create a bridge between authorities who are responsible for agriculture and rural development and those that deal with energy. The reports points out: "The agricultural sector, however, lags in this respect. A major reason for this is that problems are aggravated by the dispersed, relatively small-scale nature of energy needs in the rural sector and because, contrary to other sectors where clear and powerful counterparts exist, this sector is composed of many, often poor potential energy users. Yet, **energy is a lever which can raise agricultural and agro-industrial productivity thereby improving the level and quality of life of rural population.**" pg. iv

"The central feature of the new approach is the preparation and implementation of area-based decentralized energy plans for meeting energy needs for subsistence and development at the least cost to the economy and the environment, and linking the micro-level plans with national economic planning and development programmes, including those for the energy, agriculture and rural development sectors." pg. 5

WORLD CLEAN ENERGY CONFERENCE: The main focus of this conference, held in Geneva in November 1991, was the negative impact of conventional energy sources on the environment. It also cited the large gap between the "haves" of the world, energy wise, and the "have-nots." President Robert Mugabe of Zimbabwe was quoted: ..."the availability of abundant energy in a suitable form is not an end in itself, as it is only a means, a requirement for a whole range of other processes by which the human population is able to improve its living standards." (Zimbabwe, November 1990). In commenting on this, D.A. Davis, Executive Director of the World Energy Commission, said that as a credo for those directly involved in the energy business: "Looking at energy as a service helps: it's better than looking at energy as a god to be worshipped or as a commodity to be sold by hook or crook whatever the purpose and whatever the real cost...energy is part of a chain that ends at the human condition, well being or distress. It is a role player amongst many others contributing to that condition."

Dr. Herman Sheer, MP, from Germany and president of EUROSOLAR, provided the strongest challenge. He said we need: "solar crash programmes instead of further programmes for space technology and new weapons technologies. The alternative of Strategic Defence Initiatives are ecological SDI-Programmes - Solar Development Initiatives."

CONCERN: A concern is the tendency of developed nations to spend too much of the limited funds avaiable for energy research on activities of limited or questionable value. Too little has relevancy to help solve the real needs of mankind. Often this work supports mainly self-serving, narrow-minded research programs for academic empire builders.

Mankind cannot live on energy (nor bread) alone. However, energy has a profound effect upon one's total life. Energy can increase mankind's ability to raise food and provide safe pure water. It can improve mankind's ability to earn money and raise the standard of living. Non-polluting renewable energy can enable a person to breath clean air (instead of fossil fuel pollutants). In this regard, the three billion people in the world's rural areas have a severe problem. They consume one hundredth the energy per capita. Yet they receive the least financial support. They deserve much more attention and practical program aid to provide them energy options. Using indigenous, non-polluting renewable energy. A result will be the increase in rural households' standard of living. This is an important challenge and opportunity in the world today. Let us achieve results.

II. INTRODUCTION

The challenge, opportunity, and responsibility of helping to bring an improved standard of living to the world's rural poor deserves the highest priority. A good name for this program is **CORE: CHALLENGE, OPPORTUNITY, RESPONSIBILITY ENERGY PROGRAM.** Renewable energy often is ignored, and even opposed, by energy decision makers. They say that the technologies are not economic and cannot meet the energy requirements of an expanding economy. Many look to coal as the solution. Coal presents many problems, however, especially to the environment. When one includes all the environmental costs of coal, coal is not low cost. Dr. Nejet Veziroglu, of the University of Miami, documents that renewable energy is cost-competitive when including all costs.

In remote rural areas, where population densities are low and transmission lines impractical, renewable energy is cost-effective today. Hydropower is the most cost-effective source of non-polluting renewable energy widely in use. Biomass is the main energy source for rural areas, and it is renewable. Wind and solar power are, or are becoming, cost competitive for many applications where the wind regime and solar resource is adequate.

Rural areas can be the proving ground for a transition to a renewable energy economy. The late E.F. Schumacker coined the term, "small is beautiful" when discussing energy options. Lao Tsu, the ancient Chinese man of wisdom, 2,500 years ago said that the journey of a thousand leagues, starts where you are, with one step. **Rural areas offer the place to start the journey to a better energy future, shifting to non-polluting, renewable energy technologies.** Unfortunately, a major problem is that the people in rural areas are the least able to afford any energy. Hence, the challenge of this paper.

This paper reviews the FAO report. It summarizes a strategy paper developed by Dr. Xu Shao, Fellow at the East-West Center. The main mission of this paper is to provide some suggestions for an **ACTION PROGRAM**. The goal is to help advance a sustainable future for rural areas of the world. The contribution of renewable energy in support of responsible agricultural programs is important. A secondary mission is to promote economic development opportunities for rural areas. There must be a sensitivity to the environmental implications and results. These are key to this sustainable future. All technologies must be **technically viable, economically feasible, socially acceptable, environmentally desirable, and politically achievable.**

III. FAO REPORT

The FAO report set forth an excellent basis for meeting the challenge, opportunity, and responsibility. The major sections include the framework, implementation strategies, and guidelines. The framework for this new approach has seven suggestions. The implementation strategies address the problems and constraints. It realistically presents nine areas that may present problems. The wise man, who has a mission to serve all the people everywhere, will seek to resolve these problems. This is what Dr. Sun Yet Sen (Father of modern China) said: tien xia wei gon. The wise man effectively deals with these problems as challenges and achieves results.

The second section suggests five mechanisms to overcome problems and constraints for a successful program. They range from the involvement of the central government to the local community and people to be served. The fifth mechanism is coordination that ideally will assure that each level is appropriately involved. In some countries, the central government plays a strong role. In other countries, the grass roots or village residents may insist on their primacy. The least amount of problems is likely to result when there is cooperation between central, provincial, regional, county, and local governments and the target population. There needs to be adequate funding from the central and provincial government. Technical expertise from regional and international programs is helpful. Most important, the leaders and doers of the local community must be for the program, or it will never succeed. It must benefit their lives and achieve visible results as soon as possible.

The "Guidelines for Launching an Integrated Rural Energy Planning Program" suggests four phases. These include: awareness building, pilot projects design, implementation, and replication. Education or awareness building is basic, by those involved in the program and from the target communities leaders. Design the pilot projects to increase the chance of success. When implemented, monitoring and evaluation is important. Their cost-benefit and their acceptance by the residents or uses are vital. When successful, extend them, and make avaiable to other communities.

2631

III.   SUMMARY OF AN APPROACH TO ECO-AGRICULTURE   (BY S.H. XU)

Dr. Shao Hui Xu, a Fellow at the East-West Center in Hawaii, is from the State Science and Technology Commission in Beijing. She developed a planning paper on eco-agriculture and renewable energy. Her paper summarizes the renewable energy potential of ten provinces in China. She discusses the value of improving the agricultural production of the villages where two-thirds of the Chinese billion people live. Non-polluting, renewable energy resources can help in this expansion by bringing fertilizer, water, and electrical energy. China is already a leader in biomass technology, especially bio-gas from agricultural and human waste. This can be improved and additional co-products produced.

Dr. Xu provides a planning process to achieve greater results by a renewable-energy eco-agriculture program. She suggests targeting one or more provinces to carry out this plan. Her contacts in Gansu Province have expressed an interest in cooperating in this program.

IV.   ACTION PROGRAM FOR PROVINCIAL GOVERNMENTS TO ACCELERATE TRANSITION IN THEIR REGION

The following are seven suggestions for an **ACTION PROGRAM.** Adapt them to any given target area, county, province or regional area. Technical support should be given by those capable, experienced, and successful.

1.   RESOURCE ASSESSMENT: A good resource assessment of each of the most promising renewable technologies (biomass, hydroelectric, solar, and wind) is desirable. First, all existing useful data should be secured from national, regional, provincial, and local sources. Supplemental data or information may be needed from field stations. You should not spend a lot of money on this, however. Sometimes the local leaders or farmers have the needed information. The initial demonstration systems, if monitored, (PV or wind system), can provide good data. They also provide basic energy. One should avoid, however, putting up a wind machine where there is little wind. Yet wind power could provide the needed fresh, pure water for both human potable requirements and agriculture, where there is a strong enough wind regime. D. Neill and Z. Guo are working on a project to do this (covered in another paper).

2.   MASTER PLAN: Cooperation with, and involvement of, the local leaders, and other leaders and other knowledgeable people is needed. The three common scenarios are "business-as-usual," "moderate growth," and "optimistic growth."
o   **ENERGY NEEDS OR DEMAND:**   While the cautious approach would be the "business-as-usual," "moderate growth" is more reasonable and a worthy goal. Optimism may create added problems and possible failure and disappointments.
o   **ENERGY SUPPLY:**   Again, the business-as-usual assumes that a little more of the same energy sources will be needed each year. It includes all sources of energy (biomass - namely fire wood and biogas), petroleum products (gasoline, kerosene, bottle gas, and others), coal and peat, hydropower, and any solar or wind systems on line. The two expanded scenarios would add a greater share from the renewable energy sources. These would use the results of the resource assessment, subject to financing availability.
o   **GOALS AND PRIORITIES:**   A strong commitment to a transition to a renewable energy economy requires optimistic goals and priorities. The time frame may be near term (1 to 3 years), medium term (4 to 10 years), and long term (10 to 20 years). Some may wish to be more cautious, but there will be less achieved by setting too cautious goals and priorities.

3.   TEAM AND STRUCTURE: The team should include key leaders and doers at the local level, decision makers at the county and provincial level, and advisors from the central government and others. Technical support staff will assure sound wise decision making. The three "C's" of **communication, cooperation, and coordination** are important. There should be clear lines of authority, decision making, and responsibility. The goal is to lessen possible areas of conflicts, overlapping of authority, and maximum cooperation. Listen to key officials, respect them, and thank them for their help.

4.   LAWS OR PROCEDURES TO FACILITATE NEEDED DEVELOPMENT/ACTION: These will vary from country to country, as well as within a given province, region, and local community.

5.   INITIAL FUNDING TO HELP THE PRIORITY PROJECTS THAT HAVE PROMISING COST-BENEFIT POTENTIALS: Funding is often the major problem. Renewable energy systems may be more capital or first cost intensive. They have zero energy cost in most cases. Also their operating and maintenance costs are low. The life-cycle cost of these technologies is becoming less costly than fossil fuel systems, when including all the costs. Aid in meeting the initial cost for a project to proceed may be necessary. For example, wind and PV power today at remote locations is cost effective. Hybrid systems combine wind and PV systems.   The energy storage cost remains a problem.   D. Neill and Z. Guo have been developing better wind-PV hybrid energy systems. They are also seeking to advance a nickel-metal-hydride battery that has twice the capacity of nickel-cadmium, more than 1,000 cycles for recharging, no maintenance, no poisonous material, and no memory affect. The initial cost will be four to six times the cost of the conventional lead-acid battery. However, its life time cost and related problems will be much less than the lead-acid battery.

It is also possible to seek other sources of funding outside the country, such as the World Bank, the development banks, U.S. Agency for International Development, venture capital, and foundations.   One of the results of the world meeting on the environment held in Montreal three years ago was the **Montreal Protocol.** It established a $1.4 billion Global Environmental Facility (GEF). These funds are for projects that will improve the environment that would not normally receive financing from banks. The program encourages non-polluting renewable energy technologies.   Unfortunately, most of these funds have been obligated.   Additional funds should become available in the future. Some leaders proposed that developed nations provide financial support to the developing nations for their non-polluting sources.

6.   WORKSHOPS, PLANNING MEETINGS, TRAINING SESSIONS, AND ESTABLISHING NEEDED SUPPORT SYSTEM AT THE GRASS ROOTS LEVEL: Prepared by those who make decisions, as well as those who will be installing, operating, maintaining, and living with the various systems.

7.   IMPLEMENT SMALL SCALE, HIGH SUCCESS PROBABILITY, PROJECTS: These may be demonstration or model projects, but they should use proven technologies (not prototypes).   These projects should be successful with clear benefits.

V.   CONCLUSION

There is a great challenge, opportunity, and responsibility to help bring an improved standard of living to the rural poor of the world.   As stated in the introduction, mankind cannot live on energy (nor bread) alone. Energy has a profound effect upon a person's total life. Energy can increase mankind's ability to raise needed food, provide safe pure water, and earn more money. Non-polluting renewable energy will provide cleaner air (as opposed to energy from polluting fossil fuels), with additional benefits. It is also a good starting point for a transition to a non-polluting renewable energy economy. Small village communities can achieve total renewable energy self-sufficiency.   Again, as Lao Tsu said 2,500 years ago, "a journey of a 1,000 leagues, begins where you are." Our journey to a better energy future can begin in the villages of the developing nations. This is our **CORE PROGRAM: CHALLENGE, OPPORTUNITY, AND RESPONSIBILITY ENERGY PROGRAM.**

REFERENCES

WORLD CLEAN ENERGY CONFERENCE, Geneva, Switzerland, 11/1991
United Nations, FAO Environment and Energy Paper 12 "A New Approach to Energy
    Planning for Sustainable Rural Development."

A STRATEGY TO REDUCE GREENHOUSE GASES IN ITALY

G. Silvestrini

CNR-IEREN, National Research Council,
Via Rampolla 8 - 90142 Palermo, Italy

ABSTRACT

After the EEC decision to stabilize the carbon dioxide emissions by the year 2000 at the 1990 level, a study has been financed by the Italian Ministry of Environment in order to define what targets could be set by year 2005 and what strategies could be implemented in Italy in order to achieve consistent carbon dioxide reductions. A bottom up approach has been used in order to explore the technical potentials in different fields. The options considered have then been economically screened through a least cost procedure. Not only the energy sector has been considered but also transport, forestry and agriculture. The results indicate that there is a potential of reduction of CO_2 emissions of 25% in 2005 compared to the year 1990 and that most of the interventions analyzed could be achieved at net negative cost. In this paper some of the results of the climate stabilization scenario are presented with a special emphasis on the potential of renewable technologies that could add 39 TWh/year (17% of actual electric consumption) and avoid the annual production of 24,5 millions of tons of carbon dioxide.

KEYWORDS

Global warming, electricity, transport, renewable energies

INTRODUCTION

In 1990 the carbon dioxide emissions in Italy have been of 400 millions of tons. The National Energy Plan (NEP) prepared in 1988 forecasted an increase of CO_2 production by the end of the century of 12%. However there is a great potential to reduce the greenhoue gases emissions in all sectors. The study prepared for the Ministry of Environment has shown that it is possible to cut the carbon dioxide emissions also in sectors like transportation and electricity, that in the buisness as usual scenario are expected to grow quite rapidly (RI, 1992).

ELECTRICITY

The potential of energy saving in this sector is quite large. Even considering the high penetration rates of different appliances in the residential sector forecasted by Enel (the national electric utility), the introduction of efficient products that are already in the market could provide very large electricity and carbon dioxide reductions compared to a buisness as usual (BAU) scenario. In fig. 1 are presented the results of our analysis for major household appliances. Much stronger reduction levels could be achieved through the diffusion of high efficient appliances that could be delivered in short time with right signals from the government. The evaluated potential of lighting energy savings in residential and commercial sectors is of 8 TWh/year compared to the BAU scenario.

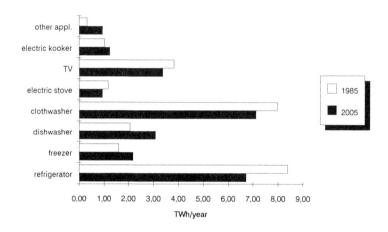

Fig. 1 Electricity consumption in 1985 for different appliances and consumption in 2005 evaluated with the introduction of efficient models

TRANSPORTATION

The study considers both passenger (urban and extraurban) and freight transportation evolution through 2005. Three different options to reduce carbon dioxide have been considered: improvement in the vehicle's fuel economy, modal shift, reduction and rationalization of mobility demand. The average fleet in year 2005 could be, according to the assumptions made in the study, 19% more efficient than the average fleet in 1988. An increase of 130% for passenger public transport and of 150% for rail and ship freight transport by year 2005 has been considered to be feasible in the italian context. The effect of many options, like urban planning, traffic management, telecommuting, biking, increase in load factors, telematic to rationalize freight transport, speed control, vehicle maintenance, driving improvement, has also been analyzed. Globally the climate stabilization scenario could lead by 2005 to a reduction in CO_2 emissions of 18% compared to 1988 values (fig. 2).

2635

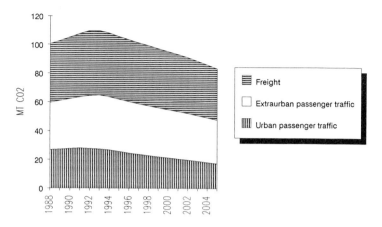

Fig. 2   Carbon dioxide emissions (millions of tons) by 2005 in the italian transportation sector in the climate stabilization scenario

RENEWABLES

The average hydroelectric production in Italy is of 50 TWh/year, while from geothermy 3,2 TWh have been produced in 1990. The attention to solar, wind and biomass energy has been till now quite low, also in relation to institutional barriers that have prevented a diffusion of these technologies in Italy. The situation is however changing with the adoption of new laws that, for example, oblige the national utility to buy the solar or wind electricity at a price of 14-15 US cents/kWh.

The potential of CO_2 reduction by 2030 from renewable sources in Italy has been estimated to be of 220 millions of tons/year or 55% of the 1990 emissions (Bernardini, 1990). In our study we have calculated that the maximum additional contribution by 2005 achieveable in the field of electric production, considered the different limitations that are particularly strong in the early phase of diffusion of these new technologies, could be of 39 TWh/year.

Analyzing the different sources of our climate stabilization scenario, almost half of the new production is supposed to come from an increase in hydroelectricity and geothermy. These values are not very far from the evaluations of the national utility that is committed to achieve an increase of 9,6 (hydro) and 6 (geo) TWh/y by 2000. As far as biomass is considered, the values considered (10 TWh/y by 2005) are slightly higher than Enel forecasted trend (4 TWh/y by 2000), but much lower than the potential suggested in EEC studies (Grassi, 1992).

Two solar technologies have been considered: thermodynamic cycles and photovoltaic cells. For the solar thermoelectric sector an installed power of 1100 MW by 2005 could be achieved assuming the same rate of diffusion that has been experienced in California (350 MW from 1984 to 1990) with the Segs plants of 30-80 MW size. The government has not considered this technology in its NEP. The penetration of photovoltaic cells in our scenario (1200 MW by 2005) is coherent with the rapid growth of this sector (30% yearly increase at European level). The government has set a target of 25 MW installed by 1995. As far as wind energy is concerned, the goal set (3000 MW) is quite ambitious compared to the targets of the NEP, 300-600 MW by the year 2000, but the rate of penetration of this technology is on line with those experienced in California, Denmark and more recently in Germany. The combination of wind and solar technologies in terms of power would be lower than 10% of the estimated demand in the electric grid, thus avoiding any possible problem related to the intermittency of these sources of energy.

In relation to low entalpy heat production, 1 million solar collectors systems in the residential sector for the production of heat water (that will substitute mainly electric systems) have been considered.

Finally in the climate stabilization scenario a goal of liquid fuels produced from sustainably grown crops (830.000 ha) equal to 5% of total energy consumption in the transportation sector has been set.

The potential of CO_2 reduction derived from this partial list of renewable technology options is presented in fig. 3. In relation to the capabilities of renewable technologies to reduce carbon dioxide emissions the following remarks should be made. For geothermic energy, the electric production is not CO_2 free. Using data on carbon dioxide releases reported in Haraden (1989) we have evaluated an average emission of 0,84 kg per kWh, a value higher than the average of the italian mix of power production that is of 0,69 kg CO_2 per kWh. For solar thermoelectric production using parabolic trough collectors, only 58% of the electricity is produced from the sun in the italian conditions, the other 42% being provided by natural gas. Finally the carbon dioxide emissions connected with the use of solar water heaters are 60-70% lower than in the case of a conventional electric heater, considering the achievable solar fraction.

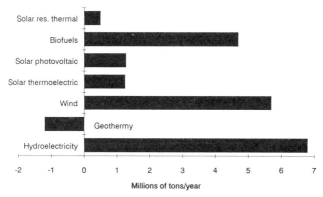

Fig. 3   Potential of carbon dioxide reduction by the year 2005 through the diffusion of renewable energy technologies in the climate stabilization scenario

COSTS

In order to compare the economic validity of different options the net cost of energy production and of carbon dioxide reduction has been calculated for the technologies considered, in the same way it is presented in Johansson (1990). For the costs of solar and wind technologies the values of the SERI (1990) study reporting the targets achievable in presence of a strong research effort have been adapted. In particular the

US goals for 2000 have been used for the year 2005 with a 30% increase in reason of the specific characteristics of the italian solar and wind resources. The cost of photovoltaic electricity derived with this approach is slightly (10%) higher than what has been estimated by italian studies (Coiante, 1990).

In fig. 4 the data for 11 different options are presented. The x- axis shows the cumulative CO_2 reduction achievable, while in the y-axis the costs of the different measures are shown (an annual discount rate of 7% has been considered). It is possible to note that large reductions of carbon dioxide can be obtained at a net negative cost.

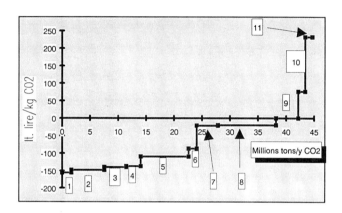

Fig. 4   Cost of conserved CO_2 for 11 options considered (1 $ = 1250 It. liras)
1 commercial lighting; 2 efficient trucks; 3 dishwashers and clothwashers; 4 Cars 10% more efficient; 5 Domestic lighting and part of the comm.; 6 Eff. water heaters ; 7 Refrigerators and freezers; 8 Cars 50% more efficient; 9 Wind energy; 10 Solar thermoel.; 11 Photovoltaic

REFERENCES

Bernardini O. (1990), Il potenziale di riduzione delle emissioni di gas di serra dalla combustione di fonti fossili in Italia, Energia n. 4

RI (1992), Barbera G., La Mantia T., Pettenella D., Picciotto F., Silvestrini G., Strategie di riduzione delle emissioni dei gas climalteranti in Italia, Rapporto intermedio, Ministero per l'Ambiente, Roma

Coiante D. (1990), La produzione fotovoltaica di idrogeno, Energia ed innovazione, Novembre/dicembre 1990

Grassi G. (1991), Produzione di elettricita' da risorse di biomassa, Agricoltura e innovazione. n. 19. Roma

Haraden J, (1989), CO_2 production rates for geothermal energy and fossil fuels, Energy vol. 14. n. 12

Johansson T., Mills E., Wilson D., (1990), Beginning to reduce greenhouse-gas emissions need not be expensive, Proc. of Second World Climate Conference, Geneva

SERI (1990), The potential of renewable energy: an interlaboratory white paper, Solar Energy Resources Institute, SERI/TP/260/3674

ENERGY AND ECONOMIC DEVELOPMENT IN ISOLATED REGIONS:
A CASE STUDY

A. Muñoz and P. Maldonado

Facultad de Ciencias Físicas y Matemáticas, Universidad de Chile,
Casilla 412-3, Santiago, Chile

ABSTRACT

High production costs and inadequate supply of electricity are severe handicaps
to economic development in remote areas. Costs are increased further by the low
level and high variability of electric loads. In order to face these problems
a simultaneous development of wind diesel energy resources and productive
projects with demand management is proposed. This paper ilustrates such a case
in Southern Chile where present electricity costs are reduced as consumption
is increased from 85 to 468 KWh/person/year. A new refrigeration plant
increases the storage capacity for the small fishing activity and effectively
uses the aleatory output of wind energy generators.

KEYWORDS

Isolated areas electrification, renowable resources, wind diesel systems.

INTRODUCTION

Electrification of remote areas in the Third World is aimed at the satisfaction
of basic domestic needs with loads concentrated in a few hours per day; the
"best solution" is obtained using small diesel units which are operated only
during the normal peak demand hours to avoid high diesel consumption.

Sustainable development in any specific community can only be achieved if at
least a certain minimum of energy supply can be provided to this group. This
minimum largely exceeds domestic load and street ligthing, and therefore, it
well provides the energetical basis for the development of some industry or
other kind of significant economic activity.

A viable way for the community's development is thus assured, through the
introduction of economic activities that both require and guarantee the
efficient operation of electric systems.

In Las Huichas island's electric system, considered as a basis for this study case, electricity is supplied by two diesel generators. Because of the island's location, fuel cost double the price in the continent.

High generation costs restrain the utility to supply what is considered only "essential" loads, during a few hours a day. To face this situation, the possibility of introducing to the system a solution based on renewable energy, is normally adopted. As the only resource available in the island is the wind, the addition of wind generators to the existant system was proposed.

SUSTAINABLE ENERGY SUPPLY FOR ISOLATED REGIONS: ELECTRICITY AND PRODUCTIVE ACTIVITIES.

To date, the island's power supply is obtained from two 100 KW diesel generators. The high variability load factor and the diesel fuel cost, force the system to operate only 6 hours/day. Table 1 shows the anual cash flow and Table 2 the related parameters.

Experience shows that an expansion of electricity supply does not necessarily mean an increase in domestic power demand (Berrie, 1983). On the other hand, an improvement in load profile is obtained when productive loads are added to fill the valleys.

In the locality studied -Las Huichas island- fishery is the only productive activity; fishermen's revenues are affected by lack of ice and preservation facilities in the island. This situation may be reversed if power supply is increased to 24 hours/day.

Figure 1 shows the foreseen electrical demand, if a 520 ton/year fish processing plant is integrated to the system. So far, there is a substantial increase of the per capita energy demand: from 85 KWh/capita/year to 468 KWh/capita/year.

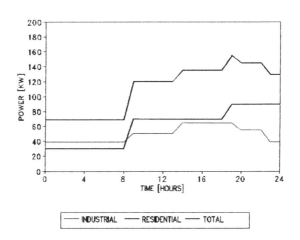

Fig. 1. Total demand new scenario.

Table 1. Main operational results.

YEARLY COSTS [US$]	PRESENT SITUATION	NEW SCENARIO
Labour and maintenance costs	29.233	67.550
Administration costs	13.617	13.617
Fuel and lubricants	18.571	86.746
Capital equipment investment(*)	—	19.113
TOTAL COSTS [US$]	61.591	187.026
YEARLY INCOME [US$]		
Energy sales	36.212	196.883
Operation surplus (loss)	(25.119)	9.857

(*) Note: Discount rate: 10%; evaluation period: 20 years.

Table 2. Main evaluation parameters.

MAIN PARAMETERS	PRESENT SITUATION	NEW SCENARIO
Energy sales [KWh/year]	168.800	937.536
Additional investment [US$]	—	162.720
Diesel energy generation [%]	100	83
Wind energy generation [%]	—	17
Diesel efficiency [litres/KWh]	0,335	0,335
Fuel price [US$/litres]	0,321	0,321
Energy sale price [US$/KWh]	0,210	0,210

Notes:
Average wind speed   : 5.1 m/sec.
Weibull distribution : c = 5.48 m/sec; k = 1.25
Diesel machines   : 2x100 KW installed capacity
Wind machines   : 3x30 KW installed capacity

WIND ENERGY OPTION FOR LAS HUICHAS ELECTRICAL SYSTEM

The project considered a wind diesel system in order to reduce substantial full costs. In Las Huichas island, wind is the most relevant renewable energy resource.

Asynchronous electric wind generators are widely used due to their high reliability and low costs. The AC voltage is fixed by the synchronous-diesel generator, but due to short term and substantial amplitude wind variations, network transients appear.

To simulate amplitude and time characteristics of the main system variables, a dynamic model of order 6 was developed. The model considers the evaluation of fuel consumption (Ughen and Oyvin, 1988), speed regulator effects, electrical transient characteristics of the machines (Concordia, 1951; Jones, 1967) and mechanical inertias. The main results (Fig. 2), show that the maximum voltage fluctuations occur during the electrical connection of the induction

wind generator. This problem is attenuated if:
1.   The capacity of each asynchronous generator is less than 20% of that o
     the diesel generator.
2.   Reactive compensation is used.
3.   Induction generators are sequentially connected.

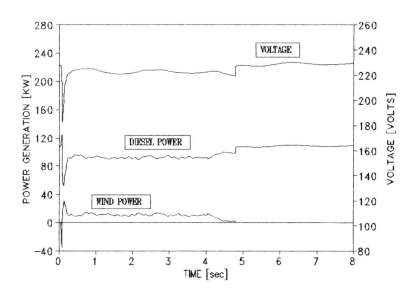

Fig. 2.   Main transients due to wind fluctuations.

Long term wind fluctuation effects may also be attenuated by developing a
demand control associated with the refrigeration plant. In fact, by using
eutectic materials and ice it is possible to store significant amounts of
energy when sufficient wind is available. It is also possible to develop a
variable voltage and frequency network, especially dedicated to supply
refrigeration systems. Table 3 shows experimental characteristics obtained when
this situation is carried out.

This option was not adopted for the Las Huichas system, because electricity
requirements are more demanding concerning network quality. When refrigeration
is the predominant load this option should be recommended.

Table 3. Variable voltage and frequency refrigeration
systems: Experimental characteristics.

FREQUENCY Hz	VOLTAGE Volts	EFFICIENCY Wmin/Kcal	ENERGY CONSUMPTION %
36,1	173	13,35	79,6
50,1	210	18,81	100,0
65,5	240	20,37	133,8

EVALUATION AND CONCLUSIONS

The main operational results are summarized in Table 1. In the new scenario, production costs are reduced from 0.36 US$/KWh to 0.20 US$/Kwh (for social reasons present price is fixed at 0.21 US$/KWh) and the internal rate of return for the project is 17%.

The project should be implemented not only because of these financial reasons but, especially, due to its socio economic impacts:

1. Both sustainable energy and industrial development can be obtained, establishing the basis for developing regional overall activity.
2. Fishing activity should improve productivity due to reductions in handling losses.
3. Commercial activity would be expanded.
4. The community could introduce essential electric equipments for medical, carpentry and maintenance purposes.
5. Local institutionality would be reinforced with the improvement of communication centres, health centres and educational activities.
6. Changes in employment structure and significant increase of fishermen's incomes would be attained.

AKNOWLEDGGEMENT

This work was supported by the Comisión Nacional de Investigación Científica y Teconólogica, Santiago, Chile as a part of the Research Projects 525-88 and 1098-92.

REFERENCES

Berris, T.W. (1983). *Power System Economics*. Peter Peregrinus, London.
Concordia, Ch. (1951). *Synchronous Machines: Theory and performance*. Wiley, New York.
Jones, Ch. (1967). *The Unified Theory of Electrical Machines*. Butterworths, London.
Ughen,K. and Oyvin, S. (1988). *A short term dynamic simulation model for wind/diesel systems*. In: Proceedings of the Tenth BWEA Wind Energy Conference, London, 22-24 March, pp. 235-242.

THE CHALLENGE AND PREJUDICE FACING RENEWABLE ENERGY IN THE UK

Rev. J.L. Leckie

Abercairny, Crieff, PH 7 3BS, UK.

ABSTRACT

The gross disproportion between government spending on nuclear energy and on renewables. The threat posed by acid rain and global warming. Claims made for nuclear power challenged. Energy efficiency will not be enough to remedy the situation owing to the anticipated rising demand aspirations of the developing nations. The trickery that has been contrived to be little renewables and hush up bad news of nuclear accidents. The cost of renewables has been falling and will continues to fall as technology advances and the benefits of large scale production are reaped.

KEYWORDS

Funding; Renewables; nuclear; energy; potential; developing nations; shenanigans

FUNDING

Funding of renewables by governments is derisory. The average IEA (1989a) member's grant has been 0.006% of GDP, the UK and USA having reached about half that. In the UK more has been accorded to commercial nuclear energy in one year than the unprincely the sum of £135.9 million (Hansard, 1989a) for all renewables in the decade 1979-89 -very little before that except for hydroelectric. The figure of £256.8 million was given by Cecil Parkinson (Hansard, 1989b) for the year's expenditure on commercial nuclear power, but it explicitly excluded BNFL, Nirex and the electricity supply industry. These are hefty spenders, so Flood's (1989) of £16.3 billion since the late 50s may well be correct. This doubles Parkinson's annual figure and is in glaring contrast with the Cavendish evidently spoke for his government in declaring: "There is no reason why spending on nuclear power should bear any proportional relationship to expenditure on renewable R&D (House of Lords, Jan 14, 91.). Why not? In justice, sauce for the nuclear goose ought to be sauce for the renewable gander. On top of all this the nuclear industry receives 99% of an amount over £1 billion annually (£1.3b in 1992) as a result of the NFFO. The remaining miserable 1% from the levy on fossil fuel goes to renewables. Between

2644

1979 and 1990 the UK government lavished a total of £30b on the coal industry (Hansard, 1991-92). Thank God, then, for those of you who are determined to develop renewables with limited funds in the private sector.

THE SOCIAL THREAT

Extensive acid rain damage and world mean temperature rise (IPCC, 1990a) have coincided with: a huge increase in industrial activity, most of it using a 90% fossil fuel energy mix; a surge in the number of vehicles, world wide; the deplorable modern practice of burning on site those 98/100 rainforest trees which cannot be economically extracted. Relatively constant for 1,000 years before 1765, the concentration of greenhouse gases in the atmosphere has increased greatly since then. Most significantly, the rate of increase of the chief culprit, carbon dioxide, trebled between 1958 and 1990 (IPCC, 1990b). A calculation from the world industrial index, published by the UN Statistical Digest (1986), shows a rise of 70% in industrial activity between 1970 and 1985. Since then IEA (1989) has indicated an increase in energy demand of about 3% per year. The number of vehicles spewing out their toxic exhaust cocktails has soared from approximately 50 million in 1950 to almost 500 million now (Walsh, 1990). The thousands of lakes and vast stretches of forest afflicted with acid are recent phenomena in the industrialized areas of Europe and North America. The incidence of respiratory disease is much above average in highly industrialized regions, especially where coal is dirtier than average (McCormick). Disturbances in nature and in people are associated in Scripture. God still the tumult of stormy waves and the tumult of the people (Psalm 65.7). The first three Gospels all record that his Son did the same. Hard on the heels of his stilling of a storm comes his cure of the man with the multiple, "Legion" personality or scrambled ego.

Global warming due to anthropogenic gases has almost certainly begun and the evidence adduced by the Intergovernmental Panel on Climate Change (IPCC c) is cogent. Some greenhouse gases can be eliminated, at a price. To take the carbon out of carbon dioxide would demand an inordinate amount of energy. Leaks of methane from coal and gas production are inevitable. They could account for much of the doubling of methane in the last century. Mounting quantities of decomposing paddy were presumably preceded by other vegetation also emitting methane. The climbing, belching cattle population was preceded by a

2645

much larger number of wild ruminants releasing methane from both ends of their bodies. It is doubtful whether these two sources ought to be blamed for the methane rise. Over all, it is imperative to minimize man-made emissions. If the "business-as-usual" scenario obtains, the IPCC (1990 d) estimates that temperature will rise by more than $3^\circ C$ by 2100. Three alternatives are open: 1) the nuclear option; 2) more efficient use of energy; 3) renewables

THE NUCLEAR OPTION

Nuclear proponents claim that commercial nuclear energy is safe, clean, cheap and peaceful. These call for scrutiny. Safe? Edward Teller, father of the hydrogen bomb, has warned about nuclear power: "...with so many simians monkeying about with things they do not fully understand, sooner or later the fool will prove greater than the proof even in a foolproof system" (Hays, 1976). Another person with the right to be heard is Pierre Tanguy (1989), Inspector General of nuclear safety in Electricite de France. He challenged the official French estimates that an accident of level 5 seriousness [the size of the 1957 Windscale fire or the near-disaster at Three Mile Island (TMI)] has a likelihood per reactor of one in every 10,000 years. He puts the chances at "a number of percent" in the next 10 years. With more than 75% of its electricity coming from nuclear energy, one can appreciate the pressure on Tanguy. He stresses on human error, confirming Teller. Sloth, inattention, carelessness are akin to the original sin, the tendency to miss the mark noted by the Psalmist when he said: "...in sin did my mother conceive me" (Ps.51.5). That is he was prone to an evil inclination from the moment of conception. Sins of omission figure as largely in Scripture as those of commission, often more largely in the New Testament. Each nuclear reactor creates 200 kg of plutonium per year, which is: "enough to inflict cancer on every single person in the world" (Carewardine, 1990a). However, it could be quite wrong to insist that nuclear energy can never be made safe. It might if safeguard were piled on safeguard, but what about the expense of piling them on?

Clean? Carwardine (1190b) gives Government figures for Britain alone at an accumulated 4,000 tonnes of high level, 16,000 tonnes of intermediate and over a million tonnes of low level waste. Vitrification can reduce the volume by two thirds, but the question is whether it can be contained in earthquake, bombing or geological shift.

Its concentration may mean that it is not comparable to naturally ocuuring radioactivity such as has lain harmless under Loch Lomond or in Oklo, Gabonfor centuries (Chapman and McKinley, 1990).

A super-microbe may emerge to devour radioactivity. Meantime we have to rely on what Academician Konstantin Sytnik called in reference to Chernobyl: "...a mighty process of dilution and dispersal of radionuclides (in nature) and this saved us" (Medvedev, 1990). It did not entirely save them as there is evidence of much increased incidence of cancer in the Kiev area and it is still too soon to know the full results.

But some such saviours may be able to cope with the immense amounts of radioactive waste that will be produced by a full nuclear scenario, or even the scenario already obtaining and planned. If so, it will be great example of the undeserved mercy of God. St.John had a vision of the earth opening to swallow the flood the dragon had spewed out to sweep away the woman, mother of the faithful (Revelation. 12.16). To what extent the earth and the oceans will continue to swallow up swelling amounts of radiation is quite unpredictable. Keepin (1990a) cites an estimate that photovoltaics and small-scale wind systems emit 10 and 20% respectively of the carbon emissions in the nuclear fuel cycle.

Cheap? The variety of costings is bewildering. Keepin (1990b) cites US estimates: the present cost of nuclear energy at 13.5 cents kWh (1987 doller); electricity from new plants will shortly cost only 5 cents kWh. This is based on what must be the economics of large scale at their zenith, mass production in a rapid switch from fossil fuel. The assumption is that capital costs will fall from the present $33,000 per kW installed capacity to $1,000 by the year 2000, surely the height of optimism. On this basis, future cost per kWh was given in pence by "The magazine of the British Nuclear Forum" may 1991, at 2.2, compared with: Coal-fired, 2.4; gas, 3.0; oil, 4.0. HMSO 205 1 (1990) records the estimates of National Power assuming a 10% CCA return and repayment of PWR construction costs over 20 years. No wonder there had been no stampede by the private sector to shoulder this industry.

Much hangs on assumptions about inflation, interest rates, pay back period, etc., but the divergence is still inexplicable to the layman. It is more than time that a definitive list of strictly comparable costings for all sources of energy was made, with proper allowance for

2647

the hidden subsidies allowed to nuclear energy and including social costs, estimated at 6.1-13.1 cents (Keepin, 1990c). Y-ARD (1989-90) provided costings for renewables in pence per kWh: large scale: tidal, 2.8; wind, 7.2; duck, 5.2; clam, 10.8; small scale: HDR, 3.9; tidal, 5.8; wind, 4.2. There is no indication of discount rate, pay back period etc. Small and large-scale are not defined. These deficiencies limit the value of the estimates. We should note that environmental costs of renewables are very low; those of nuclear energy could be very high and those of fossil fuel, especially coal, also very high.

Zweibel (1991) records how the cost of photovoltaic electricity has plummeted. In 1973 it would have been $30 kWh, if it had been translated into cost on earth; now it is 30-40 cents. It could be down to 6-10 cents by 2000 AD and will still be falling, because it will not have reached its full potential. Other renewables could reap similar benefits from advancing technology and the economies of large scale production. There is no reason why UK wind costs could not come down to the Danish level of 3.1p kW/h (2,800 turbines) or the Calafornian of 6.9c (17,000 turbines), (windirections, Winter 1991-92, Simpson, 1990). Starved of funds, renewables have often languished in retarded infancy. Or, in some cases, the placing of certain personnel in charge of their development has been like appointing King Herod as head of the RSPCC.

Between 1979 and 1990 the coal industry received £30b from the Government (*Hansard, 1991-92*). Yet the sprouting of gas fired plants recently permitted will render much of this expense vain. As with amounts lavished on nuclear power, we cannot avoid wondering what might have been accomplished if renewables had received even a quarter of what these two have enjoyed. Coal with its true environment costs added and nuclear energy stripped of its subsidies and its NFFO assistance would make a poor showing in the market place.

Peaceful? The typical light water reactor produces about 250 kg of plutonium each year, enough to make 25 nuclear weapons (Keepin, 1990d). What if a malignant narcissist such as Saddam Hussein gets hold of some? Cases of theft and highjacking of Uranium are known. Some have been on a massive scale as when a ship, carrying 200 tonnes of natural Uranium, allegedly bound for Israel, was highjacked in 1968 (Lovins and lovins, 1982).

SHENANIGANS TO DISCREDIT RENEWABLES AND COVER UP NUCLEAR
ACCIDENTS

Lovins (1977a) states that, in the mid-1970s, the case for solar power
was prejudiced by "classical" literature comparing its cost with the
pre-1973 cost of oil.

Salter (1988) observed that the firm employed by the Dept. of energy
(DEn) to investigate the wave programme in effect reduced the load
factor of his "Duck" to zero in 1983, so killing it. In 1990 this was
raised to 38%, a letter from the ETSU (23 March) to libraries directing
a revision of the report accordingly. Was it pure chance that the ETSU
letter came a month after an STV programme on the subject? In the
initial report of Nov.1980 the cable failure rate was 300 years per km
fault, reduced to 10 years in the final report of June 83. Theoretical
costings of raw materials were made at £10,000-£15,000 per tonne,
whereas manufacturer's quotations were £850 per tonne (HL paper 88).
Either the DOE paid a fee of £1.7m to support incompetence or there was
deliberate distortion. The law of Moses laid down: "Do not use
dishonest standards when measuring length, weight or quantity"
(Leviticus 19.35). To offend this statute loosens the fabric of
society. C.G. Palmer resigned as head of the Wave Programme when asked
to write its obituary, calling its closure: "a crime against nature"
(STV, "Scotish Eye" 18th feb.1990). It was also a sin against the
Creator of waves, 25-75 kW/m of which swell uncaptured off the West
coast of Britain (Lovins 1977b). Value may often be assessed by the
level of distortion to which its would - be devaluers are prepared to
stoop. It was so at the trial of Socrates. At Jesus's trial, the
chief priests manifestly lied to Pilate: "If you let this man go, you
are not Caesar's friend". Jesus had often refused the chance to be a
king, as they well knew.

The UK government kept results of the 1957 Sellafield fire secret,
because it would have jeopardized a recently forged agreement for joint
research into nuclear defence with the USA (May, 1990a). May (1990b)
also records the dismissal of Dr. Norman Macleod as Pennsylvania
Department of Health Secretary the same year that he had published data
of an abnormal number of babies in the area born with serious thyroid
defects and a dramatic increase in infant mortality following the TMI
nuclear accident. May (1989c) notes the comments of Dr.E. Sternglass,
Professor Emeritus of radiological Physics at the University of

2649

Pittsburgh that the health statistics after the accident had been: "deliberately massaged" and that a recorded: "infant mortality peak during 1979-80 had disappeared from later publications". Many shenanigans and instances of economy with the truth could be given from nations with nuclear power.

ENERGY EFFICIENCY CANNOT ALONE REMOVE THE ECOLOGICAL THREAT

The strides taken in energy efficiency are impressive. In the IEA countries, the ratio of primary energy use to GNP fell by 8% between 1973 and 1978. Denmark reduced its total use of direct fuels by 20% between 1979 and 1980. Since 1973, the USA has achieved a 25% drop in the amount of energy needed to reach a given level of GNP (Goldsmith, 1990). A DEn advertisement stated that during the years 1985-9 the manufacturing output index rose by 18.5% whilst the energy index fell by 11.8%. Savings of #500m a year were being made (*Observer Scotland, 28 June 1990*). A remarkable campaign at Newcastle Polytechnic succeeded in bringing down gas consumption from almost 28 million kWh to 20 and electricity from 6.6 million kWh to 6.0 between 1981 and 1986 (Dept of energy, 1987). However, the IEA (1989c) states that the pace of efficiency growth is slowing, though one would have thought we could all be more thrilled to thrift. It can be calculated that if every home, office and factory switched off one light bulb, which most of us could easily do, a whole power station could close down.

The IEA (1989d) also predicts a rise in energy demand of about 3% globally (2.7 in the developed; 3.8 in the developing countries, so that even if efficiency were able to maintain a 1.5% rate of increase, it would not keep up. The latest available figure for total world energy consumption is 8,013 mtoe (IEA 1989e). In terms of electricity this is 11.17 TW. Keepin (1990e) cites a reliable estimate of 21.3 TW by 2025. Almost to double consumption in so short a time does not augur well for efforts to curb pollution, especially with the present mix (90% fossil fuel) nor indeed for any mix more than 50%. Not only is world population surging by 95 million each year (in 1968, 3.5 billion people inhabited the planet; in 1990, 5.3 billion (Elrich and Elrich, 1990), but third world aspirations to a better standard of living are also galloping ahead. It is futile fondly to imagine that the poor are content with their lot. Shop window TV screens alone arouse jealously or even envy. The ancient preacher understood that:

"The eye is not satisfied with seeing, nor the ear filled with hearing" (Ecclesiastes 1.8). The eye of a starving man is not satisfied with viewing sleek First World diners nor the ear of a worker drenched in sweat filled with the hum of someone else's air conditioner. He naturally covets the combine harvesters, fertilizers, factories and energy input which bring him good dinners or a cool home. At present the less developed countries, though encompassing over 75% of the world's population, consume less than a quarter of its energy (Church of Scotland, 1991). This cannot go on. The Third World cries out for more energy. Renewables could relieve its plight, then it might heed lectures on efficiency. As First World wealth tumefies and technology advances its citizens will possess more gadgets, such as videos each one a small user of energy, but substantial in total when multiplied by millions; more fuel guzzlers, such as aircraft. They alone may tend to offset the benefits of efficient energy use.

Few are as brash as openly to affirm with Nicholas Ridley: "The fight against pollution must not be allowed to threaten business and industry's success" (Times 6 Jan.1988). Some firms, however, have gained by efficiency; others have found efficiency measures too expensive and have backslidden from ecological concern. One of the best positive examples is that of Glaxo, which has saved #4m per year by investing #2m in energy conservation (Elkington and Burke, 1989).

My concern is rooted in the belief that God became man in Christ, not a disembodied spirit or an angel in disguise. He had compassion on the sick. Therefore God must be concerned with the physical world and be all in favour of sources of energy which are a prophylaxis against disease arising from conventional energy. I admire those who do not share my belief yet have struggled to make renewables viable and put them on the market, despite much discourgement and the deplorable short-sightedness of their opponents. St.Peters prophesied that the elements would melt at the end of the age, but we have no remit to hasten the process. The men of Noah's day were oblivious to flood warnings. Are we to prove equally oblivious to modern scientists who take the role of prophet? Those of us who are persuaded that renewables can save the planet must try to stir up "people power". If it could topple the tyrants of Eastern Europe or in ancient times prevent King Saul from carrying out a wicked decree (1 Samuel 14, 45) surely it can also change the attitude of those in authority towards renewables. Politicians must be pressed into looking further than the next election; industrial magnates, further than next year's balance sheet.

REFERENCES

1. Cardwardine, M. (1990) *The WWF Environment Handbook,* Optima, London, a; 39 b; ibid.

2. Chapman, N.,I. McKinley (1990) Radioactive waste: Back to the Future. *New Scientist,* 126, No.1715, 5 May 1990, 55.

3. Church of scotland General Assembly (1991), World poverty and the response of the Church, *World Mission and Unity Reprot (Extracts),* 43.

4. Department of Energy briefing (1991). Energy expenditure.

5. EEO (1988). Energy Technology Series, No.8, Energy efficiency in buildings, EEO, p.6.

6. Elkington, J. and T. Burke (1989) *The Green capitalists,* Gollancz, London, pp.212-218.

7. Elrich, P.R. and A.H. Elrich (1990) *Population Explosion,* 4 Simon & Schuster, New York.

8. Flood, M. (1989) *The UKEA and its role in Energy Policies,* p.14, Friends of the Earth, London.

9. Goldsmith, E. and N. Hilyard (1990) *Earth Report,* 2 Beazley, London, p.66.

10. Greenpeace (1989) *Air Pollution Briefing: The big heat,* p.2, Greenpeace, London.

11. Greenpeace (1991) *Campaign Report, August 1991,* No.6, p.2, Greenpeace, London.

12. *Hansard (1989). House of Commons, Renewable energy, Written answers, 10 July 1989, a; col.384, b; col.382.*

13. *Hansard (1990-91). House of Lords, Parliamentary debate, 16 January 1991, Vol.524, Col.1158.*

14. *Hansard (1991-92). House of Commons, Parliamentary debate, 15 July 1991, col.130.*

15. Hayes, D. (1976). Nuclear Power: The fifth horseman, World watch paper No.6, p.32, World watch institute, Washington, D.C.

16. *HC 24-iv (1976) Select committee (House of Commons), Energy Resources sub-committee, 28 January 1976, Alternative sources of energy, p.xvii.*

17. HL Paper 88, (1988) House of Lords, Evidence taken before the European Communities Committee, sub-committee B, a; 199 b; 193.

18. *HMSO 205 1 (1990) House of Commons Energy committee, Fourth Report, "The cost of nuclear power", 7 June 1990, p.xxiii.*

19. IEA (1989) *International Energy Agency, Energy Policies and Programmes in IEA countries, 1989 Review,* OECD/IEA, Paris, a; 98 b;19, 53, 55 c; 35, d 54 e; 55.

20. IPCC (1990) The Intergovenmental Panel on Climate Change, *Climate change: the IPCC Scientific assessment,* J.T. Houghton et.al.,ed. Cambridge University Press, Cambridge (abbre IPCC) a; p.xii, b; pp.5-68.

21. Johnson, O. (1990) *The 1990 Almanac* (O. Johnson and W.E. Bruno, eds), ad loc. Houghton Mifflin Company, Boston.

22. Keepin, B. (1990) Nuclear power and global warming. In: *Global warming the greenpeace report (GWGR),* J Leggett, ed.,Chap 13, a; p.515 n.2, p.298, b; pp.299, 303, 305, c; p.312, d; p.314, e; p.298. Oxford University press, Oxford.

23. Lovins, A.B. (1977) *Soft Energy paths,* a; p.127, b; p.131, Gollancz, London.don.

24. Lovins, A.B. and L.H. Lovins (1982) *Brittle power,* Brick house, Andover, pp.147-8.

25. May, J.(1990) *The greenpeace book of the nuclear age,* Gollancz,; a; p.116, b; p.222.222.

2653

THE IMPACT OF ENERGY EFFICIENCY ON RENEWABLE ENERGY OPTIONS

BENT SØRENSEN

ROSKILDE UNIVERSITY CENTER, Institute for Mathematics and Physics,
PO Box 260, DK-4000 Roskilde, Denmark

Abstract. The potential for improving energy efficiency is determined, as a basis for optimizing the design of renewable energy systems.

Keywords. Energy efficiency; renewable energy.

INTRODUCTION

There is an intimate relation between the energy use density and the range of energy supply options available. The lower the density of demand, the more options become practically feasible. Decentralized energy sources such as solar and wind energy are most useful when energy use is small compared with the incident energy. Then, urban buildings may be served by collectors on rooftops or facades, detached communities by local wind turbines, etc. When energy demand is larger, renewable energy supply requires central energy plants, often placed on land not occupied by other activities, and remote from the points where energy is used. Because most renewable energy collection requires fairly large areas, compared with fuel extraction and conversion, then the fraction of energy supply, that may conveniently be covered by renewable energy, may go down if demand exceeds some critical le-vel. This makes it interesting to look at energy efficiency measures before discussing the implementation of renewable energy, and a conventional point of view will look for the break-even between saving one unit of energy and providing one unit, of course taking into account both economic and non-economic impacts of either option (Sørensen, 1979; 1991b).

ENERGY EFFICIENCY

By energy efficiency measures are meant improvements leading to provision of unaltered products and services at a lower energy spending. They can be technical improvements in each of the steps along the energy conversion chain, or they may consist in novel ways of furnishing a given energy service, e.g. through a different type of conversion process. They do not include

2654

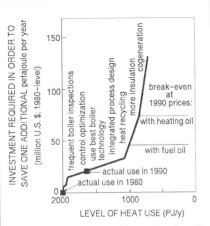

Figure 1. Marginal cost of improving the efficiency of process heat use in West German industry. The break-even levels assume a 15y depreciation period at 3% real interest (Beijdorff 1980; Sørensen 1991; IEA 1990).

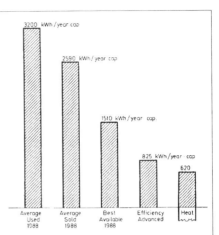

Figure 2. Total electricity consumption for appliances and lighting in Denmark, and estimated consumption if best available, advanced technology, and combined heat and power solutions are implemented (Nørgård, 1989).

changes in lifestyle or sectorial shifts. These are important because they happen all the time, but they should be treated separately from the questions of energy efficiency. A very large potential for energy efficiency improvements exist in the end-use sectors, but also in the resource and intermediate conversion steps, there is room for improvement.

Industry. In the industrial sector, substantial amounts of energy is used for motors and process heat. Adjustable-speed electronic controls of drive trains used in pumps and compressors can lower energy use by 20-35%, with investments recovered in 0.5 to 2 years (Baldwin, 1989). The efficiency of process heat systems have improved by almost a factor of two during recent decades, and further improvements of roughly 20% have been identified (Pilavachi, 1989). In addition, use of industrial cogeneration may double the efficiency (Sørensen, 1991). For example, Langley (1984) estimates the savings that may be obtained by bringing all of British industry up to current efficiency leaders, and he finds an average saving of 25% by analyzing all major industries. Figure 1 illustrates the ranking of energy efficiency measures according to economic attractiveness, for industrial process heat use in Germany.

Buildings and appliances. Building energy use depends on climate and on building use. Generally, the energy currently used for light and appliances such as refrigerators, freezers, dish and clothes washing machines, driers, cooking devices and ventilation systems, can be reduced by over 70%, with investments recovered in less than three years (Figure 2). In climates where space heating or space cooling is required, the building architecture, insulation and window types may be manipulated in such a way, that the direct energy input required becomes very modest. For new buildings, this may be achieved without noticeable increases in cost, whereas for existing buildings of lower standard, the retrofitting may be quite expensive. Average pay-back times of measures capable of halving the energy inputs have in Denmark been shown to be about 10 years. This is still a good investment, because the building residual lifetime is typically of the order of 50 years (Sørensen, 1989).

The energy efficiency of lighting is governed by the light source, the light guidance system and the arrangement of the items destined to receive the light. Presently, a mix of light sources are in use, spanning from inefficient light bulbs to highly efficient converters of watts to lumens. There is a large difference in price, too, but evaluated over the lifetime of the light source, the combined cost of light bulb and electricity to operate it turns out remarkably similar (Mc-Gowan, 1989). There used to be a difference in light quality, between incandescent, fluorescent and high pressure sodium or metal halide lamps, but new techniques have diminished such differences. Also diffe-rences in the wear caused by switching the light on and off, which used to point at some types of light sources being more useful for long sessions of lighting, while other would be preferable in case of frequent switching on and off, seem to be diminishing as the result of new technology. The end use of light is either for ambience or for specific tasks. Changes in aesthetic preferences has made it possible to increase the fraction of light being made useful, rather than being absorbed by shades and casings.

Electric appliances for home use have undergone several steps towards higher energy efficiency, and there is available technology, that could further enhance the efficiency. Most of the efficiency improvement up to the best of current standards has been achieved without noticeably increasing the cost of the items in question. The pay-back times of the extra costs associated with the most efficient appliances are estimated as follows (Nørgård, 1989):

For refrigerators, the manufacturing cost for the best available 200 litres refrigerator in 1988 is estimated at (U.S.) $25 above the production cost for an average refrigerator. To this, the consumer may wish to add another $5 due to the space taken up by the extra insulation. The pay-back time for lowering the energy consumption by 180 kWh/y is under two years. For the advanced efficiency refrigerator, a better motor and rotary compressor will add $15 to the cost, and the 40 kWh/y energy saving will have been achieved with a pay-back time of under 4 years for the additional investment. The consumer price of electricity is on average assumed to be 0.10 $/kWh, for these estimates. For freezers, the extra cost of the best available model is about $30, while the heavier heat exchanger of the advanced efficiency model will take another $20, implying pay-back times of under 2 and 6 years, respectively.

For washing machines, the most efficient model today sells for the same price as other models, so there is no extra cost implied. For the advanced efficiency savings of another 200 kWh/y, a dual water input system for both hot and cold water is required (about $40), and the pay-back period of the investment is about 2 years. The best currently available dishwasher saves 50 kWh/y with no extra cost. The advanced efficiency model is estimated to cost $25 extra and save 145 kWh/y, implying a pay-back time of under two years. For clothes driers, no additional cost is associated with the currently most efficient model, while the advanced efficiency model is expected to cost about 50$ more than current models, leading to a pay-back time of nearly 3 years. For advanced efficiency cooking systems, no additional cost is envisaged, but an industrial development effort is required. For central heating distribution systems, the best currently available technology involves control equipment adding around 40$ to the cost of the average system. The pay-back time for the associated energy saving is about 2 years. The advanced efficiency level requires new motor and pump types, estimated to add 15$ to the cost and have a

2656

pay-back time of 3 years. The best currently available ventilation systems do not involve any extra cost. The advanced efficiency systems may cost an additional 50$, corresponding to a 2.5 year pay-back time, but with a fair amount of uncertainty. For all the estimates made above, it is assumed that the more efficient equipment is introduced when a replacement is anyway called for, i.e. that no appliances are prematurely phased out in order to make way for more energy efficient types.

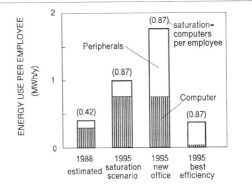

Figure 3. Energy use for computer systems in office environments. The 1988-estimate and 1995-prognosis pertains to the U.S. A 1988-1995 increase of 25% in the number of office employees is assumed (Norford et al., 1989).

For radios, TV-sets, music and video recorders and players, home computers and similar devices, there has been a substantial reduction in energy requirement, pushed ahead by the quest for increased stability of electronic components, which goes along with a reduced production of waste heat. The reduction in failure rate, and perhaps increased lifetime of the equipment, is believed to have led to near-state-of-the-art energy performance for new top-of-the-line equipment. However, there is a body of lower-quality (and lower price) equipment catering to substantial market segments. Such equipment is often characterized by use of components with lower quality, also in terms of energy efficiency. The possible effort directed at improving efficiency may thus entail ensuring a rapid dissemination of the most energy efficient techniques to all equipment, i.e. by legislation that reduces the delay in seeing this happening. Depending on the delay considered acceptable, this effort may or may not involve extra costs to the consumers.

There is a number of more rarely installed appliances (humidifiers, dehumidifiers, etc.), or more rarely used appliances (vacuum cleaners, kitchen utensils, hand tools, etc.), to which similar arguments apply. Less data is available, and for most countries, the effect on the overall picture will be small.

The service sector. The service sector comprises commercial enterprises, liberal professions and public institutions (e.g. road and airport authorities, cultural establishments and administrative units). In terms of energy use and prospects for efficiency improvements, the buildings in the service sector are covered by most of the observations made in connection with the residential sector. However, there are additional remarks related to the sales areas and display rooms, as well as related to the office environments, which will be given below. As regards transportation services, this is an important part of the service sector, which is discussed below in the transportation section.

The buildings of the service sector differ from those of the residential sector in regard to energy use, because of the different occupancy pattern. The majority of commercial buildings are primarily occupied during daytime, while residential buildings typically have their highest degree of occupancy outside working hours. This implies a higher need for air conditioning in commercial buildings (notably in those climates where cooling is required), and a higher need for space heating for residential buildings (in climatic regions where heating is relevant). One conclusion from these observations is, that the use of solar energy must exhibit a more favorable ratio between performance and cost in the case of commercial buildings, as compared to residential ones. It also indicates, that the most energy efficient window construction will not be the same for the two types of buildings. Since windows are of major importance, both in determining the heat losses from a building, and also in determining the (desirable or undesirable) solar gain of the interior, one should spent some effort in discussing the options available.

An energy efficient window must at least consist of two glass panes. Further elements may be coatings, gas fillings or solid fillings (such as foam, fibres, honeycomb structures, or microporous materials). The heat transmission through a double-pane window is basically about 3 W per square meter per degree Kelvin, but can be halved by proper coating or gas filling. An evacuated double-pane filled with silica aerogel spacer may reach around 0.5 $W/m^2/K$, but these are still in the R&D state as far as lifetime and cost is concerned. Foils, polymeric gels and liquid-crystal based materials inserted between the panes can exhibit chromogenic properties, i.e. that the transparency to radiation changes with temperature. The transition from transparency to opaqueness can take place within a temperature interval of 1.5°C, with materials costing less than 1.8 US $/m^2$ (Granqvist, 1989). This implies the possibility of self-regulating "smart" windows, that would seem extremely suitable for office environments. Currently used office windows, coated to reflect radiation from the outside, may lead to unnecessary, high energy use in case of large temperature variations (creating heating needs that would have been absent with panes transmitting radiation). The widespread installation of and regulation of shades is also an inefficient and very costly solution relative to the self-regulating windows. Smart windows may reduce the space cooling requirements by 35%, in Southern U.S. climates (Johnson et al., 1985).

Another function of windows is for display, moving now from office space to sales space. Sales

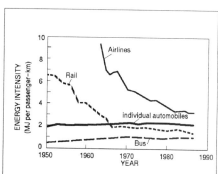

Figure 4. Development in energy intensity of passenger transportation in the U.K. (based upon U.K. Department of Energy, 1987; U.K. Department of Transport, 1987; Martin and Shock, 1989).

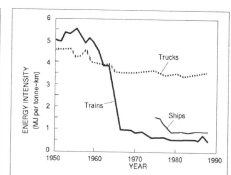

Figure 5. Development of average freight transportation energy intensity in the U.K. (based upon U.K. Department of Energy, 1987; U.K. Department of Transport, 1987; Martin and Shock, 1989).

areas and advertising displays are characterized by a high intensity of light usage, even during daytime. Furthermore the control of heat and radiation transmission through windows set other requirements than those met in office environments. This is because non-transparency of the windows can not be accepted. Not only does this rule out chromogenic control, but it may also lead to additional energy requirements for removing or preventing dew formation on the panes. Transparent gas-filled double-panes seems the best current options for such windows.

Electricity use in commercial buildings is to a large extent for lighting, and the possibilities for introducing energy efficient bulbs are at least as big as for residential buildings. The very same can be said of street and road lighting, and lights in public buildings (offices, institutions of e-ducation, hospitals, museums, stages, etc.). The actual e-nergy uses in Swedish buildings lie in the range of 80 (offices, hospitals) to 175 kWh/m^2/y (department stores)(Statens Offentliga Utredningar, 1987). The potential for improved efficiency by economically viable means is es-timated to be able to reduce the average electricity use for lighting to 15-30 kWh/m^2/y, and for climate control to 20 kWh/m^2/y (Abel, 1989).

Equipment in offices include e.g. audio, video, reproduction, telecommunication and comput-ing facilities. Their energy use has generally declined, and the best products on the market to-day have energy efficiencies that are up to 15 times better than that of the average stock (Sørensen, 1991). Figure 3 illustrates the potential efficiency improvement for office computers and computer peripherals. This segment of office electricity use is typically of the order of 40% of the total, for a saturated environment of 0.87 computers per employee. The structure of of-fice work is transforming rapidly, and the continuation of this trend is in itself going to have a substantial influence on energy use within the sector. Major areas of growth are in peripheral equipment such as laser-printers, plotters, scanners and computer-fax systems. Figure 3 as-sumes an average "on"-time of 5000 hours per year (which is low for most main-frame and workstation systems, but high for personal computers), and that the office sector grows by 25% during the 7 years considered, a feature consistent with the trend towards an "information soci-ety" seen in most industrialized countries. The penetration of software-driven technologies such as switching off idling screens is today still limited, so even the technology already in

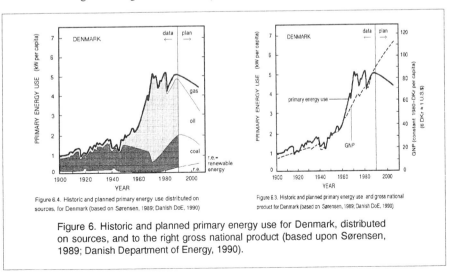

Figure 6.4. Historic and planned primary energy use distributed on sources, for Denmark (based on Sørensen, 1989; Danish DoE, 1990)

Figure 6.3. Historic and planned primary energy use and gross national product for Denmark (based on Sørensen, 1989; Danish DoE, 1990)

Figure 6. Historic and planned primary energy use for Denmark, distributed on sources, and to the right gross national product (based upon Sørensen, 1989; Danish Department of Energy, 1990).

2659

place can be used more energy efficiently. Harddisks are expected to get highly improved energy characteristics, if not being replaced by other technologies (e.g. erasable CD's). Flat screens are expected to take over completely in the near future (in computing as well as in TV and video applications). Current liquid crystal displays and plasma displays are expected to be replaced by new techniques characterized by a substantially lower energy consumption.

It is clear, that energy usage is optimized for the portable computers, which should be able to operate in stand-alone mode on batteries, but it means, that an energy efficiency of 10-15 times the current average is already available today. The portable computers are more expensive than their desktop counterparts, but for a number of reasons of which energy efficient components is only one. Still, they have other advantages which accounts for the growth rate in sales, which is believed to even rise in the next few years. The energy use in that column of Figure 3 also takes into account a shift in the composition and the energy efficiency of the peripherals: For printers, the energy use of dot matrix and typewheel printers is 10-50 W on standby and 25-100 W when printing. For inkjet printers, the corresponding figures are 3 and 10 W, while for laserprinters, they are 130 and 230 W. The trends go in the direction of replacing separate scanners, printers and fax-machines by combined units, which will reduce the idling time for each. Furthermore, the fairly energy intensive laserprinters and photocopying machines are undergoing technological changes, which could substantially reduce energy use within the next few years (use of cold-compression rollers rather than heated drums to fuse toner onto the paper). These trends are behind the indication made in the right hand column of Figure 3 (Norford et al., 1989). Electronic mail and other paperless-office trends are unclear at the moment. Potentially they could reduce both direct and indirect energy use, as well as have an impact on the forest resource problem.

Transportation. Individual passenger transportation is dominated by the passenger car, a vehicle that has undergone continuous improvements over about a century. The average new car is 35% more energy efficient than the current average stock, and prototypes with an additional 50% lowering of energy use have been constructed (Bleviss, 1988). The efficiency of collective passenger and freight transportation vehicles (including ships and planes) have improved by about a factor three since 1960, and further improvements are envisaged in prototype designs. The influence of driving styles on fuel efficiency is surprisingly large. Testing similar cars on a fixed circuit, Laker (1981) found a difference of 2 litres per 100 km between drivers being instructed to "drive normally", and the same drivers instructed to "drive economically". The cost associated with the introduction of more energy efficient automobiles is certainly not zero, but it is estimated to be of the order of the cost of other improvements made in order to stay competitive in the market (e.g. cruise control, route guidance and other electronic equipment).

Vehicles for collective transport include buses, trains (or other guided systems), airplanes and ships. For private motor cars, the present European level of energy consumption corresponds to 2.1 MJ per passenger-kilometre (one litre of gasoline being approximately equal to 32 MJ), while the corresponding numbers for bus, train and air transportation are 1.3, 1.4 and 2.7 MJ, respectively (Martin and Shock, 1989). For ships, a pure passenger estimate is more difficult to obtain, because most carriers combine passenger and freight transport. The ferries operating on major routes of passenger travel often carry trains or cars with passengers, in addition to walk-on passengers, so also here the estimate is difficult. An attempt to single out the different combinations arrives at figures of 2.1 MJ/km for walk-on passengers, 3.8 MJ/km for passenger in a car with average occupancy, and 6.1 MJ/km for passenger in train with average occupancy (Holm et al., 1980). Overall, the efficiency improvement potential is at least 15% for all types of vehicles. The most efficient among the current airline fleets has an energy consumption around half of the average, and there is a potential for a 30% improvement in energy efficiency over the next 20 years, by technical means such as improved aerodynamic efficiency and drag reduction, reducing weight of materials and improving engine efficiency. To this comes the energy savings associated with better operation of airports and air space, in order to reduce idling time at runways or in the air. The trend in the energy efficiency of passenger transport is illustrated in Figure 4.

Freight transportation include vehicles such as trucks and smaller automobiles, trains, ships and airplanes. The service sector use passenger cars, vans and light trucks. Several of the remarks made will also apply to industry use of site transportation, whereas specialized equipment used e.g. in the construction or primary resource industry is considered as covered by the discussion of industrial machinery use. Some time ago, transportation of goods by sea was the most energy efficient way, but rail transportation has taken over the leading role. Carrying freight by road is the most energy consuming way, but it should be pointed out, that the three vehicles are not interchangeable: rail and ship transport can only be used where tracks or waterways are present, and any reloading at end-points due to final destination (or point of shipping) not being served by or not being accessible by trains or ships will change the overall cost of transportation, and maybe make all-road transport economically favored. Service vehicles are usually vans or light goods vehicles (e.g. pick-ups), and the main operators are utility companies, appliance servicing companies and local merchants. The average fuel consumption is in the range of 2.5-7.1 MJ per vehicle-km, i.e. significantly higher than for passenger automobiles (Martin and Shock, 1989). The most important areas of efficiency improvements are in service work scheduling (with radio dispatch as a key technique) and correct matching of vehicle choice to task. The vehicle improvements and driver training components are similar to that of passenger cars. The trend in the energy efficiency of freight transport is shown in Figure 5.

Conversion and transmission. The efficiency of energy conversion (e.g. fuel to electricity) has gradually improved during the present century, and no radical jumps are expected for the traditional conversion processes, except for increased use of combined heat and power production and heat pumps.

DISCUSSION

Combining the numbers available for each sector of energy use, one arrives at a potential for energy efficiency improvement by known and presently economically viable technologies, which lies in the neighborhood of 40%. The time frame for achieving this amount of efficiency improvement is 10-15 years. If research and development continues during this period at the same pace as they have during the recent decades, an extrapolation of the energy intensity trend indicates, that 25 years from now, the energy use could be reduced by two thirds of the present one, if the technically available and economically sound investments are indeed made (Sørensen, 1991). The reduction in energy intensity will be compounded with any change in activity level, such as economic growth and development. An important issue in this connection is that of sectorial shifts and their implication for energy use.

We have recently seen a shift in the volume of activity, from raw materials extraction industry to manufacturing industry. This shift is characterized by a more efficient utilization of materials as well as a reduced energy intensity of products and manufacturing methods (Williams, Larson and Ross, 1987). The transformation of advanced industrialized societies to so-called information societies (i.e. societies in which the information sector is the largest sector, cf. Porat, 1985), is a reason for believing, that the per capita use of energy intensive materials will continue to decline, as the transition from agricultural, industrial and service societies to information societies proceeds and proliferates to other parts of the world. In the newly industrialized countries, industry contributes significantly to energy use, whereas in the mature industrial nations, this is no longer the case, although the industrial output per capita is not declining. In terms of human development, this trend signifies a transition from a stage, where the satisfaction of basic needs is a major effort, to societies where basic material needs can be provided with a very tiny effort, and where the big question is, what other activities are chosen by people to fill their time. In some societies, energy-intensive production-like activities are penetrating the leisure time, while in other societies, the extra free time is primarily spent on cultural and social activities. The choice has profound effects on energy requirements. Figure 6 shows one vision of the future energy demand, taken from the Danish Government energy plan (Danish Dept. of Energy, 1990).

CONCLUSIONS

The current study, which is extensively reported in Sørensen (1991), allows a number of conclusions to be drawn:
*A large reduction in energy use can be achieved through efficiency measures, that are substantially cheaper than the energy they displace. This is in contrast to the introduction of alternative sources of energy, because they are typically as expensive or more expensive than the fuels they displace (yet a full impact assessment, i.e. including indirect costs, whether expressed in monetary units or not, may still prove them preferable).
*If current average conversion methods are replaced by the most energy efficient ones among those already in the marketplace, savings of the order of 40% can be achieved. The investment cost would be depreciable over a period of zero to one or two years, except for reducing heat use in the building sector, which has a pay-back period of around 10 years. This means that if the best available technology is made mandatory upon replacement of equipment currently in use, most of the 40% reduction will take place within a period of 10-15 years.
*Research and development efforts of the past decades have improved efficiency all along the chain of energy conversions, and structural changes have generally taken a turn in the direction of less energy used per unit of activity or per unit of production. The energy intensity of most sectors has decreased by nearly 50% over the past 25 years.
*Continued improvement in energy efficiency is expected to result from ongoing research and development. If these activities are kept up, energy efficiency of the best available technology may in 5-15 years reach values, such that the energy use from the best technologies will only be a third of the present. If the use of the best available technology is penetrating the energy sector, reductions in energy intensity will be over 65% for this reason alone, some 25 years from now. This will have to be compounded with the increase in activity (perhaps a factor of two or three), but also with the effect of further structural change, where the transition from in-

dustrial to information society will affect the energy use in the down-going direction.
*The cost of continued efforts to increase energy efficiency will rise. As mentioned, the re-placement of current average technology with the best already in the marketplace entails a cost saving. The next step, from 40 to 67% reduction, is still cost effective at current energy prices, but further efficiency improvements are presently seen as associated with steeply rising costs, which can only be justified if the price of energy rises correspondingly. Of course, new - cur-rently unknown - technologies may appear, which change these estimates.
*The combined effect of pursuing energy efficiency, of economic growth and of structural change cannot be predicted with any degree of certainty, but it seems realistic to hope for a re-duction in energy use amounting to 60% (relative to the 1988 level), if governments decide to undertake the effort to make sure, that the above-mentioned energy efficiency goals are reached. A realistic time-frame for this reduction is 25 years.
*If considerations of greenhouse gas emissions leads to a decision to reduce emissions more than the 60% believed to stabilize the atmospheric concentration, i.e. to actually *decrease* the amount of greenhouse gases in the atmosphere, then further action is required, and this almost inevitably will entail the introduction of non-fossil energy sources, including renewable energy on a truly large scale.
*The cost saving associated with the first phase (the most economical efficiency measures) may be used to speed up the development of renewable energy sources on a sufficiently large scale.

REFERENCES

Abel, E. (1989). Use of electricity in commercial building, pp. 217-234 in Johansson et al., (1989).
Baldwin, S. (1989). Energy-efficient electric motor drive systems. pp. 21-58 in Johansson et al. (1989).
Beijdorff and Stuerzinger (1980). Position Paper on Energy Conservation. Round-Table Dis-cussion RT6. *Eleventh World Energy Conference.* London.
Bleviss, D. (1985). *Prospects for future fuel economy innovation.* Washington: Federation of American Scientists.
Bleviss, D. (1988). *The new oil crisis and fuel economy technologies: Preparing the light transportation industry for the 1990's.* New York: Quorum Press.
Computer Industry Abstracts (1987). La Mesa: Data Analysis Group Inc.
Danish Department of Energy (1990). *Energy 2000 - Plan of Action for a sustainable Develop-ment.* Govt. Publ.
Granqvist, C. (1989). Energy-efficient windows: Options with present and forthcoming tech-nology. pp. 89-123 in Johansson et al., 1989.
Holm, S., Jørgensen, O. and Rasmussen, H. (1980). *Energiværdier*, Report 12. København: Niels Bohr Institutet.
International Energy Agency, (1990). *Energy Policies and Programmes of IEA Countries, 1989 Review.* 608 pp. Paris: OECD Publications.
Johansson, T., Bodlund, B. and Williams, R. (eds.) (1989). *Electricity. Efficient End-Use and New Generation Technologies, and their Planning Implications.* 960 pp. Lund: Lund University Press.
Johnson, R., Connell, D., Selkowitz, S. and Arasteh, D. (1985). *Report LBL-20080.* Lawrence Berkeley Lab.

Korte et al. (1987). Possible spark-ignition engine technolog. In *Proc. Int. Conf. Vehicle Emissions and their Impact on European Air Quality*, London: Institute of Mechanical Engineers.

Laker, (1981). *Fuel economy - some effects of driver characteristics and vehicle type.* Crowthorne: TRRL Laboratory Report no. 1025.

Langley, K. (1984). *Energy use and energy efficiency in UK manufacturing industry up to the year 2000.* 2 vols. U.K. Dept. of Energy. Energy Efficiency Series, no. 3. London: Her Majesty's Stationary Office.

Martin, D. and Shock, R. (1989). *Energy use and energy efficiency in UK transport up to the year 2010.* 410 pp. U.K. Dept. of Energy. Energy Efficiency Series, no. 10. London: Her Majesty's Stationary Office.

McGowan, T. (1989). Energy-efficient lighting. pp. 59-88 in Johansson et al. 1989.

Norford, L., Rabl, A., Harris, J. and Roturier, J. (1989). Electronic office equipment: The impact of market trends and technology on end-use demand for electricity. pp. 427-460 in Johansson et al. (1989).

Nørgård, J. (1989). *Low electricity appliances - options for the future.* pp. 125-172 in Johansson et al. (1989).

Pilavachi, P. (ed.) (1989). *Energy efficiency in industrial processes. Future R&D requirements.* 416 pp. Commission of the European Community, Directorate General for Science, Research and Development, Energy Series EUR 12046. Luxembourg: Office for Official Publications of the European Community.

Porat, M. (1985). *The Information Economy.* US Department of Commerce, 5 vols., Washington DC.

Redsell et al. (1988). *Comparison of on-road fuel consumption for diesel and petrol cars.* Crowthorne: TRRL Contractor Report 79.

Statens Offentliga Utredningar (1987). *Driftel i lokaler 1987, El-husholdning på 1990-talet.* Stockholm: Report SOU 1987:69 Bilaga.

Sørensen, B. (1979) *Renewable Energy.* 685 pp. (Academic Press, London; New York, 1980, Sofia, 1989).

Sørensen, B. (1989). *The fourth International Conference on Energy Program Evaluation, Chicago*, pp. 139-142. Argonne National Laboratory.

Sørensen, B. (1991). *Selected experiences on energy conservation and efficiency measures in other countries.* 114 pp. (DASETT Series of Greenhouse Studies No. 0, Govt. Print. Canberra).

Sørensen, B. (1991b). *Energy Policy,* May, pp. 386-391; *Update* no. 46, p. 4 (United Nations, New York 1991)

U.K. Department of Energy (1987). *Digest of UK energy statistics.* London: Her Majesty's Stationary Office.

U.K. Department of Transport (1987). *Transport Statistics, Great Britain 1976-1986.* London: Her Majesty's Stationary Office.

Williams, R., Larson, E. and Ross, M. (1987). *Annual Review of Energy,* vol. 12, pp. 99-144.

THE ROLE OF RENEWABLE ENERGY SOURCES
IN THE FINNISH ENERGY POLICY

Dr John Nelson
Director - Finnpower
The Finnish Foreign Trade Association
The Finland Trade Centre
30-35 Pall Mall
London SW1Y 5lP

Since the Finnish indigenous energy resources are small, one of the main objectives of the Finnish energy policy over a number of years has been to ensure a multiplicity of energy sources. In this respect the role of the indigenous energy resources has been strengthened and the efficiency of energy production and utilisation has been emphasized. With the increasing concern over the environment, environmental impacts have become the most important element in the energy policies. In the new National Energy Strategy, which was published at the end of 1991, increased emphasis was placed on measures such as environmental protection, energy saving and the use of renewable energy sources. In the latter case, the domestic energy resources essentially comprise of water, peat and wood in its various forms. The State actively encourages the utilization of the domestic fuels and gives grants to municipalities and industry for building power plants based on domestic resources. Since 70% of Finland's primary fuels are imported a reduced energy bill has a marked effect on the national economy.

About one third of Finland's total electricity demand is covered by hydro power. This is unlikely to increase in the near future since most of the rivers offering the best potential have already been harnessed, and the remainder is protected by law to preserve the free water falls.

Finland is today one of the biggest users of bioenergy in the world. Around 19% primary energy consumption is based on biomass resources; wood in various forms accounts for 15% and peat, a slowly renewable biomass, 4%. Bioenergy is used primarily by the power plants serving the pulp, paper and other industries as well as in conventional power plants. Wood is the main renewable energy resource in Finland. Today's total energy production based on wood is around 4.1 MTOE per year. Finland has fairly large peat reserves, one third of the total area is covered by peatland. Annually, 13 million cubic metres of peat is used for energy production.

In the most optimistic calculations, is estimated that during the first decades of the 21st century at least 20% of primary energy consumption will be covered by bioenergy. The share of peat will be approximately 6% of the total energy consumption. The amount of wood that is not utilized by the pulp and paper industry is huge, 15 million cubic metres of wood is decaying in the forests. This amount of wood should be removed from the forest and utilized in energy production. It is estimated that the additional energy potential based on wood is 1.5 MTOE per year. The development of IGCC technology is an essential prerequisite for increasing utilization of bioenergy. Via this technology it is possible to double the power heat ratio and decrease CO_2 emission by replacing fossil fuels. Finland is in the forefront of the development of this technology.

The Finnish energy policy statement has emphasized the potential existing in energy conservation as well as the importance of maximising the use of bioenergy and waste materials. There is no doubt that renewable energy resources will continue to form a significant part of the National Energy Policy in Finland for decades to come.

MARKET POTENTIAL OF SOLAR THERMAL SYSTEM IN MALAYSIA

Mohd. Yusof Hj. Othman, Kamaruzzaman Sopian
and Mohd. Noh Dalimin

Solar Energy Research Group, Department of Physics
Universiti Kebangsaan Malaysia, 43600 Bangi, Malaysia

ABSTRACT

This paper reviews the market potential for solar thermal systems in Malaysia. Our study indicates that solar thermal system such as solar drying, solar water heating and process heating has a good potential for commercialization. The primary obstacle facing the utilization of these technologies is the finacial aspects.

KEYWORDS

Market potential, solar thermal systems, domestic applications, industrial applications, economic analysis.

INTRODUCTION

Malaysia experiances a high amount of sunshine throughout the year, that makes solar energy a potential candidate for energy resources (Chuah *et al.*, 1984). A market assessment was carried out for solar thermal systems. Three solar thermal technologies namely solar drying, solar hot water system and solar heating process were studied.

MARKET POTENTIAL OF SOLAR DRYING TECHNOLOGY

Solar Drying in Malaysia

There is no solar drying system installed commercially in Malaysia except demonstration systems for drying of rubber and paddy (Othman *et al.*, 1989). The rubber drying system was installed by Rubber Research Institute of Malaysia (RRIM) as a pilot system and further study on the system is still being carried out. The later was installed at one of rice mills in Kedah, Peninsular Malaysia as trial project.

Solar Drying Technology for Rubber Sheet

Malaysia is among the world largest producer of natural rubber. The annual production of natural rubber in 1987 was 1.581×10^6 tonnes (Ministry of Primary Commodities, 1988). Among the types of natural rubber produced are: Ribbed Smoked Sheet (RSS) and Air Dried Sheet (ADS). One of the important processes involved in producing both type of rubber is drying, whereby RSS requires introduction of smoke during the process. However, ADS requires only drying.

Solar Somkehouse for RSS

RRIM has constructed a pilot solar smokehouse. The specification of the system is described in Table 1. The conventional system uses rubberwood as fuel.

Table 1. Specification of RRIM's solar smokehouse.

Drying system	
Drying chamber	: 1800mm x 1500mm x 2200mm
Chamber's capacity	: 150 kg
Levels of beroties	: 2
Weight of rubber sheet	: 1 kg
Temperature	: 50°C - 55°C
Solar system	
Absorber plate	: corrugated galvanised sheet painted black
Insulator	: 12 mm polystyrene form
Panel size (each)	: 1190mm x 900mm
Top cover	: clear glass
Air circulation	: Force convection by fan

The Economic of Solar Smokehouse for RSS

Based on the system designed, economic feasibility has been calculated for both systems. A typical economic assesment based on current price of rubberwood shows that the pay back period for solar smokehouse system is only 38 month with full capacity operation of 75 kg per month (Muniandy, 1989).

In 1987 alone, Malaysia exported 404,692 tonnes of RSS rubber which is 24.7% of total natural rubber exported (Ministry of Primary Commodities, 1988). Hence, about 404,692 tonnes of rubberwood or M$12,140,760.00 (M$30.00 per tonnes) is needed to produce the RSS rubber. Futhermore, only M$3,035,190.00 is spent if a solar smokehouse is use and thus a saving of M$9,105,570.00.

Processing of ADS Rubber Sheet

In Producing ADS, currently, diesel is considered as appropriate energy source since it can be used continuously for 5 - 6 days at 58°C (140°F). At the moment, the source is considered economically feasible method of producing ADS. Standard drying tunnels are being used in the plantation. There are four tunnels in the plantation of which two are in operation.

A typical economic analysis on drying tunnel for ADS with the current price of diesel shows that the pay back period for solar drying system is only 28.3 month (Othman and Sopian, 1990). It should also be noted here that the cost calculated is only for the addition of solar panel since the cost for air circulation system (fan) is similar for both sources of energy.

Currently, there are no less than 20 such small private plantations are producing ADS drying unit using diesel fuel and of about 3 - 5 tonnes production per day per tunnel, giving a market potential of M$5.409,000.00.

MARKET POTENTIAL OF SOLAR WATER HEATER TECHNOLOGY

Domestic Solar Hot Water System in Malaysia

Domestic solar water heater has been commercially available in Malaysia since late seventies. The existance of not less than six companies involve in the manufacturing, assembling and marketing of the system tells us that the popularity and acceptibility of the system for local use is good. The companies involved are shown in Table 2. Four of the companies are fabricating the system locally or at least assembled locally and the rest import the system.

Table 2. Product availability and price with installation.

Product	Import/local	Size (gal)	Ret. price (M$)
Solarhart	import	30/60	3000/4500
Solco	import	40	
Solarmate	local	30/60	2300/3400
Aztec	local	30/60	2600/3600
Solareens	local	30/60	2500/3500
Pecol	local	30/60	2500/3500

From experience it is recommended that for domestic use, small family (3 to 4 people) should use 30 gal (136 l) and a family with more than 4 people should use 60 gal (272 l) capacity of storage tank (Ong, 1982). Based on the sale, Malaysian family prefered 60 gal system with the area of panel about 2 m². Table 3 shows the sale of domestic and industrial for solar hot water system in Malaysia in 1990. One unit is taken as a 60 gal system.

Table 3. Estimated sales of solar hot water system for Malaysia in 1990.

Product	Sales (unit)	Value (million M$)	Area (m²)
Solarhart	1450	6.525	2900
Solarmate	700	2.380	1400
Aztec	380	1.368	760
Solareens	150	0.525	300
Pecol	50	0.175	100
Total	2730	10.373	5460

Industrial Solar Hot Water System

The existing application of solar hot water system for industries varies from hotel to hospital. Table 4 shows the application which are already exist and under construction in Malaysia.

Table 4. Major projects of industrial solar hot water
system in Malaysia.

Project	Application	Cost (M$)	Status
Bintulu Fertilizer Plant	Hot water supply for hostel	510,000 ($\approx$30 m^2)	Already inst.
Sungai Buluh Golf Club	Club facilities	450,000 ($\approx$20 m^2)	Under const.
Ramley Burger	Hot water supply	60,000 ($\approx$20 m^2)	Already inst.
Hospital Univ. Sains Malaysia	Hot water supply		Under const.
Hotel Kota Kinabalu	Hot water supply	90,000 ($\approx$20 m^2)	Already inst.
Hotel in Kuching	Hot water supply	150,000 ($\approx$30 m^2)	Already inst.

Economic Aspect of Solar Hot Water Syatem

If the same 40 gal (185 l) solar hot water system uses electricity to heat the water the amount of energy consumes annually is about 4382.3 kWh and is equivalent to M$1103.40. The cost of solar water heater is M$3500.00 and the cost of electric heater is M$1200.00, hence the pay is only 25 month.

The maintenance cost for solar water heater is about M$10.00 annually for cleaning the glass covers on the absorber panels. The current tax-free interest rate is about 6% per annum and the average price of electricity for domestic consumption is M$0.23 per kWh, the economic aspect of both solar and electric water heaters is shown in Table 5 (Ong, 1982).

Table 5. Economic aspect of solar hot water system
compared with electric water heater.

	30 gal. storage capacity		60 gal. storage capacity	
	Solar	Electric	Solar	Electric
Price (M$)	2500.00	1000.00	3500.00	1200.00
Annual maint. Year 1	10.00	498.00	10.00	830.00
Year 2	10.00	498.00	10.00	830.00
Year 3	10.00	498.00	10.00	830.00
Total (M$)	2530.00	2492.00	3530.00	3690.00

From this table it can be seen that for 30 gal. water heater the pay back is a little bit more than 3 years and for 60 gal. capacity is less than 3 year. From Table 5 if the annual growth of solar water heater is 15%. Then the projected growth until 1995 can be shown as in Table 6.

Table 6. Projected potential market on solar water heater.

Year	1991	1992	1993	1994	1995
Units (60 gal.)	6279	7221	8304	9549	10981
Value (1991)	11.93	13.72	15.78	18.15	20.86

CONCLUSION

The market potential for solar thermal system exist for solar drying of agricultural produce such as rubber. A substantial saving can be achieved by using solar energy for drying of ADS and RSS rubber products. Market potential for solar water heaters are for hotels/motels, food industries, clubs and recreational clubs, housing developers for domestic application, hospitals and manufacturer of solar hot water systems.

ACKNOWLEDGEMENT

We would like to extend our sincere acknowledgement to Universiti Kebangsaan Malaysia for funding the author to attend the conference.

REFERENCE

Chuah D.G. and Lee S.L. (1984). *Solar Radiation in Malaysia*, Oxford University Press.

Ministry of Primary Industries, Malaysia, (1988). *Statistic on Commodities.*

Muniandy V, (1989). *Utilization of solar energy for rubber sheet drying.* Proc. of Seminar on Solar Drying, Dept. of Physics. UKM. 65-73.

Ong. K.S. (1982). *Use of solar water heaters in Malaysia.* Seminar on Solar Heating and Cooling. University Malaya.

Othman M.Y., Dalimin, M.N, Salleh, M.M and Yatim B, (1989). *Drying technology in agricultural produce in Malaysia: Solar option*, Clean and Safe Energy Forever Vol 2, Pergamon Press. 1540-1544.

Othman, M.Y. and Sopian K. (1990). *Energy conservation in rubber drying.* Proc. of Int. Conf. on Energy and Environment, KMIT, Bangkok. 143-152.

LARGE SCALE ACCEPTANCE OF RENEWABLE ENERGY – A RISK PERSPECTIVE

B.HARMSWORTH BSc MInstE

Data Sciences (UK) Limited, Meudon Avenue, Farnborough,
Hampshire, GU14 7NB, U.K.

ABSTRACT

This paper introduces the discipline of risk management and describes how it can support the move towards the large scale acceptance of renewable energy. The paper illustrates how risk pervades all levels of the decision making process: directing development programmes supporting system assessment and investment appraisal and providing more effective, responsive project management.

The paper concludes that the application of risk management at this critical stage in the life of the renewable energy sector offers considerable potential for smoothing its establishment as a major player in the world energy market of the future.

KEYWORDS

Risk, uncertainty, risk management, strategy, evaluation, project management.

INTRODUCTION

The characteristics of the contemporary energy market suggest that in order to achieve a breakthrough, the renewables must satisfy intense public and political scrutiny as well as demonstrating technical feasibility and cost effectiveness.

Furthermore, the impact of these market complexities is exacerbated by uncertainty and volatility. In making renewable energy decisions which impact into the medium and long term futures we stand facing what Rosenhead (1989) calls 'the funnel of uncertainty': not only is the future inherently uncertain but the further we look into the future the more uncertain it becomes. Peters (1987), in his review of the world market economy, suggests that we are now in 'an era of unprecedented uncertainty' where 'predictability is a thing of the past'.

The management of the various renewable energy programmes must also address the risk and uncertainty which is inherent in the fulfilment of technology, a significant proportion of which is novel either in terms of design or application. Success will be determined, therefore, by the ability of the programme managers to manage risk and uncertainty: preventing or reducing the impact of such factors as technical problems, unrealistic estimates and difficulties arising from socio-political and environmental influences.

This combination of factors defines the perspective for considering the role of risk management in the fulfilment of renewable energy programme objectives. It will be argued that the management of risk and uncertainty is not just good practice but is the key management strategy to achieve success in the prevailing market conditions.

PRINCIPLES OF RISK MANAGEMENT

From the project manager's perspective, there are two essential premises which justify the adoption of risk management principles. The first is that insufficient account is generally taken of events or circumstances which could prevent a project from achieving its objectives. The second is that project plans tend to be too inflexible and insufficiently robust to the consequences of changing circumstances.

Risk management aims to prepare us for the future before it arrives. This it achieves by 'testing' the project plans against the possible consequences of both damaging events or circumstances (risks) and possible variations in our perception of the future (uncertainties).

The magnitude of a risk is defined in terms of the likelihood that it will occur and the impact if it does. It is the management of risk which helps to define a low risk strategy or plan and provides the facility for rapid and cost-effective responses if risks do occur. The magnitude of an uncertainty is defined in terms of the degree of uncertainty and its significance in terms of estimating the project objectives. It is the management of uncertainty which builds the flexibility into the project which is necessary for it to respond quickly and effectively to the dynamic energy market. The vast majority of projects have performance, cost and timescale objectives. Individual risks and uncertainties will impact on these in different ways, and consequently their magnitudes are three dimensional parameters, measured against performance, timescale and cost, which need to be monitored and reviewed throughout the project's life.

A commitment to risk management is fundamentally an adjustment of the project management focus away from the conventional success-oriented approach to a problem-oriented one. Its success is heavily dependent on the setting of a climate in which every individual has a responsibility for risk identification and is encouraged to make problems visible as early as possible.

In essence, risk management recognises that the project plan should be organic; that an evolving plan is essential if the project is to be responsive to changes in internal and external circumstances. If it can be established, through a performance risk analysis, that a project is nominally feasible then a structured programme of risk and uncertainty reduction is the most effective way of ensuring that the project reaches its target.

A RISK MANAGEMENT MODEL FOR RENEWABLE ENERGY

To illustrate some of the particular applications of risk management techniques in renewable energy, consider a simple model of the energy sector decision process. The model comprises four levels: a 'national (strategic)' level overseeing the role of renewables; a 'national (tactical)' level selecting the most appropriate renewables; a 'programme' level comparing alternative technologies and a 'project' level concerned with the fulfilment of a given technology.

Level One: National (Strategic).

At the strategic level, the forecasting of energy supply and demand is a key activity. The huge uncertainties in world markets question the very validity of the long term forecasting necessary in energy policy - note the experience of the oil industry, for example, embarrassed by the 'unthinkable' fall in oil prices in the late 1980s (Marsh, 1991). The implications of accepting that the future is too uncertain for strategic decision making may be profound. Governments may feel justified in defaulting completely to market forces but equally they could pursue long term energy options not driven by short term cost-effectiveness.

Although these strategic (financial, social and environmental) factors are beyond the influence of project managers it is, nevertheless, desirable to take account of them in order to produce robust plans for the future. The most effective way to incorporate them using risk management methods is to evaluate options against a number of scenarios, each representing alternative market configurations. One could, for example, assess geothermal technology against the impact of variations in energy price, carbon taxes, market incentives, accelerating environmental concern, population growth and international conflicts. Scenarios are not used simply as a passive test but rather to provide feedback to suggest a redefinition of either programme aims or system design which could provide benefits over a wider range of future circumstances.

Level Two: National (Tactical).

In undertaking Investment Appraisals of renewable options, the assessment of risk will have a key role to play alongside the estimates of performance, cost and timescale. All estimates are, by nature, variables and to treat them as single points discards valuable information and can provide falsely optimistic results. Quantitative risk analysis provides the means to represent the variable nature of estimates and thereby provide more realistic parameters for evaluation and planning.

As an example, consider the possible future roles of offshore wind and large scale tidal power to contribute to a 2010 energy budget. One factor to be considered is the date at which the two technologies are likely to be sufficiently mature. The initial assessment is that both options could be implemented by 2010; the wind option perhaps several years earlier. A risk model of the plans to achieve the 2010 deadline might show a different picture. A thorough risk analysis would be undertaken covering all aspects both of the system and of the plan proposed to achieve the objectives. For example, the wind system analysis will reflect the risks and uncertainties of operation and maintenance in the marine environment and both options will be reviewed for the possible consequences of environmental objections. Timescale estimates for each element of the plans will then be reconfigured into probability distributions reflecting the risks and uncertainties relevant to each. Finally, a Monte Carlo simulation of the plan will be run to produce cumulative distribution functions of completion timescale as shown below.

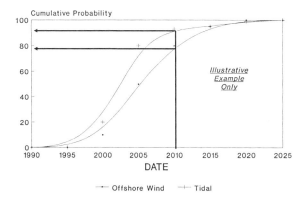

Fig. 1. Cumulative Distribution Functions of Completion Timescale

The result shows a markedly different probability of the two alternatives reaching the required target date. In these terms, the offshore wind system is more 'risky'.

A parallel exercise could be undertaken for the projected costs and once these probabilistic measures have been established they have many further uses. They can, for example, be used for analysing the sensitivity of key assumptions and changed circumstances, and tracking the progress of system development. In addition, targeting key areas of uncertainty to increase the quality of estimates and plans is essential to provide the basis for justifiable investment decisions.

Level Three: Programme.

Monte Carlo models could also be used to provide the perspective for comparing, on a common timescale basis, systems within the same programme. This is of particular value when it is necessary to take account of designs with widely disparate timeframes. (In the UK wave energy programme, for example, the Oscillating Water Column concept has already been demonstrated at sea whereas the Salter Duck is, in many respects, a second generation device still at the laboratory test stage.)

Another key role for risk analysis will be the restructuring of development programmes towards risk reduction. This will utilise a performance risk analysis to target the key risk areas for early resolution, and specify project milestones to reflect key risk reduction achievements. A performance risk analysis will consider each element of a system to identify lack of maturity and novelty of application. It will also focus on system integration, where problems can transform a set of proven sub-systems into a high risk system. Incorporating performance risk analysis into the design process should reduce risk even before management of the residual risk commences. Such an approach is particularly pertinent to the high risk devices, since priority attention to the major risks will provide an earlier 'definitive' assessment of technical feasibility.

Level Four: Project.

At the project level, the main priority will be the day to day management of risk, involving an iterative process of identification, analysis and management.

The effectiveness of risk management is founded on the comprehensiveness of the risk identification process. It should address all factors which could result in the project overspending, over-running or underachieving: uncertainty/volatility of objectives; technical feasibility; availability of resources; efficacy of tools and processes; quality of organisation and management and external factors.

The analysis stage of the process is essentially concerned with defining the magnitude of risks and considering measures which can avoid, contain, transfer or otherwise reduce them. Resulting modifications to the plans will then be assessed to establish whether they can be cost-effectively introduced to reduce risk or increase robustness.

All risk data should be held in a risk register which becomes the maintained current statement of the project's risk exposure; data discarded from the register as it is updated provide a valuable audit trail of the project decision process. Progress in managing risk can be assessed using either the output of a Monte Carlo analysis or the changing distribution of risks at discrete qualitative levels of risk magnitude.

In summary, project risk management has three objectives: firstly to reduce the risk inherent in the project by reconfiguration of risky areas; secondly to develop and implement a strategy to manage the risk that remains; thirdly, to manage the uncertainty and thereby produce more effectively directed research and plans with greater flexibility and robustness to possible futures.

CONCLUSIONS

The large scale acceptance of renewable energy requires two conditions: firstly that individual technologies achieve their potential and demonstrate credible cost-effectiveness; secondly that they are able to respond quickly and effectively to opportunities presented by the rapidly changing world market. Examples have been given of how techniques, already proven in other market sectors, can provide the rigour and flexibility of project management which will be required at all levels of decision making. This paper has demonstrated the important role which risk management will need to play if the renewable energy community is to achieve its aim.

REFERENCES

Marsh, B. (1991). Coping with uncertainty: rational investment. In: *Chaos, forecasting and risk assessment*. The Strategic Planning Society.

Peters, T. (1987). *Thriving on chaos*. MacMillan. London.

Rosenhead, J. (1989). Robustness analysis: keeping your options open. In: *Rational analysis for a problematic world*, pp. 193-218. John Wiley & Sons Ltd, Chichester.

Human energy - too valuable to be used efficiently?

R. A. Lambert

School of Development Studies, University of East Anglia
Norwich, NR4 7TJ.

ABSTRACT

In discussions on renewable energy one of the forms which is most widely used, human energy, is frequently overlooked. In poorer countries, farmers rely heavily on human energy in agriculture and in transport, with women taking the largest burden often against a background of poor nutrition. Research on renewable replacements is welcome but is unlikely to affect most of the tasks currently using human energy. There is a need for more research into improving the efficiency with which these tasks are carried out. This paper outlines some examples of the use of human energy in developing countries and the potential for improvements in efficiency.

KEYWORDS

Human energy; renewable energy; agriculture; transport; development.

HUMANS AS POWER SOURCES: UNACCEPTABLE DRUDGERY?

Despite the key role of human energy in developing countries, there has been comparatively little research on how it might be used more effectively. One reason is the preoccupation of researchers in rich countries with finding alternatives to non-renewable or mined sources. Another is the feeling that it is the moral responsibility of researchers to completely eliminate drudgery. However it is likely that the use of human energy in the rural areas of developing countries will continue for the foreseeable future. Other sources of energy, whether renewable or not, are simply not feasible. There is therefore a need to consider ways in which the existing use of human energy may be made more efficient, thereby reducing if not eliminating drudgery.

In dealing with human energy it is useful to understand how humans function as energy sources. There is a need to distinguish between the mechanical work output of the person and the food energy cost. Food energy is used for basic metabolic activities and is dissipated as heat or used to do mechanical work. The amount of energy consumed in food is typically about 8 - 13 MJ/day (2,000 - 3,000 kcal/day), with a long-term intake of less than the minimum of 8MJ/day being evidence of malnutrition. According to Grandjean (1988) the maximum amount of energy that a human can consume, averaged over a year, is 20MJ/day. For a 50kg adult, about 5MJ/day of this food energy is expended in basal metabolic activities. A further 2MJ may be expended in "everyday activities", bringing the basal-plus-everyday total to 7MJ/day. The

2677

balance remains for conversion into mechanical work. Clearly at a low food intake of 8MJ/day, there is only about 1MJ/day available for conversion into mechanical work while at the higher intake of 13MJ/day, up to 6MJ/day may be available for conversion.

The efficiency with which this metabolic energy may be converted into work is generally in the range of 10 - 20%, with a maximum of about 30%. A generally accepted value for daily work output is 250 Watt-hours (0.9MJ), which implies a food energy cost of 4 to 9 MJ/day, depending on the efficiency of conversion. This implies a fairly high total food energy intake of 11 to 16 MJ/day (2,600 - 3,800 kcal/day). The sustainable power output of a human is in the order of 1.0 W/kg bodyweight, or 50W for a 50kg person, implying an energy cost of 250-500J/s. For illustration, the mechanical work involved in climbing a flight of 12 steps, a vertical distance of 2.3m, is 1.1kJ for a 50kg person (Work = 9.81 x mass x height). At a conversion efficiency of 25% this would require 4.4kJ of food energy . If the time taken is 10 seconds (climbing slowly) then the mechanical power output is 110W.

HUMAN ENERGY USE IN AGRICULTURE

Table 1 below shows that the energy costs of much agricultural activity in Africa and Asia falls within the range of 250 - 500 J/s. The energy cost of agricultural work is not shared equally. For example, Haswell (1981) shows, for a West African region, that contribution to aggregate energy expenditure by women is 61% and men is 39%. In other parts of the continent, where more men work in towns, the burden on women can be much higher.

Table 1   Energy expenditure in subsistence agriculture in the tropics
(from Durnin and Passmore (1967) p 67)

Activity	Metabolic energy cost kcal/min	J/s
Africa		
Grass cutting	4.5	320
Clearing bush and scrub	5.8 - 8.4	410 - 590
Hoeing (women)	4.8 - 6.8	340 - 480
Weeding	3.8 - 7.8	270 - 546
Planting groundnuts	3.1 - 4.5	220 - 320
Ridging and deep digging	5.5 - 15.2	390 - 1060
Tree felling	8.4	590
Head panning, loads 20-35 kg	3.2 - 5.6	220 - 390
India		
Mowing	5.1 - 7.9	360 - 550
Watering	4.1 - 7.5	290 - 525
Weeding, digging and transplanting	2.3 - 9.1	160 - 640

Stout (1979, p55) indicates that of the gross energy input in agriculture in developing countries, in Africa over a third (35%) was supplied by humans, with the balance supplied from animal or mechanical sources. The comparable figure for Asia was 26% and for Latin America 9%. While there have been some changes since then, the use of human energy in Africa is unlikely to have decreased and, with the poor economic performance of many African countries, may have increased. A human is capable of a power output of about 0.05kW, an animal 0.2 - 0.5kW and a small tractor 20kW so the total number of farmers dependent on human energy is proportionally much higher than the gross power data suggest.

HUMAN ENERGY FOR TRANSPORT

In addition to agriculture, transport consumes large amounts of human energy. The transport burden falls particularly heavily on adult women, both in terms of time and in terms of loads carried. This transport burden has direct and indirect health implications. Curtis (1985, p10) quotes reports from Bangladesh which indicate that 50% of broken neck cases resulted from falls while carrying heavy loads. The energy required for transport is often supplied from a diet which is deficient, with severe implications for pregnant or lactating women. The need to transport water over long distances limits the quantity available for domestic use, with detrimental effects on hygiene in the home. Time spent carrying water and firewood is time that could be spent on other activities, such as childcare. A survey carried out in Makete, Tanzania (Barwell and Malmberg-Calvo, 1988, quoted by Doran, 1990 p8) showed that women spent 1,310 hours per year transporting 72 tonne-km. The comparative figures for men were 259 hours transporting 9 tonne-km. Table 4 below, summarising some of the results of a survey carried out in Chiduku, Zimbabwe (Mehretu and Mutambirwa, 1992), shows the energy cost of distance of various trips and how they affect the wife of a household. Adding the total energy cost of distance, 2.82 MJ/day, to the basal plus everyday total of 7MJ/day (Grandjean, 1988), gives a total of 9.8 MJ/day (or 2,340 kcal/day), which is higher than the mean daily supply of 8.9MJ (2132kcal) for Zimbabwe quoted by Mehretu and Mutambirwa (op cit). Clearly, there is little energy left for other activities, even if the food intake is higher than this mean value.

Table 4   Energy cost of distance with wife's share in Chiduku, Zimbabwe. (adapted from Mehretu and Mutambirwa, 1992).

| Purpose of trip | Weekly energy cost per tripmaker | | Wife's burden | |
	kcal	MJ	share %	daily energy cost MJ/day
Fetching water	1,530	6.41	62	0.57
Collecting firewood	636	2.66	63	0.24
Herding livestock	1,200	5.03	29	0.21
Local markets	2,439	10.22	48	0.70
Clinics	649	2.72	45	0.17
	Daily energy cost of these trips			1.89
	Daily energy cost of all trips made by wife*			2.82

*estimated by this author from Mehretu and Mutambirwa's data

SUBSTITUTING FOR HUMAN ENERGY

For large numbers of people in poor countries, the challenge is not to find an alternative to mined sources of energy but rather to reduce their heavy dependence on human energy. An obvious way to reduce the burden on humans is to use animals and there are many examples of the successful adoption of animal power throughout the developing world. However, frequently this may not be feasible. Wider adoption of more powerful conventional technologies, such as internal combustion engines which use mined energy, is restricted by, among other things, the large amount of capital required. It is worth noting that the skill with which motor vehicles are kept operational by those who can afford the initial capital indicates that complicated technology can be successfully adopted.

For many rural households the relevant cost comparisons for renewable energy is the cost of labour. In parts of Africa, such as Kenya and Zimbabwe, the official agricultural wage often approximates to $1.00/day. Actual wage labour rates in the informal rural sector may be much lower especially given the impact of recent structural adjustment policies and may well be as little as $0.20/day. Given a work output of 250Wh/day, this implies an energy cost of $1 - 2 per kWh, maybe even less. While this appears high compared with the cost of conventionally generated electric power (about $0.10/kWh), it is purely a recurrent cost and need only be incurred if there is a good prospect of profitable production. Furthermore, humans are very flexible and intelligent power sources. If the 250Wh/day of human energy were being used to turn a simple water pump, this could be replaced by a solar photovoltaic driven motor. With a peak Watt cost of $7.00 the amortised cost of the PV system is $1.20 - 3.00 per kWh (see Fraenkel, 1986 for method of calculation). This is close to the cost of human labour. It seems likely that costs of less than $3.00 per peak Watt PV can be expected in the next 10 to 15 years (IT Power, 1992) making such a system very competitive. The challenge is to devise some method of turning the high capital cost of the panels into a "take it or leave it" recurrent cost for the farmer, perhaps as through some form of rental arrangement.

IMPROVEMENTS IN EFFICIENCY

While it is clear that some forms of renewable energy are, on an amortised cost basis, becoming competitive with human labour in remote parts of developing countries, the widespread use of human energy for agricultural labour and transport will continue for the foreseeable future. Some examples of the potential for improvements in efficiency are outlined below.

One of the more obvious improvements is the use of the bicycle for transport. Using energy data given in Grandjean (1988, p86) the energy cost of a person travelling 1km by walking, running and cycling has been calculated at 204, 264 and 120 kJ/km - a considerable saving by cycling.

In recent research (Lambert & Faulkner, 1991) the labour time per household spent on irrigation using watering cans was reduced by the introduction of simple manually operated pumps, manufactured in the informal sector, and a piped distribution system. Figure 1 illustrates the labour time reduction for one of the adopting households. The cost of such a system could normally be recovered in the first year of production.

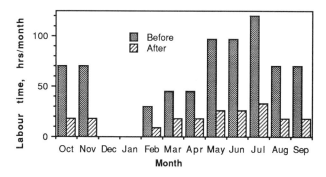

Fig 1   Time spent irrigating before and after adoption of irrigation technology

Bedale (1924) measured the oxygen consumption of women in the UK for a number of methods of carrying loads. The results are shown below in Figure 2. When converted into energy costs, the values seem low but it is notable how much variation there is between methods, indicating the scope that exists for reducing the energy cost of load carrying. Many of these technologies can be produced by the local informal manufacturing sector.

CONCLUSION

Human energy is one of the most widely used forms of renewable energy in developing countries. Technologies that use human labour, although expensive per unit of energy delivered, require no major financial capital outlay and can often be made locally. Costs are only incurred when there is the prospect of profitable production. Although the amortised costs of other renewable energy technologies, such as solar photovoltaics, are becoming competitive, the high capital costs will keep them out of reach of small farmers until an effective method is found of amortising these costs. Many tasks, particularly in agriculture and rural transport, can not be done using other forms of energy. Much of the burden of these tasks falls on women. While it is desirable to carry out research to find substitutes for human energy, there is both the need and the scope to improve efficiency with which it is used. The challenge for researchers is to find ways of making such improvements in efficiency accessible to the rural poor.

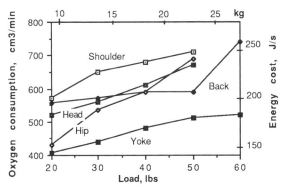

Fig 2   Oxygen consumption for various methods of load carrying
         (adapted from Bedale, 1924).

REFERENCES

Barwell, I., and Malmberg-Calvo, C. (1988). *Household travel demand in Makete district.* Village level transport survey No 192. IT Transport Ltd, Ardington UK.
Bedale, E.M. (1924). Comparison of energy expenditure of a woman carrying loads in eight different positions. *Industrial Research Paper No 29,* Medical Research Council of Great Britain.
Curtis, V. (1986). *Women and the transport of water.* Intermediate Technology Publications, UK.
Doran, J. (1990). *A moving issue for women.* Gender analysis in Development sub-series no 1. School of Development Studies, University of East Anglia, UK.
Durnin, J. V. G. and Passmore, R. (1967). *Energy, work and leisure.* Heinemann, London.
Fraenkel, P. (1986). *Water pumping devices.* Intermediate Technology Publications, UK.
Grandjean, E. (1988). *Fitting the task to the man.* Taylor and Francis, London.
Haswell, M. (1981). *Energy for subsistence.* Macmillan, London.
IT Power. (1992). personal communication.
Lambert, R. A., and Faulkner, R.D. (1991). *Simple pump technology for micro-scale irrigation.* Final Report of ODA project R4434. WEDC, Loughborough University, UK.
Mehretu, A. and Mutambirwa, C. (1992). Time and energy costs of distance in rural life space of Zimbabwe. *Social Science Medicine.* 34, 1, 17-24.
Stout, B. A. (1979). *Energy for world agriculture.* Food and Agriculture Organization, Rome.

A SOLAR REVIEW: THE TECHNOLOGIES AT A GLANCE

BY

J A Gregory
IT Power Ltd
The Warren
Bramshill Rd
Eversley Hants
RG27 0PR UK

INTRODUCTION

This paper looks at developments, projects and programmes involving the renewable energy technologies of photovoltaics, solar thermal electric, solar water heating, active and passive design, biomass and wind energy.

It does not include hydro-electric or tidal and wave energy. Nor has it included economic or price analysis, as both issues involved require much more detailed discussion than space here allows.

Rather this is an overview, at a glance, of where we are now and what is possible for today as well as for tomorrow.

SOME ENERGY PERSPECTIVES

Over the last 20 years the world price of oil has undergone many fluctuations especially during the 1973 and 1979-81 oil panics and the 1990 Gulf War.(1)

These price fluctuations in a primary energy source illustrate the rather tenuous reliance which is placed on oil. Not only is this fossil fuel finite in nature and inconsistent in pricing and supply, but it also poses environmental hazards during its exploration, extraction, refining process, transportation and end-use.(2) These include the emission of carbon dioxide, which is a major contributor the Greenhouse effect.

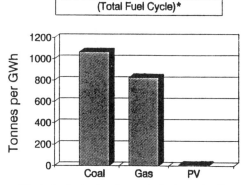

Fig.1

2682

Price fluctuations impact upon balance of trade figures, especially in countries where all/the majority of oil and oil based products are imported. This has a tremendously negative effect upon these economies in both industrialised and developing countries.(3)

Despite these uncertainties, most world economies are still dependent on oil as their major fuel resource, and many of these import over half their requirements. The United States, for instance, still imports over half its oil requirements, a figure unchanged in real terms since 1973.(4)

During the oil crises of the '70s, many world governments, especially those of industrialised countries, tried to decrease this dependency through substantially increasing their spending on renewable energy technologies. Allocations came in a variety of forms: subsidies to users, increased budgets to centres for research & development, grants to providers of energy derived from renewable energy technologies.

The sun contributes significantly to the world energy supply, with enough solar radiation being absorbed by the earth in a day to supply to world's energy needs for a year!(5) It helps also create wind, wave and tidal power, yet, until there is a crisis, we make such little use of these natural resources.

The United States initiated a number of very bold programmes. The PURPHA legislation set new possibilities for the generation of electricity outside the established utilities. Under this, Solar Tax Credit schemes were initiated whereby US taxpayers were offered a credit against tax for the installation of solar energy devices such as solar hot water systems. Wind farm developments also were covered by Tax Credits.(6,7)

The Solar in Federal Buildings programme was established in 1978 to encourage Federal Agencies to design, purchase & install solar thermal systems in federal buildings. In May 1990, contracts worth US$31 million were allocated to projects under this scheme.

However, there were problems with all these programmes. Ultimately the most negative aspect was the untimely cessation of funding, which left expensive structures without the funds necessary to maintain them. The demise of these schemes often resulted in the taxpayer bearing the financial burden for the maintenance of the installed systems, which contributed to the erosion of both public and private confidence in the technologies themselves.

Very positive aspects of the programmes included the development of many new solar and wind technologies, which have since gone onto economic commercialisation.(8)

This increase in funding for renewables also spawned many 'cowboy' operators, who entered the market to profit from the perceived 'windfall' and left it as soon as the funds (or the companies themselves) dried up, leaving in their wake many inoperative systems.

For many years overseas aid and development programmes have included a variety of renewable energy technologies. Some of the programmes have been very good and are still operating after many years whilst others, for a variety of reasons, propagated the stigma that renewable energy technologies are not reliable.

In many of the above instances the technology itself was not the root cause of any problems. Rather incidenary factors such as no provision in the budget for local training or maintenance; 'dumping' of old technologies; and incorrect technology allocation for the project were to blame.

The overall result of this headlong rush to energy independence has been that many of the renewable energy technologies have taken years to overcome the myths which have since surrounded their operation. False proclamations such as 'they don't work', there's not enough sunshine', they are too heavy for the roof', 'they are too expensive' have often spoilt the market for very real and sustainable technologies.

Which brings us to the overwhelming question: where are we today? Are there reliable solar technologies out there in the market place; do they work; are the companies reliable?

The answer is that we are poised at the threshold of a very exciting era, where lessons from the past are being applied to present and future programmes, so that technologies are not being developed by an over-extended R&D base, but primarily through market forces. The racketeers from the late '70s -'80s have dropped away, leaving players in the field who are there for the duration and not just for short-term profits. The major lending institutions such as the World Bank and the regional development bank are now looking closely at renewable energy technologies and their economic sustainability with the result that more projects which incorporate renewables are being financed.

There is a new generation of government subsidies and tax relief schemes in operation. These tend to look at the timeline needed to establish the technology/ies in the market, and, in conjunction with the operation of a phase-out scheme, these then allow the technologies to take up the full market price in a way which does not decimate the industry in the process.

The Danish government's wind energy programme is an example of how this can work (9). The British government's Non Fossil Fuel Obligation is another sustainable way of assisting fledgeling renewable energy industries to become competitive.(10)

The following is a general overview of the solar technologies currently available, various applications and initiatives with a view to stimulating future developments globally.

PHOTOVOLTAICS

Photovoltaic (PV) systems which convert sunlight directly into electricity, are extremely attractive as an energy source. This is because PV modules produce no pollution during their operation, have a life of at least 20 years for some of the material types, and require little maintenance. The two major types of PV materials available on todays' market are crystalline and amorphous silicon, with other material types currently the subject of major research and development programmes.(11)

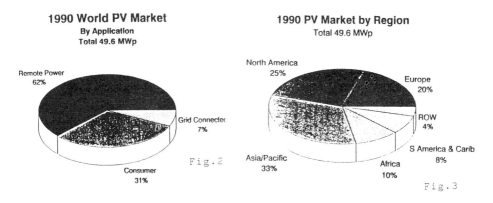

1990 World PV Market
By Application
Total 49.6 MWp

Remote Power 62%

Grid Connected 7%

Consumer 31%

Fig.2

1990 PV Market by Region
Total 49.6 MWp

North America 25%

Europe 20%

ROW 4%

S America & Carib 8%

Africa 10%

Asia/Pacific 33%

Fig.3

PV systems are used in off-grid rural areas to provide power for water pumping, lighting, vaccine refrigeration, electric livestock fencing, telecommunications, cathodic protection, water treatment, transport aids (such as highway & signal lighting and railway crossings & signals) and many others. In short, the applications are often limited more by the imagination than by the technology.

The communications sector is one of the largest markets for PVs, with applications including mountain-top microwave repeaters; rural health centres; security posts and other transceivers; radio repeaters; remote TV & radio receivers; remote weather measuring; mobile radios; remote alarms; wind sock lights; runway lighting; port navigational lighting; light houses.

Immunisation programmes for the 6 main preventable diseases in developing countries rely on vaccines being maintained at +4°C - +8°C. Vaccine refrigerators are a critical factor in these programmes, and PV-powered refrigerators are now considered both reliable and economic - especially on a life cycle cost basis where the avoided vaccine loses are taken into consideration. Over 3,000 such refrigerators have been installed world-wide in remote areas as part of immunisation programmes conducted by such organisations as the World Health Organisation (WHO) and the United Nations Children Fund (UNICEF).(12)

Other spin-off applications within healthcare programmes include the provision of lighting to rural health centres and advertisements for the prevention of AIDS shown via PV-powered video sets.

The development of PV-powered pumps was pioneered in the 1970's by primarily French companies for application in the Sahel region of Africa. In 1978 when the United Nations Development Programme (UNDP) launched a Global Solar Pumping Project, (executed by the

World Bank) the technical ability of such systems had been demonstrated only in a limited manner; it is now estimated that there are over 10,000 PV pumping systems installed world-wide.

The primary application for these PV pumping systems is village water supply and livestock watering in developing countries, with major livestock applications also in Australia and the USA. Irrigation pumps, surface water purification pumps, salt water desalination systems, circulation pumps for fisheries and ice production facilities have all used PVs successfully as their power source. Apart from the immediate economic advantage which these systems bring in the form of improved living standards and increased crop & livestock production, they also decrease the amount of time which is spent often walking several kilometres several times per day to collect household water supplies.

PV-powered village electrification systems can provide an initial phase in the provision of electricity to areas remote from the utility grid. Although inferior in amount of power supplied (at least initially) small PV remote areas power supply systems can, at a much lower price than grid extension, provide electricity for domestic lighting, electrical appliances and for charging batteries. Indeed, in Morocco a spontaneous market has sprung up for PV-powered TVs, although this may be the only electrical appliance in the house! The Navajo nation has 150 homes in Arizona which receive their electricity from PVs; and there are many remote homes across the world which rely on PVs or hybrid remote area power supply systems, including in French Polynesia where entire villages rely on solar energy.(13)

At the community level, PVs are powering mosques, government buildings, schools, village industries, street lights, battery charging stations, mountain refuge huts. Radio KTAO-FM in New Mexico USA operates solely on solar energy (14).

On a larger scale there has been significant interest during recent years in the potential for megawatt-sized PVs to feed electricity into the utility grid. This has taken the form of large central power stations, such as current PVUSA programme (15), and the new 1 MW PV power station planned for Spain (16,17).

Building-mounted distribution PV systems can generate electricity which is supplied directly into the grid. The German 2000 Roof programme is the largest government-sponsored program in the world to use PVs integrated into the roof structure to provide electricity both for the house; any excess is supplied to the utility grid.(18)

In Switzerland, two-way electricity meters are available, allowing the private generator to both buy from and sell to the local canton utility, at a predetermined price. The Swiss Federal Office of Energy has a dynamic "Energie 2000 Actionsprogramm" to install 20 MW of PV capacity by the year 2000. Installations such as the motorway sound barrier near Chur and the Mont Soleil 500 kW project are part of this programme.(19)

SOLAR THERMAL ELECTRICITY

Solar thermal electricity was the first solar electric technology to be considered for development by utilities. Put simplistically, these systems incorporate tracking reflectors which are used to concentrate sunlight onto a receiver. This receiver absorbs the reflected solar radiation converting it to heat, warming a working fluid which then drives a turbine generator.

One of the most successful systems has been developed by the Luz Corporation. Typically, a 80 MW solar electric generating system supplies the local electricity grid. The installed Luz systems operating in the Mojave Desert, California have a capacity of 350 MW peak. The electricity output replaces the equivalent of 2.3 million barrels of imported oil per year. In terms of carbon dioxide emissions, this is equivalent to replacing a typical 80 MW oil-fired plant which produces over 15,000 tonnes of CO_2 and 100 tonnes of nitrogen oxide per annum.[20]

Although the Luz Corporation ceased operation in December 1991, due primarily to problems associated with the withdrawal of the state and federal tax credits, its technologies are sound, and an international rescue bid for both the operating plants and the R&D has been launched.[21]

Another solar thermodynamic system is the central receiver. The best known of these was the test plant, Solar One, a 10 MW plant located near Barslow in California. Current demonstration plants include the Plataforma Solar de Almeria and CESA-1 plants located in Almeria Spain.[22]

SOLAR HEATING SYSTEMS

Solar drying has been used successfully in a number of countries around the world for many years, with the technologies utilised differing greatly in complexity and cost.

A very simple inflated plastic bag solar air collector, used for drying vegetables, fruits and cereal grains, has been demonstrated at the Faculty of Engineering & Technology at Mattaria, Egypt [23].

A more sophisticated grape drying process is being demonstrated in Victoria, Australia [24].

A solar kiln is being used to meet the drying needs of the small-to-medium-size wood processing enterprises in Liberia and other countries in the West African area.[25]

Loss of stored grain through insect infestation is a problem in tropical areas. A study being conducted by the Central Food Technological Research Institute in India uses flexible pouches/bags made from high-density polyurethane covered with low-density transparent polyurethane as an inexpensive solar collector. This system is capable of heating the grain to above 60°C for more than 10 minutes, which is required for exterminating the insects.(26)

A novel application for solar air collectors includes a proposal to heat an inexpensive incubator for premature/sick children in developing countries where no electricity is connected using only the power of the sun.(27)

Solar greenhouses have been used for centuries, dating back to the first recorded use by the Roman's to grow Emperor Tiberius's vegetables! Today there are many sophisticated systems around, which allow for not only heating through natural radiation, but also incorporate technologies which enhance natural daily radiation to provide warmer growing conditions.

In 1953 the first solar greenhouse in Tibet was established ; now the aperture area of greenhouses exceeds 700,000 sq metres, with an annual vegetable production rate of 225,000 kg/ha.(28) In China a unconventional application for solar greenhouses is in the production of tar for road building!(29)

Solar greenhouses (or atria) can be also designed to provide heat for buildings, as part of an active or passive solar design system.

There are a number of solar cookers on the market. Their advantage is that their only fuel requirement is the solar radiation, and there are many areas where they have been very successfully introduced.(30) Sometimes, however, they are not appropriate for the local culture: cooking by the sun requires that meals are prepared outside and during the times of peak heat.

In Zimbabwe, a country undergoing an increasingly severe drought, the Director of Energy Resources and Development has stressed the need for solar cookers as a means of off-setting increasingly scarce biomass resources.(31)

SOLAR WATER HEATING

Solar water heating is a mature technology which has been around for many years and of which many thousands of systems have been sold. There are a number of system types to choose from, ranging from simple solar batch water heaters, to thermosiphon systems, to the highly efficient and more expensive evacuated tube technology.

Applications range from solar hot water baths in Tibet to domestic hot water systems in Israel and Cyprus. High temperature steam generating solar collectors are capable of producing steam and very hot water suitable for hospitals, hotels, public buildings, apartment houses, factories, district heating and swimming pools.

In the Indian state of Gujurat the government has a programme to promote renewable energy research & applications. Solar hot water systems have been installed in dairies, hospitals and pilgrimage centres, and 2,400 domestic systems are in operation.(33) In Egypt, the New & Renewable Energy Authority has a development project where solar collectors are used to heat water for a dairy.(34)

In Australia, the 'Solar Energy Report' shows the daily summer percentage of energy which could be saved using a solar hot water system. Melbourne Metropolitan newspapers and 2 state-wide television News programmes display the Report on week days during their weather sections.(35)

The Green Paper on Renewable Energy & Energy Conservation produced by the Department of Industry, Technology & Resources in Victoria concluded that industrial process heating systems have the potential to provide 40-50% of heat needed for applications below 200°C, and could provide 30-40% of energy up to 500°C. Demonstration projects such as those at Campbellfield Hospital in New South Wales conducted by the University of Sydney, and the active solar water heating system installed at St Rose Hospital in San Antonio, Texas would seem to concur with that finding.(36) A report by the UK Energy Technology Support Unit, due to be published in late 1992, will look at the potential for water heating systems in both the UK and in Europe.

ACTIVE SOLAR DESIGN

An active solar energy system typically uses solar collectors (air or water) to absorb direct and indirect solar radiation. The heat produced is then channelled into the building structure to assist in heating either directly or indirectly. There are a variety of heat storage possibilities, such as phase-change technologies, floor storage systems (such as heating concrete) and rock beds. In Rhayadar, mid Wales, a solar village is being designed. This project is unique in that the heat storage facility is provided by 5,000 litres of water stored above the ceiling of each level!(37)

Active solar systems can also be designed to assist in cooling the building.

Active solar systems is one of the focuses of the International Energy Agency Solar Heating & Cooling Programme. Task XIII looks at building designs, whilst Task XI includes programmes in the Passive and Hybrid Solar Commercial Building programme.(38)

In Switzerland a system is being researched whereby heat collected during the summer is stored underground and released to heat the building in the winter months. This seasonal storage project comprises a 18,000 m² office building and 500 houses. Two methods of storing summer heat nearby in chalky soil are being tested. In both, the heating of the soil is done by water heated by solar panels, which is then circulated to a depth of approximately 15 meters. According to the research, this method is capable of providing up to 58% of winter heating requirements.(39)

Applications of active solar systems for buildings, either for heating or cooling, are especially appropriate in severe climatic zones or in areas where energy costs are very high. Passive design can otherwise usually satisfy the occupants' internal thermal comfort needs in more temperate climates.

PASSIVE SOLAR DESIGN

Passive low energy architecture refers to the architectural practices which utilise the local environment and the building element to naturally heat, cool and light a building: ie, in essence, the building becomes the energy system. The design of the building is the vital ingredient - designing for the climate instead of in spite of it, and designing for function not just aesthetics.

Passive low energy architecture is not a new phenomium, but rather an old one rediscovered from the Roman, Greek and other ancient civilisations such as in Tibet, where the Sunshine hall was designed and built some 3000 years ago.(28)

In 1953 the commercial Solar Building at 213 Truman NE, Albuquerque New Mexico was built. At this time, it incorporated new technologies for the internal thermal comfort of the building; these systems and the building are still in use today, and the building itself has been put on the US Historic Buildings Register.(40)

Passive low energy architecture allows the occupants to have control over the internal environment, instead of being dictated to by the programmed system. How many times when working in an air conditioned building have you been too hot or too cold or too stuffy, and wanted to throw open the windows for some fresh air? And maybe, if this were possible to do, the system would then be out of balance! Incorporating passive solar principals in either a commercial or a domestic building in turn makes the building more energy efficient, thus cutting down on the environmental cost of using fossil fuels to provide an artificial internal atmosphere which is neither comfortable nor desired.

Today the proportion of total energy use attributable to buildings generally ranges upwards of 40% in countries such as the USA, 30-35% in OECD countries and as low as 10-15% in non-industrialised countries.(41) These figures can be reduced drastically by incorporating passive low energy principals and energy efficiency measures.

Domestic, community and commercial buildings can utilise effective design, and these principals are being actively encouraged by many government programmes, as well as individual architects, engineers and builders. Schools such as the Park

Ridge Primary school in Melbourne (42), and the school at Tournal, Belgium (43) are able to heat, cool and light their interiors using the natural energy from the sun.

In the United Kingdom, the Department of Environment's Best Practice programme is the responsibility of the Energy Efficiency Office and managed by the Building Research Energy Conservation Support Unit. The Passive Solar Design Research programme is the responsibility of the Energy Technology Support Unit as part of the Department of Trade & Industry's Renewable Energy R&D programme. Whilst both programmes ultimately target builders and developers, they have addressed different aspects of energy-conscious design. The Best Practice option addresses both domestic and commercial buildings through the provision of impartial information to the market. It also supports up to 49% for research which will lead to good practice tomorrow.

The Passive Solar programme has focused on overcoming technical uncertainties surrounding design details and their potential to save energy. The majority of resources has gone to research work primarily in field trials to determine how passive solar measures perform under UK conditions.(44)

The French Agency for Energy Management (AFME, ADEME) has a research programme for passive solar design, which incorporates both domestic and commercial buildings. This is being conducted in conjunction with industry, and has been running since 1990. It also includes participation in the European Commission's PASSYS (passive solar components & system development) programme.(45)

In Denmark there has been a long-term commitment to energy efficient building design. The Danish parliament, realising that government involvement was essential as a stimuli to the implementation of the technology, offered incentives to building owners, and implemented a realistic pricing policy for energy. For the last decade Danish real estate agents have had to list the energy improvements of houses at the time of sale.

The principals of facing living areas to the sun, internal thermal mass, correct positioning and sizing of windows, the use of insulation, cross flow ventilation are elements which can be used in any climate.

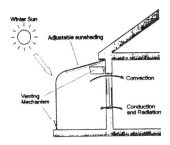

A solar sunspace/greenhouse (or atrium in commercial buildings) can be designed to provide winter heat for the building, and to assist in cooling the building in summer.

Fig.4

2691

In Middle Eastern countries, traditional house structures reflected both the climate and social traditions. The houses were designed to cope with daily temperatures averaging 45°C in the dry season to 25°C the wet season (and often below 0°C in the evenings). However, in many areas, other architectural influences and changes in social patterns have caused a move away from buildings which use their design to create a pleasant internal comfort level, to those which rely on mechanical, energy intensive means. A shift back to the traditional, whilst still incorporating modern, design methods is starting to happen.(46)

BIOMASS

Renewable energy sources currently contribute around 25% of the world's primary energy needs. Of this the two largest resources are biomass (14%) and hydro electricity (6%). This is equivalent to 1257 MTOE of biomass contributing to a world energy consumption figure of 9067 MTOE. Put another way, of the total energy resource of 400 Ej, biomass is responsible for 55Ej.(47)

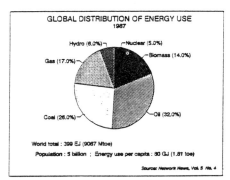

Fig.5

In developing countries, estimates show that around 35% of energy needs are provided from biomass, with countries such as Ethiopia, Rwanda & Sudan receiving over 90% of their primary energy from biomass. In rural China and India, over three quarters of the energy comes from biomass.(48)

In the United States, biomass is responsible for 3 - 4% of the total primary energy consumption, which is roughly equivalent to the energy produced in the States by nuclear energy. The biomass energy is produced from small scale biomass to electricity plants operated by the utilities, energy-waste plants, and a large proportion is produced by domestic wood-burning stoves.(47)

Biomass can be burnt directly for heat to generate steam or gases for the production of electricity, fermented to make alcohol fuels, extracted as oils from plants as diesel substitutes, anaerobically digested to biogas.

In many developing countries wood, the traditional source of biomass fuel, is becoming scarce. In regions of Africa, women have to spend many hours each day foraging for enough wood with which to cook and to provide heat for the colder evenings.(49)

In Myanmar (Burma), firewood, charcoal and biomass residue produce almost 80% of total energy consumed in 1988/89. In 1989, this still left a need to import 890,000 barrels of oil. The result of this reliance on biomass has been deforestation; and reliance on imported fuels means that the money spent on importing cannot be used to assist in the creation of an indigenous and wealth-creating sustainable energy resource.(50)

Governments, NGOs and international agencies are initiating programmes in developing countries to increase the availability of biomass, such as through the implementation of agroforestry systems.(51)

There are a number of fuel-efficient stove programmes in operation (52). In Kenya, ITDG has a one-pot Maendelo stove project. This stove is fuel efficient, has low smoke emissions and includes local manufacture by women's group. This not only adds to the participants' economic well-being, but also benefits the community as a whole by reducing fuel consumption and costs, mitigating health problems caused by the inhalation of smoke, and freeing up the women's time from cooking and fuel gathering.(53)

In two south Indian villages, biogas and wood gasifier electricity generation add to the socio-economic well-being of the whole community; similar benefits are being gained in Nepal, where a community forestry for energy project is successfully employed.

In Brazil an alcohol fuel programme has been operative since 1975. Ethanol is produced from sugar cane and used as a fuel substitute in vehicles. In 1991 over 4 million vehicles ran on pure ethanol; the rest, some 9 million vehicles, drove on a blend of 20% ethanol with petrol.

Stillage, a by-product of the ethanol production process, is being used as fertiliser, and sugar cane bagasse is being used as animal feed and for the generation of electricity.(54)

Zimbabwe also has a sugar cane - alcohol programme, with over 40 million litres of ethanol being produced annually. As in Brazil, the stillage is being used as fertiliser, cattle fodder and as an electricity generating source. Kenya and Malawi have since established similar programmes.(46)

In industrialised countries, the problem is not so much the loss of biomass wood resources so much as the over-abundance of an increasing source: garbage. In the past, house-hold, industrial and agricultural waste was usually either put into landfill tips and forgotten, or inefficiently burnt, releasing harmful Greenhouse gases (especially methane form landfill tips) and other pollutants into the atmosphere.

There are now a number of biomass conversion technologies coming onto the market which are able to produce heat, electricity, gaseous and liquid fuels and liquid and solid by-products from biomass materials.(56)

Coarse refuse fuel has been used in the United Kingdom, United States and Sweden largely as a partial coal substitute.(57)

In Eye (UK), a 14MW power station is being built, at a cost of £22 million. This plant is designed to burn poultry litter in an environmentally-acceptable way, and will burn up to 130,000 tonnes of litter a year.(58)

In Sweden, around 13% of all primary energy comes from biomass, and there are discussion underway proposing to increase this to 50%. (59) Many of the 150 district heating plants now run on biofuels not oil. Some use wood, other rely on refuse or peat.(60)

In the United States, the biomass industry now generates over 9,000 MW (compared to only 250 MW in 1980), resulting primarily from the PURPA legislation. Ethanol as a fuel extender has also increased to an estimated 5.6 billion litres with 60 plants in 22 states, and is projected to increase to 10.6 billion litres in 1995.(61)

After the 1992 harvest, the burning of surplus agricultural straw in the field will be banned in the UK. This amounts to some 5 - 7 million tonnes, even though straw is employed for farm uses and in industry. Straw as a fuel is already being utilised in the UK (170,000 tonnes of straw is used to fire boilers on farms); in Denmark, the world leader in the application of this technology, 54 straw-fired district heating systems are in operation.(62)

WIND ENERGY

Today there are many wind technologies to chose from, ranging from small water pumping systems to large megawatt electricity turbines.

1 - 2 million mechanical water pumping windmills are installed world-wide, with primary markets being the USA, Australia and with over 680,000 having been installed in Argentina. Windmill pumping systems have been used for many decades and applications include irrigation and drainage (low lift, high volume), livestock watering (small volumes) and domestic water supply.

There are a number of types of electricity-generating wind turbines, varying in size from 20 W to 2 MW.

Small turbines can be used to charge batteries, power household appliance (as part of a stand alone power system) and as the energy source for telecommunications relay stations. (63,9)

Around the world, most larger wind turbines are grid-connected and are either configured to sell electricity directly to the utility or to provide electricity to the local load. Some 1500 MW of wind turbines have been installed in California, and over 500 MW have been installed in Europe.

A report undertaken by the European Wind Energy Association (and partially funded by the EC Directorate for Energy) found that 'studies indicate that Europe's existing and well-dimensioned electricity grid can absorb and distribute wind power to meet at least 10% of the Community's electric energy needs by 2030'. (64) Some EC countries have quite extensive programmes in operation; other are not so advanced.

In the Netherlands there is over 150 MW of installed capacity, and a target of 1000 MW by the year 2000. Recent changes to the regulations governing wind power subsidies are likely to increase the number of machines in the market.(65)   Greece plans to have an installed capacity of 18 MW by 1993; the Spanish utility, Union Fenosa, is set to install a 3MW wind farm at Camarison in Galacia, making it the country's largest (66); and Germany has subsidies to encourage the installation of 200 MW of wind plant over the next few years, with 4 new small wind farms being granted financial support in January 1992.(67)

In Denmark, there are in excess of 2000 turbines, with a total installed capacity of 360 MW, producing 1 - 2% of the country's needs. Of this there are 62 listed wind farms, with an average installed capacity in excess of 1 MW and the majority of which are owned by the electricity utilities.(9,64)

The wind industry in Denmark has been assisted by government support in two areas. In research & development, the programme has been updated every year since 1980 for the following 3 years, thus maintaining a continuous effort in R&D over 12 years, and avoiding the 'stop-go' programmes of other countries.

The introduction of a subsidy in 1979 precipitated the emergence of small turbine manufacturers. In 1979 the subsidy was 20%; in 1980 it was increased to 30%; it was subsequently was decreased as the market become able to cover the costs, and was completely removed in 1989.(9)

The United Kingdom is supporting the development of local wind farms through the Non Fossil Fuel Obligation. The 1990 Order included 9 wind projects, representing 12 MW; the 1991 Order contained 49 wind projects with an overall capacity of 82 MW. Britain's first commercial windfarm, located at Delabole in Cornwall, commenced operation in December 1991. This will supply more than 10 million kWh of electricity per year, which is enough to provide energy for 3,000 homes.(10,68)

In the South East Asia region, India currently has 16 MW of wind power in the state of Gujarat, including the newly inaugurated 10MW wind farm, 16 MW in the state of Tamil Nadu, and a further 6 - 8 MW spread around the rest of the country, bringing the installed capacity in the country to around 40 MW.(69)

Offshore wind farms are beginning to attract attention. The Danish utility Elkraft has recently commenced production from its offshore wind farm at a site near Vindeby. This will produce 12 million kWh per year. A Danish utility, Elsam, has announced plans to establish another offshore farm. Although production of an offshore facility is twice as expensive as that of a land facility, better performance due to increased wind regimes and therefore higher potential profits are the quoted driving forces.(70)

TOWARDS A SUSTAINABLE SOLAR FUTURE

Over the past 10 - 20 years, many advances have been made in the demonstration of technology feasibility in renewable energy systems. Performances have improved and prices generally decreased, though the development of economic assessment procedures has lagged well behind. This has meant that renewable energy systems have not been costed on a playing field level with more conventional energy sources. On this basis, many potential projects have been put aside as 'uneconomic'.

However, critical capital costing mechanisms are starting to be applied to renewable energy projects.(71) Factors such as economic sustainability and environmental impact are beginning at last to be applied to conventional energy project. The Global Environment Facility is one international fund which is taking these factors into positive account.(72)

Successful, environmentally-sustainable, development-orientated projects which incorporate renewable energy technologies have a number of features in common: the technology/ies are matched to the users' needs; the systems are installed correctly; there are adequate maintenance programmes; and they are financed in such a way so that financing of the project does not hinder its completion or continuance.

Government programmes which incorporate long term planning for the technology and its place within the overall economy (from small water pumping systems to large grid connect projects or energy efficiency building programmes) impact in a positive manner on the development of the technologies and their market development.

Today there are many successful examples of institutional, private and public sector programmes in both developed and developing countries; where this is not the case, one or more of the above ingredients is missing.

Solar energy technologies are available for a variety of end-uses, from the simple to the complex. There are many more projects and programmes than the ones briefly outlined above.

It takes perhaps a dash of imagination added to the above ingredients, and a belief that we are only guardians of this world for tomorrow's children, to generate globally sustainable development through the use of renewable energy technologies.

REFERENCES

1. '1991 Idris Lecture: Primary Energy - Regional Consumption Patterns 1989', N C Coleman, Energy World, No. 193, November 1991
2. Environmental Considerations in Energy Development, Asian Development Bank, May 1991.
3. 'Hawaii condo calm as utility rates increase', Solar Industry Journal, 4th quarter 1990, Vol. 1, Issue 4.
4. Solar Industry Green Plan, A call for Action, Solar Energy Industry Association, 1991; 'The Renewable Energy Plan. A Research Plan for the Future', Office of Renewable Energy, US Department of Energy, Washington.
5. 'Large Scale Solar Power Generation', B McNelis, proceedings of Solar Energy in the 80s, London, 1980.
6. 'The Renewable Energy Plan. A Research Plan for the Future', Office of Renewable Energy, US Department of Energy, Washington.
7. 'Solar Tax Credits: The US Experience', Laura J Sallman-Smith, World Renewable Energy Conference, Reading 1990.
8. 'Overcoming Institutional Barriers', Solar Industry Journal, First Quarter 1991, Vol. 2, Issue 1.
9. Wind Energy, Research & Technological Development, 1990, in Denmark, Ministry of Energy, Danish Energy Agency, Copenhagen, 1990.
10. PAEG Bulletin, Parliamentary Alternative Energy Group, Special Issue, December 1991, UK.
11. 'Solar photovoltaic power: a US electric utility research & development programme', Electric Power Research Institute, Palo Alto, USA, 1990; 'Electricity from Sun and Wind', Electric Power Research Institute, Palo Alto, USA, 1991.
12. 'Photovoltaics in Developing Countries', SunWorld, Vol 15, No. 5, 1991.
13. 'Renewable Energies for Australasia', Solar Progress, Vol 12, No.1.
14. 'Radio station goes Solar', SunWorld Vol 15, No.5 1991.
15. '1989-90 PVUSA Progress Report', PVUSA Team, Department of Energy, USA.
16. 'German-Spanish solar station', Independent Power Generation, January 1992.
17. Independent Energy, March 1992.
18. 'Technology regulations and consultation concepts for small grid-connected PV plants in the German 2000 Roof program', 10th EC PV Conference, Lisbon, 1991.
19. 'Energie 2000, Actionsprogramm', Federal Office of Energy, Switzerland, February 1991.
20. SunWorld, Vol. 14, No. 4, 1990; 'Electricity from Sun & Wind', ibid.
21. 'Major solar power company runs into trouble', Independent Power Generation, January 1992.
22. 'Electricity from Sun and Wind', Electric Power Research Institute, Palo Alto, USA, 1991; Guidelines for the Economic Analysis of Renewable Energy Technology Applications, International Energy Agency, 1991.
23. 'Design of a combined system of solar drying and anaerobic digestion for the treatment of food processing residuals', H A Heikal, M N Metwally, S A El-Shemi, Appropriate technology for rural development: Solar dryers & photovoltaic systems, papers presented at a workshop in Rabat, Morocco organised by the International Foundation for Science in collaboration with the Institut Agronomique et Veterinaire Hassan II.
24. 1989/90 Annual Report, Energy Victoria, Melbourne Australia
25. 'Design of lumber dry kiln using solar energy for small and medium size woodworking enterprises', I K A Okoh, T Debey Sayndee, Appropriate technology for rural development: Solar dryers & photovoltaic systems, ibid.
26. 'Disinfesting grain with solar heat', Appropriate Technology, Vol 18, No.3, Dec 1991.
27. Proposal by IT Power Ltd, The Warren, Eversley, UK
28. 'Tibet shares the sun', SunWorld, Vol 15, No.5, 1991.

29. 'Paving the way: the use of solar energy in China', SunWorld, Vol 14, No.4, 1990.
30. Heaven's Flare: A Guide book to Solar Cookers, Joseph Radabaugh, Home Power Inc., Ashland, USA.
31. 'Zimbabwe calls for more solar power', Independent Power Generation, January 1992.
32. 'Solar hot water heating: a guide for home builders', Solar Industry Journal, Vol 1, Issue 4, 1990; 'Power from Sunlight', International Power Generation, January 1992.
33. 'Reviewing India's Gujurat: How a state in western India is serious energy shortfalls & supporting developments by using renewable energy technologies', SunWorld, Vol 15, No.1, 1991.
1990.
34. 'An Overview of Egyptian Renewable Energy Programs & the Renewable Energy Field Testing project', New & Renewable Energy Authority, Nasir City,, Egypt, 1990.
35. 'Solar Energy Report - media exposure for the solar resource, J A Gregory, K Guthrie & E Stamotopolous, paper presented at the Solar '88 conference, University of Melbourne, 1988; 1989/90 Annual Report, Energy Victoria, Melbourne.
36. Green Paper, Renewable Energy & Energy Conservation, Department of Industry, Technology & Resources, Melbourne, 1990; 'Innovative solar water heating system installed on Texas Hospital, Solar Industry Journal, Vol 2, Issue 1, 1991.
37. 'The Sun-powered Village', Review, Issue 16, Summer 1991.
38. Solar Update Newsletter, International Energy Agency Solar Heating & Cooling Programme, No.17, May 1991. **ADDITIONAL**
39. 'Concept Energetique de l'Office Federal de la Statistique a Neuchatel, stockage saisonnier d'energie solaire', P Chuard, D Aiulfi & P Jaboyedoff, Proceedings from the CISBAT conference, Lausanne, October 1991.
40. 'World's First Solar Office Building', Solar Industry Journal, Vol. 1, Issue 4, 1990.
41. 'Energy Consumption in Office Buildings'. Energy Management, March/April 1992; 'Energy Economics and the environment in the Building Sector', Solar Industry Journal, Vol.1, Issue 4, 1990.
42. 'Lights out, School is in', Park Ridge Primary School, Taylor Oppenheim, Melbourne.
43. Passivsolare Gemeinschafts-und-Geschäftsbauten, Bundesamt für Energiewirtschaft Forschungsprogramm Solararchitektur, S Robert Hastings (ed), Brugg, 1990.
44. 'Turning Passive Solar Design into Best Practice', David Vincent, Department of Environment, Review, Issue 17, Autumn 1991.
45. La Lettre de l'Architecture (Bio) Climatique, No. 6, Automme 1991; 'Solar energy applications to buildings; Solar radiation data', Energy Research & Development Contractors' Catalogue 1987, T C Steemers & G Caratti (Eds), Commission of the Heating & Cooling Programme, No.17, May 1991.
46. 'Energy Conscious Design Concepts for the Arabian Gulf areas', SunWorld, Vol. 14, No. 3, 1990; Solae Efficient Design Manual for Housing, J A Gregory & F Darby (Eds), Enery Victoria, 1990.
47. D Hall & J Scurlock, Biomass Journal, 1990
48. Network News, Vol 5, No.4. 1991, Biomass Users Network, 84 Soi Rajakroo Paholyothin Road, Bangkok 10400.
49. Women & Environment in the Third World, Irene Dankelman & Joan Davidson, Earthscan Publications, London, 1989.
50. Environmental Considerations in Energy Development, Asian Development Bank, May 1991; 'Energy & People: A Dossier on Woodfuel in the Developing World', Commission of the European Communities.

51. Women & Children First, Recommendations from the UNCED symposium of the impact of environmental degradation and poverty on women & children, Geneva May 1991; Network News, Vol 5, No.4. 1991, ibid; Biologue & the Regional Eneeergy Program Report, Vol 8, No.3, Sept. 1991

52. 'Build an oven, Cook a meal': How solar energy empowered women in Costa Rica', SunWorld, Vol. 14, No.4, 1990; Dankelman & Davidson, ibid.

53. 'Benefits all round for women stove producers in West Kenya', & 'Cooking stoves: extra utility at the right price', Appropriate Technology, Vol 18, No.3, Dec 1991.

54. Network News, Vol 5, No.4. 1991, ibid.

55. 'Distillate production of a simple direct solar still', SunWorld, Vol 14, No.3, 1990.

56. 'Energy from waste: Clean, Green & Profitable', Energy World, November 1991; 'Go-ahead for SELCHP' & 'Solid waste to energy', Independent Power Generation, January 1991.

57. Network News, Vol 5, No.4. 1991, ibid.

58. 'Financing the Renewables' Review, Issue 15, Spring 1991.

59. Network News, Vol 5, No.4. 1991, ibid.

60. 'The Swedish Energy Dilemma', Review, Issue 15, Spring 1991.

61. Network News, Vol 5, No.4. 1991, ibid

62. 'What next for straw?', Review, Issue 17, Autumn 1991.

63. Time for Action: Wind Energy in Europe, European Wind Energy Association, October 1991.

64. 'Wind Energy - the right approach', Independent Power Generation, March 1992. Time for Action: Wind Energy in Europe, European Wind Energy Association, October 1991.

65. 'Private Developers granted larger share of subsidy cake', 'Wind Power Monthly, Vol 8, No.2, February 1992..

66. 'Union Fenosa starts joint venture company', Independent Energy, March 1992.

67. 'Four new wind farms and a positive public survey', 'Wind Power Monthly, Vol 8, No.2, February 1992.

68. 'The first of many', Independent Power Generation, March 1992; 'First windfarm running at last', 'Wind Power Monthly, Vol 8, No.2, February 1992.

69. 'Private sector catches on fast to power shortage solution', Wind Power Monthly, Vol. 7, No.6, June 1991; 'Indian Inauguration', 'Wind Power Monthly, Vol 8, No.2, February 1992.

70. 'First off-shore windpark now operating', Independent Energy March 1992.

71. Professor Olav Hohmeyer, Fraunhofer Institut für Systemtechnik und Innovationsforschung, Karlsruhe (D); Professor Robert Hill, Newcastle Polytechnic, Newcastle upon Tyne (UK); Dr Richard Ottinger, Pace University School of Law, New York (USA).

72. 'The Global Environment Facility', IBRB, Geneva Office, August 1991; UNCED PrepCom III, Statement by the World Bank, 1991.

CLIMATIC VARIATIONS IN THE NORTHEAST OF THE
IBERIAN PENINSULA

J. E. SEGARRA and J. J. RIVERA

Department of Meteorology and Soil Science.
Polytechnical University of Catalonia
Av. Bases de Manresa, 61-73
08240-MANRESA (Spain)

ABSTRACT:

In this task is shown the change suffered, by some meteorological variables, especially in the latest years. In concrete the minimum and maximum temperatures, although in different degrees. The number of days with frost have decreased appreciably, in the same manner as those with snowfall. On the other hand, the number of hours of sunlight has increased, and there haven't been great changes in the annual precipitation. As a whole, these alterations demonstrated give an impression of a notable climatic change. If this is due to an increment of the "greenhouse effect", it seems a more probable hipothesis each time.

KEY WORDS:

-Mean values of the minimum, maximum and average temperatures.
-Annual rainfall. Hours of sunlight.
-Greenhouse effect.

INTRODUCTION:

The zone chosen to carry out this task, is situated northeast to the Iberian Peninsula, and is named "Central Catalonia".

It's on a region of 1500 Km², with an average altitude over the sea of about 300-500 m, separated from the coastal zone by two mountain ranges and 40 Km from the sea, and from another side at about 50 Km from the limits of the Pirinees. It has its own unique climate. In its capital, Manresa (65000 inhab.) you can find the meteorological station with abundant datum, which also consists of a network of observers whom inform every

2700

day. The station is built in an Engineering School dependant on the Polytechnical University of Catalonia.

Quite acceptable meteorological values have been obtained since 1915. But since 1979 the datum has been registered with great care and precision, being completely trustworthy.

From the industrial point of view, the human activity has been quite even and with minimum development. Therefore the influence of the urban contamination is minor, and apart from that, the instruments of observation are far away from the urban centres.

The results offered can be considered to be precise and affected little by the increment of atmospheric contamination. The same can't be said if datum was to be registered from the city of Barcelona and its industrial surroundings.

PRESENTATION AND ANALYSIS OF METEOROLOGICAL DATUM:

In table 1 we present the datum concerned for the months of winter, in this zone. In them you can see an increase in the minimum and maximum temperatures, in consideration with the latest periods of time. These increments become especially notable in the last period considered, since 1987 to and including 1991. Generally the increment in the minimum temperatures is 25% higher than that of the maximum temperatures. If this heating was a cause of the "greenhouse effect", it seems logical to imply that it is a mayor increment of the minimum temperatures, such as it has lately been announced.

In table 2 the datum referred to appears for all the months of the year. If we state them, its to point out that the increment of temperature is a little higher in the winter months (November-December-January-February-March) than in those of summer (June-July-August-September) and has less relevance in those months in between (April-May-October). This aspect can reaffirm the previous assumptions made about the "greenhouse effect".

In table 3 there obviously appears a great decrease with the number of days with frost; with consequence to everything previously mentioned. It also must be pointed out that the amount of snowfall has diminished notably. From an average of 2 days per year, it has gone by at zero snowfalls in the winters 86-87, 87-88, 88-89, 89-90 and 90-91.

In figure 1 is shown the increase in the mean annual temperature by decades of the zone subjected to the study.

In figure 2 appears the variation in the rainfall suffered by the zone.

On the contrary, figure 3 offers the logical increase in the number of hours of sunlight per year, since 1986. This increase may have importance because of the profitable use of solar energy as an alternative source of energy.
With relation to the energy datum, we only have with total precision that of the last

three years which are the following:

Year 1989: 5025.6 MJ/m²
Year 1990: 5151.6 MJ/m²
Year 1991: 5357.2 MJ/m²

Last of all, figure 4 analyses the different growth in minimum and maximun temperatures; aspect which we believe is essential to attribute the phenomenon to the entitled "greenhouse effect".

CONCLUSIONS:

1.- There is an evident heating of the atmosphere in the studied zone.

2.- The increment is larger in the minimum temperature than in the maximum, in the winter months than in those of summer.

3.- There is an appreciable disminution in the rainfall, and an increase in the number of hours of sunlight.

4.- Since a notable increase in the atmospheric contamination caused by human activity has not been produced in the zone, it can therefore be concluded with enough probability that the datum added to that from other regions, contribute at any rate to believe that an increment is being produced in the "greenhouse effect".

Table 1. Mean temperatures of winter months in degrees Celsius (by periods of years).

	1930-1991			1980-1991			1987-1991		
	Min.	Max.	Ave.	Min.	Max.	Ave.	Min.	Max.	Ave.
NOVEMBER	3	14	8.5	5.0	14.1	9.6	4.8	14.3	9.6
DECEMBER	-0.1	9.5	4.7	0.9	10	5.5	1.5	10.8	6.1
JANUARY	-1.2	9.3	4.1	-0.3	9.7	4.7	0.3	10.6	5.4
FEBRUARY	-0.2	12.3	6.1	1.3	12.2	6.7	1.6	14	7.8
MARCH	2.6	16.3	9.5	3.6	16.4	10	4.4	17.8	11.1

Table 2. Mean temperatures in degrees Celsius (by periods of years).

	1930-1991			1980-1991			1987-1991		
	Min.	Max.	Ave.	Min.	Max.	Ave.	Min.	Max.	Ave.
JANUARY	-1.2	9.3	4.1	-0.3	9.7	4.7	0.3	10.6	5.4
FEBRUARY	-0.2	12.3	6.1	1.3	12.2	6.7	1.6	14	7.8
MARCH	2.6	16.3	9.5	3.6	16.4	10	4.4	17.8	11.1
APRIL	5.1	19.2	12.2	5.6	18.2	11.9	5.8	18.7	12.3
MAY	9	23	16	8.9	22.5	15.7	9.4	23.5	16.4
JUNE	13	27.4	20.4	13.4	27.6	20.5	13.5	27.8	20.7
JULY	16	31	23.5	16.4	31.8	24.1	16.9	32.2	24.6
AUGUST	16	30	23	16.9	30.7	23.8	17.5	31.8	24.7
SEPTEMBER	13.4	26.6	20	14.1	27.2	20.7	14.7	27.9	21.3
OCTOBER	8.4	20.8	14.6	9.2	21.0	15.1	9.8	21.2	15.5
NOVEMBER	3	14	8.5	5.0	14.1	9.6	4.8	14.3	9.6
DECEMBER	-0.1	9.5	4.7	0.9	10	5.5	1.5	10.8	6.1
Average	7.1	20	13.6	7.9	20.1	14	8.3	20.9	14.6

Table 3. Number of days with minimum temperature less than or equal to zero degrees Celsius.

1930 - 1991	1980 - 1991	1987 - 1991
61.3	45	43.2

2703

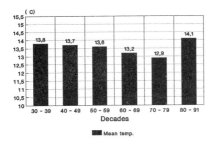

Fig. 1. Variation of the annual
mean temperature

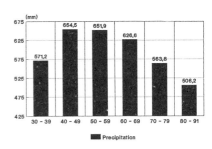

Fig. 2. Annual precitation by decades

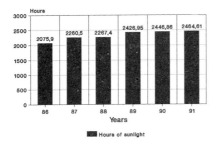

Fig. 3. Hours of sunlight per year

Fig. 4. Increment of the average
values of the minimum and
maximum temperatures with
respect to the mean in the
period 1980-1991

Consequences of Climatic Variability to Design Methods

R. Aguiar

LNETI - Departamento de Energias Renováveis
Estrada do Paço do Lumiar, 1699 Lisboa Codex - PORTUGAL

ABSTRACT

Climatic variability within a year has been considered in many studies of solar radiation, but the year-to-year variability is seldom dealt with. The present investigation shows that typical 5% interannual fluctuations of the yearly average reflect 10% fluctuations of the monthly averages and up to 100% probability fluctuations for some daily radiation ranges. And, as can be seen with the help of the utilizability concept, the monthly fluctuations amplified further when estimating system performance. A χ^2 type distribution with 3 degrees of freedom is proposed for the estimation of climatic variability in the case of only one year of data available. Specially for sizing systems of with high solar fractions or minimum solar fraction required each for year/month, this paper demonstrates that climatic variability must be taken into account.

KEYWORDS

Solar radiation; climatic variability; system design.

INTRODUCTION

It is seldom possible, or practical, for a system designer to use daily or hourly radiation data sequences with the length of the system's expected lifetime – some 20 years. Most systems really don't need to be designed with the sequences of hourly or daily values, but just with average monthly averages; or with probability/utilizability fun-

2705

ctions which very often can be parametrised just by the first. In any case, the values used are supposed to correspond to the 20 years of the system's lifetime. There are at least three factors which bring problems to this picture. First, networks for radiation measurements are still quite sparse. Most stations are recent and thus there are not many places for which one can obtain radiation data with the record length desired (not to mention orientation, tilt and time step). Let's take monthly averages for simplicity; the values one can read in Atlas, or obtain from the National Meteorological Services correspond most of the times to just a few years, and not as desirable to long-term averages. This could be a problem even if the radiation sequence would be stationary, but – and this is the second factor – it is known that the climate system has shown visible trends in this century; and that it is expected to continue do so in the future, namely as a result of an increasing atmospheric greenhouse effect. Finally, and even using the correct long term averages/functions, to design a system, this is not the same as using the whole historic sequence, in particular when a minimum solar fraction should be attained every year.

The objective of this paper is to estimate the effect of interannual climate variability in system design procedures – eg. the errors committed when a designer uses short-term information to estimate long-term solar fractions. A short overview will be given of the variation of descriptive statistics with the number of years used in the computation. Next, a specific example of DHW solar system dimensioning will illustrate how the effect of interannual climatic variation amplifies expected deviations of the system performance. Finally, a method based on a χ^2 distribution will be proposed for when it is necessary to estimate climatic variability from just a few years of data.

DESCRIPTIVE STATISTICS VS. RECORD LENGTH

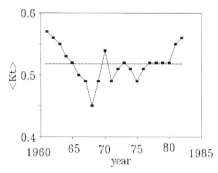

Fig. 1 - Yearly K_t averages; Athens, 1961-1982.

The data base available to the authors consists in hourly sequences of radiation for several locations in Europe, and for the Atlantic islands of Azores and Madeira. Monthly averages, probability functions and autocorrelation at lag one day for the daily clearness index (K_t) sequences were computed for these places for the *horizontal* plane; results for *tilted* planes are expected to be similar.

Due to lack of space, only the statistics for Athens, Greece, will be presented, as a typical example of the results obtained. Figure 1 depicts the yearly averages of K_t for the 22 year period. It can be seen that even at this (smoothed) scale, variations of up to 5% can show up, in respect to the long-term (22 year) mean.

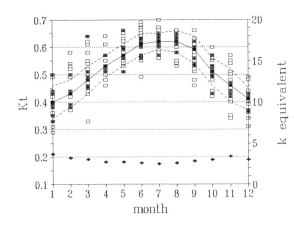

Figure 2 shows monthly averages of K_t ($\overline{K}_t$) for the entire 22 year period, along with the observed standard deviation range (broken lines). Variations of up to 25% can seen to be possible, relative to the long-term means, although errors in the range 5% - 10% are the most frequent.

Fig. 2 - Monthly averages of K_t, Athens 1961-82 ($\square$). ⁕ - multiple observations.
$\square$ - long-term averages; + - equivalent χ_k^2 degrees of freedom.

Figure 3 shows the probability function of K_t. It can be seen that interannual variations of 100% can be possible for many K_t bins. Thus the variability in yearly K_t stands amplified in terms of monthly K_t values, and much more so in terms of probability values for specific K_t ranges. Conceivably, this can produce large errors for system design procedures using average values/functions.

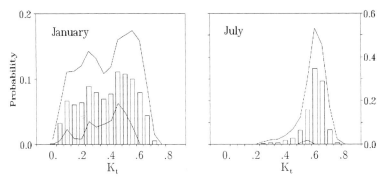

Fig. 3 - Probability function for K_t, Athens 1961-82. Bars show long-term average values for each bin; lines show range of ± standard deviation.

The monthly autocorrelation at lag one day, r_1, and its variance, was also computed (not shown). It varies strongly, perhaps not only due to climatic noise, but also because of the short sequence lengths (30 days) used for its computation.

CLIMATIC VARIABILITY vs. ACCURATE SYSTEM DESIGN

Suppose one wants to design a DHW system for Athens using the utilizability technique (eg. T.A. Reddy, 1987). Let's take the case where 22 years or just 5 years of daily radiation data are available. Figure 4 shows the utilizability function $\phi(H_c)$ values for each critical daily radiation level H_c for the long-term and 5-year cases. This function is usually represented in terms of critical adimensional levels, but when dimensionally represented one can see that for a certain month the reduction in performance effect stemming from a lower average hourly radiation $\bar{H}$ is amplified by a corresponding lower utilizability. As a specific example, let's take the case of Fig. 4 and a system with a threshold radiation of $H_c = 0.2 \ MJ \cdot m^{-2}$.

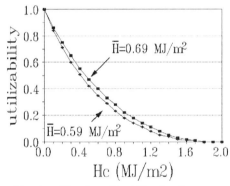

Fig. 4 . - Hourly utilizability for January, Athens: □- 1961-82; +- 1968-72.

When computing the expected collected energy, a reduction of 14% in $\bar{H}$ from the long-term value is amplified by a reduction in ϕ to a final value of 19%. This shows that: i) a significant reduction of the system's solar fraction can be expected for periods up to a few years; ii) the reduction is more than proportional to the interannual climatic variability (i.e. its non-linear).

A SIMPLE METHOD FOR ESTIMATING CLIMATIC VARIABILITY

For the (common) case when a solar system designer only has a few years of data, and wants to estimate the true long-term mean and variability, classical statistical theory can offer an approximate answer (S.H. Crandall, 1963). In fact, χ_k^2 distributions with k degrees of freedom have long been appreciated as good models for the probability distribution of essentially additive, Gaussian, stationary random variables. The mean m of the chi-square distribution is just k, and its variance σ^2 is 2k. The ratio (m/σ) equals $\sqrt{2/k}$. For a variable z with average $\bar{z}$ related with a chi-square distributed variable y by $z = \bar{z} \cdot y/k$, the ratio is the same. If one supposes that $\bar{K}_t$ has a chi-square distribution, than the number of equivalent degrees of freedom k_{eq} can be estimated by $k_{eq} = 2 \cdot <\bar{K}_t>^2/\sigma^2$, where $<\bar{K}_t>$ is the long-term mean of the monthly clearness index. Figure 2 shows that the estimate of k_{eq} for Athens is very low: around 3, for any month. If for example a single $\bar{K}_t$ value is available to the system designer, say K_t^*, confidence limits for the true long-term mean $<\bar{K}_t>$ can be obtained with the help of the χ^2 distribution. As a specific case, if 80% confidence limits are desired, one could expect that $<\bar{K}_t> \in [\max(0.1, 0.02 \cdot K_t^*), \min(0.8, 2. \cdot K_t^*)]$; but if K_t^* was already an average for several years, this range could be shortned.

CONCLUSIONS AND FINAL COMMENTS

This paper shows that the (monthly) statistical characteristics of solar radiation are subjected to a substantial interannual climatic variability. If a solar system designer attempts to use information for just a few years, he may obtain values which will be significatively different from those obtained with the long-term averages; furthermore, the differences seem to be non-linear in the average radiation level. A χ^2 type method is proposed for estimating climatic variability in the absence of detailed data.

A number of other questions related to interannual climatic variability are prompted by this work. For instance, in particular for high solar fraction systems, a user can be interested on obtaining a minimum solar fraction for each year (or month) from his system, rather than on having a certain long-term value. The current work also shows that the use of Typical Meteorological Years as a substitute for long-term radiation data can be questioned as it does not take into account interannual variability. Finally, in these days of concern with the increase of the greenhouse effect, significant trends on several meteorological variables, including cloudiness – and thus ground-level solar radiation – can be expected. For systems which are built to last for 20 years in the future, these trends are of no minor importance for their accurate sizing.

REFERENCES

Reddy, T. A. (1987). *The Design and Sizing of Active Solar Thermal Systems.* Clarendon Press, Oxford.

Crandall, S. (1963). Measurement of stationary random processes. In: *Random Vibration* (S.Crandall, ed.),Vol 2, Chap. 2, pp. 35-44. M.I.T. Press, Cambridge, Mass.

THE ESTIMATE OF GLOBAL AND DIFFUSE RADIATION AT QUETTA, PAKISTAN

S.Z.Ilyas, S.M.Nasir and S.M.Raza
Department of Physics
University of Balochistan
Quetta.

ABSTRACT

Estimate of global radiation is obtained by using monthly mean daily fraction of possible sunshine i.e. $\bar{s}$ and the monthly mean extraterestrial i.e. $\bar{H}_o$ radiation. We computed the regression coefficients a and b, using least square technique for the period 1975-1985 and then determined the monthly mean daily fraction of possible sunshine ($\bar{s}$) using Angstrom-page i.e. $\bar{H} = \bar{H}_o$ (a+b$\bar{s}$); the values of which are in close agreement with experimentally determined values $s = \dfrac{\bar{n}}{\bar{N}_d}$ where $\bar{n}$ is the monthly average number of instrument recorded bright sunshine hours per day and $\bar{N}_d$ the average day length. We estimated the diffused component of radiation by employing Liu and Jordan (1) Page (2) and Iqbal (3) relationships from diffused radiation following $\bar{K}_T$ i.e. clearness index values.

An annual variation of the ratio of the daily diffused radiation to the daily global radiation for Quetta is obtained by Liu and Jordan, Page and Iqbal relationships for diffused radiation using $\bar{K}_T$ values. On comparison of the annual variation of monthly average $\bar{K}_T$ and $\bar{H}_d/\bar{H}$ values, we find Liu and Jordan relationship more efficient for predicting diffuse component of radiation at Quetta while the ratio $\bar{H}_d/\bar{H}$ employing correlations (1,2,3) plotted against $\bar{K}_T$ values shows that and Jordan (1) correlation generally underestimates diffuse radiation. This underestimate is perhaps due to non-

correction of shadow band effect, different values of solar
constant and other climatic variables such as the optical
properties of clouds and ground albedo

INTRODUCTION

The Angstrom-Page formula (2,4) correlates global solar
irradiation to sunshine duration. In this work the constants a
and b, usually referred as regression coefficients, have been
determined using least square technique for the period 1975-
1985, using the relationship.

$$\bar{H}/\bar{H}_O = a + b \ (\bar{n}/N_d) \qquad \ldots\ldots\ldots\ldots \ (1)$$

where a and b are regression coefficients, the values of which
are 0.42 and 0.37, respectively, $\bar{n}$ the monthly average number
of instrument recorded bright sunshine hours per day and $\bar{N}_d$ the
average day length. Eq.(1) is a modified from of Angstrom's
relation, developed by Prescot (5). The values of $\bar{H}_O$ were
determined by the relation

$$\bar{H}_O = (24/\pi) \ I \ E_O \ Sin \ L \ Sin \ D \left[(\frac{\pi}{180^o}) w - \tan w \right] \ \ldots (2)$$

where I is solar constant (4.921 $MJ/m^2/hr$), E_O the eccentricity
correlation factor, L the latitude ($30^o 11'$), D the declination
(degrees) and w the sunset hour angle (degrees). The extra
terrestrial radiation $\bar{H}_O$ were obtained by taking the average
over the whole month. We then determine the monthly mean daily
fraction of possible sunshine ($\bar{s}$) using Angstrom Page formula
(2,4)

$$\bar{H} = \bar{H}_O \ (a + b.\bar{s}) \qquad \ldots\ldots\ldots\ldots\ldots \ (3)$$

The monthly mean daily fraction of possible ($\bar{s}$) was also
experimentally determined by using

$$s = \bar{n}/\bar{N}_d , \qquad \ldots\ldots\ldots\ldots\ldots \ (4)$$

the values of which were found in close agreement with the
theoretically determined values employing Angstrom-Page formula.
Theoretical and experimental values of $(\bar{s})$ are given in table (1).

The diffused component of radiation is estimated by
using Liu and Jordan (1), Page (2) and Iqbal (3) relationships
following clearness index values i.e. $\bar{K}_T$. The clearness index
is the ratio of the measured monthly average global to the
monthly average daily extraterrestial irradiation on a horizontal surface. In order to calculate solar irradiation on a
tilted collector plane one needs to know, besides the data on
global radiation on horizontal plane, the percentage of
diffused radiation in global.

This paper presents a comparision of annual variation of
the ratio of the daily diffused radiation to the daily global
radiation for Quetta by Liu and Jordan, Page and Iqbal relationships for diffused radiation using $\bar{K}_T$ values.

RESULTS AND DISCUSSIONS

The monthly average values of clearness index i.e. $\bar{K}_T$
are shown in Fig.(1) which represents two peaks i.e.for March-
April and September-October months and a minimum in June-July
for the period 1975-1985. Quetta is an oval valley at an
altitude of 1799 meters, surrounded by dry mountains. The minimum
in the values of clearness index in the months of June-July
(usually summer) is due to dust storm which is a regular
climatological feature every year in Quetta. The maximum values
of $\bar{K}_T$ in the months of November, December and January could be
attributed to clouds and snow fall every year in winter at
Quetta.

For the estimate of diffuse radiation from global,
there are various empirical correlations. We employed only
three relationships for comparision, Liu and Jordan (1) and
Klein (6) developed the following correlation.

$$\bar{H}_d/\bar{H} = 1.39 - 4.027\ \bar{K}_T + 5.531(\bar{K}_T)^2 - 3.108(\bar{K}_T)^3 \quad \ldots \text{(5)}$$

Page (2) developed a correlation in a linear form

$$\bar{H}_d/\bar{H} = 1.0 - 1.13\ \bar{K}_T \qquad \ldots\ldots\ldots\ldots\ldots \text{(6)}$$

2712

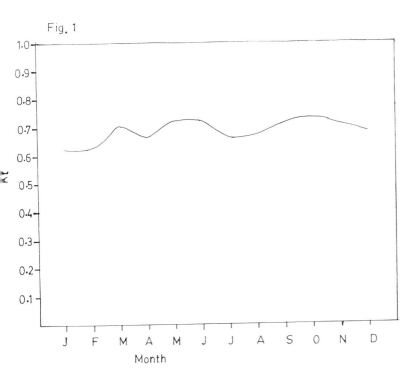

Fig. 1

Kt

Month

J F M A M J J A S O N D

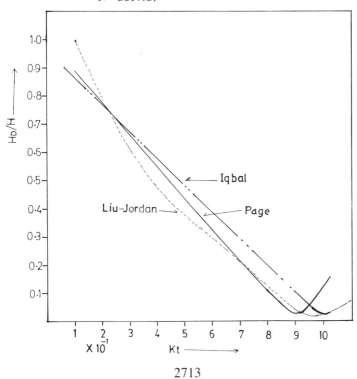

Fig. 2   Plot of Kt against Hᴅ/H
          for Quetta.

Hᴅ/H

Iqbal

Liu-Jordan

Page

X 10⁻¹   Kt

2713

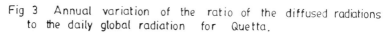

Fig 3   Annual   variation   of   the   ratio of   the   diffused radiations
        to   the   daily   global   radiation   for   Quetta.

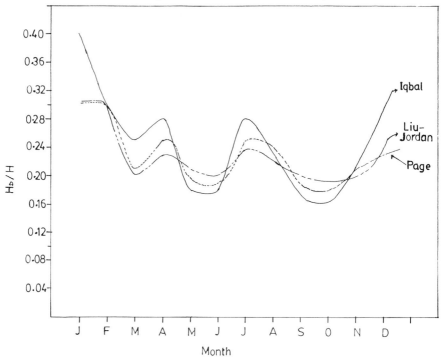

Iqbal (3) used the following correlation for estimating the diffused component of radiation

$$\bar{H}_d/\bar{H} = 0.958 - 0.952\, \bar{K}_T \qquad \ldots\ldots\ldots (7)$$

After estimating diffused radiation from above relationships, we computed $\bar{H}_d/\bar{H}$, the values of which are plotted as a function of $\bar{K}_T$ and monthly average as shown in Figs.(2) and (3), respectively.

We conclude that Liu and Jordan (1,6) relationship from diffused radiation is relatively better as compared to Page and Iqbal correlations (2,3). Liu and Jordan (1,6) correlation underestimates diffuse radiation and requires modifications.

ACKNOWLEDGEMENTS

We are thankful for the finaicial support of British Council.

REFERENCES

1. Liu, BYH and Jordan R.C. "The interrelationship and Characteristic distribution of Direct, diffuse and total solar radiation" Solar Energy 4, Vol.3,1 (1960).

2. Page, J.K. "Estimation of monthly mean values of Daily total short wave radiation on vertical and inclined surface from sunshine records for latitudes 40 s - 40 N". Proc UN New sources of Energy. Paper No.598, Vol.4, 378 (1961).

3. Iqbal, M. "Correlation of Average Diffuse and Beam radiation with Hons of Bright Sunshine" Solar Energy.23, No.2, 169 (1979).

4. Angstrom, A. "Solar and Terrestrial Radiation", Q.J.R.Meteor Soc. 50 (1924) 121.

5. Prescot, J.A. "Evaporation from water surface in relation to solar Radiation" Trans. R. Soc. Aust. Vol.64, 114 (1940).

6. Klein, S.A. "Calculation of monthly average insolation on tilted surfaces" Solar Energy, Vol.19, No.4. 325(1977).

SOLAR RADIATION AVAILABILITY IN SOUTH-EAST OF ROMANIA

V. BADESCU

Mecanica, Termotehnica,
Polytechnic Institute of Bucarest,
Bucarest 79590, ROMANIA

ABSTRACT

Previous studies emphasized that the most sunshined regions of
Romania are situated in south-west (Oltenia) and south-east
(Dobrogea). In this paper we report about the amount of direct,
diffuse and global solar radiation available in eleven
localities of Dobrogea.The values of the global and diffuse
solar daily irradiation were computed by using the Angstrom-
Page and Liu-Jordan formulae, respectively. Also, the paper
give some informations about the monthly average values of
relative sunshine and air temperature.

KEYWORDS

Global radiation, direct radiation, diffuse radiation, relative
sunshine, Angstrom-Page formula.

INTRODUCTION

As is well known, the spatial distribution of solar radiation
may be strongly non-uniform. Indeed, the amount of solar
radiation at ground level depends on both the latitude and
some other factors, as the relief and the air mass circulation.
Consequently, studies concerning the meteo-climatological
features of the site must be performed before a solar
installation is designed.

Previous studies emphasized that the most sunshined regions
of Romania are situated in south-west (Oltenia) and south-
east (Dobrogea) (see, e.g. Bădescu (1989)). Among these two
regions the last one seems to constitute the best alternative
for solar energy applications. Indeed, Dobrogea has many arid
zones and less productive agricultural areas. In this paper
we report about the amount of direct, diffuse and global

solar radiation available in eleven localities of Dobrogea, selected to give a broader coverage of the region in both latitude and longitude.

GEOGRAPHICAL DATA ABOUT DOBROGEA

Figure 1 shows the three historical provinces of Romania : Moldavia, Valahia and Ardeal. Dobrogea is a region situated in south-east of Valahia. Its surface is around 15500 km^2. The relief of Dobrogea consistes of hills and plateaus with average altitude around 150m asl. However, the Macin Mountains from the north of Dobrogea have a maximum altitude of 451 m asl.

Table 1 shows the main geographical data of the eleven localities of Dobrogea that we selected to compute solar radiation.

COMPUTATION OF SOLAR RADIATION

In order to estimate the availability of solar global radiation we used the Angstrom-Page formula. This relationship supposes to be known the value of relative sunshine, which is the ratio between the hours with bright sunshine and the daily light hours during the period. In case of the localities Adamclisi, Gorgova, Jurilovca and Sulina we used the multi-year monthly average values of relative sunshine σ reported in Oancea and Constantinescu (1982). For the other localities we computed the relative sunshine by using the multi-year monthly average values of bright sunshine hours reported in Agenda Meteo (1985). For lack of space we show here the values of relative sunshine only for May (Table 2). The same Table 2 contains the multi-year values of air temperature for six localities of Table 1.

The monthly average values of the daily global irradiation at ground level, H, were computed by using the Angstrom-Page formula (Page, 1961) :
$$H = (a+b\,\sigma)H_o \qquad (1)$$
where a and b are regression coefficients and H_o is the daily average irradiation outside the atmosphere. The monthly values of a and b were determined in a previous work (Oancea and Constantinescu, 1982) by using the actinometric measurements performed at Constanța. The daily diffuse solar irradiation D was computed with the Liu-Jordan formula (Liu and Jordan,1960) modified by Page (1961) :
$$D/H = 1.0-1.13(H/H_o) \qquad (2)$$
Then, the daily direct irradiation I was evaluated with :
$$I = H-D \qquad (3)$$

RESULTS AND DISCUSSIONS

By using the eq.(1) we computed the monthly average values of the global solar irradiation H for the eleven localities of Table 1. For lack of space we show here only the values

2717

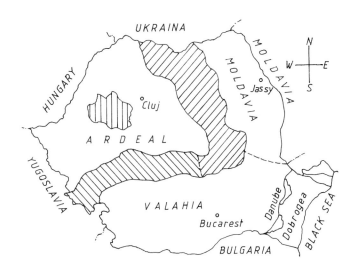

Fig.1. Geographical placement of Dobrogea on the
Romanian territory. The dashed area shows
the Carpathians Chain.

Table 1. The localities of Dobrogea where solar
radiation was computed.

Letter	Localities	Longitude (E)	Latitude (N)	Altitude (m)
a	Adamclisi	$28^{\circ}00'$	$43^{\circ}08'$	158
b	Calaraşi	$27^{\circ}21'$	$44^{\circ}12'$	19
c	Constanţa	$28^{\circ}38'$	$44^{\circ}13'$	13
d	Galaţi	$28^{\circ}01'$	$45^{\circ}30'$	71
e	Gorgova	$29^{\circ}12'$	$45^{\circ}11'$	3
f	Hîrşova	$27^{\circ}57'$	$44^{\circ}41'$	30
g	Jurilovca	$28^{\circ}53'$	$44^{\circ}46'$	27
h	Medgidia	$28^{\circ}16'$	$44^{\circ}15'$	64
i	Sf.Gheorghe	$29^{\circ}36'$	$44^{\circ}54'$	1
j	Sulina	$29^{\circ}40'$	$45^{\circ}09'$	3
k	Tulcea	$28^{\circ}49'$	$45^{\circ}11'$	4

computed in May (Fig.2b). These values could be compared with
those obtained during a study performed at the level of the
whole country (Bădescu, 1989)(Fig.2a). The present smaller

2718

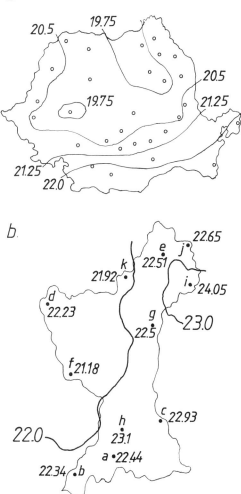

Fig.2. Solar global irradiation in May
(MJ m⁻²day⁻¹). a.Distribution over the
Romanian territory (Bădescu, 1989).
b. Distribution over Dobrogea (present
study).

scale study gives new valuable details concerning the spatial
distribution of solar global radiation.

Table 2. The monthly average values of some
meteorological and actinometric
parameters (May).

Letter	Localities	Relative sunshine	Direct Irradiation $(MJm^{-2}day^{-1})$	Diffuse Irradiation $(MJm^{-2}day^{-1})$	Air Temperature (^{o}C)
a	Adamclisi	0.52	12.96	9.48	–
b	Călăreşi	0.55	13.31	9.03	17.0
c	Constanţa	0.58	14.03	8.90	16.5
d	Galaţi	0.55	13.23	9.00	15.9
e	Gorgova	0.56	13.61	8.90	–
f	Hîrşova	0.50	11.99	9.19	–
g	Jurilovca	0.56	13.53	8.97	16.0
h	Medgidia	0.58	14.24	8.86	–
i	Sf.Gheorghe	0.63	15.46	8.59	15.9
j	Sulina	0.57	13.73	8.92	–
k	Tulcea	0.53	12.85	9.07	16.6

The eqs.(2) and (3) were used to compute the monthly values of
daily direct and diffuse irradiation, respectively. Table 2
shows the results obtained in May.

A sound analysis of the whole set of data allowed some general
conclusions. During the warm season (April-October) the global
solar irradiation is higher in north of Dobrogea. In the winter
months a larger amount of solar radiation is received in the
south of the region. Note that the same remarks apply also for
relative sunshine.

REFERENCES

Agenda Meteo (1985). Institutul de Meteorologie şi Hidrologie.
    Ed.Tehnică, Bucureşti.
Bădescu, V. (1989). Observations sur la distribution du rayon-
    nement solaire en Roumanie. La Meteorogie, 30, 9-16.
Liu, B.Y. and Jordan, R.C. (1960). The inter-relationship and
    characteristic distribution of direct, diffuse and total
    radiation. Sol. Energy, 4, 1-19.
Oancea, C. and Constantinescu, M. (1982). Sur l'estimation du
    gisement solaire dans les sites ou n'existe pas des mesures
    directes de l'irradiance solaire. Proc. Natn. Conf. on New
    Sources of Energy, Jassy, Romania, 1, pp 41-46.
Page, J.K. (1961). The estimation of monthly mean values of
    daily shortwave radiation. Proc. United Nations Conference
    on New Sources of Energy, Rome, 4, 378.

SOLAR ENERGY RADIATION OVER UYO (5°2'N, 7°50'E) NIGERIA

N. M. EKPO

Department of Physics,University of Uyo,
P. M. B. 1017, Uyo. Nigeria.

Abstract:

Estimation of daily global solar radiation in Uyo (5°2'N, 7°50'E) Nigeria has been done for the period 1982 - 1990. The global radiation was estimated from sunshine hours, using Amstrom's regression equation. The regression constants used are the average values of Turton which could be used to predict global radiation for all the tropics (Massaquoi 1988, Turton 1987). The monthly average global solar radiation in Uyo ranges from 10MJ/M²day to 16.35MJ/M²day during the decade. The study shows that global solar energy is available in abundance throughout the years in Uyo. The radiation is largely diffuse due to low clearness index of 0.35 - 0.45. The yearly cumulative average solar energy ranges from 4500MJM⁻² to 5050MJM⁻². More than 48,180MJM⁻² of solar energy has been dissipated in Uyo during the decade.

KEYWORDS

Energy; Solar Energy; Global Radiation; Extraterrestrial Radiation; Clearness Index

ENERGY

Energy means so many things to so many people. It is linked to politics, health, and economy. It affects the development and stability of entire industries and even governments as well as the feasibility of a family. For all countries, a cheap, abundant and clean supply of energy is very vital as there is a close relationship between the amount of power generated and the gross national product.

Solar Energy

The world's sources of energy with attendant disadvantages are finite in quantity and are depleting at an ever-growing alarming rate. The almost undepletable, cheap and clean source of energy is the solar energy which is abundant in this part of the world (tropics) though untapped.

According to Gierman (Gierman 1985) developing countries, are in a particular good position to utilize solar energy because the solar energy in these parts is abundant and the inhibitants are scattered over vast areas making access to electricity or conventional fossil fuel difficult and expensive. Moreso, there is acute shortage of conventional sources of energy in these countries. Generally, solar radiation data is still scarce especially in the developing countries though solar radiation climatology of variuos countries has been studied (Abdalla 1987,Ahmed et al., 1983, Bahhah et al., 1985, Bennet 1965

Danneshyer 1978, Excell 1976, Flocas 1980, Ibrahim 1985, Khogali 1983, Massaquoi 1988, Paltridge et al., 1976, Sambo 1986, Sahara et al., 1986).

Detailed information on the availablility of solar radiation on the surface of the earth is essential not only for understanding many physical and biological processes, but for the optimum design and study of solar energy conversion systems, the prediction of their performances and estimation of efficiencies of existing systems.

Solar energy at the earth surface depends on many factors of non-global character. A study of solar radiation under local climatic conditions, preferaby over a long period, should be useful to both the locality where the radiation is collected, and the wider world community (Massaquoi 1988). According to Ibrahim (Ibrahim 1985), to develop a global radiation map requires knowledge of radiation data in various countries and for world-wide marketing of solar equipment, designers and manufacturers need to know the average global radiation available in different regions.

This work provides for the first time such information for the past decade in Uyo (5°2'N, 7°50'E) Nigeria.

Solar Energy Estimation

The average number of hours of sunshine n were obtained from daily measurements covering a period of 9 years (1982-1990) using sunshine recorder.

The monthly average of daily global radiation $\bar{H}$ on a horizontal surface was estimated using Amstrom regression equation (Page 1979).

$$\bar{H} = H_o(a + b \, n/N) \tag{1}$$
where
H_o - extraterrestrial solar radiation on a horizontal surface on an average day of each month (Klein 1979).
a,b - regression constants.
$\bar{n}$ - monthly average of hours of bright sunshine.
N - mean daily number of hours of day light in a given month between sunshine and sunset.

Values of N were calculated from the Cooper's formula (Cooper 1969).

$$N = 2/15\cos^{-1}(\tan\delta\tan\phi) \tag{2}$$
where
ϕ - latitude (ϕ=05°02'N for Uyo).
δ - solar declination.

$$\delta = 23.45°\sin360[(284+D)/365]. \tag{3}$$
where
D - day of the year from January 1.
The extraterrestrial solar radiation was calculated from the equation of Duffie and Beckman (Duffie et al 1985).
$$H_o = (24/\bar{n})I_{sc}[(1+0.33\cos(360D/365)][\cos\delta\cos\phi\sin\omega+ (2\pi\omega/360)\sin\delta\sin\phi]. \tag{4}$$
where
I_{sc} - the solar constant (I_{sc} = 1365WM²).
ω - sunshine or sunset hour angle (Sukara et al; 1986, Massaquoi 1988).
$\omega = \cos^{-1}(-\tan\delta\tan\phi). \tag{5}$

The regression constants used in this work are the average values (a = 0.3, b = 0.4) of Turton (Turton 1987) for humid tropics which could be used to predict global radiation for all the tropics as supported by the works of Massaquoi (Massaquoi 1988) for serra Leone and Sambo (Sambo 1986) for Nothern Nigeria.

RESULTS

Global Solar Energy Radiation

Fig.1 shows the monthly variation of global solar energy in Uyo for the years 1982 - 1990. Values of the monthly average global solar radiation range from 10MJ/M²day to 16.35MJ/M²day during the decade. The minimum values characterize the cool-dry (December - January) and rainy (June - August) seasons of the year. Maximum values were obtained during the hot-dry (March - May) seasons. The maximum of 16.35MJ/M²day was recorded during the hot-dry period (April) of 1982, a year that witnessed draught in this part of the world.

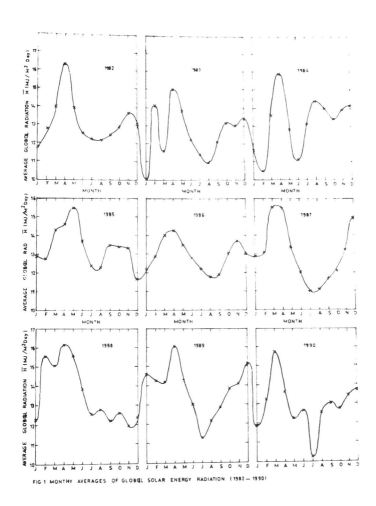

FIG 1 MONTHY AVERAGES OF GLOBAL SOLAR ENERGY RADIATION (1982 — 1990)

Extraterrestrial Solar Energy Radiation

A-part of sahara dust during the cool-dry seasons and water molecules in the rainy seasons that attenuate the extraterrestrial radiation, the monthly variation of the extraterrestrial radiation itself over Uyo (Fig.2) contributes also to the fluctuations of the monthly average global radiation. The monthly average of extraterrestrial radiation over Uyo ranges from a minimum of 33MJ/M²day to a maximum of 37.20MJ/M²day. The maximum values in April and September seems to reveal the effect of the nearness of the sun when directly over latitude 05°2'N on its journey to and fro the nothern or southern hemisphere while the minima in December and July indicate how far away the sun is from latitude 05°2'N when in either of the hemispheres. This goes to confirm the fact that according to Massaquoi (Massaquoi 1988) the extraterrestrial radiation H_o is a function only of latitude and is independent of other locational parameters.

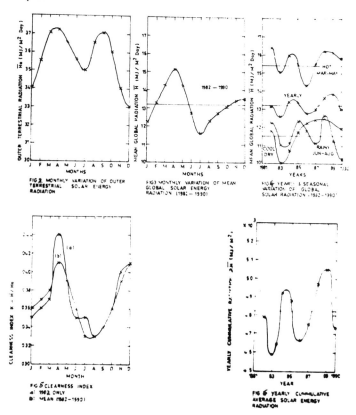

FIG 2 MONTHLY VARIATION OF OUTER TERRESTRIAL SOLAR ENERGY RADIATION

FIG 3 MONTHLY VARIATION OF MEAN GLOBAL SOLAR ENERGY RADIATION (1982 - 1990)

FIG 4 YEARLY & SEASONAL VARIATION OF GLOBAL SOLAR RADIATION (1982 - 1990)

FIG 5 CLEARNESS INDEX a) 1982 ONLY b) MEAN (1982-1990)

FIG 6 YEARLY CUMMULATIVE AVERAGE SOLAR ENERGY RADIATION

Global Radiation Pattern

The monthly mean of global radiation varies between a minimum of 11.62MJ/M²day for the months of July and 15.10MJ/M²day for the months of April during the decade (Fig.3). The global radiation has characteristic values for the different seasons of the year (Fig.4). During the rainy seasons, the values fluctuate between 10.25MJ/M²day (1983) and 12.58MJ/M²day (1987) about a mean value of 11.51MJ/M²day. In the hot-dry seasons, the values range from 14.25MJ/M²day to 16.35MJ/M²day with a mean value of 15.5MJ/M²day. The yearly average fluctuates between 12.56MJ/M²day and 13.61MJ/M²day about a mean value of 13.2MJ/M²day. Low global radiation was observed throughout all seasons in 1983 while maximum radiation was obtained in all seasons of 1988.

2724

Clearness Index

Apart from the monthly variation of extraterrestrial radiation due to latitude, the seasonal fluctuation of the global radiation is consequent of the low clearness index $K = \overline{H}/H_o$ (Fig.5) which shows that the atmosphere over Uyo is partly cloudy throughout the years. It is more cloudy during the cool-dry seasons of November - January, and most cloudy during the rainy seasons of June - August. The monthly values of K range from 0.35 to 0.45. A large part of the global radiation is therefore diffuse.

CONCLUSION

This study shows that solar energy is available in abundance throughout the years in Uyo. The radiation is largely diffuse. The yearly cumulative average solar energy (Fig.6) ranges from 4500MJM^{-2} to 5050MJM^{-2}. A total of more than 48,180MJM^{-2} (about 30075GeV/M^2) of solar energy has been dissipated in Uyo during the decade which could be harnessed to supplement the energy needs of Uyo and its environs.

ACKNOWLEDGEMENT

The author wishes to thank the Department of Metheorological Services, Uyo, for assistance in data collection.

REFERENCES

Abdalla, Y.A.G.(1987). Solar radiation over Doha (Qatar). Int. J. Solar Energy, 5, 1-5.

Ahmed I.,M.Al-Hamdani and K.Ibrahim (1983).Solar radiation maps for Iraq. Solar Energy 31,29-33.

Bahhsh,H.R.Srinivasan and V.Behel (1985). Correlation between Hourly Diffuse and Global radiation for Dhahran, Arabia. Solar and Wind Technology 2, 59-63.

Bennet,I.(1965). Monthly Maps of mean daily insolation for the United States. Solar Energy 9, 145 (1965).

Cooper,P.I. (1969). Absorption of Solar Radiation in Solar Stills. Solar Energy 12,333-337.

Daneshyer,M.(1969). Solar radiation statistics for Iran. Solar Energy 21,345-349.

Duffie,J.A. and W.A.Beckman (1980). Solar Engineering of Thermal Processes P.22 John Wiley,   New York.

Excell,R.H.B.(1976). The solar radiation climate of Thailand. Solar Energy 21,345-349.

Flocas,A.A..(1980). Estimation and prediction of global solar radiation over Greece. Solar Energy 24,63-67.

Gieman,K.(1985). Understanding of Solar Energy - A general Review - VITA Arlington, Virginia 1985.

Ibrahim,S.M.A(1985). Predicted and Measured Global Solar Radiation in Egypt. Solar Energy 35, 185-189

Khogali,A.,M.R.T.Ramdem, Z.E.H. Ali and Y.A. Fattali(1983). Global and diffuse solar irradiance in Yemen. Solar Energy 31,55-59.

Klein,S.A.(1979).Calculation of monthly average Insolation on titled surfaces. Solar Energy 19,325-328.

Massaquoi,J.G.M.(1988). Global Solar radiation in Sierra Leone(West Africa). Solar and Wind Technology Vol.5,3.281-284.

Page,J.K.(1979). Methods for the estimation of Solar Energy on vertical and inclined surfaces. Solar Energy Conversion (Editor A. E. Dixon and J. D. Leslie) Pergamon Press, Canada p.37-99.

Paltridge, G.W. and D. Proctor (1976). Monthly mean solar radiation statistics for Australia. Solar Energy 18,235-238.

Sambo,A.S.(1986).Emperical models for the correlation of Global solar radiation with Meteorological data for Nothern Nigeria, Solar & Wind Technology 3,89-92.

Sukhera,M.B.,M.A.R. Pasha and M.S.Naveed (1986). Solar radiation over Pakistan-comparison of measured and predicted data. Solar and Wind Technology,Vol.3,No.3, 219-221.

Telahun,Joseph (1987).Estimation of Global solar radiation from sunshine hours, Geographical and Meteorological Parameters. Solar and Wind Technology Vol.4 No.2, 127 (1987).

Turton,S.M.(1987).The relationship between total irradiation and sunshine in Humid Tropics. Solar Energy 38,353-354.

ON THE RANDOM MEASUREMENTS OF ROBITZSCH PYRANOGRAPHS

Bulent G. Akinoglu

Middle East Technical University
Dept. of Physics
06531, Ankara, TURKEY

ABSTRACT

Robitzsch type bimetallic pyranographs have been recording solar radiation for the last few decades in most of the meteorological stations all over the world. However, the recordings are incorrect due to the climate and time dependence of the calibration constants. Present report gives the results of the work to correct the data of these instruments for three locations in Turkey. The results show that, neither the corrected nor uncorrected data of Robitzsch type pyranographs should be used in solar energy research and development.

KEYWORDS

Solar radiation, bimetallic actinographs, bimetallic pyranographs, Robitzsch pyranographs

INTRODUCTION

Robitzsch type bimetallic pyranographs use the difference in the thermal expansions of two metals painted black and white. Their calibration constants depend on the ambient temperature and other climatological parameters of the location (Duffie and Beckman; Esteves and Rosa, 1989). Therefore, these instruments are now replaced with the Eppley types (which use thermoelectric effect to record solar radiation) in the developed countries. However, in most of the stations all over the world, old ones are still the only instruments recording data, and a bulk of data have been collected for many years.

There are some works to correct the data of Robitzsch pyranographs (e.g. Esteves and Rosa, 1989). This report presents the results of the work to correct the old data of the Robitzsch pyranographs for the three locations in Turkey. State Meteorological Office of Turkey installed new instruments and recorded simultaneous data for the year 1983, in these locations. The analysis are carried for these three locations and the results show that it is better to use the estimation models rather than to correct the data of a location using the correction factors of another location.

ANALYSIS

The work is performed for the monthly-average daily, monthly-average hourly, daily and hourly values of solar radiation. Only a brief presentation of the analysis for the monthly-average values are given in this report.

The work is carried using standard statistical error analysis. Only two of the statistical errors are given in the tables. These are

$$MBE = [\sum_{i}^{N} (H_{im} - H_{ic})] / N \qquad (1)$$

and

$$RMSE = [(\sum_{i}^{N} (H_{im} - H_{ic})^2) / N]^{1/2}, \qquad (2)$$

where H_{im} and H_{ic} are the measured and corrected values of the data, respectively.

Monthly-average of daily solar radiation data are widely used in the performance calculations of solar energy systems. There exist quite a large number of models to estimate these values. Among them, sunshine based models are easy to apply and estimate with acceptable accuracy. Therefore, the quality of the corrected data is tested using a sunshine based model which was found to be one of the best in a recent article (Akinoglu and Ecevit, 1990). This is a quadratic correlation between the monthly-average daily solar radiation and monthly-average daily bright sunshine hour

$$H/H_o = 0.145 + 0.845 \ (s/S) - 0.280 \ (s/S)^2 , \qquad (3)$$

where H and H_o are monthly-average daily and extraterrestrial daily solar radiation, s and S are the monthly average bright sunshine hours and daylength, respectively.

The linear relations between Robitzsch and Eppley readings of monthly-average daily solar radiation are obtained for the three locations. The relation of a location is used to correct the Robitzsch data of the other. The results show that the corrected values are not better than the estimated values of eqn. (3).

The second approach is to use the monthly correction factors of one location (defined as: $F_m = H_e / H_a$, where H_e and H_a are the readings of Eppley and Robitzsch, respectively) to correct the monthly data of the other. The estimated values are still better than the corrected values, and such a result can also be demonstrated from Fig. 1. This figure shows the monthly variation of correction factors for the locations. The difference in the correction factors of two locations for the same month of the year can be as large as 25 %.

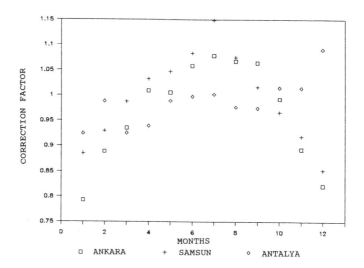

Fig. 1. Monthly variation of F_m

These three locations are from the north, mid and south regions of Turkey and have different climates (latitudes are 41.28N for Samsun, 39.95N for Ankara and 36.88N for Antalya). Hence, one may think that the correction factors for these locations are functions of the climatological parameters and hopefully have the same behavior for all instruments. If one can obtain a correlation between the correction factors and monthly climatological parameters using the data of one location, it may be used for the other locations. Therefore, linear correlations between the correction factors and monthly climatological parameters are obtained using multiple regression analysis for three locations. These climatological parameters are monthly averages of noon temperature (T), relative humidity (RH) and bright sunshine hours (s). The correlations for Ankara are

$$F_m = 0.812 + 0.010 \ T \qquad , \qquad (4)$$
$$F_m = 1.070 + 0.0072 \ T - 0.0033 \ RH \qquad , \qquad (5)$$

and

$$F_m = 1.077 + 0.0062 \ T - 0.0033 \ RH + 0.0051 \ s \ . \quad (6)$$

The correlations of one location are used to correct the data of the other locations and the corrected data are compared with the estimation of the model (eqn. (3)). As it can be seen from Table 1, the corrected data are not better than the estimations of the model.

Table 1. Errors for the monthly calculations

SAMSUN

error	uncorrected	eqn.(4)	eqn.(5)	eqn.(6)	estimated
MBE	-7.9	10.2	2.2	10.1	2.0
RMSE	24.8	16.1	14.4	17.0	12.7

ANTALYA

error	uncorrected	eqn.(4)	eqn.(5)	eqn.(6)	estimated
MBE	7.3	32.6	21.4	16.8	24.5
RMSE	14.5	48.2	37.3	31.4	26.7

Monthly-average of hourly data are often used for the hourly analysis of solar energy systems. After examining these data, it is observed that the relations between the readings of the two types of instruments depend both on the hour of the day and month of the year. Hence, correction factors are defined for each hour as

$$f(i,j) = I_e(i,j) \, / \, I_a(i,j) \quad , \qquad (7)$$

where I_e and I_a represent the readings of Eppley and Robitzsch type pyranographs and, i and j stands to indicate the hour of the day and month of the year, respectively. Using hourly correction factors, the monthly-average of hourly data of the next year (the available data for the next year is only for 6 months of 1984) were corrected and tested using uncorrected data. Hourly and monthly defined MBE and RMSE (Akinoglu, 1990), are used for comparison. Table 2 gives the results for Ankara. The hourly data of 1984 (6 months) are corrected using hourly correction factors (eqn. (7)) of hourly data for 1983. As it can be seen from this table, the errors before the correction are less than the errors in the corrected data. In this table subscripts m and h stands to indicate monthly and hourly calculations. Although uncorrected data give smaller errors, this does not mean that they can be used in the performance calculations. They still possess larger errors than the estimations of standard model (Duffie and Beckman) of monthly-average hourly radiation (Akinoglu, 1990).

CONCLUSION

A detailed work is carried to correct the data of Robitzsch pyranographs. One year simultaneous data of the Robitzsch and Eppley pyranographs, for the three locations, are used in the analysis. In conclusion, readings of the Robitzsch pyranographs are functions of climatological parameters and this dependence is different for different instruments. Instead of correcting the old Robitzsch data, estimation models are recommended for the solar radiation calculations.

Table 2. Errors for the hourly calculations

Ankara	uncorrected			eqn. (7)	
hours	MBE_h	$RMSE_h$		MBE_h	$RMSE_h$
7.5	1.61	2.91		2.54	3.66
8.5	0.24	3.00		0.49	2.57
9.5	-1.48	3.15		-2.15	4.54
10.5	-3.07	4.39		-4.49	7.63
11.5	-2.70	3.32		-3.98	6.16
12.5	-2.73	3.64		-3.49	6.47
13.5	-2.36	2.74		-2.88	5.41
14.5	-1.87	2.91		-1.12	3.91
15.5	-0.22	2.38		1.16	2.41
16.5	-0.43	2.67		1.56	2.95
17.5	-0.66	1.06		0.45	0.76
months	MBE_m	$RMSE_m$		MBE_m	$RMSE_m$
1	-2.39	4.20		-0.04	3.00
2	-1.28	2.03		0.84	1.46
3	-1.02	1.54		0.35	1.09
4	-0.92	1.73		-1.35	2.57
5	0.53	2.60		-0.56	4.77
6	-2.37	4.61		-5.83	9.40

REFERENCES

Akinoglu, B.G. (1990). The effect of using different monthly-based estimation models on the calculations of hourly radiation values for Turkey, Proc. WREC, ed. A.A.M.Sayigh, Pergamon Press.
Akinoglu, B.G.; A.Ecevit (1990). Construction of a quadratic model using modified Angstrom coefficients, *Solar Energy,* 45, 85-92.
Duffie, J.A., W.A.Beckman (1980). *Solar Engineering of Thermal Process,* Wiley Interscience, New York.
Esteves A., Rosa C., (1989). A simple method for correcting Robitzsch type pyranometers, *Solar Energy,* 42, 9-13.

A METEOROLOGICAL MODEL FOR SOLAR ENGINEERING APPLICATIONS

I. FARKAS

Institute for Mathematics and Computer Science
University of Agricultural Sciences
Gödöllő, H-2103, HUNGARY

ABSTRACT

On the basis of long term measurements a simplified hourly solar climatic model has been elaborated. For ambient air temperature and absolute humidity a sifted cosinusoidal and for radiation on horizontial surface a combined cosinusoidal and a second order exponential approximation functions were derived. Using the daily distributions referring to a given month an extended model has been established valid for all the year round. The number of day in a year is the only parameter to this entire model. To identify the parameters of the approximation functions adaptive simplex method was applied to avoid the necessities of calculation the derivates of a multivariable function. The model was tested with the measured data for Budapest and it has been found that a fairly good prediction can be reached through the simplified equations.

KEYWORDS

Ambient temperature; absolute humidity; direct radiation; diffuse radiation; approximation equation; parameter identification.

INTRODUCTION

To analyse the dynamical operation of a solar system requires meteorological data. In case of on-line processes these values can be directly gathered and used in the system, but for simulation purposes an appropriate model is needed providing their hourly values. In the latter case the data originated from long term measurements can be used succesfully. For simplification of computations approximating equations can be introduced for ambient air temperature, for absolute humidity of ambient air and for solar radiation on horizontal plane in clear and cloudy days. Since their measured values are available in discrete time points, the most easily applicable methods for the approximation are which have no necessity of calculation the derivates of a function and are also good for multivariable problems. In the optimization program developed adaptive

2731

simplex method developed by Nelder and Mead (1965) was used to meet the requirements mentioned above. The computer programs carried out for the optimization in Farkas (1990) was coded in Pascal.

AMBIENT TEMPERATURE

For the daily temperature distribution a cosinusoidal function was fitted:

$$\hat{t}_w(\tau) = t_m \cos\left[\pi/12 \ (\tau - \tau_o)\right] + t_a \ . \tag{1}$$

The parameters of the Eq. (1) were calculated for average clear and cloudy days in every month for Budapest region on the basis of long term measured data in Szabó and Tárkányi (1969). The results are listed in Tables 1.

Table 1. Estimation results of ambient temperatures

Month	N_{day}	t_m, $^\circ$C clear	cloudy	t_a, $^\circ$C clear	cloudy	τ_o, h clear	cloudy
1	18	3.494	2.005	-4.500	-2.000	14.000	14.000
2	47	4.080	2.501	-2.975	0.500	14.000	14.000
3	75	5.521	2.497	5.516	3.483	14.000	14.000
4	103	6.501	2.998	13.491	11.000	14.000	14.000
5	136	6.501	2.992	15.491	12.987	14.000	13.969
6	163	6.584	2.972	21.450	16.950	14.000	14.000
7	199	6.501	3.507	22.491	18.483	14.000	14.017
8	229	6.501	2.501	23.491	16.458	14.000	13.999
9	259	6.501	3.507	17.491	15.484	14.000	14.000
10	289	3.503	2.998	12.492	12.000	14.000	14.000
11	319	4.021	2.005	3.000	5.000	14.000	14.000
12	346	3.494	2.005	0.500	0.000	14.000	14.000

The approximating functions give a really good estimation because the maximum temperature difference less then 0.87 $^\circ$C for clear days and less then 0.52 $^\circ$C in case of cloudy days comparing to the measured values.

An idea is arrising now how to construct only one function for estimation the year round measured data instead of the function defined by Eq. (1) and its parameters for average day in every month listed in Tables 1. It can be observed from the results that these parameters can be estimated with a function versus day in a year. Several functions were studied then for possible approximation of temperature parameters. Finally it has been found that for amplitude a parabolic, for average a combined parabolic and exponential and for time delay a constant can be reasonably used:

$$\hat{t}_m = a_1 \ N_{day}^2 + a_2 \ N_{day} + a_3 \ ; \tag{2a}$$

$$\hat{t}_a = \left[a_4 \ N_{day}^2 + a_5 \ N_{day} + a_6\right] \exp\left[-\left[\frac{N_{day} - a_7}{182.5}\right]^2\right] \ ; \tag{2b}$$

$$\hat{\tau}_o = a_8 \ . \tag{2c}$$

To get only one approximation function these patterns should be substituted into Eq. (1). Simultaneous identification of the parameters in Eqs (2a-c)

with the same data base for Budapest yields:
for clear days:

$a_1 = -0.0001298$ $a_2 = 0.04555$ $a_3 = 2.689$ $a_4 = -0.001107$

$a_5 = 0.4396$ $a_6 = -19.86$ $a_7 = 195.0$ $a_8 = 14.00$, (3)

and for cloudy days:

$a_1 = -0.00004218$ $a_2 = 0.01571$ $a_3 = 1.778$ $a_4 = -0.0008132$

$a_5 = 0.3122$ $a_6 = -10.48$ $a_7 = 227.3$ $a_8 = 13.97$. (4)

The maximum differences between the measured and the here calculated values by Eqs (3-4) falled into the same range as the calculated ones by Eq. (1). The maximum differences, only in some exceptional cases, reached as 1-2 $^{\circ}$C.

ABSOLUTE HUMIDITY OF AMBIENT AIR

For daily distribution of absolute humidity of air the same process, as it was used for ambient temperature, can be followed:

$$\hat{x}_w(\tau) = x_m \cos\left[\pi/12 \ (\tau - \tau_o)\right] + x_a \ . \qquad (5)$$

The results of parameter identification are shown in Table 2.

Table 2. Estimation of absolute humidities

Month	N_{day}	x_m, g/kg		x_a, g/kg		τ_o, h	
		clear	cloudy	clear	cloudy	clear	cloudy
1	18	0.2801	0.2865	2.0583	2.8738	14.1045	14.0536
2	47	0.3154	0.4699	2.0658	3.6162	14.0000	14.0000
3	75	0.6705	0.3852	3.5321	3.9150	14.0000	14.0124
4	103	0.9590	0.7216	5.3129	6.5137	14.0000	14.0000
5	136	1.1682	0.7438	6.4321	7.2921	14.0000	14.0496
6	163	1.8479	0.9652	9.8121	9.2079	14.0000	14.0000
7	199	1.9236	1.3306	10.4525	10.6075	14.0000	13.9877
8	229	2.0803	1.3480	11.2421	11.2771	14.0000	14.0000
9	259	1.5542	1.1131	8.0329	8.9129	14.0000	14.0000
10	289	0.6328	0.8219	5.8883	7.4900	14.0000	14.0000
11	319	0.5990	0.4836	3.6212	5.1108	14.0000	14.0255
12	346	0.4496	0.4098	3.1092	3.5933	14.0000	13.9505

The approximating functions, also here, give a good estimation because the absolute humidity difference less then 0,244 g/kg for clear days and less then 0.208 g/kg in case of cloudy days comparing to the measured values.

Similarly to the Eqs (2a-c) for the estimation of absolute humidity parameters it can be written that:

$$\hat{x}_m = b_1 \ N_{day}^2 + b_2 \ N_{day} + b_3 \ ; \qquad (6a)$$

$$\hat{x}_a = \left[b_4 \ N_{day}^2 + b_5 \ N_{day} + b_6\right] \exp\left[-\left[\frac{N_{day} - b_7}{182.5}\right]^2\right] \ ; \qquad (6b)$$

2733

$$\hat{\tau}_o = b_8 . \tag{6c}$$

Identifying the parameters of Eqs (6a-c) it has been obtained:
for clear days:

$b_1 = -0.00005608$ $b_2 = 0.02175$ $b_3 = -0.4613$ $b_4 = 0.0001966$

$b_5 = -0.2264$ $b_6 = 56.79$ $b_7 = 350.8$ $b_8 = 13.98$,   (7

and for cloudy days:

$b_1 = -0.00003073$ $b_2 = 0.01222$ $b_3 = -0.1145$ $b_4 = 0.0002525$

$b_5 = -0.2510$ $b_6 = 59.92$ $b_7 = 346.8$ $b_8 = 14.01$.   (8

The maximum differences between the measured and the here calculated value
by means of Eqs (7-8), similarly to the temperature prediction, falled int
the same range as the calculated ones by Eq. (5). The maximum differences
only in some exceptional cases, reached as 0.5-1 g/kg.

SOLAR IRRADIANCE

For the approximation of daily solar direct radiation on a horizonta
surface Garg et al. (1981) suggested a combined cosinusoidal and a fourt
order exponential function. During some preliminary study in Farkas (1988
it has been found that a cosinusoidal function with an extension of a
second order exponential function fits well, too:

$$\hat{I}(\tau) = \begin{cases} I_m \cos\left[\pi/2 \, \tau_h \, (\tau - \tau_o)\right] \exp\left(-\dfrac{\tau - \tau_o}{\tau_h}\right)^2, & \text{if } |\tau - \tau_o| \le \tau_h \\ 0, & \text{if } |\tau - \tau_o| > \tau_h , \end{cases}$$ (9

The results of identification by using Eq. (9) for direct and diffuse
radiation components are illustrated in Table 3.

Table 3. Estimation of solar radiation components

Month	N_{day}	I_m, W/m^2		τ_h, h		τ_o, h	
		direct	diffuse	direct	diffuse	direct	diffuse
1	18	192.79	82.31	4.520	5.240	11.515	10.551
2	47	288.17	96.01	5.536	6.393	11.478	11.447
3	75	424.17	136.73	5.997	7.722	11.427	11.365
4	103	523.16	140.72	6.915	8.458	11.491	11.342
5	136	595.39	145.61	7.635	9.601	11.488	11.873
6	163	606.52	166.41	7.873	9.742	11.449	11.551
7	199	571.98	161.43	7.600	9.494	11.520	11.672
8	229	557.92	139.77	7.167	8.531	11.580	11.472
9	259	471.33	112.08	6.365	7.761	11.568	11.426
10	289	335.20	104.31	5.610	6.759	11.487	11.487
11	319	239.93	80.31	4.812	5.899	11.563	11.525
12	346	171.41	65.97	4.621	5.268	11.533	11.448

The approximating functions give a reasonable estimation because the
radiation difference less then 36.0 W/m^2 for direct and less then 33.0 W/m^2
for diffuse radiation components comparing to the measured values.

Estimating the radiation parameters for amplitude a combined parabolic and exponential function, for halving time interval a second order polinom and for time delay a constant fit well:

$$\hat{I}_m = \left[c_1 \, N_{day}^2 + c_2 \, N_{day} + c_3 \right] \exp\left[- \left(\frac{N_{day} - c_4}{182.5} \right)^2 \right] ; \tag{10a}$$

$$\hat{\tau}_h = c_5 \, N_{day}^2 + c_6 \, N_{day} + c_7 ; \tag{10b}$$

$$\hat{\tau}_o = c_8 . \tag{10c}$$

Identifying the parameters of Eqs (10a-c) it has been obtained:
for clear days:

$c_1 = -0.007084$ $c_2 = 3.7375$ $c_3 = 216.8$ $c_4 = 134.0$

$c_5 = -0.0001319$ $c_6 = 0.04587$ $c_7 = 3.657$ $c_8 = 11.50$, $\qquad$ (11)

and for cloudy days:

$c_1 = -0.00008100$ $c_2 = -0.1748$ $c_3 = 193.6$ $c_4 = 186.9$

$c_5 = -0.0001649$ $c_6 = 0.05732$ $c_7 = 4.451$ $c_8 = 11.47$. $\qquad$ (12)

In comparision to the identical results in Table 3 it can be stated that the maximum differences between the measured and the here calculated values by means of Eqs (11-12) were a bit higher as the calculated ones by Eq. (9). But, the estimation still stays in the range of engineering applicability.

RESULTS

By means of the approximating functions introduced in this paper it is very convinient to handle the daily course of climatic parameters, because the number of day in a year is the only input parameter. In spite of their simplicity a reasonably good accuracy was achived. So, the model can be used as a tool in simulation of solar engineering problems.

REFERENCES

Farkas, I. (1988). On the estimation possibilities of solar radiation and ambient temperature used for solar systems, Technical-Agricultural Academy, Bydgoszcz, Poland, Agriculture, 27(158), 5-10.
Farkas, I. (1990). Block oriented system simulation program (BOSS), IBM implementation, Institiute for Mathematics and Computer Science, University of Agricultural University, Gödöllő, Hungary.
Garg, H.P., B. Bandyopadhyay and V.K. Sharma (1981). Investigation of rock bed solar collector cum storage system, Energy Conversion and Management, 21, 275-282.
Nelder, J.A. and R. Mead (1965). A simplex method for function minimization, Computer Journal, 7, 308-313.
Szabó, Gy and Zs. Tárkányi (1969). Solar radiation data for the planning in building industry, Institute for Building Sciences, Budapest, (in Hungarian).

MATCHING SILICON CELLS AND THERMOPILE PYRAMOMETERS RESPONSES

Alados-Arboledas, L; Castro-Díez, Y.; Batlles, F.J. and Jiménez, J.I.

Grupo de Física de la Atmósfera Dpto de Física Aplicada, Facultad de Ciencias, Universidad de Granada, 18071, Granada, Spain.

ABSTRACT

Silicon photovoltaic sensors provide a cheap alternative to standard thermopile sensors in piranometry. However, their use implies some problems related with their spectral response. Global and diffuse radiation, with a polar axis shadowband, both by means of thermopile sensor and silicon photovoltaic sensor, has been measured from 1990 summer until 1991 fall in Almeria, Spain, covering a complete range of atmospheric conditions. These coincident 10-minute data sets have been used to define a correction factor for the silicon detector measurements. The comparison of both data sets, has been made by applying the Kolmogorov-Smirnov test. Before the correction procedure was applied, severe discrepancies were found specially for the diffuse irradiance. Results of the correction method applied over an independent data set show an improvement on the measurements from the silicon detector, with R.R.M.S.D. of about 5% for both, global and diffuse irradiance.

KEYWORDS Pyranometers, silicon cells, diffuse radiation, global radiation, measurements.

INTRODUCTION

The accurate assesment of global, diffuse and direct solar radiation is of primary interest in the estimation of radiation interception by the diverse surfaces of hills, solar collectors, building, vegetation, animals, greenhouses, and thus in the energy balance of a region on the earth's surface, ecology, plant responses and other problems, including those related with turbidity.

A monitoring station typically measures only two of the insolation components and calculates the third. However, the measurement of diffuse horizontal irradiance and direct normal irradiance are not widespread due to the high cost of instruments and periodical maintenance. The use of a pyranometer with a polar axis shadowband for measuring the diffuse horizontal irradiance side by side another one for measuring the global horizontal irradiance provides the cheapest alternative for measuring the three solar radiation components, obtaining the third by means of their relationship

$$D = G - I_n \cos Z \qquad (1)$$

where G is the global total horizontal irradiance, I_n is the direct normal irradiance, D is the diffuse horizontal irradiance and Z is the zenith angle of the sun.

Since few years ago it has surged the alternative of measuring the solar radiation by means of photovoltaic silicon cells used as sensors in low-cost pyranometers, (Kerr et al., 1967, LICOR, 1984, Michalsky et al, 1987, Pereira et al. 1990). This fact may lead to an increase in solar radiation stations. Among the interesting features of photovoltaic pyranometers are their short time response to solar radiation fluctuations, and their stability and ruggedness that in conjunction with its tolerance to soiling make them suitable for use in severe environments.

However, there are several problems that must be solved before photovoltaic pyranometers can be considered as an useful alternative to thermopile pyranometers. The photodiode wavelength response is selective and not broad enough to cover all the solar spectrum. Furthermore, the silicon response presents temperature dependence (about 0.2%), that could cause a relevant change in sensitivity on mid-latitude regions. Finally, as is it the case with all pyranometric detectors, including those based on thermopile sensors, there are problems associated with their cosine response at high angles of incidence, only partially corrected by the use of a diffusing plastic disk.

In this work we discuss the limitations of silicon photodiodes in performing irradiance measurements traditionally made with thermopiles, and correction procedures that improve the intercomparability of the two radiometric techniques. The goal is to suit silicon cell for radiometric observations by deriving empirical corrections to simulate thermopile sensors. We present the results of an intercomparison between a set of photodiodes with a polar axis shadowband and a conventional complement of first-class thermopile instruments.

DATA

Experimental data were recorded at Almeria, Spain, a seashore location (36.83° N, 2.41° W). Two photovoltaic pyranometer, Licor 200-SZ, one with a polar axis shadowband and another without it, were used to perform the silicon sensor measurements for diffuse and global components, while an identic set of class I radiometers, Kipp & Zonenn CM-11, were used as reference. Other mesurements included in the radiometric station are the air temperature at 1.5 m and inner temperature of an Eppley pyrgeometer. Data from all instruments were averaged and stored every 10 minutes. Measurements used in this work began in early 1990 and finished after 1991 fall. Thus, a complete range of temperatures and solar angles were included among the samples taken. After analytical checks for measurements consistency were applied in order to eliminate problems associated with shadowband misalignements, and other questionable data. In fact we have used only cases at solar zenith angle less than 85^0. About 35000 ten minutes values are available in total.

CORRECTION PROCEDURE AND RESULTS

The first problem that must be adressed is the temperature dependence, Michalsky *et al.* (1987) proposed to normalize photodiode response at a standard temperature of 30^0 C using an empirically developed expression

$$R / R_{30} = 0.9815 + 0.0006 \, T \, (^\circ C) \qquad (2)$$

where T is the temperature measured inside the photvoltaic pyranometer by means of a thermistor. We have used this correction expression with the measured air temperature at 1.5 m and the inner temperature of a pyrgeometer operated side by side the pyranometers. The use of the inner pyrgeometer temperature is more effective but the results for the 1.5 m air temperature show that this could be an interesting alternative.

Following the correction method proposed by Michalsky *et al.* (1991) the next step is deriving a "silicon efficacy", that is, the ratio of thermopile pyranometer response to silicon pyranometer response, in order to macht silicon pyranometer with thermopile pyranometers measurements. To carry it out we have categorized the data according to the clearness of the sky, the brightness of skylight, and the solar zenith angle as it is done in Perez *et al.* (1990). The sky clearness parameter, denoted ϵ, is a function of the cloud and aerosol amount, defined by

$$\epsilon = ((D_u + I_{nu}) / D_u + 1.041 \, Z^3) / (1 + 1.041 \, Z^3) \qquad (3)$$

where subindex u denotes uncorrected values measured with the silicon pyranometer, I_{nu} is uncorrected direct normal irradiance evaluated with diffuse and global uncorrected values. The skylight brightness parameter, denoted Δ, depends on the aerosol burden and the cloud thickness and is defined by the relationship

$$\Delta = D_u / (I_o \, \cos Z) \qquad (4)$$

where I_o is the extraterrestrial solar irradiance in W/m^2. The third parameter in the model is the solar zenith angle, that defines the sun's position in the sky hemisphere.

Table 1. Boundaries for the correction parameters

Parameter	Boundaries			
ϵ	1.45	2.70	4.30	6.40
Δ	0.007	0.015	0.040	0.075
Z (degrees)	75	60	50	35

Table 1 contains the boundaries for the five bins of each parameter. This boundaries are chosen in such way that real cases of each parameter were evenly distributed. The boundaries in ϵ and Δ are different from those considered by Michalsky *et al.*(1991) using their rotating shadowband radiometer, due to the inherent differences between the two measuring techniques with respect to diffuse irradiance. The rotating shadowband radiometer performs a realtime correction for the amount of skylight blocked by the shadowband when making diffuse measurement while we do not correct shadowband measurements at this stage.

2738

he ratio of the thermopile irradiance measurement to the corresponding silicon cell irradiance measurement is calculated for each member of a bin. The mean of these ratios is used as a correction factor for any subsequent silicon cell irradiance measurement with parameters falling within the bounds of the associated bin. In this way we have obtained two sets of correction factors, one for global and another for diffuse horizontal irradiance. We have used two thirds of the data for this purpose, applying the correction model to the remaining data.

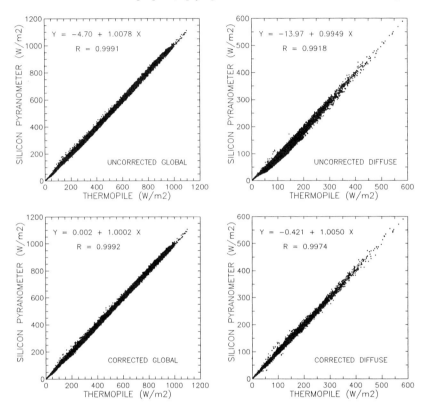

Fig. 1. *Scatter plots of silicon pyranometer vs. thempile responses.*

Figure 1 shows the scatter plots of the thermopile versus the silicon pyranometers after temperature correction by means of equation 2, using the inner pyrgeometer temperature. Global measurements shows a good agreement, while diffuse shows a marked tendency to underestimation, specially in clear conditions, low values of diffuse irradiance. Figure 2 shows the discrepancies between the two data sets, specially relevant is the distribution of bias for diffuse irradiance, showing the aforementioned underestiamation tendency.

The application of the correction procedure greatly improves the matching of the two kinds of sensors measurements, specially for the diffuse irradiance. Figure 1 also shows the improvement in the scatter plots, the experimental points lies close to the 1:1, line of perfect matching. Figure 2 shows the new apparence of bias distribution, mean bias deviations are

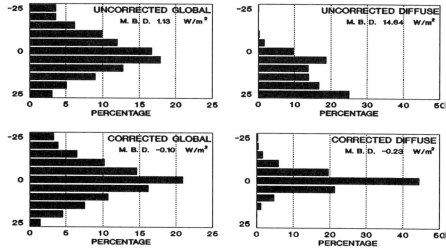

Fig. 2. *Mean Bias Deviation distribution between the two data sets.*

close to 0.0, both for global and diffuse irradiance. Table 2 shows the results that we have obtained. The Kolmogorov-Smirnov test, showing the probability of the null hypothesis that the two data sets were subset of the same distribution, indicates a relevant improvement in the matching of the two distribution, thermopile and silicon cell data. The Relative Root Mean Square Deviation, R.R.M.S.D., shows a discrepancy between the two kind of sensor measurements about 5%, both for global and diffuse irradiance, a close value to the experimental errors associated with the thermopile measurements.

Table 2. R.R.M.S.D. and Kolmogorov-Smirnov test between silicon cells and thermopiles.

		R.R.M.S.D %	K-S Test %
GLOBAL	Uncorrected	4.9	90.17
	Corrected	4.6	100
DIFFUSE	Uncorrected	17.26	11.54
	Corrected	5.1	35.48

REFERENCES

Kerr, J.P., Thurtell G.W. and Tanner C.B. (1967). An integrating pyranometer for climatological observer stations and mesoscale network. *J.Appl.Meteorol.*, 6, 688-690.

LICOR, Inc, (1984) *Radiation Measurements Instruments*, LMI-584.

Michalsky J.J., Harrison, L. and LeBaron, B.A. (1987). Empirical radiometric correction of a silicon photodiode rotating shadowband pyranometer. *Solar Energy*, 39, 87-96.

Michalsky J.J., Perez, R., Harrison, L. and LeBaron, B.A. (1991). Spectral temperature correction of silicon photovoltaic solar radiation detectors. *Solar Energy*, 47, 299-305.

Pereira, A.C. and Souza Brito, A.A. (1990). A microprocesor-based semiconductor solar radiometer, *Solar Energy*, 44, 137-141.

Perez, R., Ineichen, P., Seals, R., Michalsky, J.J. and Stewart R. (1990). Modelling daylight availability and irradiance components from direct and global irradiance, *Solar Energy*, 44, 271-289.

DISCRETIZATION OF SOLAR RADIATION FOR SOLAR ENERGY APPLICATIONS

A. ADANE and A. MAAFI

Université des Sciences et de la Technologie d'Alger
(U.S.T.H.B.), Institut d'Electronique, BP N° 32,
El Alia, BAB EZZOUAR (ALGIERS), ALGERIA

ABSTRACT

A method for estimating the loss of information in discrete models is derived in this paper,using the calculation of the relative entropy function.This method is used to divide daily sunshine duration and global irradiation data into different states (these data were recorded in four locations of Algeria, i.e.,Algiers,Batna,Oran and Setif).It is found that the loss of information involved by a two-state discretization is less than 12 % and the obtained results confirm that two-state models are suitable for sizing photovoltaic systems or solar power plants.For higher power systems,a higher number of discrete states must be used for modelling solar radiation,in order to reduce the loss of information.

KEYWORDS

Solar radiation; entropy function; information theory; photovoltaic systems; modelling.

INTRODUCTION

In the case of locations for which solar radiation has not been measured,the calculation of the size of solar power systems requires the simulation of the statistical features of collected solar energy.The most common procedure of generating time sequences of solar radiation is by means of autoregressive moving average (ARMA) models.Another way is to use Markov models in discrete time (see for example, G.F. Lameiro and W.S. Duff, 1979,M.J. MEJON et al., 1980).So,the first-order two-state Markov chains have recently been used to fit the daily sunshine duration and global irradiation data for Algiers.Then.the obtained model has successfully been applied for sizing stand-alone photovoltaic (PV) systems (see, A. Maafi and A. Adane, 1989) .In 1988,A.J. Aguiar et al.,divided solar radiation for Lisbon into ten states and also found that the better solution for modelling and generating radiation sequences was to use first-order Markov models in dicrete time.In the papers mentioned above,the discretization of sunshine duration or solar

radiation into two,three,four or ten states has been discussed.However,no
quantitative study was reported about the loss of information involved by
such a discretization.To overcome this problem,a method for estimating the
loss of information in discrete models,is developed in this paper.This method
which consists in computing the entropy function for different types of discretization,is then applied to the decomposition of sunshine duration and
global irradiation (measured in four locations of Algeria) into different
states.

METHODOLOGY

Consider a meteorological variable G,recorded every day in a given location
during a number of years sufficiently large to make the obtained data statistically meaningful.If the G data are normalized,then,they become $0 \leqslant g \leqslant 1$.
The method of estimation of the loss of information firstly consists in dividing the scale of g data into n equal intervals or meteorological states ϕ_k
which correspond to:

$$(k - 1)/n \leqslant g < k/n \qquad \text{for } k = 1, 2, ..., (n - 1) \qquad (1)$$

$$(n - 1)/n \leqslant g \leqslant 1 \qquad \text{for } k = n$$

The integer k labels the order of the states in the g-scale.In practice,the
possible values of n (which are also integers) must not exceed ten intervals
of the g-scale.In addition,when the number n increases too much,there is an
important augmentation of the bias error,the number of data decreases in the
intervals and the conditions for long-term situations are not achieved.For
each of the values of n,an histogram of g-data is then obtained and gives the
probability P_k of having a state ϕ_k.This classification gives rise to discrete models which are equivalent to coding systems producing messages by combination of n different symbols.Therefore,it can be said that the amount of
incertainty or information involved by the reception of a message or the outcome of an experimentation,is measured by the entropy function defined as:

$$Q(n) = - \sum_{k=1}^{n} P_k \log_2 P_k \qquad (2)$$

In this expression,the correlation between states is neglected (the latter are
considered as independent in the first approximation).When the number n varies,the background of the obtained entropy notably changes from a model to
another.So,it is better to consider the relative entropy $Q(n)/Q_M$ to make
possible the comparison between each type of discretization.The maximum entropy Q_M is obtained when all the probabilities are equal.Then:

$$Q_M = \log_2 n \qquad (3)$$

The discrete probability distribution of g for n = 10 is almost identical to
the continuous ones and can be taken as reference.In the sense of information
theory,the rate of loss of information due to the discretization into n states,is expressed as:

$$p = \left| \frac{Q(10) - Q(n)}{Q(10)} \right| \qquad (4)$$

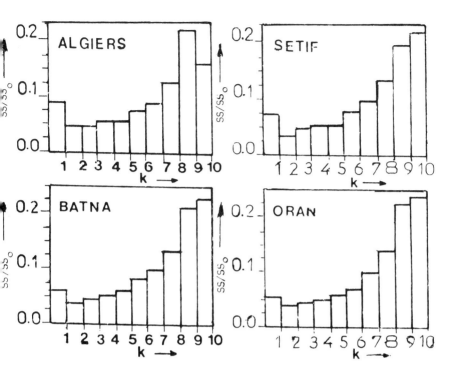

Fig. 1. Histograms of SS/SS_o data for Algiers,Batna,Oran and Setif.

EXPERIMENTAL DATA

The data used to implement this method are daily sunshine duration SS for Algiers(1972/82) ,Batna(1971/84) ,Oran(1970/84) and Setif(1981/87) .Daily global irradiation H for Algiers(1972/82) are also considered.these data are firstly divided by their extra-terrestrial values SS_o and H_o,in order to get normalized data,i.e.,$g = SS/SS_o$ and H/H_o .Then,they are sorted into two, three,four,five,six,seven,eight,nine and ten states respectively.this clas- sification yields nine types of histogram which differ from one to another by the number of intervals.For example,fig. 1 exhibits the histograms of SS/SS_o for Algiers,Batna,Oran and Setif on a yearly basis.These histograms which consist of ten intervals,have the same general shape and show a con- centration of data around the high values.Such distributions are essentially due to the. mediterranean features of the climate of the North of Algeria.

ESTIMATION OF THE LOSS OF INFORMATION

In the first step,the relative entropy and the loss of information are com- puted for SS/SS_o and H/H_o data on a monthly basis.The same kind of results are obtained for the four locations.To illustrate this calculation,the dif- ferent distributions of Q/Q_M obtained for Algiers in the case of SS/SS_o da- ta,are presented in fig. 2. The obtained loss of information is less than 5 % for each month of the year,except July and August for which p can rise

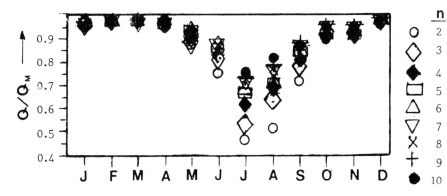

Fig. 2. Monthly distributions of relative entropy for different
types of discretization.

to 39 %.The high values of loss of information are explained by the high
persistence of fine weather in Summer involved by the redundancy of these
states. p has also been estimated on a yearly basis for SS/SS_0 and H/H_0 da-
ta.Fig. 3 and table 1 show that the same kind of results are obtained for
the four locations.The loss of information is maximum and does not exceed
12 % for n = 2 states.When n is greater than five states,p is less than 4 %.
The lowest values of p are obtained for Algiers (p is less than 5 % for
n = 2).

DISCUSSION AND CONCLUSION

In general,two fundamental parameters of solar radiation are required for
sizing stand-alone photovoltaic systems.These are the yearly average of dail

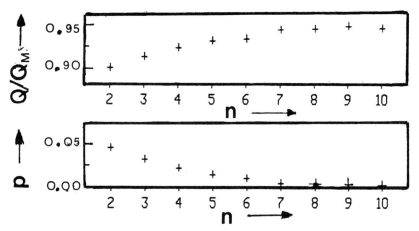

Fig. 3. Relative entropy Q/Q_M and loss of information p against
the number n,on a monthly basis for Algiers.

2744

Table 1. Relative entropy for the four considered locations on a yearly basis (when $n = 2$, $p = 5\%$ for Algiers, 12 % for Oran, 11 % for Batna and Setif).

n	Q/Q_M			
	Algiers(H/H_o)	Batna SS/SS_o)	Oran(SS/SS_o)	Setif(SS/SS_o)
2	0.894	0.817	0.795	0.829
5	0.929	0.888	0.871	0.888
10	0.944	0.918	0.907	0.932

solar radiation and the deficit of solar energy (which is often expressed as number of days of storage).In Algiers,the energy deficit occurs during the February/May period.Then,the two-state approximation is largely sufficient . to account for the main features of solar radiation.It will be noticed that the loss of information must not be considered as a modelling error.But,this parameter measures the ability of a discrete model to reduce the redundancy of states when the number n increases.In addition,the increase in number of states requires too elaborate calculations and leads to no significant improvement for sizing solar power systems working in a fixed position .The simplicity and efficiency of the two-state discretization explain why the first-order two-state Markov chains was successfully applied for sizing stand-alone photovoltaic systems in Algiers (see, A. Maafi *et al.*, 1990).Sequences of days of bad and fine weather have recently been simulated for Algiers using the first-order two-state Markovian model.Synthesized and measured sequences have been compared,giving rise to satisfactory results.

The method presented before seems to be an efficient tool for estimating the loss of information in discrete models and the above analysis confirms that the two-state discretization is suitable for sizing PV systems and solar power plants.As seen in fig. 2,a more detailed description of solar radiation requires its discretization into more than five states and can be used for sizing high power systems like solar concentrators,since the latter are more sensitive to the variations of solar radiation.In addition,it will be interesting to extend this method to the discretization of hourly solar radiation in order to find the number of states required for sizing low inertia solar systems.

REFERENCES

Aguiar, R.J., M. Collares-Pereira and J.P. Conde (1988), *Solar Energy*, 40, 269-279
Lameiro, G.F. and W.S. DUFF (1979), *Solar Energy*, 22, 211-219
Maafi, A. and A. Adane (1989). A two-state Markovian model of global irradiation suitable for photovoltaic conversion. *Solar Wind Technol.* 6, 247-252
Maafi, A., A. Adane and M. Benghanem (1990). A two-state Markovian estimation of the size of photovoltaic systems. *Proc. WREC (Reading)*, 1, 284-288
Mejon, M.J., Ph. Bois et R. Lestienne (1980), *Rev. Phys. Appl.*, 15, 113-122

DETERMINATION OF PEREZ SOLAR DIFFUSE IRRADIANCE MODEL COEFFICIENTS FOR VALENCIA (SPAIN)

V. GOMEZ, A. CASANOVAS, M.P. UTRILLAS AND J. MARTINEZ-LOZANO

Departament de Termodinàmica. Facultat de Física. Universitat de València
46100 Burjassot (València), Spain

ABSTRACT

The Perez solar diffuse irradiance model has been tested against measured irradiance data at Valencia, Spain. The model performs satisfactorily using the coefficients given by Perez et al.(1990). Yet some improvement is achieved by fitting the model to the measured Valencia irradiance data.

KEYWORDS

Perez model; solar diffuse irradiance; solar irradiance models.

THE PEREZ MODEL

Over the last year several models have been developed to estimate the diffuse solar irradiance over tilted and arbitrary oriented surfaces. Several authors (Ineichen et al., 1985, Bourges, 1986, Chowdhury and Raman, 1987, Hay and McKay, 1987, Ineichen et al., 1988, Reindl et al., 1990, Utrillas et al., 1991, Feuermann and Zemel, 1992), have reported better agreement of Perez's model predictions compared to experimental data.

Several versions of Perez model are available, being the formers more elaborate and the later the more simplified. Simplifications consider a zero-width horizon and a point circumsolar zone, together with drastic reduction in the number of coefficients to fit and linearization of equations. In spite of simplifications the loss of performance of the model is not noticeable. Details of Perez's model can be found elsewhere (Perez et al., 1983, Perez et al., 1987, Perez et al., 1988, Perez et al., 1990).

EXPERIMENTAL SETUP AND DATABASE

The measured data used cover the January 1 to December 31 1990 period and include values of direct irradiance, horizontal global irradiance and vertical global irradiance on the North-, South-, East- and West planes. The vertical plane pyranometers are provided with artificial horizons. Measurements have been performed at the Campus of Burjassot, Valencia, Spain (latitude: 39.5° N, 40 m height over the sea level). Horizon obstructions are less than 4° except in a 30° sector on the Northwest where they reach 6°.The experimental setup has been described in a previous paper (Utrillas et al., 1991).

PEREZ'S MODEL COEFFICIENTS DERIVED FROM VALENCIA DATA

Only the last version of Perez's model has been considered for fitting. The circumsolar brightness F_1 and horizon brightness F_2 of this model show a linear dependence on the zenithal angle θ_z and the sky's brightness Δ defined through:

$$\Delta = mI_{do}/I_{on} \qquad (1)$$

where m is the optical air mass, I_{on}, the extraterrestrial normal irradiance and I_{do}, the diffuse horizontal irradiance. The tabulated data of Kasten and Young (1989) have been used for m. The circumsolar and horizon brightness show an additional dependence on the clearness index ϵ defined through:

$$\epsilon = (I_{do} + I_n)/I_{do} \qquad (2)$$

where I_n is the normal irradiance.

This dependence on the clearness index is not lineal and is described therefore by eight sets of coefficients each belonging to a different sky category defined by an ϵ range.

A modified clearness index ϵ' independent from zenithal angle has been proposed additionally by Perez et al. (1990):

$$\epsilon' = (\epsilon + \kappa\theta_z^3) / (1 + \kappa\theta_z^3) \qquad (3)$$

where k = 1.041.

The used categories correspond to the last version of model (Perez et al., 1990). The circumsolar and horizon brightness are given for ϵ' range by:

$$F_1 = F_{11} + F_{12} \Delta + F_{13} \theta_z \qquad (4)$$
$$F_2 = F_{21} + F_{22} \Delta + F_{23} \theta_z \qquad (5)$$

given a total of 48 different coefficients F_{ij} to be adjusted.

Table 1. Generic circumsolar (F_1) and horizon brightening (F_2) coefficients developed for Valencia.

ϵ' category	UPPER LIMIT	CASES (%)	F_{11}	F_{12}	F_{13}	F_{21}	F_{22}	F_{23}
1	1.065	21.27	0.012	0.369	-0.024	-0.074	0.039	0.019
2	1.230	7.83	0.483	0.047	-0.250	0.044	0.049	-0.049
3	1.500	8.41	0.706	0.447	-0.477	0.194	-0.091	-0.089
4	1.950	10.74	0.994	0.109	-0.588	0.400	-0.512	-0.104
5	2.800	15.83	1.069	-0.298	-0.582	0.574	-1.280	-0.054
6	4.500	19.63	1.073	-0.898	-0.526	0.749	-2.759	0.064
7	6.200	9.33	1.431	-6.135	-0.338	1.106	-7.316	0.274
8	------	6.97	0.878	-7.618	0.068	1.164	-12.35	0.610

$$F_1 = F_{11}(\epsilon') + F_{12}(\epsilon') \Delta + F_{13}(\epsilon') \theta_z$$
$$F_2 = F_{21}(\epsilon') + F_{22}(\epsilon') \Delta + F_{23}(\epsilon') \theta_z$$

Table 1 shows the coefficients obtained from the fit. Cases where $\varepsilon' > 15$ or zenithal angle θ_z is grater than 85° have been not preserved. A total of 3474 hourly cases distributed among categories accordingly to Table 1 have been obtained.

DISCUSSION OF RESULTS

Both the proposed coefficients by Perez *et al.* (1990) obtained from 13 USA and European locations, and the derived coefficients for Valencia (Table 1) have been used in comparing the Perez model predictions with the measured data.

Although Perez is a diffuse irradiance model, only global irradiance values on tilted surfaces are normally required. Therefore is usual to give the error of the calculated global irradiance instead of giving the error in the diffuse solar irradiance. Yet is better to operate with the later on discussing model results. Both kinds of error are therefore presented in this paper with separate comments for each of them.

Three different statistical estimators have been used as a measured of the model deviation from experimental values: the mean bias deviation MBD, the mean absolute deviation, MAD, and the root mean square deviation, RMSD. Experimental values are also affected by errors, and we avoid therefore to use error as a synonym for model deviation from measured data.

We have also given the relative values of the estimators to allow for some normalisation in comparing results from different sources.

Estimation of Global Solar Irradiance

Table 2 shows the obtained MBD, MAD and RMSD global irradiance relative values using the Perez's model. The model performs well showing a 10.6% MAD for the model provided with Perez parameter set (Perez *et al.*, 1990) and 8.2% MAD for the model fitted to Valencia data. Yet the improvement is basically in the North-, East-, and South planes, with little variation on the West plane. This improvement obtained by using local coefficients is similar to that achieved by Harrison and Combes (1989) for Calgary, Canada.

Table 2. Deviations in the estimation of the global solar irradiance by using Perez model

	GLOBAL SOLAR IRRADIANCE				
	AVERAGE (Wm^{-2})	MODEL COEFF.	RELATIVE DEVIATION (%)		
			MBD	RMSD	MAD
NORTH 90°	78.69	USA-EUR.	-13.7	24.3	18.5
		VALENCIA	1.6	17.6	12.5
SOUTH 90°	267.65	USA-EUR.	-6.4	10.8	8.1
		VALENCIA	-1.8	8.0	5.9
EAST 90°	188.16	USA-EUR.	-9.0	17.8	11.8
		VALENCIA	-2.3	13.9	8.7
WEST 90°	193.59	USA-EUR.	-2.0	13.9	9.5
		VALENCIA	4.2	13.7	9.4
ALL	182.02	USA-EUR	-6.7	15.2	10.6
PLANES		VALENCIA	0.0	12.4	8.2

Estimation of Diffuse Solar Irradiance

Table 3 shows the obtained MBD, MAD and RMSD diffuse irradiance relative values using the Perez' model. The North plane gives as expected similar results as for global solar irradiance. The deviation of the model in evaluating diffuse solar irradiance doubles the global irradiance evaluation in the other planes.

Table 3. Deviations in the estimation of the diffuse solar irradiance by using Perez model

DIFFUSE SOLAR IRRADIANCE					
	AVERAGE (Wm^{-2})	MODEL COEFF.	RELATIVE DEVIATION (%) MBD	RMSD	MAD
NORTH 90°	73.59	USA-EUR.	-14.6	26.0	19.8
		VALENCIA	1.7	18.8	13.3
SOUTH 90°	114.32	USA-EUR.	-15.0	25.3	19.0
		VALENCIA	-4.2	18.8	13.8
EAST 90°	100.52	USA-EUR.	-16.8	33.4	22.1
		VALENCIA	-4.4	26.0	16.2
WEST 90°	94.61	USA-EUR.	-4.2	28.4	19.4
		VALENCIA	8.5	28.0	19.2
ALL	95.76	USA-EUR	-12.7	28.8	20.1
PLANES		VALENCIA	0.0	23.6	15.7

The model uses as input horizontal diffuse irradiance, evaluated through direct normal and horizontal global irradiance. Therefore the model may be biassed if direct and global irradiance shows similar values, as in the case of clear skies. In addition to this, the Eppley NIP pyrheliometer, with a 5°43' aperture, includes circumsolar diffuse irradiance in the measurements. The Gueymard (1986) pyrheliometer correction factor can be applied to modify the measurements of direct irradiance, but this correction has no be performed in this paper.

CONCLUSIONS

The overall performance of the Perez' model is satisfactory in the case that we use the set of coefficients proposed by Perez *et al.* (1990). This set has been obtained from the data of 13 different locations. Using the Perez' model with coefficients fitted to Valencia data increases slightly the performance, yet this latter test lacks, for the moment, of strict independence from the data set. Atypically the West plane does not present any improvement using the local set of coefficients. This behaviour of the applied model may be related.

REFERENCES

Bourges, B. (1986). Le calcul de l'éclairement solaire sur planes inclinés. *La Météorologie*, 7 Sér., 11, 58-68.

Chowdhury, B. H. and Raman, S. (1987). Comparative assessment of plane-of-array irradiance models. *Solar Energy* , 39, 391-398.

Feuermann, D. and Zemel, A. (1992). Validations of models for global irradiance on inclined planes. *Solar Energy* , 48, 59-66.

Gueymard, C. (1986). Une paramétrisation de la luminance énérgetique du cel clair en foction de la turbidité.*Atmosphere-Ocean*, 24, 1-15.

Harrison, A. W. and Coombes, C. A. (1989). Performance validation of the Perez tilted surface irradiance model. *Solar Energy* , 42, 327-333.

Hay, J.E. and McKay, D. C. (1987). Calculation of solar irradiance for inclined surfaces:

Verification of models which use hourly and daily data. *I.E.A. Solar Heating and Cooling Program, TASK IX Final Report.* Atmospheric Environment Service, Downsview, Ont., Canada.

Ineichen, P., Guisan, O. and Razafindraibe, A. (1985). *Mesures d'ensoleillement a Genève: Modèles de transposition du rayonnement solaire.* Groupe de Physique Appliquée, Université de Genève, CUEPE Pub. No. 25.

Ineichen, P., Zelenka, A., Guisan, O. and Razafindraibe, A. (1988). Solar transposition models applied to a plane tracking the sun. *Solar Energy*, 41, 371-377.

Kasten, F. and Young, A.T. (1989). Revised optical air mass tables and approximation formula. *Applied Optics*, 28, 4735-4738.

Perez, R., Stewart, R. and Scott, J. (1983). *An anisotropic model of diffuse solar radiation with application to an optimisation of compound parabolic concentrators.* ASRC-SUNY, Pub. No. 870. Albany, NY.

Perez, R., Seals, R., Ineichen, P., Stewart, R. and Menicucci, D. (1987). A new simplified version of the Perez diffuse irradiance model for tilted surfaces. *Solar Energy*, 39, 221-231.

Perez, R., Stewart, R., Seals, R. and Guertin, T. (1988). *The development and verification of the Perez diffuse radiation model.* Report SAND88-7030.

Perez, R., Ineichen, P., Seals, R., Michalsky, J. and Stewart, R. (1990). Modeling daylight availability and irradiance components from direct and global irradiance. *Solar Energy*, 44, 271-289.

Reindl, D. T., Beckman, W. A. and Duffie, J. A. (1990). Evaluation of hourly tilted surface radiation models. *Solar Energy*, 45, 9-17.

Utrillas, M. P., Martinez-Lozano, J. A. and Casanovas, A. J. (1991). Evaluation of models for estimating solar irradiation on vertical surfaces at Valencia, Spain. *Solar Energy*, 47, 223-229.

ESTIMATION OF THE MONTHLY MEAN HOURLY SOLAR IRRADIATION AT VALENCIA (SPAIN):THE HELIOS MODEL

J. MARTINEZ-LOZANO, M.UTRILLAS, V. GOMEZ AND A. CASANOVAS

Dpto. de Termodinámica. Facultat de Física. Univ. de Valencia
46100 Burjassot (Valencia), Spain.

ABSTRACT

European Helios model calculated values have been tested with global irradiance data measured on vertical planes at Valencia (Spain). CEC's and Perez' versions of the Helios model give similar deviations from measured data, except on the North plane, where the Perez' version performs better than the CEC model version.

KEYWORDS

Helios model; tilted surfaces; global solar irradiance.

THE HELIOS MODEL

The Helios Solar Irradiation Model (Koussis et al., 1989) has been developed within the European Solar Microclimate Programme of the CEC DG XII Solar Energy R & D Programme. Helios code computes monthly mean values of the direct, diffuse and reflected components of solar irradiation on a surface of arbitrary orientation, as well as the total values, for every hour of a mean monthly day, taking into account the shading and the reflection from nearby topographic features.

The Helios code is based in the CEC radiation model for tilted surfaces (Page, 1986), that gives the monthly mean of hourly global solar irradiation based on the values for clear- and overcast skies. Two different diffuse solar irradiation model are used by Helios: the CEC- and Perez'25° circumsolar model (Perez et al., 1987).

The Helios model needs only very common climatic data as input: site latitude, longitude and altitude, monthly mean of insolation (hours), tilt and azimuth of the surface, and local variables as atmospheric turbidity, specified by the Linke turbidity factor, albedo and topography. It is applicable therefore to locations where no radiation data are available.

Linke turbidity factor is estimated by the Doigniaux and Lemoinde (1983) method from the Angstrom-Prescott formula relating hourly global irradiation and insolation. The effect of topography is estimated from computation of the skyline, reduction of sky dome and reflection

2751

from the surrounding terrain.

EXPERIMENTAL SETUP AND DATABASE

The measured hourly values of irradiation used cover the January 1, 1990, to December 31, 1991, period. Measurements have been performed at the Campus of Burjassot, Valencia, Spain (latitude: 39.5° N, altitude:40 m above the sea level). Horizon obstructions are less than the 4°, except in a 30° sector on the Northwest where they reach 6°, and the topographic aptitudes of Helios have not been used in the calculations.

Measurements include direct irradiation, horizontal global irradiation and vertical global irradiation on the North-, South-, East-, and West planes. The vertical plane pyranometeres have been provided with artificial horizons and zero value of ground albedo have been used therefore in calculations. The experimental setup has been described in a previous paper (Utrillas et al.,1991).

Input values for the model are, therefore, the sunshine hours and atmospheric turbidity. Table 1 gives the monthly-mean hourly values of the former obtained from the 1961-1990 period (Centro Meteorológico Zonal de Valencia, 1992). Monthly values of the coefficient of the Angstrom-Prescott formula estimated for Valencia (Martínez-Lozano, 1984) are included in Table 1. Done input data, the model predicts an average value over thirty years. This values have been compared with experimental hourly irradiation data obtained during the years 1990 and 1991.

Table 1. Sunshine hours and Angstrom coefficients for Valencia, Spain

Month	Angstrom Coefficient	Insolation 1961-1990	Insolation 1990	Insolation 1991
January	0.74	5h 34'	5h 41'	5h 28'
February	0.76	5h 43'	6h 49'	4h 53'
March	0.76	6h 16'	6h 28'	6h 0'
April	0.76	7h 6'	7h 21'	8h 29'
May	0.74	8h 5'	9h 19'	9h 24'
June	0.75	8h 47'	9h 45'	8h 45'
July	0.70	9h 45'	10h 9'	10h 14'
August	0.69	8h 58'	9h 17'	9h 39'
September	0.72	7h 41'	6h 37'	8h 43'
October	0.69	6h 24'	6h 21'	6h 37'
November	0.71	5h 17'	6h 46'	6h 18'
December	0.69	4h 55'	4h 39'	3h 28'

RESULTS AND DISCUSSION

Two different statistical estimators have been used as a measure of the model deviation from experimental values: the mean absolute deviation, MAD, and the root mean square deviation, RMSD. Experimental values include errors, and we avoid therefore to use error as a synonym for model deviation from measured data. Relative values of the estimators have also been given to normalise results from different sources.

Tables 2 and 3 show the monthly deviation of the model for each plane during the 1990 and 1991 years.The results obtained from Perez'- and CEC versions of Helios model are similar, except in the North and West planes. Perez' model perform better in the North plane and the CEC model in the West plane. This behaviour of the model may be related to the moderate

West obstructions of the measuring location.

Table 2. Deviations of mean values estimated by Helios model from 1990 experimental data for Valencia, Spain. CEC- and Perez' version of Helios model are compared.

MODEL		CEC		PEREZ	
Plane	Mean value (W.m-2)	MAD (%)	RMSD (%)	MAD (%)	RMSD (%)
North 90°	74.9	16.3	19.5	12.8	16.5
South 90°	271.5	18.2	26.3	18.3	25.7
East 90°	189.3	17.0	22.6	16.5	24.2
West 90°	186.3	13.9	20.7	13.9	27.3
Total	180.6	16.6	25.7	16.9	27.3

Table 3. Deviations of mean values estimated by Helios model from 1991 experimental data for Valencia, Spain. CEC- and Perez' version of Helios model are compared.

MODEL		CEC		PEREZ	
Plane	Mean value (W.m-2)	MAD (%)	RMSD (%)	MAD (%)	RMSD (%)
North 90°	75.9	18.2	23.5	16.0	21.5
South 90°	277.6	22.3	28.6	22.7	28.6
East 90°	202.6	20.9	32.0	20.6	33.0
West 90°	195.4	18.5	27.9	21.5	33.0
Total	187.9	20.6	31.2	21.2	33.1

Year 1991 values are less accurate than the 1990 values, due to the atypical insolation values during some months as April and December. In addition to this the 1990 year is included in the period of insolation input data. Simplification performed on topography doesn't affect obviously the year by year difference.

Table 4. Deviations of values estimated by Helios model and actual 1990 sunshine hours from 1990 experimental data for Valencia, Spain. CEC- and Perez' version of Helios model are compared.

MODEL		CEC		PEREZ	
Plane	Mean value (W.m-2)	MAD (%)	RMSD (%)	MAD (%)	RMSD (%)
North 90°	74.9	18.9	22.5	14.8	18.7
South 90°	271.5	14.5	18.4	14.6	18.6
East 90°	189.3	14.8	19.1	14.1	19.1
West 90°	186.3	12.2	18.9	11.2	18.9
Total	180.6	14.4	20.2	13.6	19.6

Tables 4 and 5 show that the deviation of Helios model using the actual sunshine hours improve the ability of the model on estimating the hourly monthly means of irradiation. The performance of Perez' model in the West plane is increased remarkably using the actual sunshine hours as input to the model. Yet the global difference between the 1990 and 1991 years is not reduced in the process.

We can conclude therefore that the most of the observed deviations are of a random nature and greater that the differences between models. The choice of an Helios' model version is therefore not decisive on results.

Table 5. Deviations of values estimated by Helios model and actual 1991 sunshine hours from 1991 experimental data for Valencia, Spain. CEC- and Perez' version of Helios model are compared

MODEL		CEC		PEREZ	
Plane	Mean value (W.m-2)	MAD (%)	RMSD (%)	MAD (%)	RMSD (%)
North 90°	75.9	19.4	24.6	15.9	21.7
South 90°	277.6	18.4	24.1	18.7	24.3
East 90°	202.6	18.0	24.7	17.5	25.0
West 90°	195.4	16.0	23.9	15.4	22.9
Total	187.9	17.7	25.9	17.3	25.8

REFERENCES

Centro Meteorológico Zonal de Valencia. (1992). Unpublished data.
Doigniaux, R. and Lemoine, M. (1983). Classification of radiation sites in terms of different indices of atmospheric transparency. In: *Solar Radiation Data.* (W. Palz, ed.), Solar Energy R&D in the European Community , Ser. F, Vol. 2, D. Reidel Publishing Co., Dordrecht.
Koussis, A., Lalas, D. P. and Papadopoulos, C. (1989). *Helios Solar Radiation Modelling Package. User's Guide (Version 2.01).* Lamda Technical Ltd., Athens.
Martínez-Lozano, J. A.. (1984). Global solar irradiation and sunshine duration in Valencia (Spain). *Rev. Geofísica,* 40, 279-290.
Page, J. K. (1986). *Prediction of Solar Radiation on Inclined Surfaces.* Solar Energy R&D in the European Community, Ser. F, Vol. 3, D. Reidel Publishing Co., Dordrecht.
Perez, R., Seals, R., Ineichen, P., Stewart, R. and Menicucci, D. (1987). A new simplified version of the Perez diffuse irradiance model for tilted surfaces. *Solar Energy,* 39, 221-231.
Utrillas, M. P., Martinez-Lozano, J. A. and Casanovas, A. J. (1991). Evaluation of models for estimating solar irradiation on vertical surfaces at Valencia, Spain. *Solar Energy,* 47, 223-229.

REGRESSION ANALYSYS AND BIVARIATE FREQUENCY DISTRIBUTIONS

OF DAILY SOLAR RADIATION DATA

Manuel Vazquez

Area de Maquinas y Motores Termicos. Universidad de Vigo
E.T.S. Ingenieros Industriales. Apartado 62, 36200 Vigo, Spain

ABSTRACT

Regression analysis of daily diffuse and global solar radiation data of a five year period from Madrid has been done. Data was fitted with third order polynomials. The correlations for each month and for the periods Feb-May, Jun-Sep, and Oct-Jan are given together with that for the entire year. Bivariate (global-diffuse) frequency distributions of the data have also been studied. This study has shown they are multimodal and thus the usual unimodal or bimodal empirical functions can not adequately fit the data. It is suggested that the frequency distribution can be modelled as a set of statistical functions centered on the modes.

KEYWORDS

Solar radiation; global radiation; diffuse radiation; regression analysis; probability distributions

INTRODUCTION

Before optimally exploiting a renewable energy resource it is of importance to know the potential of that resource at the place of exploitation. In the case of Solar Energy, that potential is both site and time dependent. It would be necessary therefore, to have at each studying site a measurement station recording direct, diffuse and global solar radiation. Since both capital and operation costs of the stations are relatively expensive, they are scarce. Moreover, many of the stations record only daily values and global radiation. To overcome this lack of information, several studies have been devoted to analyze the relationship between the diffuse and the global components for stations where both components are measured. Then, after having seen a quasi-universality in the relationship, several authors have proposed to use the regression correlations found to determine the diffuse and the direct components in places where only global radiation is measured. Complementary with this type of study another interesting statistical procedure to analyze Solar Radiation data is studying the probability density distributions of the data. Since in this case, a quasi-universality in the distributions has also been seen, they are applied to stations with very little data. In this paper, both kinds of research are undertaken with data from Madrid: Regression Analysis for daily diffuse and global solar radiation data, and bivariate (diffuse-global) probability distributions.

DATABASE

The data analyzed in this article is the daily values of horizontal global and diffuse irradiations over a five year period, from Nov. 1980 to Oct. 1985, measured by the Instituto Nacional de Meteorologia (I.N.M) of Spain at the station of Madrid-Ciudad Universitaria. The data was supplied on a half-hourly basis and daily values were obtained by integration. Both components of solar radiation were measured using Kipp and Zonen CM5 pyranometers and the values of diffuse radiation include an isotropic shade ring correction.

REGRESSION ANALYSIS

Daily clearness indices K_t (horizontal global radiation/horizontal extraterrestrial radiation) and diffuse coefficients K_d (horizontal diffuse radiation/horizontal extraterrestrial radiation) were obtained from the data. Details of the calculations are given in Vazquez, Ruiz and Perez (1991). One graph of K_d versus K_t was made for each month of the year over the period in study. Next, the regression analysis for the data in each plot was undertaken. The models chosen to fit the data are, as in earlier works (Liu and Jordan (1960), Collares-Pereira and Rabl (1979), Erbs, Klein and Duffie (1982)), polynomials of the form:

$$K_d = a_o + a_1 * K_t + a_2 * K_t^2 + \ldots + a_n * K_t^n \qquad (1)$$

Polynomials up to the tenth-order were tested. The parameters in the polynomials were obtained by least squares estimates. The only condition imposed on each curve is that its left-hand side had to start at a point of the $K_d = K_t$ line and to have in it the same slope of that line (i.e., for $K_d = K_t$, $\delta K_d / \delta K_t = +1$) in accordance with Vazquez and Ruiz (1990) and Vazquez, Ruiz and Perez (1991). It was seen that the higher the order the better a polynomial fits the data. Nevertheless that trend weakens markedly as the order is increased. So, it was apparent that the increase in accuracy obtained with the use of polynomials of more than third order did not compensate the complexity of the model. Therefore, third-order polynomilas were those finally fitted.

Figure 1 (left) shows the diffuse-to-global correlation curves fitted for several months. Their respective equations appear in Table 1 together with the minimum and maximum values of K_t for which the equations are valid. For values of K_t smaller than those minimum values, the correlations coincide with the straight line $K_t = K_d$. Displacement of the curves in Fig. 1 confirms the seasonal effect (absorption and air mass effect), explained in Vazquez, Ruiz and Perez (1991). Observing the relative position of the curves, one can deduce that the data can be grouped in three seasonal periods: Fall-Winter including September, October, November, December and January corresponding with the curves on the left; Spring including March, April and May corresponding with the curves on the right; and Summer including June, July and August with their curves in the middle. This last group also includes February which is a transitional month. Nevertheless, in order to simplify the results, the data was grouped in the following three periods: February-May, June-September, and October-Jannuary. Their respective correlation curves are plotted in Fig. 1 (right). Their corresponding equations appear also in Table 1 together with that for the entire year.

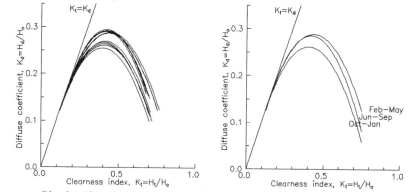

Fig.1. Plot of the daily diffuse-to-global correlations for Madrid. Left.- Correlations for each month; Right.- Correlations for the periods Feb-May,Jun-Sep,Oct-Jan.

TABLE 1. DAILY CORRELATIONS FOR MADRID

$$K_d = a_o + a_1*K_t + a_2*K_t^2 + a_3*K_t^3 \qquad (2)$$

Period	a_o	a_1	a_2	a_3	$K_t\geq$	$K_t\leq$
January	-0.02884	1.47076	-1.94749	0.21606	0.1233	0.7208
February	-0.05824	1.70434	-2.16796	0.23324	0.1669	0.7035
March	-0.03967	1.54776	-1.92089	0.20845	0.1460	0.7562
April	-0.03231	1.47097	-1.73040	0.10230	0.1377	0.7660
May	-0.03852	1.53432	-1.87729	0.17145	0.1452	0.7525
June	-0.05370	1.65994	-2.04913	0.13282	0.1636	0.6961
July	-0.06172	1.72225	-2.13633	0.13692	0.1718	0.6818
August	-0.05797	1.69818	-2.12506	0.13886	0.1670	0.6802
September	-0.03239	1.49338	-1.89498	0.12511	0.1319	0.7074
October	-0.04015	1.58053	-2.13706	0.27855	0.1396	0.6993
November	-0.03408	1.52184	-2.02751	0.23032	0.1316	0.7071
December	-0.02859	1.47574	-2.00613	0.22878	0.1210	0.7011
Feb-May	-0.04124	1.56059	-1.93542	0.20610	0.1483	0.7517
Jun-Sep.	-0.04813	1.62158	-2.02743	0.13427	0.1557	0.6944
Oct-Jan.	-0.03132	1.49741	-2.00368	0.22951	0.1268	0.7097
Year	-0.07012	1.84754	-2.79359	0.84992	0.1614	0.7650

BIVARIATE PROBABILITY DISTRIBUTIONS

Complementary with regression analyses, another statistical way of analyzing solar radiation data is looking at the probability density distributions of the data. Most works relative to this objective have studied frequency distributions of only one variable (either the global radiation or less commonly the direct radiation). References to bivariate frequency distributions are very scarce. In the present case we have looked at the frequency density distributions of pairs of values K_t, K_d. Due to the limited data, the plane K_t-K_d was divided according to a grid by lines plotted at intervals of 0.02 in the graph units. For each cell(K_t,K_t+0.02; K_d, K_d+0.02) the number of occurrences (data) was counted and the values K_t and K_d were averaged. Then a computer programme was used to take the irregularly spaced data (cell frequency, cell average K_t, cell average K_d) and convert it to regularly spaced data (the programme computes the frequency in each grid intersection). The interpolation method used is based on inverse distance.

Following the above procedure, graphs of frequency vs K_d vs K_t were plotted for each month of the year over the period of study. But, because of limited data in each plot, the three- dimensional surfaces appeared with discontinuities. Consequently, data was grouped according to the three yearly periods refered above: Feb-May, Jun-Sep, and Oct-Jan. A plot of f vs K_d vs K_t was made for each one of those periods. They appear in Fig. 2 together with the plot for the entire year. Fig. 3 is a projection on the K_d-K_t plane of this last plot.

It can be seen in the plots of Fig. 2 that there are two well-defined clusters (concentrations) of frequency. The one on the right corresponds to clear or almost clear days, and the one on the left corresponds to very cloudy or overcast days. But whereas both clusters are important in the Spring and Fall-Winter periods, the Summer period is characterized for having the frequency density nearly exclusively concentrated on clear day values. Besides the above mentioned clusters, in Fig. 2 there appear other intermediate clusters (associated to peaks of frequency) corresponding to partly cloudy days. So, any attempt to model the density frequency distribution of K_t and K_d as a unimodal or bimodal function, such as those that have been done up to now, were unsuccessful (see the comments by Vignola and Daniels (1991). Our proposal is that the density frequency distribution can be modelled by a set of statistical functions centered on the modes (peaks). Possible functions are the bivariate Normal statistical function, or even cones, the base of both being either circular or elliptic. I plan to present that work in another paper.

2757

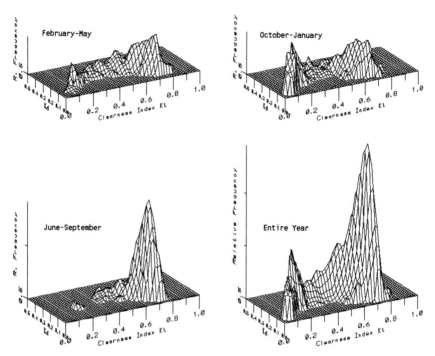

Fig. 2. Plot of relative frequency *vs* K_d *vs* K_t for Madrid

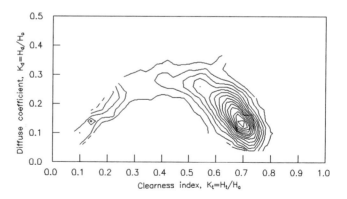

Fig. 3. Projection on a K_d-K_t plane of the surface of Fig. 2 (entire year), showing lines of constant frequency

CONCLUSIONS

Regression analysis of daily diffuse and global solar radiation data of a five year period from Madrid has been done. Polynomials up to tenth order were tested to fit the data. Third order polynomials were those finally chosen since they represent a balance between simplicity and accuracy. Comparisons of the regression curves made for each month of the year has shown that the data can be grouped in three periods: Fall-Winter (Sept-Oct-Nov-Dec-Jan), Spring (March-April-May), and Summer (June-July-Aug). February can be included in the last group. In the paper, the polynomial equations for each month and for the periods Feb-May, Jun-Sep, and Oct-Jan are given together with that for the entire year. Bivariate (global-diffuse) frequency distributions of the data have also been studied. This study has shown they are multimodal and thus the usual unimodal or bimodal empirical functions can not adequately fit the data. It is suggested that the frequency distribution can be modelled as a set of statistical functions centered on the modes.

AKNOWLEDGEMENTS

I would like to thank the Instituto Nacional de Meteorología of Spain, especially Mr. Almarza, for providing me with data in diskette.

REFERENCES

Collares-Pereira, M. and A. Rabl (1979). The average distribution of solar radiation—Correlations between diffuse and hemispherical daily and hourly insolation values. *Solar Energy* **22**, 155.
Erbs, D.G., S.A. Klein and J.A. Duffie (1982). Estimation of the diffuse radiation fraction for hourly, daily and monthly-average global radiation. *Solar Energy* **28**, 293
Vazquez, M. and V. Ruiz. (1989). Scattering and absorption of direct solar radiation stopped by clouds on days of high clearness index. Proceedings of the Congress of the International Solar Energy Society, Kobe, Japan.
Vazquez, M., V. Ruiz and R. Perez. (1991). The roles of scattering, absorption, and air mass on the diffuse-to-global correlations. *Solar Energy* **47**, 3, 181.
Vignola, F. and D.K. McDaniels (1991). Statistical properties of hourly beam radiation. Proceedings of the Congress of the International Solar Energy, Denver, USA.

A GENERAL NEW SIMPLE RELATIONSHIP BETWEEN SOLAR IRRADIANCE AND SUN HEIGHT

Olmo, F.J., Vida, J, Foyo-Moreno, I. and Jiménez,J.I.

Grupo de Física de la Atmósfera, Dpto de Física Aplicada, Facultad de Ciencias, Universidad de Granada, 18071, Granada, Spain.

ABSTRACT

A great amount of global and direct irradiance data have been obtained from different measuring campaigns carried out with the Swiss Mobile System place at ten different mid-latitudes sites in Europe, ranging from 0 to 3250 m.s.l, covering solar elevation angles from 0 to 77 degrees. The Linke Turbidity Factor as well as those of Angströn, Schüepp and the Clearness index, have been compiled for each data series. The possible relationships among global irradiance, solar altitude and the turbidity indexes have been analyzed afterwards.

The three-dimensional statistics investigation carried out for the three parameters show a strong dependency of solar irradiance on solar altitude for any given turbidity index. Though measurements were taken in quite different places and throughout different time periods too, it was evident the absolute absence of strange or unexpected values, being the only common aspect the slight modification with turbidity. Bearing in mind these results, we have made a regression analysis involving the whole data set, for every turbidity index. From here, we get that global irradiance in the absence of clouds, can be expressed as a simple solar altitude depending function. This is an extremely important result, mainly because of its general validity.

KEYWORDS: Global irradiance, Sun height, Turbidity indexes, Graphical analysis, regression analysis.

INTRODUCTION

Global and diffuse irradiance measurements are of great importance from the energetic point of view, to build general models as well as to use the solar energy in a specific locality. Pyranometric data from horizontal and inclined surfaces provided information with respect to the urban heat budget and to the solar plane collector design.

Many authors have tried to parameterize the global and diffuse irradiance and irradiation in terms of the angular parameters and also from turbidity. The main subjet of this research work was the generalization and extension of the data base. Nevertheless, experimental data bases of global and diffuse irradiance are few, specially for its spatial and angular distribution, probably due to the difficulties inherent to the calibration and maintenance of this radiometric stations.

The irradiation incident on a horizontal surface can be considered as the contribution of three components: The direct flux, the sky diffuse flux and the flux reflected by the ground and other underlying surfaces. All the components depend on the solar elevation, the latitude, the slope of the surface, the cloudiness and on the optical properties of the atmosphere and the underlying surface.

In this work we analyse the dependence of global irradiance on a horizontal surface with the solar elevation and the atmospheric conditions, determined by the turbidity indexes computed for clear skies from data provided by the Swiss Mobile System during several measuring campaigns carried out by Valko et al (1989) distributed all aver Europe.

DATA BASE

In the late spring and early summer of 1988 a measuring campaign was carried out with the Swiss Mobile System, in collaboration between the authors and the Swiss Meteorological Institute (Zurich). Later on, the data base was completed with those of other campaigns, covering a wide range of mid-latitudes in Europe. The data base also covers a wide range of altitudes, from sea level up to hight mountain zones, Pic of Veleta (Granada, Spain) which stands at 3000 m.s.l. The total number of clear sky measuring series were 1084, more than 173.000 single point global irradiance measurements, distributed as indicated in Table I.

TABLE I. Number of clear skies series classified by solar elevation.

Sun h.	-> 0	-> 5	->10	->15	->20	->25	->30	->35	->40
Nº Ser	22	27	35	53	68	84	91	93	97

Sun h.	->45	->50	->55	->60	->65	->70	->75	->80	Total
Nº Ser	87	77	82	71	71	43	55	28	1084

The estructural characteristics and operation of the Swiss Mobile System are well known from the publication by Valko (1982). We will make some comments on the global irradiance measurements, made with a CM5 pyranometer, which takes measurements every 20 s in each series. In addition the Mobile System is provided with a "fish eye" objective, to take pictures of the sky dome and, after a proper digital proccesing, determine the cloudiness. The selection of the clear skies series used in this work, has been realized after such proccesing of the slides.

In order to complete the data base, we have included the turbidity indexes obtained from the

2761

measurements of direct irradiance in the ten campaigns. Many authors have tried to determine the relationships between the scattering, atmospheric turbidity and solar radiation. Until now the best results come from the turbidity indexes that include the study of monochromatic irradiance as it gets through the different atmospheric layers, such as the Linke, Angström and Schüepp turbidity indexes (Page, 1986).

In a particular place the turbidity depends strongly on the moisture, the temperature, the aerosol distribution, etc., in summary, on all the factors that directly affect the atmospheric properties in that place. From the energetic point of view, these variations are of most importance, i.e., the variation on the spectral solar distribution can produce variations of 30 % in the outputs of silicon solar cells (Abdelrahman, 1988).

In this work, to evidence these effects from the clear skies series, the turbidity indexes mentioned above have been computed, using the results of the LOWTRAN 4B program for the atmospheric transmittance, for mid-latitudes atmospheres (Heimo, 1984), except for the Spanish campaigns, in which, due to the special design of the measurements only the clearnes index (D_h/G_h) has been computed.

We have classified the global irradiance on a horizontal surface in intervals of 5 degrees of solar elevation, for each turbidity index adequate interval. In this classification there exists many classes with null frequency, probably due to the fact that this solar elevation-turbidity index is impossible or no value corresponding to such combination exists. In order to eliminate the classes with frequency zero, we took the intervals of 10 degrees in solar height and other combination of turbidity indexes. Table II shows the mean values and the frequencies of global irradiance for each interval. As it can be seen in the tables, the null frequencies correspond now to absolutely improbable combinations.

TABLE II. Classification of global irradiance: a) Frequencies, b) Mean values.

Sun h. D_h/G_h		5.0	15.0	25.0	35.0	45.0	55.0	65.0	75.0
.125	a	0	504	960	1612	1414	1568	784	808
	b		268	448	614	783	886	969	1042
.275	a	132	460	1276	1372	728	1032	372	52
	b	128	205	354	546	724	788	869	935
.425	a	32	736	1092	654	248	280	288	32
	b	110	192	311	471	668	755	828	900
.575	a	236	506	296	0	0	0	0	0
	b	95	186	297					
.725	a	156	152	36	0	0	0	0	0
	b	197	128	300					
.875	a	456	32	0	0	0	0	0	0
	b	32	100						

RESULTS AND DISCUSSION.

The three-dimensional representation of global irradiance versus the two input variables, solar elevation and turbidity index, show for any of the turbidity indexes, the same behaviour: there is a strong dependency with the solar elevation, strongly increasing the irradiance for any of the turbidity indicators. On the other hand, global irradiance presents little sensibility to the variation of the turbidity index. It is remarkable that the values analised are those obtained experimentally by the Mobile System, without any special filtered treatment. This result can be explained as a consequence of the great number of data analised and, specially, for the adequate selection of the turbidity and sun height intervals. The zero value corresponds to those combinations for which either it is impossible to find data, or we do not have any experimental values. Figure 1 shows this behaviour for the clearness index, as a case study of any turbidity index.

To make evident the variation of global irradiance with sun height, it is shown in Figure 2 these variation parameterized with the clearness index too. From the curves, it can be seen that there exists a good correlation between global irradiance and sun height, for any value of the clearness index. In addition, the little sensibility of global radiation suggests as as first approximation that the regression analysis can be made once for all the values, ignoring the variation with turbidity. At this point, we find it necessary to remark that the regression analysis has been carried out with only eight points, but, as it can be seen in Table II, any of this points correspond to more than 32 points in the worse situation.

The results are similar for all the turbidity indexes. Having in mind the spatial and temporal distribution of the data, we can conclude that global irradiance under cloudless skies can be expressed as a linear function of sun height, (in degrees). The regression equation that we have obtained is:

$$G_h = 13.40\ \alpha + 27.60\ (W/m^2)$$

where α is the solar elevation in degrees, and the coefficients corresponds to the average values for all the data used. The correlation coefficient is .989 for a confidence level of 95%, and r.m.s.e is always less than 6%. This result is very important, both for its simpli- city, and specially for its great

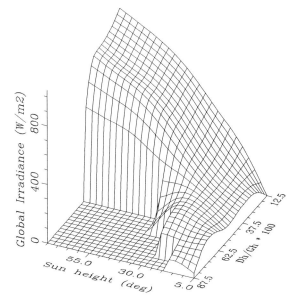

Fig. 1. *Three dimensional plot.*

generality, derived from the fact that the data base hasn't been filtered, except for the quality control of the measurements.

Some more relevant information can be inferred from the analysis of the data when studying the variation of global irradiance with one of the selected turbidity indexes, parameterized with solar elevation. From the three-dimensional surface we find that global irradiance strongly grows with sun elevation, and softly for each turbidity index. Both tendencies have been confirmed by the graphs in figure 2. There exist a good sensibility to the parameter, showing the same porcentual modification in irradiance for a given turbidity variation at any solar altitude. Fig. 2 shows this behaviour for the clearness index, but for other indexes the results are also very similar.

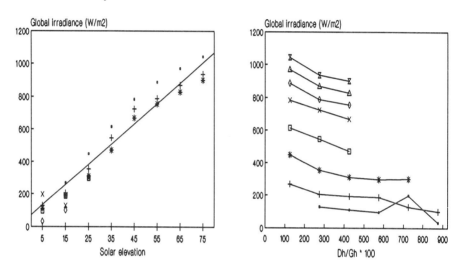

Fig. 2. *Mean Global Irradiance vs Sun height (left) and Clearness index (right).*

Finally, we conclude that these results are of great importance and have to be taken into consideration in any attempt of modelling the global and diffuse irradiance on inclined surfaces.

REFERENCES

Abdelrahman, M.A. and Said, S.A.M. (1988). Comparison between atmospheric turbidity coefficients of desert and tempeature climates. *Solar Energy*, 40, 3, 219-225.

Heimo, A. (1984). Diagnosis of the atmosphere through optical measurements performed with a Mobile Station. *Thesis*, ETH ZÜrich.

Page, J.K. (1986). Solar Energy R & D in the European Community. Serie F. Vol 3. Solar Radiation Data. *Reidel Pub. Co.*

Valko, P. (1982). Empirical study of the angular distribution of sky radiance and of ground reflected radiation fluxes. Report n° 3 within Project F of the CEC Solar Energy R & D Programme.

Valko, P., Merlo, G., Olmo, F.J. and Wild, M. (1989). Analysis of Sky Radiance and Slope Irradiance. Angular distribution measured at different sites in Europe. European Solar Microclimates. Zürich. Final Report. *CEC (in press).*

DEPENDENCE ON SUNSHINE AND WAVELENGTH OF THE RATIO OF MONTHLY MEAN HOURLY TO DAILY GLOBAL RADIATION

ALFONSO SOLER* AND ROSA RAMIREZ**

*Departamento de Física e Instalaciones.ETS Arquitectura.28040 Madrid.Spain.
**Instituto Nacional de Meteorología.Paseo de las Moreras s/n.28040 Madrid.

ABSTRACT

Plots of $\bar{r}=\bar{I}/\bar{H}$ vs. $\bar{S}_o$ for the hours centered at 0.5 h.,1.5h.,...from solar noon,are used to estimate values of $\bar{I}$ when values of $\bar{H}$ are available or can be estimated for a certain location,$\bar{I}$ and $\bar{H}$ respectively being the mon thly mean hourly and daily global radiation.The dependence on sunshine of these plots has been studied for different ranges of values of the daily sunshine fraction σ .The dependence on σ is found to be smaller than for $\bar{r}_\lambda$ vs. $\bar{S}_o$ plots
The dependence of the $\bar{r}_\lambda = \bar{I}_\lambda /\bar{H}_\lambda$ vs. $\bar{S}_o$ plots on wavelength λ is also studied,$\bar{I}_\lambda$ and $\bar{H}_\lambda$ respectively being the monthly mean hourly and daily global radiation for a certain wavelength.It is shown that $\bar{r}_\lambda$ vs. $\bar{S}_o$ plots are dependent on wavelength.

KEYWORDS

Hourly global radiation.Sunshine hours.Wavelength.

INTRODUCTION

From experimental or estimated values of the monthly mean daily global radi- ation,it has become customary to estimate values of the monthly mean hourly global radiation $\bar{I}$, using plots of $\bar{r}= \bar{I}/\bar{H}$ vs. $\bar{S}_o$,$\bar{S}_o$ being the montlhy mean daily maximum possible number of sunshine hours[1-5].The dependence of $\bar{r}$ vs. $\bar{S}_o$ plots on turbidity and cloudiness (clear/overcast skies) has been studied for a specific location [6] .The dependence on latitude has recently been investigated [7].Similar plots have been obtained for global illuminan- ces [8].
In the present work,using Uccle´s data we present some of the results obtai- ned relating :1)the dependence of $\bar{r}$ vs.$\bar{S}_o$ plots on the values of the daily sunshine fraction $\underline{S}/\underline{S}_o$, and 2) the dependence of the $\bar{r}_\lambda$ vs. $\bar{S}_o$ plots on wa- velength λ ,being $\bar{r} =\bar{I}_\lambda /\bar{H}_\lambda$.

DEPENDENCE ON THE DAILY SUNSHINE FRACTION OF THE RATIO OF MONTHLY
MEAN HOURLY TO DAILY GLOBAL RADIATION

The $\bar{r}$ vs. $\bar{S}_o$ plots are usually established using all the measured values of
global radiation,independently of the state of the sky.these plots may be
referred to,as plots for "average skies".in the present work we study the
dependence of $\bar{r}$ vs $\bar{S}_o$ plots on the values of the daily sunshine fraction
$\sigma = S/S_o$,where S is the number of daily sunshine hours,and S_o the astronomical daylength.The dependence on σ of the $\bar{r}$ vs. $\bar{S}_o$ plots is studied using
values of $\bar{I}$ obtained by Dogniaux [9] at Uccle for the period 1958-1975
for the following classes of days:

        -days with duration of sunshine S=0
        -days with sunshine fraction σ comprised between 0 and 0.1,
         0.1 and 0.2 ..,0.8 and 0.9.
        -days with sunshine fraction σ higher than 0.9
        -days with sunshine fraction σ comprised between 0 and 1

Values of $\bar{H}$ are obtained by adding the monthly average hourly values.
For the hours centered at 0.5h, 1.5h ,...6.5h,from the solar noon,and
each class of days,best fits (in a least squares sense)were obtained for
the $\bar{r}$ vs. $\bar{S}_o$ plots using second degree polynomials.Correlation coefficients
were mostly higher than 0.95.Using the equations for the best fits,values
of $\bar{r}$ are computed for different values of $\bar{S}_o$.
In Table 1 we give $\bar{r}$ values for 0.5h.from solar noon and different ranges
of values of σ ,from $0.5 < \sigma < 0.6$ to $0.9 < \sigma$,and also for $0 < \sigma < 1$.It is
seen that for each value of $\bar{S}_o$,$\bar{r}$ tends to diminish with increasing values of σ (for $0 < \sigma < 1$ values are close to those obtained for $0.8 < \sigma < 0.9$)
The decrease of $\bar{r}$ with increasing σ is about 5% for $\bar{S}_o$=9h,if we compare $\bar{r}$
values for $0.5 < \sigma < 0.6$ and $0.9 < \sigma$,and 1% for $\bar{S}_o$=15 h. and the same ranges
of values of σ.

Table 1 $\bar{r}$ values for 0.5h. from the solar noon,$9h < \bar{S}_o < 16h$
          and different ranges of values of σ .

	$\bar{S}_o$							
	9	10	11	12	13	14	15	16
$0.5 < \sigma < 0.6$	0.191	0.174	0.159	0.145	0.134	0.124	0.117	0.111
$0.6 < \sigma < 0.7$	0.192	0.175	0.160	0.147	0.135	0.125	0.118	0.112
$0.7 < \sigma < 0.8$	0.188	0.169	0.153	0.140	0.129	0.120	0.114	0.111
$0.8 < \sigma < 0.9$	0.186	0.168	0.153	0.140	0.129	0.120	0.114	0.110
$\sigma > 0.9$	0.181	0.165	0.151	0.139	0.129	0.120	0.115	0.110
$0 < \sigma < 1$	0.184	0.167	0.153	0.141	0.130	0.122	0.115	0.110

In Table 2 the corresponding $\bar{r}_d = \bar{I}_d / \bar{H}_d$ values are given,and have been computed as $\bar{r}$ values in Table 1,using data in [9].($\bar{I}_d$ and $\bar{H}_d$ respectively
being the monthly mean hourly and daily diffuse radiation).

The corresponding figure has been shown in [10] .For each value of $\bar{S}_o$, $\bar{r}_d$ values diminish with increasing values of σ (for $0<\sigma<1$ values of $\bar{r}_d$ are close to those obtained for $0.8<\sigma<0.9$).The decrease with σ is higher than 10% for all values of $\bar{S}_o$,if we compare $\bar{r}_d$ values for $0.5<\sigma<0.6$ and for $\sigma>0.9$.

Table 2 . $\bar{r}_d$ values for 0.5h. from the solar noon,$9h \leqslant \bar{S}_o \leqslant 16h$ and different ranges of values of σ .

	$\bar{S}_o$							
	9	10	11	12	13	14	15	16
$0.5<\sigma<0.6$	0.174	0.159	0.146	0.134	0.124	0.116	0.109	0.105
$0.6<\sigma<0.7$	0.170	0.154	0.141	0.129	0.120	0.112	0.106	0.102
$0.7<\sigma<0.8$	0.162	0.148	0.135	0.125	0.116	0.109	0.104	0.100
$0.8<\sigma<0.9$	0.159	0.146	0.134	0.124	0.115	0.108	0.103	0.099
$\sigma>0.9$	0.155	0.141	0.129	0.119	0.110	0.103	0.098	0.094
$0<\sigma<1$	0.174	0.158	0.145	0.133	0.123	0.115	0.109	0.104

Using fits for 1.5h.,...,6.5h., from solar noon,it is seen that as rule,the variation in $\bar{r}_d$ values with increasing values of σ is larger than for corresponding $\bar{r}$ values.

DEPENDENCE ON WAVELENGTH OF THE RATIO OF MONTHLY MEAN HOURLY TO DAILY GLOBAL RADIATION.

In this section,results are reported on the dependence of the $\bar{r}_\lambda = \bar{I}_\lambda/\bar{H}_\lambda$ vs. $\bar{S}$ plots on wavelength λ ,where $\bar{I}$ and $\bar{H}$ are respectively the monthly mean hourly and the monthly mean daily global radiation for wavelength λ. To our knowledge the dependence has never been studied before,perhaps due to the few reliable continuous measurements available.Data used are those obtained by Dogniaux for Uccle in 1980.The published data include one year of continuous measurements of total blobal radiation ,and of global radiation for 315 nm;400 nm ;446 nm;545 nm ;646 nm;730 nm ;816 nm and 914 nm ($\Delta\lambda$ =10 nm) [1]. For each wavelength,best fits for the $\bar{r}_\lambda$ vs. $\bar{S}_o$ plots were obtained with second degree polynomials for each of the hours around the solar noon. Correlation coefficients were usually higher than 0.95.The fits are given in Fig .1 for the hours centered at 3.5h., from solar noon.It is seen in Fig. 1 that for λ = 545 nm,646nm,730nm and 816 nm the $\bar{r}$ vs. $\bar{S}_o$ curves are practically coincident.The curve for λ = 400 nm lies close to the curve for global radiation.The curves for 400 nm,446 nm and 914 nm lie far from each other.The curve $\bar{r}_\lambda$ vs. $\bar{S}_o$ for 315 nm not only does not lie close to any other,but also presents a dependence of $\bar{r}_\lambda$ on $\bar{S}_o$ different than that obtained for the other wavelengths. As similar results are obtained for 0.5h,1.5h,etc from solar noon,we must conclude that the $\bar{r}_\lambda$ vs. $\bar{S}_o$ plots for each of the hours are not coincident:for each of the hours the same plot can not be used for all wave - lengths.

As for $\bar{H}$,if $\bar{H}_\lambda$ is known,$\bar{I}_\lambda$ could be computed using the $\bar{r}_\lambda =\bar{I}_\lambda /\bar{H}_\lambda$ vs. $\bar{S}_o$ plots.If the relation between $\bar{H}$ and $\bar{H}_\lambda$ was known or estimated for a certain location,$\bar{I}_\lambda$ could be computed from $\bar{H}$ values.Unfortunately this does not appear to be an easy way to predict $\bar{I}_\lambda$ taking into account the results in Fig.1 ,and those for other hours from solar noon.
(Note:the relation between instantaneous values of total global radiation and instantaneous values of global radiation for different wavelengths has been given in [12]).

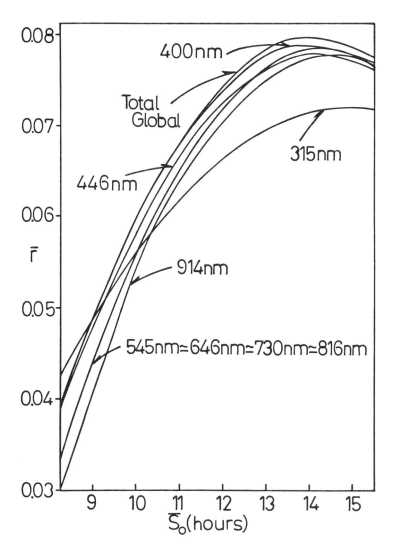

Fig1. $\bar{r}_\lambda$ vs. $\bar{S}_o$ plots for 3.5 h. from solar noon

2768

CONCLUSIONS

The following conclusions were reached using Uccle's data.
1) The dependence of $\bar{r}$ vs. $\bar{S}_o$ plots on σ is smaller than that of $\bar{r}_d$ vs.$\bar{S}_o$ plots.
2) The $\bar{r}_\lambda$ vs $\bar{S}_o$ plots are dependent on wavelength for each hour around solar noon.

REFERENCES

1.Whillier,The determination of hourly values of total solar radiation from daily summations.Arch.Meteor.Geophys.Bioklimtol.Ser.B.7 ,197-204 (1956)
2.B.Y.Liu and R.C.Jordan.The interrelationship and characteristic distribution of direct,diffuse and total solar radiation.Solar Energy 4 ,1-19 (1960).
3.M.Iqbal,A study of Canadian diffuse and total solar irradiation.II Monthly average hourly horizontal radiation.Solar Energy 22,87-90 (1979)
4.A.Soler,Estimation of the monthly average hourly global ,diffuse and direct radiation.Solar & Wind Tech 4 ,192-194 (1978).
5.A.Soler,Solar radiation correlations for Madrid.Solar & Wind Tech. 5 293-297 (1988)
6.A.Soler,Dependence on turbidity and cloudiness of the distribution of the monthly average hourly diffuse,global and direct radiation.Solar & Wind Tech. 4 ,81-93,(1987)
7.A.Soler and K.K.Gopinathan.Estimation of monthly mean hourly global radiation for latitudes in the 1°N-81°N range.Solar Energy.Communicated.
8.A.Soler,Global and diffuse illuminances:estimation of monthly average hourly values.Lighting Research and technology 22 ,193-196 (1990)
9.R.Dogniaux and M.Lemoine,Variations horaries et distribution de frequences des composants du rayonment solaire et des temperatures de l'air et du sol nu a Uccle.Miscellanea Serie B,43.Institut Royal Meteorologique de Belgique.
10A.Soler, The dependence of the distribution of the monthly average hourly diffuse radiation on the values of the daily sunshine fraction. Solar & Wind Technology 7,545-547 (1990).

11.R.Dogniaux et coll.Distribution spectrale du rayonment solaire a Uccle. Institut Royal Meteorologique de Belgique.(1981)
12.D.Crommelynck and A.Jaukoff,A simple algorithm for the estimation of the spectral radiation distribution on a horizontal surface based on global radiation measurements.Solar Energy 45 ,131-137 (1990)

ESTIMATION OF MONTHLY MEAN DAILY VALUES OF DIFFUSE RADIATION FOR SPAIN

K.K.GOPINATHAN[*]AND ALFONSO SOLER

Departamento de Física e Instalaciones.ETS Arquitectura.28040 Madrid.Spain
* Permanent adress:The National University of Lesotho.Roma.Lesotho.

ABSTRACT

Several years of measured data of global and diffuse radiation,together
with sunshine duration for 5 locations in Spain,are used to establish empirical relationships to connect monthly mean daily diffuse irradiation
with clearness index and percent possible sunshine.a correlation connecting sky radiation with both,clearness index and percent possible sunshine
together is found to be the most accurate.the selected correlation is used
to estimate the monthly mean daily diffuse radiation for the 5 locations
used to establish it,for 2 other locations in Spain with measured diffuse
radiation data,and for 2 Portuguese locations for which data are also avai-
lable.The deviations between measured and estimated values rarely exceed
10%,and in general are very low.the selected correlation is used to estimate monthly mean daily values of diffuse radiation for 51 widely spread
locations in Spain using measured or estimated values of global radiation
and sunshine duration,and some features of diffuse radiation over Spain
are described.

KEYWORDS

Monthly mean daily diffuse radiation.Estimation for Spain.

INTRODUCTION

For stations where no measured data on diffuse radiation are available,a
common practice is to estimate them from other measured parameters like
global radiation and sunshine duration.a host of empirical correlations
are available in the literature [1-6] for estimating monthly mean daily
diffuse radiation from various parameters.
Although there are 7 locations in Spain for which long term hourly data
on diffuse radiation are available,a systematic analysis of these data
has never been carried out.No attempt has yet been made to obtain any
correlation from the combined data that could be employed all over the
country to evaluate the monthly mean daily diffuse radiation for those
locations for which measured data are not available.

Different investigators have suggested different correlations to be employed for European locations.However none of these studies included stations
from Spain,and as climate conditions of Spain,with lots of sunshine,are
different from other parts of Europe,the observations valid for other
European regions may not hold true for Spain.
A separate study of diffuse radiation for Spanish locations was thus found
necessary.the main objectives of the present study are as follows:
1) To obtain values,suiting Spanish conditions,of the regression
   parameters,for the different types of correlations available.
2) To find among the different types of correlations the most
   accurate one for Spain.
3) To check the validity of the selected correlation using it to
   estimate monthly mean daily values of diffuse radiation for lo-
   locations for which experimental values are available,whether
   or not these locations have been used to develop the correlation.
4) If results from 3) are satisfactory,to obtain estimated values
   of monthly mean daily diffuse radiation for up to 51 locations in
   Spain for which values of sunshine duration and/or values of global radiation are available.

DIFFUSE RADIATION CORRELATIONS FOR SPAIN

The following symbols are adopted:
$\bar{H}$:Monthly mean daily global radiation on a horizontal surface
$\bar{H}_d$:Monthly mean daily diffuse radiation on a horizontal surface
$\bar{H}'$:Monthly mean daily global radiation on a horizontal surface before
   striking the ground.
H_d':Monthly mean daily diffuse radiation on a horizontal surface before
   striking the ground.
K_T:Clearness index $(\bar{K}_T=\bar{H}/\bar{H}_o)$
$\bar{S}$:Monthly average daily number of bright sunshine hours.
$\bar{S}_o$:Monthly average daily number of possible sunshine hours.

Measured data (4 to 7 years) of diffuse and global radiation obtained
with Kipp & Zonen model CM-5 pyranometers,together with daily sunshine
data for 5 locations in Spain are used in linear,multilinear and poly-
nomical analysis to obtain the constants in different correlations available in the literature[1-6].The values of the regression coefficients
obtained,the correlation coefficients r ,and the standard error estimates S.E. are given below.

$$\bar{H}_d/\bar{H} = 0.9760 - 1.0828\ \bar{K}_T \quad ;r=0.913\quad S.E.=0.042 \qquad (1)$$

$$\bar{H}_d/\bar{H} = -0.3320 + 6.3605\ \bar{K}_T - 13.8363\ (\bar{K}_T)^2 + 8.4174\ (\bar{K}_T)^3 \qquad (2)$$
$$r=0.916;\ S.E= 0.042$$

$$\bar{H}_d/\bar{H} = 0.8030 - 0.7137\ (\bar{S}/\bar{S}_o) \quad ;r=0.917\ ;S.E=0.041 \qquad (3)$$

$$\bar{H}_d/\bar{H}_o = 0.1884 + 0.2049\ (\bar{S}/\bar{S}_o) - 0.2967\ (\bar{S}/\bar{S}_o)^2 \qquad (4)$$
$$r=0.637;\ S.E.=0.060$$

$$\bar{H}_d'/\bar{H}' = 16.0252 -110.32\ (\bar{H}'/\bar{H}_o) + 229.40\ (\bar{H}'/\bar{H}_o)^2 +24.439\ (\bar{H}'/\bar{H}_o)^3$$
$$-556.33\ (\bar{H}'/\bar{H}_o)^4 + 445.51\ (\bar{H}'/\bar{H}_o)^5. \qquad (5)$$
$$r=0.914\ ;S.E.=0.043$$

$$\bar{H}_d/\bar{H} = 0.9017 - 0.3912\ (\bar{S}/\bar{S}_o) - 0.5291\ \bar{K}_T \quad ;r=0.933;S.E=0.037 \quad (6)$$

2771

The results above ,at first glance,indicate that eqn.(6) should be the
most accurate.the accuracy of the estimated diffuse radiation for eqns.
(1) to (6) was tested by calculating the mean biass error (MBE) and the
root mean square error (RMSE).This was done for the 5 stations used to
develop eqns.(1) to (6) and for 2 other stations in Spain for which measured data are also available.the lowest MBE and RMSE values for maximum
number of stations are obtained for eqn.(6).
An attempt was also made to find the ranking of the correlations relating
the values of MBE and RMSE obtained for each station,the first ranking
being the one which gives the lowest MBE or RMSE values and so on.For each
MBE or RMSE values a weighting of 6 was assigned to the eqn with the best
performance,5 to the second best and so on,as in 10 .Eqn. (6) clearly
ranks the top and can be considered as the best.
The diffuse radiation estimated with eqn.(6) was compared with measured
data for the 7 Spanish locations and 2 Portuguese locations.Results for
Madrid and Málaga (this last location was not used to establish eqn.6)
are given in Figs.1 and 2.For all locations the deviation between measured
and estimated values rarely exceeded 10%.

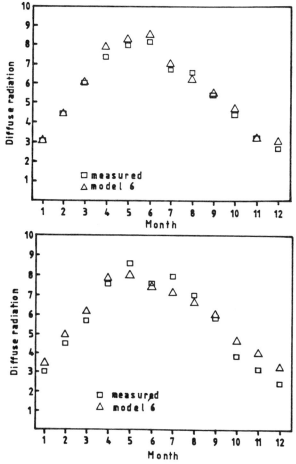

Figs. 1 and 2. $\bar{H}_d$ for Madrid and Málaga

ESTIMATING MONTHLY MEAN DAILY DIFFUSE RADIATION OVER SPAIN

There are 51 stations in Spain for which measured values of the daily sunshine fraction are available.for 15 of these stations,measured data on
global radiation are also available.for the remaining 36 stations,global
radiation values were estimated from an empirical Angstrom type model,
developed for stations with reliable data of global radiation.finally,
values of $\bar{H}_d$ were estimated with eqn (6) using measured or estimated
values of $\bar{H}$.
Some features of the estimated diffuse radiation over Spain are as follows:
    1)The diffuse radiation values show clear north-south and east-west
      gradients during summer months,with northern and esatern regions
      receiving more diffuse radiation than the south and west respec-
      tively.
    2)A mild north-south gradient can be observed in winter months,
      with the southern stations receiving more diffuse radiation than
      the north.
    3)The yearly mean daily diffuse radiation shows mild north-south
      and east-west gradients.

ACKNOWLEDGMENT

The authors are greatful to the Dirección General de Investigación Científica y Técnica (DGICYT) for financial support for K.K.Gopinathan durig
his sabbatical.

REFERENCES

1.B.Y.H.Liu and R.C.Jordan ,The interrelationship and characteristic distribution of direct,diffuse and total solar radiation.Solar Energy 4
  1-19 (1960).

2.J.K.Page,The estimation of monthly mean values of daily short wave
  radiation on vertical and inclined surfaces from sunshine records
  for latitudes 40°N-40°S.Proc.U.N. conference on new sources of ener-
  gy.Paper S.98,Vol.4,378-390.(1961).

3.A.Soler,Dependence on latitude of the relation between the diffuse
  fraction of solar radiation and the ratio of global to extraterrestrial
  radiation for monthly average daily values.Solar Energy 44,297-302.
  (1990)

4.J.E.Hay,A revised method for determining the direct and diffuse compo -
  nents of total short wave radiation.Atmosphere,14,278-287,(1976).

5.M.Iqbal,Correlation of average diffuse and beam radiation with hours
  of bright sunshine.Solar Energy,23,169-173 (1979).

6.K.K.Gopinathan,Computing the monthly mean daily diffuse radiation
  from clearness index and percent possible sunshine.Solar Energy
  41,379-385,(1989).

QUADRATIC VARIATION OF GLOBAL SOLAR RADIATION WITH BRIGHT SUNSHINE HOURS

Bulent G. Akinoglu

Middle East Technical University
Dept. of Physics
Ankara, Turkey

ABSTRACT

A physical formalism is developed for the relation between monthly-average daily global solar radiation and bright sunshine hours. Formalism yields quadratic relation if the effect of first reflection cycle between ground and the sky is included.

KEYWORDS

Solar radiation, bright sunshine hours, solar radiation models.

INTRODUCTION

Ångstrom's (1924) well-known empirical linear formulation between fractional global solar radiation and bright sunshine hours is the starting point for the sunshine-based models. This formulation is widely used to estimate the monthly-average global solar radiation. Higher order correlations between monthly-average global solar radiation and bright sunshine hours also exist in the literature (Bahel et al., 1987; Ogelman et al., 1984) and recently the global applicability of these correlations have been verified (Akinoglu and Ecevit, 1990).

Davies and McKay (1984) stated that the models which use bright sunshine hours have the same physical basis as the models using cloud data. A recent attempt by Jain (1990) showed that a physical formulation can be established between monthly-average bright sunshine hours and monthly-average diffuse and global solar radiation. His formalism yields a linear relation.

In this report, a description of a physical formalism between the global solar radiation and bright sunshine hours is given.

The effect of reflection between ground and the sky is included. The formalism leads to a quadratic relation, the coefficient of which can be described by monthly-varying atmospheric and climatic parameters. The parameters and the coefficients of the quadratic correlation are calculated using the monthly-average data of a location in Turkey.

FORMALISM

The direct component, G_D , of solar radiation is only present when there is a bright sunshine period within a time interval. If n_i is the fractional period of bright sunshine within a time interval, the direct radiation reaching the earth can be expressed as

$$G_D = G_0 n_i \tau \tag{1}$$

at that interval, where G_0 is the extraterrestrial solar irradiance and τ is the transmission coefficient of the atmosphere during clear-sky condition.

Diffuse component during a bright sunshine period can be expressed as

$$G_{d1} = G_0 n_i (1-\tau) \beta' \tag{2}$$

where β' is atmospheric forward scattering coefficient.

Diffuse component during a cloudy sky period is

$$G_{d2} = G_0 (1-n_i) \tau \tau' \tag{3}$$

where τ' is the transmission coefficient of the clouds. Adding eqns. (1), (2) and (3), and including the first reflection cycle between sky and the ground, global solar radiation reaching the earth can be obtained in terms of the fractional period of bright sunshine within a time interval, viz.

$$G = G_0 [n_i \tau + n_i (1-\tau) \beta' + (1-n_i) \tau \tau'] (1 + \alpha \beta) \tag{4}$$

where α and β are the ground reflectance and atmospheric back-scattering coefficient, respectively. Note that the total diffuse component can be derived from eqn.(4) as

$$G_d = G_{d1} + G_{d2} + G_{d3} \tag{5}$$

$$= G_0 n_i (1-\tau) \beta' + G_0 (1-n_i) \tau \tau' + G_0 [n_i \tau + n_i \tau (1-\tau) \beta' + (1-n_i) \tau \tau'] \alpha \beta$$

where G_{d3} is the diffuse part coming from the first reflection cycle between ground and the atmosphere.

The ground albedo α varies from 0.1 for forest and grass up to 0.7 for fresh snow (Houghton, 1954). However, for semi-urban and cultivation sites 0.2 is a reasonable value as measured by Ineichen et al. (1990).

2775

Cloud albedo depends on the type, height and amount of the clouds but an average value can be assigned in the long-term radiation estimations. The average value of this parameter was determined to be 0.5 by several authors (Fritz, 1948; Houghton, 1954; Monteith, 1962).

Using the results obtained by Houghton (1954) a value around 0.25-0.40 can be derived for the average forward scattering coefficient of the clear atmosphere.

Equations (4), (5) and (1) are for a small time interval in which the irradiations and the parameters have constant values. To obtain the monthly-average daily values, these equations should be integrated from sunrise to sunset and averaged. Assuming that the form of the equations does not change after this integration and averaging procedures, from eqns. (1) and (5), monthly-average daily direct and diffuse solar radiation can be written as

$$H_D = H_0 \, (n/N) \, \tau_e \tag{6}$$

and

$$H_d = H_{d1} + H_{d2} + H_{d3} \tag{7}$$

$$H_0 \, [n/N(1-\tau_e) \, \beta'_e + (1-n/N) \, \tau_e \tau'_e + [\, (n/N) \, \tau_e + n/N(1-\tau_e) \, \beta'_e + (1-n/N) \, \tau_e \tau'_e] \, \alpha_e \beta$$

where n/N is monthly-average daily fractional bright sunshine hours and the subscript e stands to indicate that the parameters are monthly-effective parameters. H_0 is monthly-average daily extraterrestrial solar radiation on horizontal surface.

Without the aerosol effect, Davies and McKay (1982) expressed the atmospheric back-scattering β as

$$\beta = \alpha_R (1 - CA) + \alpha' \, CA \tag{8}$$

where α_R is molecular scattering, α' is cloud base albedo and CA is the fractional cloud amount. Assuming that $CA = 1 - n_i$ (Davies and McKay, 1982) and $\alpha_R = 0.0685$ (Lacis and Hansen, 1974), and replacing n_i by n/N, and α' and β by their monthly-effective counterparts, α'_e and β_e,

$$\beta_e = 0.0685 n/N + \alpha'_e (1 - n/N) \tag{9}$$

can be written.

From eqns. (6), (7) and (9) a quadratic variation between the monthly-average daily global solar radiation and monthly-average daily bright sunshine hours can be obtained which has the form

2776

$$H/H_0 = a_0 + a_1 n/N + a_2 (n/N)^2 . \qquad (10)$$

The coefficients are

$$a_0 = \tau_e \tau_e' (1 + \alpha_e \alpha_e') , \qquad (11)$$

$$a_1 = \tau_e (1 + \alpha_e \alpha_e') (1 - \beta_e') + \beta_e' (1 + \alpha_e \alpha_e') \tau_e \tau_e' (0.0685 \alpha_e - 2\alpha_e \alpha_e' - 1) \qquad (12)$$

and

$$a_2 = \tau_e \alpha_e (0.0685 - \alpha_e') (1 - \beta_e') + \beta_e' \alpha_e (0.0685 - \alpha_e')$$

$$+ \tau_e \tau_e' \alpha_e (\alpha_e' - 0.0685) . \qquad (13)$$

Knowing the value of H_D the monthly-effective parameter τ_e can be calculated. From τ_e and H_d the parameter τ_e' can be obtained for each month, by assigning appropriate values to the parameters α_e, α_e' and β_e'. Hence, the monthly parameters τ_e and τ_e' can be used to determine the monthly coefficients of the quadratic variation eqn. (10), with the help of eqns. (11), (12) and (13). Table 2 gives the results of the calculations for the data of Ankara (Table 1).

Table 1. Data for Ankara (latitude=39.95N)
(H in MJ/m^2-day , n in hours)

	Jan	Feb	Mar	Apr	May	Jun	Jul	Aug	Sep	Oct	Nov	Dec
H	4.4	8.3	12.5	15.6	19.1	23.4	21.5	21.6	18.2	11.4	5.8	4.5
n	1.5	3.1	5.4	6.2	8.0	10.2	9.4	10.5	9.3	6.5	2.8	2.4

Table 2. Parameters and the coefficients

	τ_e	τ_e'	a_0	a_1	a_2
Jan	0.72	0.22	0.177	0.709	-0.057
Feb	0.63	0.34	0.234	0.584	-0.047
Mar	0.56	0.31	0.189	0.581	-0.047
Apr	0.57	0.27	0.173	0.608	-0.049
May	0.53	0.29	0.172	0.578	-0.046
Jun	0.50	0.50	0.270	0.448	-0.037
Jul	0.57	0.20	0.125	0.657	-0.052
Aug	0.57	0.11	0.070	0.715	-0.057
Sep	0.59	0.15	0.100	0.701	-0.056
Oct	0.55	0.25	0.148	0.617	-0.049
Nov	0.42	0.48	0.222	0.445	-0.036
Dec	0.49	0.34	0.185	0.534	-0.043

CONCLUSION

If one takes into account the first reflection cycle between the ground and the atmosphere, the variation of the fractional monthly-average daily global solar radiation with respect to fractional bright sunshine hours is quadratic rather than linear. The coefficients of this variation can be derived in terms of some climatic and atmospheric parameters. Using the measured monthly-average daily global, diffuse and bright sunshine hours, it is possible to determine these coefficients without empirical approaches. These coefficients can be used to estimate the solar radiation at the locations with the same climatological and geographical conditions. Equation (5) can be used to establish a quadratic formulation between the diffuse solar radiation and bright sunshine hours. A linear formalism can be obtained using the same procedure and neglecting the multiple reflections between ground and the sky (Akinoglu, 1991).

REFERENCES

Akinoglu, B.G. and A.Ecevit (1990). Construction of a quadratic model using modified Ångstrom coefficients to estimate global solar radiation, *Solar Energy*, 45, 85-92.
Akinoglu, B.G. (1991). A physical formalism for the modified Ångstrom equation to estimate solar radiation. *Turkish J. Phys*. (in press).
Ångstrom, A. (1924). Solar and terrestrial radiation. *Q. J. Roy. Met. Soc.*, 50, 121-126.
Bahel, V., H.Bakhsh and R.Srinivasan (1987). A correlation for estimation of global solar radiation, *Energy*, 12, 131-135.
Davies, J.A. and D.C.McKay (1982). Estimating solar irradiance and components. *Solar Energy*, 29, 55-60.
Davies, J.A., M.Abdel-Wahab and D.C. McKay (1984). Estimating Solar Irradiation on Horizontal Surfaces. *Int. J. Solar Energy*, 2, 405-424.
Fritz, S. (1949). The albedo of the planet earth and clouds, *J. Meteor.*, 6, 277-282.
Houghton, H.G. (1954). The annual heat balance of the northern hemisphere, *J. Meteor.*, 11, 1-9.
Ineichen, P., O.Guisan and R.Perez (1990). Ground-reflected radiation and albedo, *Solar Energy*, 44, 207-214.
Jain, P.C. (1990). A model for diffuse and global irradiation on horizontal surfaces, *Solar Energy*, 45, 301-308.
Lacis A.A. and J.E.Hansen (1974). A parametrization for the absorption of solar radiation in the earth's atmosphere. *J. Atmos. Sci.*, 31, 118-133.
Ogelman, H., A.Ecevit and E.Tasdemiroglu (1984). A new method for estimating solar radiation from bright sunshine data, *Solar Energy*, 33, 619-625.
Monteith, J.L. (1961). Attenuation of solar radiation: a climatological study, *Q. J. Roy. Met. Soc.*, 87, 171-179.

SIMULATION OF THE DAILY GLOBAL HORIZONTAL SOLAR IRRADIATION BY A FIRST ORDER TWO STATE MARKOVIAN PROCESS

A. KHTIRA and J. BURET-BAHRAOUI

Faculté des Sciences-Laboratoire d'Energie Solaire
B.P. 1014-Rabat -Morocco

ABSTRACT

Five years of daily solar irradiation data for Rabat (Morocco) are analyzed. By the comparison of the clearness index $K=H/H_0$ and of a sequence of threshold values K_S, we have found that the solar radiation in Rabat can be well represented by a first-order two-state Markovian process for $K_S =0.65$. Then we have used the model to generate synthetic values. The comparison between the generated series and the real series shows that the statistical characteristics of global irradiation are faithfully reproduced.

KEY WORDS

Solar irradiation - Markov model - Probabilities - Densities - Simulation

INTRODUCTION

A series of daily measurements of global solar radiation on a horizontal surface recorded during the period 1983-1987 in Rabat have been analyzed on a statistical basis . This information is essential to give the input for analytical models to evaluate the long-term performance of solar energy conversion systems and to generate synthetic daily radiation sequences which can be used for numerical simulations of any solar system.

THE MARKOV MODEL APPLICATION

Firstly, the data of the global irradiation H have been normalized and made independent from the seasonal variation; then we calculated the clearness index $K=H/H_0$ (H_0 is the extra-terrestial irradiation on a horizontal surface). The clearness index has been described by a first-order two-states Markov process. The K values are compared to a series of reference numbers K lying between 0 and 1. The threshold K_s which separates the two states is choosen to give the better Markov process. With this method, the clearness index is separated into two particular states. The first one contains all the elements E_0 for which $K < K_s$, where E_0 corresponds to a "cloudy sky". The second one contains the elements E_1 corresponding to a "clear sky" ;for $K > K_s$.

Transition matrix

The transition between the two states E_0 and E_1 is characterized by 4 transition probabilities:

$E_0 \longrightarrow E_0$ (Probability P_{00})

$E_0 \longrightarrow E_1$ (Probability $P_{01} = 1-P_{00}$)

$E_1 \longrightarrow E_1$ (Probability P_{11})

$E_1 \longrightarrow E_0$ (Probability $P_{10} = 1-P_{10}$)

So, we can write the transition matrix as:

$$P = \begin{bmatrix} P_{00} & P_{01} \\ P_{10} & P_{11} \end{bmatrix}$$

The transition probabilities P_{ij} are calculated for each month (Fig.1).

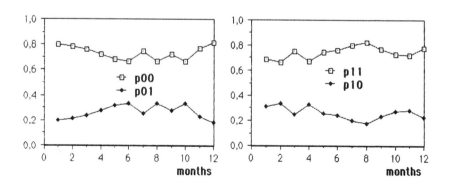

Fig.1. Monthly transition probabilities P_{ij} calculated from clearness index data

SIMULATION OF THE SYNTHETIC VALUES

Transformed variable

The generation of synthetic values is easy when the variable can take all the values between 0 and 1. For this reason, we should transform K into a new variable X which satisfies this condition. X is given by a linear equation:

$$X = \frac{K - 0.1}{0.55} = X_0 \qquad (1)$$

where $0.1 < K < 0.65$ (state E_0)

and

$$X = \frac{K - 0.65}{0.15} = X_1 \qquad (2)$$

where $0.65 < K < 0.8$ (state E_1)

The probability densities of X are fitted by a three degree polynomial function for E_0 and E_1 states, for any season. We give one example for the three months (December, January, February) (Fig. 2).

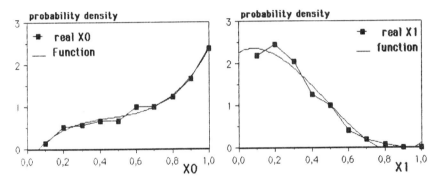

Fig. 2 The seasonal probability densities for X_0 and X_1 adjusted by a polynomial function (for months Dec, Jan, Feb)

Simulation of the global irradiation synthetic sequences

We can simulate the values of the clearness index for a day $J+1$ by the following equations:
if the day J belongs to the state E_0:

$$K = P_{00} K_0 + P_{01} K_1 \qquad (3)$$

if the day J belongs to the state E_1:

$$K = P_{10} K_0 + P_{11} K_1 \qquad (4)$$

The two clearness index values K_0 and K_1 have been calculated by the equations (1) and (2); X_0 and X_1 are two values taken from two synthetic sequences generated by the computer.These values have the same probability density as the real data.

Then,we calculate the daily global horizontal solar irradiation by the following equation:

$$H = H_0 \, K \qquad\qquad (5)$$

THE MODEL VALIDATION

In order to test the validity of the model, we have compared the statistical characteristics of simulated data with the experimental ones. The parameters used for the test are:

1-the monthly mean value $\overline{K}$ and the standard deviation σ for the clearness index (table1).

2-the monthly marginal probabilities P_0 and P_1(Fig.3).

3-the monthly probability densities for the real values and the simulated ones (Fig.4). The comparison using χ^2 (Khi 2) test with a 0.05 signification, demonstrates that there is a good concordance for 67 % of the months.

Table 1. The monthly mean clearness index values and the standard deviation for the real and the simulated values.

month	1	2	3	4	5	6	7	8	9	10	11	12
$\overline{K}_{real}$	.542	.545	.603	.606	.632	.636	.644	.663	.627	.633	.540	.550
$\overline{K}_{model}$	.520	.523	.592	.596	.609	.621	.620	.635	.604	.613	.561	.524
$\sigma \times 10^3\,_{real}$	13.0	13.8	12.4	10.6	9.6	9.6	6.0	6.3	8.9	9.0	14.2	13.6
$\sigma \times 10^3\,_{model}$	10.2	11.3	9.9	9.7	8.3	7.4	6.0	5.9	8.0	7.9	10.1	10.3

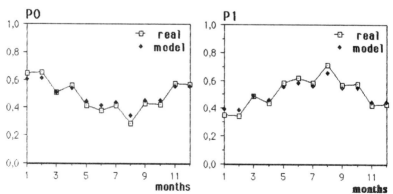

Fig. 3. Monthly comparison between the real marginal probabilities P_m (m=1 or 2) and the simulated ones.

2782

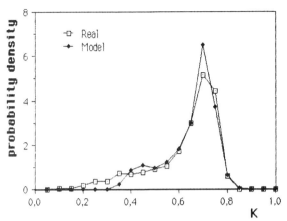

Fig.4. Annual comparison between the real probability density and the simulated one.

CONCLUSION

In conclusion,this model allowed us to generate a series of daily global solar irradiation data on a horizontal surface , which have the same statistical characteristics as the basic data. Those series may be utilized as input data for any solar energy system .

REFERENCES

R.J.Aguiar , M.Collares-Pereira , and J.P.Conde (1988). Simple procedure for generating sequences of daily radiation values using a library of markov transition matrices. Solar Energy, 40 , N°3, 269-279.

A.Maafi,Adnane et A.Ouabdesselam (1987). Ajustement des données d'insolation d'Alger par un modèle Markovien du premier ordre . Revue de Physique Appliquée, vol22 , 425-430.

J.P.Pelletier (1982). Techniques numériques appliquées au calcul scientifique.. Editions Masson , Paris.

U.Amato,A.Andretta,B.Bartoli,B.Coluzzi,V.Cuomo (1986). Markov processus and Fourier analysis as a tool to describe and simulate daily solar irradiance. Solar Energy , 37 , N°3, 1799-194.

R.Lestienne (1979). Application d'un modèle markovien simplifié à l'étude du comportement du stockage d'une centrale solaire. Revue de Physique Appliquée , 14 , 139-148.

COMPARISON OF CORRELATIONS FOR ESTIMATION
OF GLOBAL SOLAR RADIATION

S.K. Srivastava and O.P. Singh
Applied Sciences Department
(Energy Lab)
Institute of Engineering & Technology
Sitapur Road, Lucknow-226 020
(INDIA)

ABSTRACT

Empirical correlations for estimating the monthly average daily global solar radiation on horizontal surface have been compared and the validity of different correlations have also been tested. All the models provide satisfactory results with Bahel et al. model (Bahel et al., 1986) giving more accurate results. Gopinathan model (Gopinathan, 1988) shows a maximum deviation of 34 percent while in the case of Rietveld model (Rietveld, 1978) maximum deviation of 16 percent has been observed. Bahel et al. model (Bahel et al., 1987) and Louche et al. model (Louche et al., 1991) also yield satisfactory results with maximum deviation of 12 percent.

KEY WORDS

Solar energy conversion; global radiation; empirical correlations; sunshine; extraterrestrial; regression.

INTRODUCTION

Utilization of solar energy requires detailed informations concerning its availability. To achieve an optimal designed solar energy conversion system, we require knowledge about the solar radiation obtainable at particular locations. With long term performance as a goal, the design of a solar system requires a knowledge of monthly average data of solar radiation for the locality under consideration. The best radiation information is obtained from experimental measurements of solar radiation. For locations where these data are not easily available, empirical correlations developed by various researchers are available to estimate the solar radiation values from meteorological parameters.

2784

Among the various empirical correlations used so far, for estimation of solar radiation, the most widely used one is due to Angstrom (Angstrom, 1924) modified by Page (Page, 1964). The only meteorological parameter needed in Angstrom correlation is the sunshine duration. There are few correlations (Liu & Jordan, 1960; Whillier, 1965) in which declination angle and latitude have been used while Sayigh (Sayigh, 1977) used sunshine duration, relative humidity and temperature to calculate the availability of daily total radiation.

In the present work calculated values of global solar radiation using five different empirical correlations have been compared with the experimental values of global solar radition and validity of different correlations have also been tested.

METHODOLOGY

Measurement of global solar radiation on horizontal surface has been made at Lucknow (latitude-26.75°N, longitude-80.50°E, altitude-120m above sea level) using precision pyranometer (No.0317) manufactured by National Instruments Ltd., Calcutta (India), with calibration factor 5.50 mV/cal/cm^2/min. A potentiometric chart recorder with a range of 0-10 mV and speed of 20 mm per hour has been used to record the output of the pyranometer. Measured data of global solar radiation on horizontal surface has been taken from October 1989-March 1990 and April 1991-September 1991.

MATHEMATICAL MODELS

The various correlations used for comparison are as follows:

1. $(H/H_o) = 0.18 + 0.62 \, (n/N)$ (Rietveld, 1978)

2. $(H/H_o) = 0.175 + 0.552 \, (n/N)$ (Bahel et al.,1986)

3. $(H/H_o) = 0.16 + 0.87 \, (n/N) - 0.61 \, (n/N)^2$
   $+ 0.34 \, (n/N)^3$ (Bahel et al.,1987)

4. $(H/H_o) = a + b \, (n/N)$ (Gopinathan,1988)
   $a = -0.161 + 0.568 \, h - 0.184 \, h^2$
   $b = 0.963 - 0.639 \, h + 0.217 \, h^2$

5. $(H/H_o) = 0.206 + 0.546 \, (n/N)$ (Louche et al.,1991)

where

H   = monthly average daily global solar radiation on horizontal surface

H_o = monthly average daily extraterrestrial solar radiation on horizontal surface

n   = monthly average daily hours of sunshine[*]

N   = monthly average daily hours of bright sunshine[*]

a&b = regression constants

2785

h   =   altitude of the site in kilometer

(here h=0.128 km)

*Values of n and N have been taken from handbook of Solar
Radiation Over India (Mani & Rangarajan, 1982).

RESULTS AND DISCUSSION

A comparison of calculated values of global solar radiation,
obtained by using five different correlations, with the
measured values has been made and the results are shown in
Fig.1.   The correlation given by Bahel et al.(Bahel et al.,
1986) generally yields the best results with a maximum
deviation of 11.27 and 9.28 percent during April and June

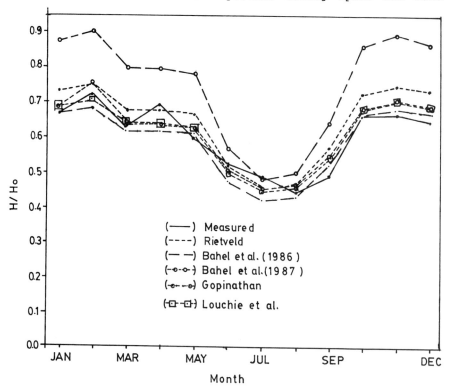

FIG. 1. VARIATION OF(H/Ho) WITH THE MONTH OF THE YEAR

respectively.   The second correlation of Bahel et al.(Bahel
et al., 1987) which is a non-linear fit for the third order,
also provides good results with maximum deviation of 12
percent.   However the predictability of first correlation
given by Bahel et al. (Bahel et al., 1986) is superior to
second one (Bahel et al., 1987).   Louche et al. model provides
better results than the model of Rietveld and Gopinathan.

Louche et al. model shows a maximum deviation of 12 percent while in general the deviation of 2-7 percent has only been observed. Reitveld correlation also gives satisfactory results with maximum deviation of 16 percent and seems better than the correlation of Gopinathan in which the regression constants are correlated to the altitude of the site. Gopinathan model provides deviation upto 34 percent and thus shows that this model is not much suitable for the places which are situated at lower altitudes.

CONCLUSION

A comparison of empirical correlations for estimation of the monthly average daily global solar radiation on horizontal surface suggests that Bahel et al. model (Bahel et al., 1986) yields better results. The model given by Gopinathan for evaluation of monthly average daily global solar radiation may not be considered suitable at lower altitudes.

REFERENCES

Angstrom, A.K. (1924). Solar and atmospheric radiation. Q.J.R. Met.Soc., 50, 121-126.

Bahel, V., R. Srinivasan and H. Bakhsh (1986). Solar radiation for Dhahran, Saudi Arabia. Energy, 11, 985-989.

Bahel, V., H. Bakhsh and R. Srinivasan (1987). A correlation for estimation of solar radiation. Energy, 12, 131-135.

Gopinathan, K.K. (1988). The distribution of global and sky radiation throughout Lesotho. Solar and Wind Technology, 5, 103-106.

Liu, B.Y.H. and R.C. Jordan (1960). The interrelationship and characteristic distribution of direct, diffuse and total radiation.Solar Energy, 4, 1-19.

Louche, A., G. Notton, P. Poggi and G. Simonnot (1991). Correlation for direct normal and global horizontal irradiation on a French Mediterranean site. Solar Energy, 46, 261-266.

Mani, A., and S. Rangarajan (1982). Solar Radiation Over India (Allied Publishers Private Ltd., Ed.). pp. 352.

Page, J.K. (1964). The estimation of monthly mean values of daily total short-wave radiation on vertical and inclined surfaces from sunshine records for latitudes 40°N-40°S.Proc.UN Conference on New Sources of Energy, 4, 378-389.

Rietveld, M.R. (1978). A new method for estimating the regression coefficients in the formula relating solar radiation to sunshine. Agric-Met., 19, 243-252.

Sayigh, A.A.M. (1977). Estimation of total radiation intensity a new formula, 4th Course on solar energy conversion ICTP, Trieste.

Whillier, A. (1965). Solar radiation graphs. Solar Energy, 9, 164-169.

ANALYSIS AND FREQUENCY DISTRIBUTION OF HOURLY AND DAILY BEAM AND GLOBAL RADIATION VALUES FOR THE NORTHERN NEGEV REGION OF ISRAEL

A. I. KUDISH[1] and A. IANETZ[2]

[1]Solar Energy Laboratory, Department of Chemical Engineering, Ben-Gurion University of the Negev, Beer Sheva 84105, ISRAEL, [2] Israel Meteorological Service, Research and Development Division, P.O.B. 25, Bet-Dagan 50205, ISRAEL

ABSTRACT

Average hourly and daily values and frequency distributions for both global and beam radiation are reported for the semi-arid, northern Negev region of. Israel. The results of the analysis are reported in the form of clearness index, beam transmittance and beam fraction.

KEYWORDS

Global radiation; beam radiation; frequency distribution; clearness index; beam transmittance; beam fraction; monthly average hourly and daily values.

INTRODUCTION

Information concerning the average hourly and daily values for both global and beam radiation are critical for the optimum design of solar energy conversion systems. In addition, knowledge of the statistical properties of hourly and daily beam radiation is of utmost importance with regard to concentrating solar collectors systems, since they utilize only the beam radiation. These systems have very short time constants and require information on the frequency distribution of beam radiation intensities, time intervals and correlation parameters.

The meteorological station of the Solar Energy Laboratory at the Ben-Gurion University of the Negev, in Beer Sheva (31°15'N, 34°45'E, 315 m MSL), measures concurrently both normal incidence beam and global radiation on a horizontal surface (as well as other meteorological parameters). Beer Sheva is located in the northern Negev region of Israel, a semi-arid zone, and is blessed by relatively high average daily irradiation rates (Kudish *et al.,* 1983, 1992).

The normal incidence beam radiation is measured using an Eppley Normal Incidence Pyrheliometer, Model NIP, and the global radiation is measured using an Eppley Precision Spectral Pyranometer, Model PSP. This station is part of the national network of meteorological stations and the instruments' calibration constants are checked at regular intervals by the Israel Meteorological Service.

Recently, we have reported upon the correlations between values of daily beam, diffuse and global radiation (i.e. the beam and diffuse fractions of global radiation as a function of the clearness index) (Ianetz and Kudish, 1992) and also the frequency distribution of the average monthly daily clearness index (Kudish and Ianetz, 1992) for Beer Sheva, Israel.

In the present work we analyze both the hourly and daily frequency distributions of the horizontal beam radiation (both as beam transmittance, $k_b = I_b/I_0$, and fraction, I_b/I) and the clearness index, $k_t = I/I_0$. The data base utilized in this analysis consists of a minimum of four and a maximum of seven months for each month of the year and includes only those days for which a complete set of measurements exist for both the normal incidence and global radiation. The validity of of the individual hourly radiation values were checked in accordance with WMO recommendation (WMO, 1983). Those values which did not comply with the WMO recommendations were considered erroneous and rejected. The analysis is based upon solar time.

RESULTS

The monthly average hourly values for k_t and k_b (the corresponding values for I_b/I are thereby defined, since $I_b/I = k_b/k_t$,) are summarized in Tables 1-2, respectively. Only those hours for which the average zenith angle is less than 84° (viz. the solar altitude is greater than 6°) are included in this analysis. The corresponding monthly average daily values are presented in Fig. 1.

Table 1. Monthly average hourly k_t values.

Month	6-7	7-8	8-9	9-10	10-11	11-12	12-13	13-14	14-15	15-16	16-17	17-18
J		0.408	0.473	0.530	0.553	0.550	0.526	0.510	0.463	0.397	0.294	
F		0.432	0.520	0.567	0.595	0.588	0.579	0.542	0.526	0.477	0.365	
M		0.490	0.563	0.598	0.622	0.627	0.616	0.593	0.550	0.490	0.388	
A	0.461	0.560	0.609	0.658	0.684	0.690	0.669	0.644	0.615	0.570	0.492	0.363
M	0.494	0.598	0.655	0.686	0.714	0.720	0.702	0.691	0.659	0.615	0.538	0.407
J	0.535	0.642	0.720	0.752	0.773	0.776	0.766	0.744	0.720	0.679	0.611	0.481
J	0.470	0.581	0.682	0.732	0.757	0.762	0.756	0.740	0.716	0.678	0.613	0.490
A	0.475	0.588	0.678	0.729	0.752	0.756	0.740	0.712	0.681	0.634	0.543	0.373
S	0.474	0.579	0.651	0.706	0.732	0.744	0.722	0.697	0.654	0.591	0.478	0.276
O		0.638	0.680	0.692	0.681	0.681	0.662	0.624	0.565	0.457	0.309	
N		0.409	0.530	0.578	0.590	0.575	0.546	0.511	0.484	0.390	0.348	
D			0.530	0.578	0.590	0.575	0.546	0.511	0.448	0.350		

The hourly frequency distributions for k_t, k_b and I_b/I are reported in Tables 3-5 (the position of the monthly maximum values are indicated by underlining). Due to space limitations, only the hourly frequency distributions for Jan. and June (the months

2790

possessing the minimum and maximum average hourly k_t and k_b values, respectively, cf. Table 6) are shown in Fig. 2. We will describe verbally the salient features of the hourly frequency distributions in the next section.

Table 2. Monthly average hourly k_b values.

Month	6-7	7-8	8-9	9-10	10-11	11-12	12-13	13-14	14-15	15-16	16-17	17-18
J		0.171	0.279	0.342	0.382	0.366	0.337	0.319	0.292	0.239	0.121	
F		0.224	0.335	0.394	0.394	0.370	0.390	0.365	0.339	0.300	0.182	
M		0.231	0.320	0.366	0.400	0.419	0.411	0.387	0.362	0.310	0.230	
A	0.169	0.304	0.372	0.437	0.471	0.480	0.456	0.425	0.400	0.350	0.280	0.136
M	0.211	0.346	0.423	0.475	0.512	0.518	0.500	0.482	0.449	0.415	0.344	0.212
J	0.252	0.408	0.527	0.575	0.610	0.624	0.611	0.585	0.566	0.529	0.466	0.343
J	0.179	0.317	0.463	0.554	0.591	0.608	0.596	0.576	0.556	0.524	0.461	0.339
A	0.144	0.290	0.433	0.533	0.584	0.598	0.580	0.543	0.524	0.492	0.411	0.243
S	0.113	0.270	0.391	0.493	0.548	0.575	0.561	0.540	0.497	0.442	0.332	0.114
O		0.272	0.389	0.443	0.460	0.472	0.471	0.444	0.393	0.295	0.149	
N		0.157	0.339	0.409	0.450	0.451	0.434	0.392	0.323	0.234	0.113	
D			0.300	0.380	0.398	0.391	0.375	0.355	0.299	0.207		

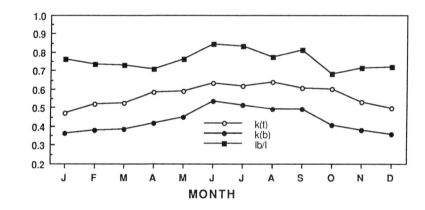

Fig. 1. Monthly average daily values of k_t, k_b, and I_b/I.

Table 3.Hourly frequency distribution of k_t (%).

k_t	Jan	Feb	Mar	Apr	May	June	July	Aug	Sept	Oct	Nov	Dec
0.05	3.78	1.45	2.55	0.75	0.29	0.00	0.07	0.00	0.00	0.30	1.60	1.43
0.15	8.46	6.61	4.50	2.91	1.17	0.07	0.07	0.14	2.43	3.44	4.52	6.79
0.25	11.01	8.47	6.27	5.06	2.39	0.35	0.64	2.51	3.72	4.64	6.96	8.58
0.35	12.47	10.17	7.66	8.77	5.43	1.62	1.70	5.79	6.76	5.91	10.59	9.89
0.45	14.03	13.63	12.72	9.28	10.46	7.40	10.33	8.79	8.81	11.64	15.68	13.71
0.55	16.49	16.73	14.83	16.46	15.05	10.16	11.18	13.05	13.74	17.43	17.90	19.31
0.65	21.82	22.77	23.71	24.14	25.46	22.99	24.63	25.26	27.41	24.25	28.53	27.53
0.75	11.81	20.03	26.87	31.32	36.66	54.37	51.10	44.45	36.75	25.15	11.08	9.42
0.85	0.14	0.15	0.89	1.31	3.08	3.03	0.28	0.00	0.38	7.24	3.15	3.34

Table 4. Hourly frequency distribution of k_b (%).

k_b	Jan	Feb	Mar	Apr	May	June	July	Aug	Sept	Oct	Nov	Dec
0.05	26.94	21.67	19.34	16.34	10.55	3.26	3.49	6.02	9.87	0.00	0.00	0.00
0.15	15.71	10.86	11.51	11.76	7.55	3.42	4.45	5.66	9.06	8.27	15.20	20.69
0.25	11.88	12.01	10.18	11.00	12.10	9.42	6.97	9.43	8.26	11.51	13.27	9.09
0.35	10.94	12.82	13.36	13.24	14.06	12.30	11.35	10.74	12.69	11.96	11.92	12.10
0.45	10.24	14.60	15.97	14.87	15.25	11.31	15.88	17.05	19.03	16.87	13.21	12.31
0.55	13.88	14.31	14.95	17.92	19.23	25.89	30.42	30.77	28.00	21.01	18.12	16.18
0.65	10.06	13.28	14.19	13.83	20.58	32.88	27.15	20.32	13.09	22.17	22.79	20.69
0.75	0.35	0.46	0.51	1.03	0.67	1.52	0.30	0.00	0.00	8.21	5.49	8.95

Table 5. Hourly frequency distribution of I_b/I (%).

I_b/I	Jan	Feb	Mar	Apr	May	June	July	Aug	Sept	Oct	Nov	Dec
0.05	17.02	13.53	13.22	9.14	5.90	1.24	1.71	3.43	3.36	2.54	8.02	13.42
0.15	7.61	6.25	5.42	6.20	4.18	1.56	1.71	2.61	4.08	4.32	6.45	7.24
0.25	8.17	6.67	7.73	6.81	3.86	1.56	1.86	2.38	4.38	4.93	6.27	7.49
0.35	9.98	7.65	6.81	8.14	6.38	1.87	3.12	3.50	4.99	7.27	7.19	6.67
0.45	8.67	9.47	8.33	8.03	8.05	3.65	4.31	4.69	6.22	10.69	8.85	8.23
0.55	10.54	9.41	11.70	12.08	12.02	6.45	6.54	8.12	10.91	13.98	10.23	10.86
0.65	9.60	13.77	13.22	14.18	14.06	12.75	13.38	13.04	18.96	19.40	15.85	12.02
0.75	10.60	15.53	16.79	19.78	20.01	31.73	32.42	30.92	26.30	21.73	22.12	14.81
0.85	17.83	17.72	16.79	15.62	25.54	39.19	34.94	31.30	20.80	15.15	15.02	19.26

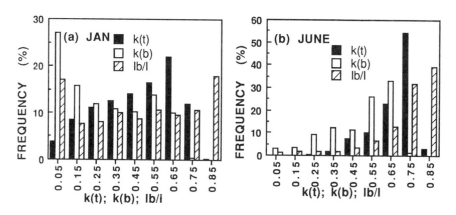

Fig. 2. Hourly frequency distribution of k_t, k_b, and I_b/I.

DISCUSSION

As mentioned above, we have limited our analysis to those hours for which the average zenith angle is less than 84° and have thereby truncated the 'solar day' as indicated by the hourly values reported in Tables 1-2. The monthly average hourly k_t and k_b values for these truncated days are reported in Table 6. The hourly data have been further divided on the basis of average zenith angle (or air mass), which essentially separates them into two groups, viz. midday and morning/afternoon hours, as reported in Table 6. We observe that the average hourly zenith angle is never less than 40° during the months Jan., Feb., Oct., Nov. and Dec., for this site.

The frequency distributions of k_t for all months possess a single maximum value and are skewed towards the higher k_t values. This maximum occurs at $k_t = 0.65$ for Jan., Feb., Nov. and Dec. and at $k_t = 0.75$ for the remaining months and the highest peak, ~ 54%, occurs in June. The situation is quite different with regard to k_b. Firstly, six months (Jan., Feb., Mar., Apr., Nov. and Dec.) exhibit a distinct bimodal distribution (May and Sept. appear to be bimodal also). The bimodal distributions are characterized by one maximum in the low k_b range (0.05-0.15) and a second in the high range (0.45-0.65). The four remaining months are characterized by a single maxima located in the range of 0.55-0.65. Secondly, the k_b distributions are not as sharply peaked as the case of k_t.

Table 6. Average hourly k_t, k_b, and I_b/I values.

	Jan	Feb	Mar	Apr	May	June	July	Aug	Sept	Oct	Nov	Dec
k(t)-a	0.468	0.519	0.554	0.585	0.623	0.683	0.665	0.638	0.609	0.599	0.524	0.516
k(t)-b			0.615	0.672	0.695	0.755	0.744	0.728	0.724			
k(t)-c	0.530	0.576	0.550	0.613	0.602	0.663	0.639	0.611	0.651	0.651	0.580	0.556
k(b)-a	0.289	0.329	0.344	0.357	0.407	0.508	0.480	0.448	0.406	0.379	0.330	0.338
k(b)-b			0.404	0.458	0.489	0.595	0.580	0.560	0.556			
k(b)-c	0.361	0.380	0.340	0.390	0.382	0.483	0.441	0.407	0.456	0.447	0.410	0.380
lb/l-a	0.485	0.529	0.525	0.538	0.606	0.724	0.699	0.678	0.630	0.602	0.536	0.550
lb/l-b			0.574	0.631	0.674	0.783	0.776	0.766	0.762			
lb/l-c	0.541	0.562	0.523	0.575	0.593	0.715	0.675	0.654	0.688	0.662	0.638	0.591

a - includes only those hours for which the zenith angle < 84°
b - includes only those hours for which the zenith angle < 40° (air mass < 1.3)
c - includes only those hours for which the 40° < zenith angle < 61.6° (1.3 < air mass < 2.1)

CONCLUSIONS

Hourly and daily beam transmittance and clearness index data for the northern Negev region of Israel have been analyzed. In the next phase of our study, we will attempt to develop frequency distribution models for these parameters.

REFERENCES

Ianetz, A. and A. I. Kudish (1992). Correlations between values of daily beam, diffuse and global radiation for Beer Sheva, Israel. *Energy* 17, in press.
Kudish, A. I., D. Wolf and Y. Machlav (1983). Solar radiation data for Beer Sheva, Israel. *Solar Energy* 30, 33-37.
Kudish, A. I. and A. Ianetz (1992). Analysis of the solar radiation data for Beer Sheva, Israel, and its environs. *Solar Energy* 48, 97-106.
World Meteorological Organization (1983). World Climate Program Report, WCP-48.

SEASONAL PARTITIONING OF DATA FOR IMPROVED ESTIMATION
OF GLOBAL RADIATION FROM THE ANGSTROM FIT

M. HUSSAIN

Physics Department and Renewable Energy Research Centre,
Faculty of Science, University of Dhaka,
Dhaka-1000, Bangladesh.

ABSTRACT

A division of the year has been made by grouping together
months around summer and winter and seasonal fits to the Angstrom relation for thirteen stations in Europe, Africa and Asia
lying between $48°N$ and $20°S$ latitudes have been obtained. It
is found that the partitioning of data leads to better estimates of monthly averaged daily global radiation G than from the
annual fits.

Data of two and three of the stations have also been used to
obtain satisfactory collective bi-annual fits for a region.

KEYWORDS

Global radiation; sunshine duration.

INTRODUCTION

For locations for which data on global radiation are not available one often estimates them from the Angstrom relation

$$\frac{G}{G_o} = a+b(\frac{n}{N}) \quad \ldots \ldots \ldots \ldots \ldots (1)$$

where G is the monthly averaged daily global radiation, G_o is
the corresponding extraterrestrial radiation, n is the monthly
averaged daily duration of bright sunshine hours and N is the
corresponding day length while a and b are regression constants. Usually data of all months of the year for a station are
used to determine a and b from a least-square fit. It is found
that both a and b vary from station to station. This introduces uncertainties in estimating G for a location where only
sunshine data have been recorded.

2794

Attempts have been made by different workers (Sayigh,1977; Garg and Garg,1982) to obtain station independent correlations by introducing in equation1 variables other than sunshine duration on which G/G_0 may depend. Latitude, humidity, precipitable amount of water, temperature, cloud cover and surface albedo of the earth have been used as extra parameters. This results in station independent formulae but the rms errors of estimates are at times large (Garg and Garg,1982) and the computations are harder while for some correlations data on extra variables are non-existent for many locations.

In the present work a different approach is made. The direct as well as the diffuse components of the global radiation must vary with changes in the turbidity of the atmosphere, the precipitable amount of water, the ozone content of the atmosphere, and the albedo of the earth's surface and of the sky. These quantities vary from month to month but the aggregate effect of such variations tends to cancel out for a location leading to a fairly satisfactory fit to equation1. But the variables generally differ in their summer and winter values and the cancellation may not be exact. Hence the year is divided into two with months around summer and around winter grouped together and an investigation is made in this paper to find if a suitable biannual partitioning of data may be made in order to obtain better fits to the Angstrom relation than the annual fits for locations in Europe, Africa and Asia.

For stations not close to a pyranometer installation one may use data of two or more locations in the same climatic region to obtain a collective region dependent fit for the Angstrom relation. It is considered worthwhile to investigate the seasonal effect on collective fits too in this paper.

To look for seasonal dependence of the Angstrom fit one requires good quality data over many years. Only recently such data are becoming available. Castro-Diez et al.(1989) observe that even now only at a few places records for G and n are available for the same duration of a good number of years. In this work the long-term data presented by them for Trappes, Carpentras, Madrid, Bulawayo and Salisbury have been utilized. Several years' data of Ibadan (Ideriah and Suleman,1989) and Enugo (Eze and Ododo,1988) have been available in the literature. Indian monthly averaged data (Mani,1980,1981) are not for the same years but G records are for ten years or longer and n records are for much longer periods for the six stations studied — Ahmadabad, Bhavnagar, New Delhi, Calcutta, Bombay and Nagpur. As both G and n are of very long durations it is considered that they represent climatological values.

The thirteen stations considered in this work (Table 1) belong to Europe, Asia and Africa covering a latitude range between 48^ON and 20OS.

ANALYSIS OF DATA

For each of the stations least square analyses were made for

2795

equation 1 using monthly averaged daily data of G and n for all
months of the year and also for months around summer and winter.
It is possible to divide the year into two using various groupings of months for summer and winter. Four combinations have
been examined after preliminary trials with different groupings
March to August (months 3-8), March to September (months 3-9),
April to August (months 4-8) and April to September (months 4-
9) have been grouped as summer months and the rest as the winter months for the northern hemisphere. The rest of the year
consists of months around the winter. For the southern hemisphere the reverse is the case.

Table 1. RMS errors for estimates of G/G_0 over the year for
each month using annual and seasonal fits to Angstro-
m's relation

Station	Average G/G_0	RMS error		
		Fits, months 1-12	Fits, months 4-9 & 10-3	Fits, months 3-9 & 10-2
Trappes	0.406	0.011	0.007	0.009
Carpentras	0.543	0.023	0.022	0.011
Madrid	0.574	0.022	0.015	0.008
New Delhi	0.626	0.037	0.028	0.026
Ahmadabad	0.631	0.023	0.014	0.014
Calcutta	0.542	0.012	0.012	0.008
Bhavnagar	0.628	0.018	0.015	0.015
Nagpur	0.591	0.013	0.011	0.010
Bombay	0.579	0.018	0.012	0.008
Ibadan	0.472	0.016	0.014	0.013
Enugo	0.472	0.014	0.008	0.010
Salisbury	0.630	0.009	0.008	0.007
Bulawayo	0.645	0.008	0.007	0.007

The rms error over the year for each station was evaluated for
annual and the biannual fits. Table 1 shows the rms errors for
annual and two of the biannual fits together with the annual
average value of G/G_0. The rms errors for the stations have
only marginal differences for the four types of groupings.How-
ever for most stations it appears that better fits are obtained if the summer like period is considered to start from March.

Collective fits have been obtained for eight groupings of the
stations using data over the year and also for months 10-2 and
3-9 (Table 2). The rms errors for the fits for each station
over the year are shown in Table 2 for the different groupings.

Table 2. RMS errors for estimated G/G_0 over the year from the Angstrom fits of data for groups of stations

Group	Station	Fits, months 1-12 rms error	bias error	Fits, months 10-2 & 3-9 rms error	bias error
1.	Trappes	0.015	-0.007	0.010	-0.000
	Carpentras	0.027	0.007	0.013	0.000
2.	Carpentras	0.025	0.019	0.020	0.017
	Madrid	0.029	-0.018	0.020	-0.017
3.	Ahmadabad	0.026	0.001	0.016	-0.001
	Bhavnagar	0.018	0.001	0.016	0.000
4.	Ahmadabad	0.027	0.006	0.017	0.006
	Bhavnagar	0.021	0.009	0.020	0.007
	New Delhi	0.040	-0.017	0.034	-0.013
5.	Bombay	0.016	-0.003	0.010	-0.004
	Calcutta	0.018	0.006	0.012	0.003
6.	Bombay	0.019	-0.004	0.010	-0.004
	Calcutta	0.019	0.007	0.012	0.004
	Nagpur	0.014	0.001	0.013	-0.001
7.	Enugo	0.014	0.006	0.013	0.005
	Ibadan	0.015	-0.006	0.016	-0.005
8.	Bulawayo	0.009	-0.001	0.008	-0.004
	Salisbury	0.009	0.002	0.008	0.004

RESULTS AND DISCUSSION

For each station one clearly finds from Table 2 that biannual fits for both months 3-9, 10-2 and 4-9, 10-3 have lower rms errors than the annual fits. The same conclusion holds for the other two types of partitions considered.

A close look at the rms errors and correlation coefficients for the four type of partitions (which could not be accomodated in the paper)show that March to August or March to September should be considered as the summer group of months. For Ahmadabad, Bhavnagar and Ibadan fits for months 3-8 and 9-2 show a big improvement with r values greater than 0.92. However rms errors are hardly affected. It appears that either months 3-8, 9-2 or 3-9, 10-2 may be used for dividing the year into two.

Table 2 shows rms errors and bias errors for collective biannual fits for months 3-9 and 10-2 using data of two or more stations. For the different groups of stations, the rms errors for biannual fits are evidently smaller than for fits over the year. It may be seen that for many cases annual fits for the individual stations have higher rms errors than the collective biannual fits. The bias error should also be small for good fits. For the pair of stations Carpentras and Madrid the collective fits have large bias errors and are unsatisfactory. It appears that it is not advisable to use a group fit for Carpentras and Madrid, stations in France and Spain having different physiographical features. Other three types of partitions of months lead to similar results.

The group for Ahmadabad, Bhavnagar and New Delhi shows a bit
large bias error for New Delhi for the partition of the year
into months 3-9, 10-2. The situation improves for an alternative partition into months 3-8, 9-2 with a bias error of-0.007
in place of -0.013 but rms errors remain hardly affected
(Please see Appendix).

REFERENCES

Sayigh, A.A.M. (1977). Solar energy availability prediction
    from climatological data. Solar Energy Engineering (Edited
    by A.A.M. Sayigh), pp. 61-82. Academic Press, New York.
Garg, H.P. and Garg, S.N. (1982). Prediction of global solar
    radiation from bright sunshine hours and other meteorologi-
    cal parameters. Solar-India 1982, Proc. of National Solar
    Energy Convention, pp. 1.004-1.007. Allied Publishers, New
    Delhi.
Castro-Diez, Y., Alados-Arboledas, L. and J cinenez, J.L.(1989)
    A model for climatological estimations of global, diffuse
    and direct solar radiation on a horizontal surface. Solar
    Energy, 42, 417-424.
Ideriah, F.J.K. and Suleman, S.O. (1989). Sky conditions at
    Ibadan during 1975-1980. Solar Energy, 43, 325-330.
Eze,A.E. and Ododo, J.C. (1988). Solar radiation prediction
    from sunshine duration in eastern Nigeria. Energy Convers.
    Mgmt., 28, 69-73.
Mani, A.(1980). 'Handbook of Solar Radiation Data for India'.
    Allied Publishers Limited, New Delhi.
Mani, A. and Rangarajan, S. (1983). 'Solar Radiation Over
    India'. Allied Publishers, New Delhi.

APPENDIX

Table 3. Sample fits to individual stations and their group fit

Station	Months 9-2 a b r			Months 3-8 a b r			rmse	bias
Ahmadabad	0.349	0.386	0.97	0.247	0.569	0.99	0.014	
Bhavnagar	0.367	0.373	0.98	0.280	0.497	0.99	0.014	
New Delhi	0.369	0.379	0.99	0.200	0.715	0.92	0.025	
Ahmadabad							0.018	0.007
Bhavnagar	0.370	0.369	0.98	0.275	0.534	0.96	0.020	0.001
New Delhi							0.034	-0.007

SEASONAL PARTITIONING OF DATA FOR IMPROVED ESTIMATION OF
DIFFUSE RADIATION FROM PAGE'S CORRELATION $\frac{D}{G} = c + d\frac{G}{G_o}$

Physics Department and Renewable Energy Research Centre,
Faculty of Science, University of Dhaka,
Dhaka-1000, Bangladesh.

ABSTRACT

Data on D and G for eleven locations in Europe, Asia and Africa lying between $48°N$ and $20°S$ latitudes have been used to determine the correlation parameters c and d for Page's correlation for fits of monthly averaged daily D and G over the year and also for seasonal fits for summer and winter periods. A significant improvement in estimates of D/G are obtained by partitioning the data.

To obtain station independent fits for different geographical regions data of two or more stations have been used to look for collective correlations. It is found that region dependent fits give satisfactory estimates when collective correlation constants c and d are used which are determined using data of a pair or a group of stations in the same climatic region for biannual fits.

KEYWORDS

Diffuse radiation, global radiation, extraterrestrial radiation.

INTRODUCTION

Most often Page's correlation formula

$$\frac{D}{G} = c + d\frac{G}{G_o} \quad \ldots \ldots \ldots \ldots \ldots (1)$$

is used to estimate the monthly averaged daily diffuse radiation D for a place for which monthly averaged daily G is known. The extraterrestrial radiation G_o is computed from solar geometry and the regression constants c and d are determined from least square fits between monthly D/G and G/G_o using data of all months of the year for a station. In this paper it is inv-

2799

estigated whether seasonal partitioning of data leads to improved estimates of D/G and hence D. To this end data of eleven
stations belonging to Europe, Africa and Asia have been divided
into summer and winter groups of months. Correlation constants
for Page's relation have been obtained for each station for
summer and winter groups of months and the rms error over the
year for the estimates of D/G using the constants are compared
with the estimates for annual fits of the stations.

An attempt has also been made to obtain biannual collective
regression fits by grouping the stations into similar or nearly
similar climatic region.

DATA AND THEIR ANALYSES

Long term quality data of monthly averaged G and D for more
than ten years for six stations in India — Ahmadabad, Bhavnagar, New Delhi, Bombay, Calcutta and Nagpur have been obtained from Mani's tables (Mani,1980,1983). Values of extraterrestrial radiation G_o were available in the compilations.
For five other stations data have been found in the work of
Castro-Diez et al.(1989). These stations situated between 48^oN
and 20^oS latitude are Carpentras and Trappes in France, Madrid
in Spain and Salisbury and Bulawayo in Zimbabwe.

Least square fits have been obtained for the individual stations using data over the year and also for months around
summer and winter by partitioning the year. Four types of partitions were made with March to August(months 3-8), March to
September(months3-9), April to August(months4-8) and April to
September(months4-9) as summer months for the northern hemisphere. The rest of the months of the year correspond to the
winter for each partition. Summer and winter in southern
hemisphere correspond to winter and summer, respectively for
the northern hemisphere. The regression constants c and d and
the correlation coefficient r were obtained for each fit.
Estimated values of D/G and the rms errors over the year have
been computed for annual and the pairs of biannual fits.

Finally, Carpentras and Trappes, two stations in France were
grouped together and collective fits for data of the stations
were obtained for summer as well as for winter periods. Similarly, Bulawayo and Salisbury in Zimbabwe were used to obtain
collective fits. Three stations in north India,— Ahmadabad,
Bhavnagar and New Delhi belonging to the semi-arid climate
were grouped together and collective biannual fits were obtained. Similarly, Calcutta, Bombay and Nagpur belonging to wet-
and-dry region were employed to obtain collective fits.

RESULTS

Table 1 contains the correlation coefficients for annual fits
and a pair of biannual fits corresponding to a particular
partitioning of months (4-8,9-3). All the correlation coefficients r are excellent. For the partitions into months 4-9,
10-3 and 3-9,10-2 it is found that some of the r values (-0.15,
-0.36 and -0.48) are small. Fits for 3-8,9-2 have a bit low r

2800

Table 1. Individual station fits to the relation
$$\frac{D}{G}=c+d\frac{G}{G_o} \quad \text{for groups of months}$$

Station	Months 1-12 r	Months 4-8 r	Months 9-3 r
1. Carpentras	-0.97	-0.99	-0.95
2. Trappes	-0.95	-0.92	-0.98
3. Madrid	-0.92	-1.00	-0.91
4. Bulawayo	-0.98	-1.00	-0.96
5. Salisbury	-0.99	-0.98	-0.98
6. Ahmadabad	-0.98	-1.00	-0.92
7. Bhavnagar	-0.99	-1.00	-0.96
8. New Delhi	-0.91	-0.92	-0.92
9. Calcutta	-0.95	-0.99	-0.97
10.Bombay	-0.98	-0.99	-0.98
11. Nagpur	-0.99	-1.00	-0.99

value of -0.70 in one single case only.

Table 2 shows that monthly estimates of D/G for the biannual fits have generally smaller rms errors over the year compared to the annual fits. Large rms errors of around 10% are present among predictions from annual correlations but there is no such case for biannual fits. The errors from different partitioning are of similar magnitude. However, a consideration of r values lead to the general choice of the partition 4-8,9-3. That is the summer group of months consists of April to August and the winter group is for September to March.

Table 2. RMS errors in D/G for individual station fits to the
relation $\frac{D}{G}=c+d\frac{G}{G_o}$ for groups of months

Station	Average $\frac{D}{G}$	rmse months 1-12	months 3-8	months 3-9,10-2	months 4-8,9-3	months 4-9,10-3
1.Carpentras	0.404	0.013	0.009	0.010	0.009	0.010
2.Trappes	0.590	0.023	0.013	0.018	0.012	0.015
3.Madrid	0.355	0.029	0.021	0.024	0.021	0.025
4.Bulawayo	0.302	0.014	0.014	0.015	0.014	0.014
5.Salisbury	0.317	0.016	0.016	0.016	0.015	0.015
6.Ahmadabad	0.370	0.037	0.019	0.017	0.023	0.017
7.Bhavnagar	0.380	0.030	0.014	0.018	0.018	0.018
8.New Delhi	0.378	0.046	0.028	0.031	0.025	0.028
9.Calcutta	0.471	0.041	0.020	0.022	0.018	0.021
10.Bombay	0.415	0.039	0.027	0.025	0.025	0.025
11.Nagpur	0.394	0.020	0.012	0.016	0.011	0.014

Table 3 presents the rms errors in D/G and mean bias errors for collective fits.

Table 3. RMS errors in estimated D/G for collective fits
to $\frac{D}{G} = c + d\frac{G}{G_0}$ for different partitioning of data

Group of Station	months 3-8,9-2 rmse	bias	months 3-9,10-2 rmse	bias	months 4-8,9-3 rmse	bias	months 4-9,10-3 rmse	bias
Carpentras	0.022	0.010	0.016	0.006	0.021	0.006	0.017	-0.002
Trappes	0.024	-0.009	0.023	-0.005	0.014	-0.005	0.017	0.000
Bulawayo	0.016	-0.003	0.016	-0.003	0.016	-0.002	0.016	-0.002
Salisbury	0.018	0.003	0.018	0.003	0.017	0.002	0.017	0.002
Ahmadabad	0.020	0.002	0.022	0.003	0.024	0.002	0.023	0.004
Bhavnagar	0.021	-0.008	0.025	-0.002	0.025	-0.004	0.026	0.000
New Delhi	0.040	0.002	0.045	-0.001	0.043	0.000	0.048	0.002
Calcutta	0.024	0.002	0.025	0.000	0.024	0.001	0.026	-0.003
Bombay	0.030	-0.003	0.029	0.001	0.029	-0.003	0.027	0.001
Nagpur	0.017	0.000	0.022	0.006	0.015	0.001	0.019	0.003

DISCUSSION

The eleven stations studied belong to different parts of the
world and have different physiographical features. Every individual station gives excellent biannual fits. When those lying
within a limited region having similar climates are grouped
together the collective biannual fits obtained are found to be
happy. The first group (Table 3) covers France, the second is
for Zimbabwe, the third scans the wet-and-dry zone of northern
India while the fourth represents the wet-and-dry region of
northern and central India. It appears then that biannual fits
are to be used for estimating D from G in preference to annual
fits. When a location has a nearby station where both G and D
are measured one may use the regression constants for that
station to estimate D for the location. Otherwise one may use
the parameters for collective fits.

It may be mentioned that Collares-Pereira et al.(1979)proposed
a season dependent correlation. But this does not give satis -
factory estimates for locations having the monsoon climate. For
example, computations show that for Calcutta the formula gives
low values with errprs of 20% or more for the rainy months May,
June, July, August and September. The present work gives a
thumb rule for seasonal partitioning of data in order to
estimate D with G good accuracy using Page's relation.

REFERENCES

Castro-Diez, Y., Alados-Arboledas, L. and Jcinenez, J.L.(1989). A model for climatological estimations of global, diffuse and direct solar radiation on a horizontal surface. Solar Energy, 42, 417-424.

Collares-Pereira, M. and Rabl, A. (1979). The average distribution of solar radiation — corelations between diffuse and hemispherical and between daily and hourly insolation values. Solar Energy, 22, 155.

Mani,A. (1980). 'Handbook of Solar Radiation Data for India'. Allied Publishers Limited, New Delhi.

Mani, A. and Rangarajan, S. (1983). 'Solar Radiation Over India'. Allied Publishers, New Delhi.

NEW CORRELATIONS FOR GLOBAL SOLAR
RADIATION FOR LATITUDES 40°N-40°S

G.M.Singh

Solar Energy Lab.,Department of Physics,
Guru Nanak Dev University,
Amritsar-143005 (India)

ABSTRACT

A prior requirement to the design of a solar energy conversion system and for the assessment of its potential is the knowledge of the insolation data for a given location.for estimation of global solar radiation on horizontal surface, linear and quadratic regression relations have been developed at Amritsar (31.63°N, 74.87°E) using measured data for the period of three years. The values obtained from the existing correlations and the one's obtained during the present study are compared with the actually observed values using Root Mean Square Error (RMSE) and Mean Bias Error (MBE) tests. The calculated and observed values are also compared using Sub-divided bar diagrams.these statistical tests indicate that out of 20 correlations studied, correlations developed during the present study are most sutable for estimation of global solar radiation.the use of these correlations is recommended for estimation of global solar radiation on horizontal surface.

KEY WORDS

Global solar radiation; correlations; sunshine duration

METHODOLOGY

The data used in this study are daily global solar radiation on horizontal surface (H) and sunshine duration (n),which were measured at Amritsar. The instrumentation used for the measurements of H include thermoelectric pyranometer along with a radiation integrator and Campbell Stokes sunshine recorder for n. These instruments were calibrated from the National Radiation Centre of the Indian Meterological Department (IMD) at Pune. The allied meterological parameters such as ambient temperature (T), relative humidity (R)and rain fall, were measured by conventional methods.

GLOBAL SOLAR RADIATION CORRELATIONS

Black et al (1954) gave relation from their stations data
$\overline{H}/\overline{H}_O$ = 0.23 + 0.48 (n/N)   (1)
(H and N are monthly average daily values of extraterrestrial radiation and maximum possible   sunshine duration).
Glover and McCulloch (1958), while attempting to improve upon the insolation-sunshine correlations,have included the latitude effect and presented the following formulation
H/H_O = 0.29 cos L + 0.52 (n/N)   (for L <60°)   (2)
Swartman and Ogunlade (1967) reported the following relationships   correlating solar radiation intensity   with sunshine and relative humidity for tropical conditions

$$H = 490 \ S^{0.357}R^{-0.262}$$ (3)

$$H = 460 \ e^{0.607(S-R)}$$ (4)
$$H = 464 + 265 \ S - 248 \ R$$ (5)
where S is the ratio of hours of bright sunshine to 12 hours
Lof et al (1966) gave the following correlation for two different ranges of percentage of possible sunshine (n/N) for Pune (India)
H/H_O = 0.30 + 0.51 (n/N)   for   25-49
H/H_O = 0.41 + 0.34 (n/N)   for   65-89   (6)
Rietveld (1978) obtained from his studies the relationship
H/H_O = 0.18 + 0.62 (n/N)   (7)
which he believed to applicable every where in the world.
Gupta et al (1979) have reported the linear relationship
$\overline{H}/\overline{H}_O$ = 0.079 + 0.878 (n/N)   (8)
using which they have the standard deviation of 0.037
Garg et al (1968) presented the following linear correlation for Delhi
H/H_O = 0.10 + 0.82 (n/N)   (9)
Dogniaux and Lamoine (1983) arrived at another latitude dependent correlation while incorporating average Linke turbudity factor
H/H_O = [0.00506(n/N)- 0.00313]L + 32029(n/N)+ 0.37022 (10)
Garg and Garg (1982) gave the following correlations based on sunshine duration and atmospheric water vapour content
H/H_O = 0.414 + 0.400(n/N)- 0.0055W_{at} (11)
where W_{at} is atmospheric water vapor content/unit volume of air
Hussain (1984) re-examined the above correlation by considering the data from North India stations only and presented the relationship with the replacement of N by N'
H/H_O = 0.394 + 0.364(n/N')- 0.0035 W_{at} (12)
where N' is given by
cos 7.5 N' = (cos 85°- sin L sin $\boldsymbol{\delta}$)/(cos L cos $\boldsymbol{\delta}$)   (12a)
Hussain (1985) pesented set of empirical constants for linear relationship of the type
H/H_O = a + b(n/N)   (13)
He reported it to be station independent. In these he used the data according to periods of monsoon.
Garg and Garg (1985) presented a linear relationship after analysing the data of 11 widely separated stations which include all varied type of meterological conditions
H/H_O = 0.3156 + 0.4520(n/N)   (14)
Reddy (1971) has developed a new formula for computing H
$H = K[(1+0.8s)(1-0.2t)/R^{\frac{1}{2}}]cal \ cm^{-2}day^{-1}$ (15)

where letters have usual notations given in the reference cited. Modi and Sukhatme (1979) has presented a set of values for empirical constants corresponding to the 12 months of the year

$$H/H_O = a + b \ (X) \tag{16}$$

the coefficients of this equation are in respect of three meterological parameters, viz. percent of possible sunshine, precipitation and all clouds.

Hay (1979) correlated n/N' with H'/H , where H' is the global solar radiation that first strikes the ground before undergoing multiple reflections with the clear atmospheric and cloud base

$$H-H' = H \ \rho \ [0.25(n/N')+0.60(1-n/N')] \tag{17}$$

where ρ is the surface albedo, H' is given by the equation

$$H'/H_O = a + b \ (n/N')$$

and N' is given by equation (12a)

For Amritsar $\rho = 0.20$, a=0.30, b=0.40 as given by Mani and Rangarajan (1980)

Gopinathan (1988) developed a formula recently incorporating elevation of location (A) above sea level along with L,n,T &R

$$H/H_O = 0.801- 0.378 \ \cos L + 0.012A + 0.316(n/N)$$
$$-1.215 \times 10^{-3} T -1.049 \times 10^{-3} R \tag{18}$$

New correlations (Singh,1990)

Linear $\quad\quad H/H_O = 0.2765 + 0.4392(n/N) \tag{19}$

Quadratic $\quad H/H_O = 0.2694 + 0.4597(n/N)-0.0147(n/N)^2 \tag{20}$

RESULTS AND DISCUSSION

Monthly average daily values of global solar radiation on horizontal surface are calculated using the different existing models including the one's developed during the present study. The results are presented in Table 1 along with the corresponding measured (observed) values.To illustrate the different values, the results in the Table 1, are divided into four figures (1 to 4), each figure containing six subdivided bars (five of calculated values & one of measured value). In this study; two statistical tests RMSE and MBE are used to evaluate the accuracy of correlations described above:

$$RMSE = [\{(H_{iC}-H_{iM} \ )\}/n]^{\frac{1}{2}} \tag{21}$$

where H is the ith calculated value, H is the ith measured value and n is the total no. of observations. This test provides information on short term performance of the correlations.

$$MBE = (H_{iC} - H_{iM})/n \tag{22}$$

This test provides information on the long term performance and also low MBE is ideal.

The performance of twenty global radiation correlation were evaluated using RMSE and MBE tests. Results are presented in Table 2 . Values of RMSE and MBE indicate that present correlations (equations 19 and 20) are most suitable . Among others, previously available correlations those given by Black et al (1954), Dogniaux and Lemoine (1983),Glover and McCulloch (1958), Rietveld(1978),Reddy (1971), Hussain(1985), Garg and Garg(1985) and Hay(1979) are in decending order of accuracy according to MBE test. The remaining correlations give overestimated values whereas the Black et al correlation gives very slightly underestimated values. The order of suitability

according to RMSE test is as following:Black et al
(1954),Dogniaux and Lemoine(1983), Glover and McCulloch(1958),
Garg and Garg (1982), Hussain (1984), Rietveld(1978), Hussain
(1985),Reddy(1971). The others give considerable overestimated
values for the estmation of monthly average daily total solar
radiation.as the performance of existing correlations from
diffrent parts of the world (particularly from 40^ON to 40OS)
are evaluated using the measured data, therefore being the most
accurate, the use of new equations (No.19 & 20) is recommended.

REFERENCES

Black,J.N.,C.W.Bonython and J.A.Prescott(1954).Solar radiations
  and the duration of sunshineQuart.J.Roy. Met. Soc.,84, 344.
Dogniaux,R. and M.Lemoine(1983).Classification of radiation
  sites in terms of different indices of atmospheric transparency
  Proc. Int.Daylighting Conf.,Phoenix,Arizona
Garg,H.P.,R.Ganguli and C.L.Gupta(1968).Computation of average
  hourly solar radiation intensity from sunshine hours
  data,Indian Chemical Manufacturer,6,83-88.
Garg,H.P. and S.N.Garg(1982).Solar radiation from bright
  sunshine hours and other meterological variables, Proc. NSEC,
  Allied Publishers, New Delhi, 1.004-1.007.
Garg,H.P. and S.N.Garg(1985).Correlation of monthly average
  daily global,diffuse and beam radiation withbright sunshine
  hours, Ener.Conv. & Mgmt.,25, 409-417.
Glover,J. and J.S.G.McCulloch(1958).The empirical relation
  between solar radiation and hours of sunshine,Quart.J.Roy. Met.
  Soc.,84, 172-175.
Gopinathan,K.K.(1988).A new model for estimating total solar
  radiation, Solar and Wind Tech., 5, 107-109.
Gupta,C.L.,K.Usha Rao and T.A Reddy(1979).Radiation design data
  for solar energy applications, Energy Mgmt.,Oct.-Dec.,299-313.
Hay,J.E.(1979).Calculation of monthly mean solar radiation for
  horizontal and inclined surfaces,Solar Energy,23, 301-307.
Hussain,M(1984),Estimation of global and diffuse irradiation
  from sunshine duration and atmospheric water vapour content,
  Solar Energy,33, 217-220.
Hussain,M(1985). An assessment of solar and wind energy
  availability in Bangladesh, Proc.Conf.on Physics and Energy,
  Dhaka, 37-53.
Lof,G.O.G.,J,A.Duffie and C.O.Smith(1966).World distribution of
  solar energy,Solar Energy,10, 27-37.
Mani,A.and S.Rangarajan(1980). Hand Book of Solar Radiation,
  Allied Publishers,New Delhi, 12-14.
Modi,V.and S.P.Sukhatme(1979).Estimation of daily total and
  diffuse insolation in India from weather data,Solar Energy,22,
  414-416.
Reddy,S.J.(1971).An empirical method for estimation of total
  solar radiation, Solar Energy, 13, 289-290.
Rietveld,M.R..(1978).A new method for estimating the regression
  coefficients in the formula relating solar radiation to
  sunshine, Agric. Met., 19, 243-252.
Singh, G.M.(1990).Investigation of solar terrestrial radiation
  and correlations at Amritsar, SESI Journal, 4, 149-150.
Swartman,R.K. and O.Ogunlade(1967).Solar radiation parameters
  from common parameters, Solar Energy, 11, 170-172.

Table 1 : Monthly average daily values of Global Solar Radiation on Horizontal Surface (kWhm^{-2})

Eqn.	Jan.	Feb.	Mar.	Apr.	May	Jun.	Jul.	Aug.	Sep.	Oct.	Nov.	Dec.
Obs.	3.104	3.989	4.876	6.259	6.230	6.596	6.323	5.303	5.453	4.466	3.714	3.011
1	3.048	3.845	4.553	5.823	5.960	6.637	5.926	5.610	5.161	4.397	3.631	3.020
2	3.290	4.151	4.914	6.286	6.433	7.165	6.396	6.055	5.572	4.747	3.921	3.260
3	5.068	5.388	5.454	6.190	6.170	6.409	5.613	5.419	5.629	5.740	5.628	5.315
4	4.884	5.253	5.350	6.264	6.226	6.576	5.481	5.190	5.515	5.684	5.548	5.158
5	5.074	5.434	5.521	6.300	6.264	6.547	5.656	5.392	5.683	5.825	5.704	5.346
6	3.536	4.417	5.309	6.603	6.939	7.503	6.916	6.531	5.889	4.941	4.045	3.414
7	3.284	4.165	4.888	6.346	6.411	7.245	6.351	6.034	5.608	4.813	3.993	3.296
8	3.670	4.692	5.432	7.222	7.145	8.264	7.036	6.724	6.351	5.516	4.606	3.760
9	3.574	4.562	5.295	7.008	6.961	8.016	6.863	6.551	6.168	4.346	4.858	3.647
10	3.279	4.129	4.904	6.238	6.416	7.107	6.387	6.038	5.536	4.703	3.879	3.234
11	3.515	4.417	5.202	6.415	6.559	6.906	6.065	5.664	5.401	4.730	3.972	3.442
12	3.489	4.372	5.134	6.376	6.592	7.000	6.242	5.843	5.485	4.751	3.973	3.414
13	3.433	4.275	5.258	6.549	6.862	7.444	6.471	6.120	5.618	4.751	3.879	3.289
14	3.422	4.299	5.126	6.476	6.702	7.372	6.683	6.307	5.755	4.872	4.010	3.354
15	3.297	3.663	4.077	5.328	5.844	6.443	5.677	5.263	5.684	4.786	4.011	3.203
16	3.610	4.665	5.798	6.741	7.037	7.223	6.703	6.296	5.685	4.827	3.956	3.377
17	3.457	4.480	5.282	6.375	7.077	7.163	6.380	6.116	5.647	4.782	4.007	3.273
18	3.747	4.667	5.593	6.828	7.153	7.645	7.141	6.717	6.063	5.101	4.199	3.606
19	3.153	3.985	4.714	5.993	6.196	6.812	6.156	5.823	5.329	4.521	3.709	3.113
20	3.153	3.986	4.714	5.994	6.196	6.812	6.156	5.823	5.330	4.520	3.709	3.113

Table 2 : MBE and RMSE values of 20 correlations for estimating H

Eqn.	MBE	RMSE
1	− 0.005	0.249
2	0.015	0.316
3	0.164	1.202
4	0.148	1.124
5	0.164	1.230
6	0.036	0.616
7	0.015	0.337
8	0.047	0.977
9	0.039	0.773
10	0.015	0.293
11	0.034	0.317
12	0.032	0.327
13	0.027	0.451
14	0.027	0.477
15	0.016	0.458
16	0.042	0.602
17	0.029	0.460
18	0.054	0.801
19	0.004	0.199
20	0.004	0.199

Eqn. = Equation No. : Obs. = Observed values

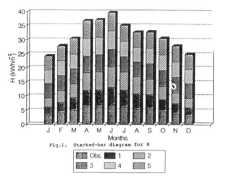

Fig.1. Stacked-bar diagram for H

| Obs. | 1 | 2 |
| 3 | 4 | 5 |

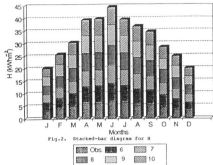

Fig.2. Stacked-bar diagram for H

| Obs. | 6 | 7 |
| 8 | 9 | 10 |

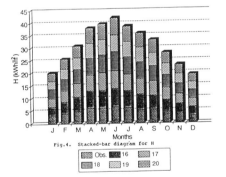

Fig.4. Stacked-bar diagram for H

| Obs. | 16 | 17 |
| 18 | 19 | 20 |

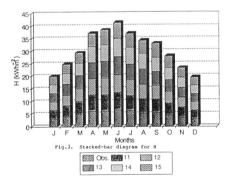

Fig.3. Stacked-bar diagram for H

| Obs. | 11 | 12 |
| 13 | 14 | 15 |

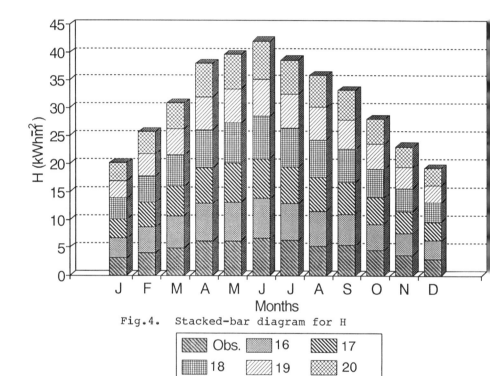

Fig.4.   Stacked-bar diagram for H

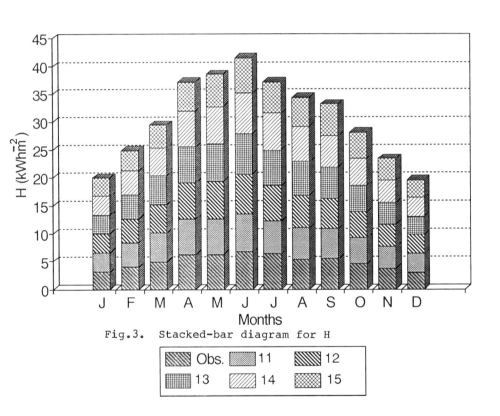

Fig.3.   Stacked-bar diagram for H

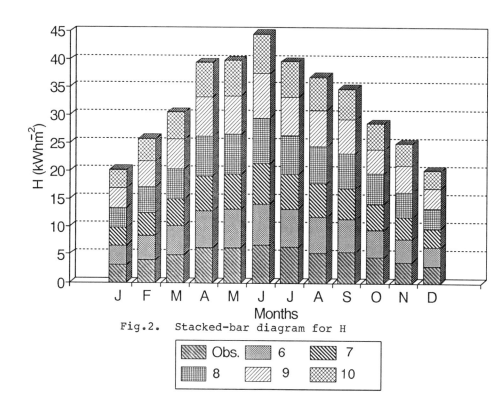

Fig.2. Stacked-bar diagram for H

Legend: Obs. | 6 | 7 | 8 | 9 | 10

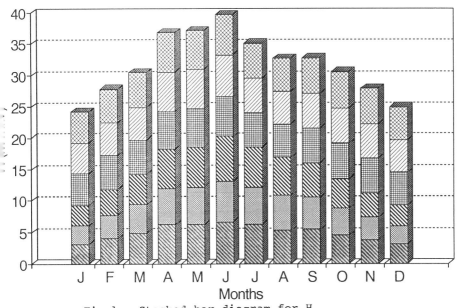

Fig.1. Stacked-bar diagram for H

ESTIMATION OF GLOBAL IRRADIATION IN DRY SEASON
FROM EVAPORATION AMOUNTS

SYDNEY M. CHAGWEDERA

Department of Science and Mathematics Education
University of Zimbabwe
P. O. Box MP 167
Mount Pleasant
Harare
ZIMBABWE

ASTRACT

A correlation formula

$$H/H_o = A + Bx + Cx^2 + Dx^3$$

has been found to give good estiamtes of global irradiation, H, from mean
monthly evaporation amounts, x. A, B, C, D, are regression coefficients.
These are particularly true for dry seasons where the percentage error for
Harare (Zimbabwe) ranges from 0 - 13.5% and for Bulawayo (Zimbabwe) ranges
from 1,8 - 16,2% for eight drier months. The data used is standardized
over 20 years for the irradiation and over 15 years for the evaporation
amounts.

KEYWORDS

Irradiation; evaporation; regression; global; correlation.

INTRODUCTION

A number of meteorological parameters such as sunshine hours, cloud cover,
and precipitation affect the magnitude of global solar radiation incident
on a given location (Lewis, 1989). Since a knowledge of global irradiation
is required in the setting up of solar energy systems, and since it is more
difficult to measure it than the many other meteorological parameters, a
host of correlation equations have been developed. Perhaps the popular
are Angstrom-type equations which give linear regression coefficients
(Chagwedera and Sendezera, 1990; Sayigh, 1977; Zungu and Sendezera, 1989).
However, the computer revolution has led to other polynomial regressions.
For example, Hussain and others (1990) have found a correlation formula:

$$n/N = A + Bx + Cx^2$$

to estimate sunshine duration (n) from precipitation amount (x).

2814

Closely related to precipitation (which is clearly suitable for wet areas) is evaporation. Some meteorological stations such as Harare (17° 50'S) and Bulawayo (20° 09'S) in Zimbabwe have been keeping evaporation records for over 15 years. Not ing that about eight months in a year in Zimbabwe are considered dry it would be appropriate to search for a correlation formula for the estimation of global irradiation from evapration amounts.

PROCEDURE

Detailed records of global solar radiation for two Zimbabwe cities, Bulawayo (20° 09'S) and Harare (17° 50'S) were obtained form Zimbabwe Meteorological Services (Chagwedera and Sendezera, 1990). Mean monthly evaporation totals in millimetres (mm) for the two stations were also obtained from the Department of Meteorological Services (Zimbabwe 1981). The evaporation pan used is the American Weather Bureau Class A pan which is made of galvanised steel 1 800 mm in diameter and 300 mm in height. A protective wire mesh that prevents animals from drinking the water is installed although this reduces evaporation by about 10%. Allowance is also made for rain falling in the pan.

Polynomial regression fits (up to the cubic term) between the mean monthly evaporation amount (x) and the monthly average H/Ho were obtained for the two stations. These had the form:

$$H/Ho = A + Bx + Cx^2 + Dx^3$$

where A, B, C, D are the regression coefficients and H is monthly average daily global irradiation on a horizontal surface (MJ m^{-2} day^{-1}) and H_0 is monthly average daily irradiation on top of the earth's atmosphere (extra terrestrial) – obtained from Butera (1982).

TABLE 1 : DATA RESULTS FOR BULAWAYO (20° 09'S)

MONTH	Ho (MJm-2d-1)	Hm (MJm -2d-1)	H/Ho	HC (MJm-2d-1)	EVAP (mm)	$\frac{Hm-Hc}{Hm}$ x 100 (Δ)
JAN	41.97	23.4	.6759	28.5	182	− 21.4
FEB	40.14	22.1	.7041	28.3	150	− 28.1
MAR	36.44	21.6	.6873	25.1	169	− 16.2
APRIL	31.37	20.2	.7219	22.6	130	− 11.9
MAY	26.69	17.7	.7254	19.4	126	− 9.4
JUNE	24.26	15.8	.7415	18.0	108	− 13.9
JULY	25.14	17.4	.7254	18.2	126	− 4.6
AUG	28.91	19.5	.6926	20.0	163	− 2.6
SEPT	33.91	19.5	.6480	21.9	214	− 1.8
OCT	38.26	22.8	.6284	27.4	191	1 − 5.3
NOV	41.09	22.0	.6680	27.4	191	− 24.5
DEC	42.23	21.9	.6794	28.7	178	− 31.1

2816

TABLE 2: DATA AND RESULTS FOR HARARE (17° 50'S)

MONTH	Ho $(\text{MJm}^{-2}\text{d}^{-1})$	Hm $(\text{MJm}^{-2}\text{d}^{-1})$	H/Ho	HC $(\text{MJm}^{-2}\text{d}^{-1})$	EVAP (mm)	$\dfrac{\text{Hm}-\text{Hc}}{\text{Hm}} \times 100$ (Δ)
JAN	41.46	20.6	.6617	27.4	163	-33.0
FEB	40.01	20.2	.6584	26.3	137	-30.2
MAR	36.85	21.5	.6617	24.4	163	-13.5
APRIL	32.24	20.4	.6588	21.2	140	- 3.9
MAY	27.87	18.6	.6576	18.3	129	- 1.6
JUNE	25.55	16.4	.6562	16.8	108	-13.9
JULY	26.38	18.0	.6577	17.4	112	- 3.4
AUG	29.92	19.8	.6628	19.8	130	- 3.3
SEPT	34.46	22.7	.6676	23.0	171	- 0
OCT	38.40	23.6	.6659	25.6	227	- 1.3
NOV	40.73	21.1	.6649	27.1	257	- 8.5
DEC	41.62	20.3	.6623	27.6	188	-28.4

RESULTS

For Bulawayo (20° 09'S) a correlation formula:

$$H/Ho = A + Bx + Cx^2 + Dx^3$$

has cofficients:
$$A = 0,8389$$
$$B = -0,9003 \times 10^{-3}$$
$$C = -0,5961 \times 10^{-7}$$
$$D = 0,4657 \times 10^{-9}$$

whereas Harare (17° 50'S) has the regression cofficient given by:

$$A = 0,6836$$
$$B = -0,6714 \times 10^{-3}$$
$$C = 0,4947 \times 10^{-5}$$
$$D = -0,1013 \times 10^{-7}$$

2817

Table 1 and 2 give the results for Bulawayo and Harare respectively.
Hc and Hm are the calculated and measured irradiation values respectively.
is the percentage difference

$$\frac{Hm - Hc}{Hm} \times 100$$

It must be pointed out that the main rainy season in Zimbabwe is mid-
November to mid - March, that is about 8 months of the year are Δ
relatively dry. Thus looking at table 1 (Bulawayo) it is seen that Δ
varies Δ from 1,8 - 6 - 6,2 for the relatively dry months while for Harare
ranges from 0 - 13,4 , for the drier months. For the wet months Δ has
much higher values for both stations.

CONCLUSION

From these results it is clear that a correlation formula

$$H/Ho = A + Bx + Cx^2 + Dx^3$$

gives satisfactory prediction of global irradiation, H, from evaporation
totals. If the dry months are considered the correlation is even better
pointing out to the usefulness of the formula for estimation of global
irradiation in dry seasons (areas?) from evaporation amounts.

REFERENCES

Butera, F. (1982). In Non-Conventional Energy Sources ACIF Series, Vol.3
                    27 World Scientific.

Chagwedera S. M. and Sendezera E. J. (1990). Prediction of Global Solar
                    Radiation at two Locations in Zimbabwe: A Comparative
                    Analysis of Three Angstrom-Type Equations, in Press
                    Renewable Energy Journal.

Hussain, M.; Ali, G. and Aditya, S. K. (1990). On Estimation of Sunshine
                    Duration form Precipitation   Amounts, in Energy and
                    the Environment into the 1990s, 3087-1090,
                    Proceedings of the 1st world Renewable Energy Congress,
                    Reading, UK, 23-28 September 1990. Pergamon Press,
                    Oxford.

Lewis, G. (1989).   The Utility of the Angstrom-Type Equations for the
                    Estimation of the Global Irradiation, Solar Energy
                    43, 297-299.

Sayigh, A. A. M. (1977). Solar Energy Availability Prediction from
                    Climatological Data in Solar Energy Engineering,
                    61-82, Academic Press, New York.

Zimbabwe Department of Meteorological Services (1981). Climate Handbook
                    of Zimbabwe, Harare (Salisbury), Zimbabwe.

Zungu, M. B. and Sendezera, E. J. (1989). Global Solar Radiation in
                    Kwaluseni (Swaziland), Swaziland Journal of Science
                    and Technology, 10, 49-54.

On the Utilization of Wave Energy
for the Restoration of the Natural Coastal Environment

Hitoshi Hotta, Takeaki Miyazaki and Yukihisa Washio

Marine Development Research Department
Japan Marine Science and Technology Center
2-15, Natsushima-cho, Yokosuka, 237, Japan

ABSTRACT

Wave energy is a renewable and clean energy resource, and its total amount is infinite in spite of low density of energy, and it can be utilized anywhere in the ocean coastal zone. Authors made a proposal for a system to utilize the wave energy in order to resolve the environmental problems of contermination of seawater and the sea bed in the enclosed bay. The "Mighty Whale" is a floating wave power device, which has high performances on the absorption of wave energy and the dissipation of waves, furthermore which can convert the wave energy to the energy of compressed air. Compressed air is transported into the bay to be spent for the aeration for the refreshment. In this paper, the cost for construction of the "Mighty Whale" and the system for refreshment of the coastal environment.

UTILIZATION OF WAVE ENERGY

Characteristics of Wave Energy

The annual average of wave energy per unit length of shoreline around Japan is about 6 kW/m, and it is about 7 kW/m at the coast of the Pacific Ocean. At the coast of the Japan Sea in winter, wave power is more than 20 kW/m. Since the total length of shoreline of Japan is about 30,000 km, we can select the optimal location for the wave power utilization.

By the way, wave energy changes every second, and mean value of it in a day, a week and a month are change, also. Furthermorem, the density of energy of wave is not so high. In spite of these shortcomings of wave energy, we need to develop for the conventional utilization of it, because it is clean and renewable energy resources ,if we don not need to consider the cost for energy conversion.

2819

Conversion from Wave Energy to Compressed Air Power

If we can store compressed air in the ocean, we can utilize it
not only for the generation of electric power but also for the
refreshment of the coastal zone and "aquaculture." It is
neccessary to increase the dissolved oxygen in sea water or to
exchange sea water by the transportation of sea water from the
offing for the refreshment.

Regarding this system, seawater is transported with compressed
air as a mixed fluid of liquid and air just the same as the
"air lift" technology. As for the air transported to near
shoreline, it is discharged at the sea bottom for the
generation of air bubbles to increase oxygen in sea water, that
is, "aeration."

The compressed air is generated by the compressor coupled to
the air turbine which is installed on the wave power device and
rotates by the air flow generated by the air chamber of the
Oscillating Water Column (OWC) type wave power device. The
optimal matching between the compressor and the air turbine is
one of the themes of the research and development on this
system.

Multi-Purpose Utilization of the Wave Power Device

Since the wave power device can absorb wave energy, it means
that the wave power device can dissipate the wave and create
the calm sea area behind the device also. However, the
performance for wave dissipation of the floating wave power
device, such as "the Mighty Whale" shown in Fig.1 [1], is not
so good in comparison with the fixed type device, especialy in
waves of long period such as storm waves.

From the point of view of the effect on coastal enviroment,
such a shortcoming turns into a merit. Naturally, big waves
change the sea water in an enclosed sea. However, a man-made
offshore structure to protect the waves is used to prevent the
change of the sea water, and it causes some pollution problems.
Of course, it is very important to keep our lives safe.
Authors guess that we will need to consider not only safety but
also amenity at the seashore in the very near future.

Consequently, the floating wave power device such as the Mighty
Whale has possibility to be utilized as a floating break water.
And, it creates the calm sea area behind the device where
"aquaculture" or many marine sports are available. It means
that such a device can create conventional space in the ocean
near the sea shore.

THE SYSTEM FOR REFRESHMENT OF THE COASTAL ZONE

Structure of the Mighty Whale

The scheme of the shapes of the Mighty Whale is shown in Fig.1.
This device is considered from three parts: Air chambers, a
floating chamber, and a stabilizer slope. Wave energy is

2820

converted to air power by OWC of the air chamber, and drive the air turbine and the air compressor. The stabilizer slope fulfils the function for reduction of pitching motion in waves and for the space in the ocean.

Performances of the Mighty Whale

Fig.2 shows the efficiency of conversion of wave power into air power (wave power absorption) versus wave length (λ) divided by length(L) of the device [1]. According to the results of the two dimensional scale model test in regular waves, the maximum efficiency of wave power absorption is about 60% and the performance of dissipation of incident wave is about 80% within the range of normal sea state at the coast of Japan. These results indicate that the Mighty Whale has excellent performances of in wave power absorption and dissipation of incident waves, and better than former types of such wave power devices [2].

By the way, the mooring force in the waves is very small also, and the minus values of it were measured in the condition of shallow water. It means the Mighty Whale will be able to go ahead against incident waves sometimes, so that the mooring force is very small in general. Consequently, it was made clear that it will not be difficult to moor the Mighty Whale in waves. Authors guess that this propulsion effect in waves is one of the ways to utilize the wave energy.

The Compressed Air System

The Mighty Whale will have a unique function to compress the air utilizing the wave power. Several units of the air turbine, such as Wells turbine, will be installed on the Mighty Whale, rotate by the air flow through the orifice of the air chamber. If we can convert the axial torque of the air turbine to the axial torque of the air compressor directly, we can generate the compressed air, which may be very useful for the aquaculture and other acivities in the coastal area.

ABILITY FOR REFRESHMENT OF COASTAL ZONE

The compressed air produced by the compressed air system is transport to the coastal area which has some problems on the contamination of sea water and mud of sea bed, and utilized by the aeration. It must increase the oxygen in the seawater and the mud, and it makes every lives active to spend some nutrient components. Consequently, the environmental situation in the coastal area may be refreshed by this system.

When we assume that the Mighty Whale will be installed at the site of 2m in the significant wave height and 8 seconds in the significant wave period, the total input wave power for one unit of the Mighty Whale is about 1,400 kW. Since the average efficiency in regular waves of the wave power absorption is about 50%, and the average efficiency of the turbine is about 40%, the input power for 9 units of air chamber to the air compressors is about 280 kW. Consequently, the rated power of the air compressor should be about 30 kW.

When the air must be compressed up to 4 kg/cm , the air compressor of 30kW in rated power can generate the compressed air of about 1400Nm /day. Since the Mighty Whale is designed in this paper has 9 units of the air chamber, air turbine and compressor, the total volume of the compressed air for the Mighty Whale in a day will be about 13,000Nm /day, that is about 0.15Nm /sec.

By the way, when this compressed air will be utilized for aeration at the point of 8m in water depth, it is estimated that the discharge of air of 0.01Nm /sec can make some effects for the increase of dissolved oxygen in sea water for the area in the coastal zone with a radius of about 50m. Furthermore, when we consider the horizontal diffusion of air bubbles, and the effect of breaking of the layer of sea water by vertical current caused by the movement of bubbles from the sea bottom to the surface, the effective area by the airation may widen about double in radius.

Consequently, we can design the distance between each air pipe line on the sea bed for the airation at about 250m, and this system has 16 nozzles for blow out of the air bubbles. It means authors estimate this system can supply air for the dissolvement of oxygen to the sea of about 1 km around for the purification of coastal sea and aquaculture.

CONCLUSION

In this paper, the following conclusions were obtained.

(1)The possibility and the performance of a system for the refreshment of sea water in coastal zone utilizing wave power
   is discussed and authors reach a possible positive solution.

(2)It was estimated that the system using the Mighty Whale of
   100m in length as a floating OWC type wave power device can
   make refreshment for the sea water of the area of about 1 km
   around by airation.

(3)Authors believe that with this new way to develop and
   utilize coastal space, energy can be proposed to keep nature
   in the coastal zone.

REFERENCES

[1]Miyazaki,T., Hotta,H. and Washio,Y.,1990, Design and Performance of the Mighty Whale Wave Energy Convertor, Proceedings
   of 1st World Renewable Energy Congress,Volume 5,pp.2894-2898
[2]Hotta,H., Miyazaki,T., Washio,Y. and Ishii,S., 1988, On the
   Performance of the Wave Power Device Kaimei -The Results on
   the Open Sea Tests-, Proceedings of the 7th Int. Conf. on
   Offshore Mechanics and Arctic Engineering, Volume 1,pp.91-96

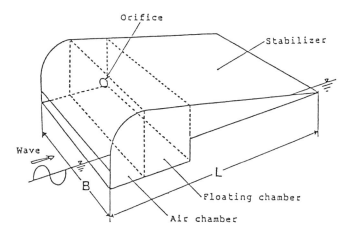

Principal dimension of the scale model

Length	0.70 m
Breadth	0.49 m
Draft	0.09 m
Depth	0.20 m
Depth of curtain wall	0.025m
Weight	5.5 kg

Fig.1 Scheme of the Mighty Whale

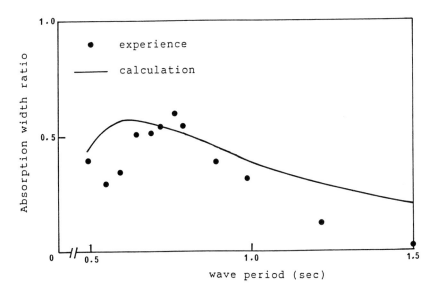

Fig.2 Performance of wave energy absorption in regular waves
by the Mighty Whale

Optimum Design of the Darrieus-Type Cross Flow
Water Turbine for Low Head Water Power

A. Furukawa*, Y. Takamatsu*, K. Okuma* and K. Takenouchi**

* Department of Mechanical Engineering, Faculty of Engineering, Kyushu
  University, 6-10-1 Hakozaki, Higashi-ku, Fukuoka, 812, JAPAN
** Department of Industrial Design, Kyushu Institute of Design, 4-9-1
  Shiobaru, Minami-ku, Fukuoka, 815, JAPAN

ABSTRACT

Model tests of a ducted Darrieus-type water turbine with straight-bladed runner have been
conducted for utilizing extra-low head water power. The guiding principles for designing
high efficiency runner and operating under optimum control are clarified as a summary of
those results.

KEYWORDS

Darrieus-type water turbine; blade performance; operating characteristics; cavitation limit;
extra-low head water power

INTRODUCTION

Darrieus-type water turbines have been developed for free stream of river, tidal current and
extra-low head water power (Faure, 1985; ITDG, 1980; Kihoh et al., 1990). In the case
of extra-low head utilization, the merit of simple structure of the Darrieus-type might not
outweigh the cost problem of conventional types at present because of its inferior efficiency.
It will, however, be useful for the energy problem in future to develop various types of water
turbines. In this paper, guiding principles for designing the ducted Darrieus-type turbine
and useful data of its characteristics are described and its operating range is also discussed.

DESIGN OF HIGH PERFORMANCE RUNNER

A conceivable view of a ducted Darrieus-type water turbine system with straight-bladed
runner is illustrated in Fig.1, where the geometry of the runner should be determined
according to the following principles of the design obtained from experimental results
(Takamatsu et al., 1991). However, it should be noted in Fig.1 that optimum wall shapes

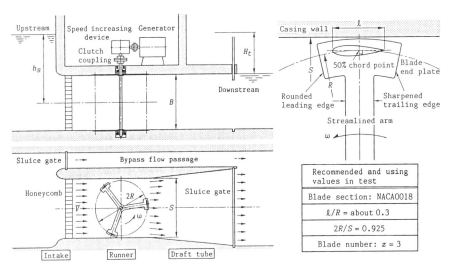

Recommended and using values in test
Blade section: NACA0018
ℓ/R = about 0.3
$2R/S$ = 0.925
Blade number: $z = 3$

Fig.1 Conceptual view of a ducted Darrieus-type water turbine system

of the intake, the runner casing and the draft tube must be examined as a future study to make the runner performance maximum and all hydraulic losses minimum.

A symmetric Darrieus blade in a circular motion along the runner pitch circle is equivalent to a cambered one in a linear motion in still water by a flow curvature effect. Therefore, by use of a symmetric blade section, a greater proportion of power can be generated in upstream half of the blade rotation where flow oncoming to the runner is less disturbed. And as a thin blade section has high lift slope and high lift/drag ratio while a thick one has a wide no-stall region in its characteristics, a symmetric blade section with reasonable thickness ratio from 0.15 to 0.18 is preferable, for example NACA0018, as a high efficiency Darrieus blade. The dynamic stall effect also appears in the Darrieus blade characteristics due to the circular motion. The blade chord length, ℓ, of about 0.3 times the radius of runner pitch circle, R, would be recommended since the no-stall region in the Darrieus blade characteristics becomes wider with the blade chord length longer due to the dynamic stall effect. A blade setting attitude on the runner pitch circle affects the runner performance sensitively. The blade should be set tangent to the pitch circle at its midchord point, considering that an instantaneous attack angle to a Darrieus blade, α^*, should be evaluated at the midchord point of the blade and the Darrieus runner yields a high efficiency when α^* varies within a no-stall region widely in a revolution of the runner.

The runner efficiency and the tip speed ratio, $U/\overline{V}$, at the maximum power generation decrease as an increase of number of blades, z, where U is the peripheral velocity of the runner pitch circle and $\overline{V}$ the average oncoming velocity to the runner. On the other hand, a torque ripple, defined as the ratio of the peak to peak amplitude to the average torque, also decreases with number of blades so that the impact to the gear train of the increasing speed system can be weakened. The compromise between both results yields the use of triple-bladed runner. The Darrieus blades should be supported parallel to the rotating shaft by streamlined arms, with taking notice of reducing an additional loss, which decreases the

2825

shaft power from the runner. The opened side-edges of the blade also deteriorate the output power. However this effect can almost disappear by attaching the end plates, the lateral area of which takes about seven times the blade cross-sectional one, to the blade edges in the case of the runner blades with low aspect ratio, B/ℓ, where B is the span length of the blade. To attain higher power generation for ducted Darrieus-type water turbines, the casing gaps between the runner pitch circle and parallel duct walls should be kept minimum as $2R/S \simeq 1.0$ when S denotes the duct width because a leakage flow through the gaps contributes no power generation.

USABLE DATA OF RUNNER PERFORMANCE

Figure 2 (a) shows performance of the Darrieus-type runner with recommended geometry. These data were measured by experiments and can also be estimated (Furukawa et al., 1992). The power and head coefficients are defined as $C_p = P/(\rho \overline{V}^3 B R)$ and $C_h = 2gH/\overline{V}^2$, respectively, where P is the output power generated by the runner, H the supplying head through the runner, ρ the density of water and g the gravitational acceleration. The runner efficiency is evaluated from $\eta = (2R/S) C_p/C_h$. The maximum power coefficient $C_p = 2.83$ is obtained at $U/\overline{V} = 5.4$ and the maximum efficiency $\eta = 0.62$ at $U/\overline{V} = 3.3$. Then this triple-bladed runner gave measured torque ripple of less than 0.58 over the range of $3.0 < U/\overline{V} < 6.1$. Figure 2 (b) shows, as examples, generated output power, P, in cases of $\overline{V} = 1.25$ and 1.75 m/s and shaft power loss, P_ℓ, of the runner with $B = 1.0$ m and $R = 0.185$ m against the runner speed, U. Since this runner generates positive power over all range to $U/\overline{V} = 7.0$, it is possible to self-start from rest to an operating point under the condition of $(P - P_\ell) > 0$.

In the operation in higher oncoming flow velocity, cavitation occurs on the Darrieus blade and deteriorates the performance. The cavitation criteria $[\,k\,]$ of this runner against $U/\overline{V}$ are also shown in Fig.2 (a), where $[\,k\,]$ is the dimensionless form of the instantaneous maximum head drop on the blade in one revolution of the runner (Takamatsu et al., 1992). As

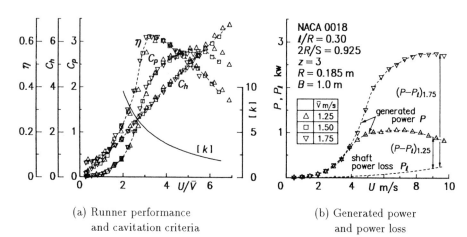

(a) Runner performance
and cavitation criteria

(b) Generated power
and power loss

Fig.2 Performance, power loss and cavitation criteria for designing runner

a magnitude of $\overline{V}$ becomes higher with total head, H_t, upper limit of H_t due to cavitation is estimated by the next approximate formula:

$$H_t < C_h \, (p_a - p_v) \, / \, (\rho g) \, / \, \{ \, 1.0 - C_h + [\,k\,] \, (U/\overline{V})^2 \, \} \, , \qquad (1)$$

where p_a is the atmospheric pressure and p_v the vapor pressure. This formula gives, for instance, $H_t < 0.74$ m for $U/\overline{V} = 5.4$ suggesting the restriction on usable range of the Darrieus-type water turbine.

OPERATING CHARACTERISTICS AND OPERATING RANGE

An operating point of the water turbine is determined as an intersection of curves of shaft power from the runner, $(P - P_t)$, and the load, P_G. Hence both curves must be made clear in the design so as to extract power as effectively as possible from extra-low head water power. Figure 3 (a) shows curves of $(P - P_t)$ of the optimum designed runner, which has a performance shown in Fig.2, for various oncoming flow velocities against the runner speed. Two kinds of conceivable load curves are also illustrated in Fig.3 (a). One is the almost constant loads as driving an electric generator with automatic voltage regulator (AVR) and the other is the loads proportional to U^3 as driving a pump directly. In the former case, a stable operating point is in higher runner speed range than that of the maximum power point of the runner, U_m. Therefore some devices, for example as a sluice gate to regulate $\overline{V}$ or a clutch to decrease the load as shown in Fig.1, may be installed to increase the runner speed higher than U_m and to move an operating point from rest to the stable one, according to the runner speed at which the AVR works. Besides, the runner efficiency at the stable point is significantly lower than the maximum one as can be found from Fig.2. In order to operate stably at a point near the maximum efficiency one, an auxiliary device is necessary to decrease the shaft power in the higher range of the runner speed than that of the maximum efficiency point as an additional shaft power loss, ΔP_t, for instance as shown in Fig.3 (a). In the latter case, it is able to operate stably even at the maximum

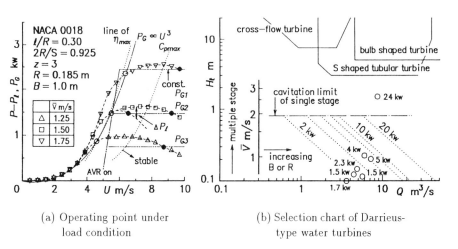

(a) Operating point under load condition

(b) Selection chart of Darrieus-type water turbines

Fig.3 Operating characteristics of Darrieus-type water turbine

efficiency point so that the specification of a power extractor or a pump is selected easy by matching both the shaft power and the load at that point.

Figure 3 (b) shows a selection chart for the Darrieus-type water turbines with recommended geometry. Dotted lines represent the shaft power available. These data were obtained with some proper assumptions on hydraulic loss, H_t, and power loss, $(P_t + \Delta P_t)$, by using values of $C_p = 2.54$, $C_h = 3.89$, $\eta = 0.60$ and $[k] = 3.45$ at the operable point of $U/\overline{V} = 4.0$ just higher than the maximum efficiency tip speed as shown in Fig.2. Published test data for ducted and unducted ones (Faure, 1985; ITDG, 1980; Kihoh et al., 1990) are also plotted as a symbol of white circle in Fig.3 (b) under an assumption of $H_t = \overline{V}^2/2g$ for unducted ones. A Darrieus-type turbine should be selected essentially within extra-low head power region compared with conventional types, because of upper limit of total head due to cavitation. It would be emphasized that this type has an adaptability to wide flow rate range by increasing the straight blade-span, B, or the radius of the runner pitch circle, R, independently without changing total head.

CONCLUDING REMARKS

When an available head and a flow rate are given, a ducted Darrieus-type water turbine is selected from a selection chart provided for utilizing extra-low head water power. Then a high efficiency runner can be designed in accordance with proposed guiding principles and usable data, and the configurations of duct-components are also decided. An optimum method to operate this type of water turbine should be considered with a loading system as described. Thereafter the economic problem will have to be discussed.

This research was financially supported by the Grant-in-Aid from the Ministry of Education, Science and Culture, Japan (No.02555040).

REFERENCES

Faure, T. D. (1985). Experimental Results of a Darrieus Type Vertical Axis Rotor in a Water Current. NRC Canada., Div. Mech. Eng., TR-HY-005.
Furukawa, A., K. Takenouchi, K. Okuma and Y. Takamatsu (1992). An Approximate Method for Estimating the Blade Performance of Darrieus-Type Cross-Flow Water Turbines. Mem. Fac. Eng., Kyushu Univ., 52-2, 82-94.
Intermediate Technology Development Group (1980). River or Canal Current Water Pumping Turbine. Project Bulletin.
Kihoh, S., K. Suzuki and M. Shiono (1990). Power Generations from Tidal Currents. Proc. 4th Pacific Cong. Marine Sci. Tech., 1, 479-485.
Takamatsu, Y., A. Furukawa, K. Okuma and K. Takenouchi (1991). Experimental Studies on a Preferable Blade Profile for High Efficiency and the Blade Characteristics of Darrieus-Type Cross-Flow Water Turbines. JSME Int. Journ., Ser.II, 34-2, 149-156.
Takamatsu, Y., A. Furukawa, K. Okuma, K. Takenouchi, T. Sasaki and G. H. Zhao (1992). Studies on Cavitation Occurring on the Runner Blade of a Darrieus-Type Water Turbine. JSME Int. Journ., Ser.II, 35-1, 46-52.

APPILICATION OF THE NEW COST ANALYSIS
TO WAVE POWER EXTRACTION AT BREAKWATERS

H. KONDO*, I. SUGIOKA*, T. WATABE**, S. OSANAI*** and S. OZAWA****

* Center for Cooperative R & D, Muroran Institute of Tech.
  27-1 Mizumoto-Cho, Muroran-Shi, Hokkaido, 050 Japan
** Department of Mechanical Systems Engineering, ditto
*** Port & Harbor Division, Hokkaido Development Bureau
  N-18, W-2-1-1, Kita-Ku, Sapporo-Shi, Hokkaido, 060 Japan
**** Cold Region Port & Harbor Engineering Research Center
  Kita Bl. N-7, W-2, Kita-Ku, Sapporo-Shi, Hokkaido, 060 Japan

ABSTRACT
Cost of wave power extracted with systems installed in breakwater has been
analysed with the approach which includes environmental cost. A case study
shows the power extraction at breakwaters is now promising one from the
viewpont of cost effectiveness.

KEY WORDS
Breakwater; cost analysis; Pendulor system; wave power.

Background
Now that we are clearly aware of several facts that excess consumption of
fossil fuel in the 20th century has damaged environment, conventional
approaches for the cost of natural energy resource is hardly justified. A
new approach which can evaluate the total cost of energy including
environmental and/or social costs must be authorized as soon as possible,
and should be applied to practical cases.
Fig. 1 illustrates environmental changes due to extracting wave energy and

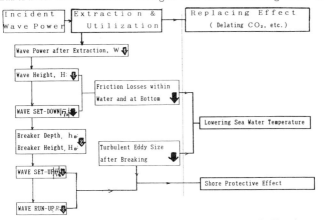

Fig. 1 Effect of Power Extraction to Coastal Environment

2829

the resulting major effects nearshore: when the energy is extracted at water depth h in sea waves decrease their height H. That brings subsequently less breaker depth hb and height Hb, and then smaller wave setdown $\overline{\eta}$d, set-up $|\overline{\eta}u|$ (KONDO & WATABE, 1991a). As we can see in the figure, most of effects of power extraction are favorable for environment in addition to no emission of gases such as SOx, NOx and CO_2.

NEW COST ANALYSIS
An analytical approach was proposed by two of us which could take into account the environmental effects on cost of natural energy utilization (KONDO & WATABE, 1991b). A simplified explanation of the analysis is as in the following.
The true cost of a energy Pt consists of the apparent or market cost Pm, the environmental cost and the social cost Ps as,

$$Pt = Pm + Pe + Ps \qquad (1)$$

Cost of Pm is dominated with many parameters and they are different for each kind of energies. For the case of wave energy it may be approximated as,

$$Pm = f(h, H, tw, v, \ldots Sp, \ldots\ldots) \qquad (2)$$

where tw is water temperature, v is flow velocity and Sp denotes the kind of extracting system employed.
Pe represents the increased cost to get rid of environmentally bad effects by consuming the energy such as disposal cost of harmful gases from fossil fuel power plants, and in a generalized expression it may be expressed as,

$$Pe = \Sigma \, \Delta Px,i = \Sigma \, \frac{\partial Pe}{\partial x,i} \, \Delta x,i \qquad (3)$$

Most of renewable energies gives no harmful effect and so the cost, and even better effects to improve environment which brings the minus Pe.
Ps is the costs to be payed socially and politically to get energy such as the emergent war expense payed by the allied countries during the Persian Gulf War in 1990-91.

UTILIZATION SCHEME OF WAVE ENERGY
One of major problems in wave energy utilization is inconvenience in energy transmission from a power plant to users, which considerably restricts common power uses. Possible usage of the wave power extracted at ports located in higher latitude seas is shown in Table 1.

CASE STUDY
A case study to which the present cost analysis is applied to proposed wave power plant of 1500 kW (rating) at a small island in Japan Sea (Kondo, et. al, 1989). Pendulor system which has proved to have an excellent efficiency (Kondo, et. al, 1985; Yano, et. al, 1985; Watabe, et. al, 1989) is employed in the study.

2830

TABLE 1   POSSIBLE UTILIZATION SCHEME OF WAVE ENERGY EXTRACTED AT PORT

FORM OF ENERGY	SCALE OF POWER PLANT		
	Small (<1000kW)	Medium (1000-10000kW)	Large (>10000kW)
Electricity	○ Lighthouse & Bouy ○ Lightening port area ○ Cathodic Protection ○ Power supply for ship	○ Power for remote island ○ Lightening shoreside road ○ Power for sewage station ○ Artificial reef growing	○ Hydrogen Production ○ Power for artificial and remote islands ○ Commercial power supply
Thermal	○ Ice making & Refregr. ○ Hot water & heating supply for houses ○ Heating for fishfarm	○ Power for Heat-Pump ○ Hot water supply ○ Snow melting power for quay and road	○ Snow melting for nearshore road and airport ○ Hot water and heating power the houses
Mechanical	○ Aeration for fishfarm ○ Simple sewage work ○ Pumping for lowland drainage	○ Pumping seawater ○ Sand bypassing power ○ Antifreezing of water area at cold region port	○ Desaltization of sea water ○ Power supply for offshore fish farm and factory ○ Power for seawater pumping

Design Condition
Extracting Water Depth: 10 m below LWL
Design Tidal Range: 0.3 m
Mean Incident Wave Power: 10 kW/m
Maximum Design Wave: Hs = 8 m, Ts = 10 sec.
Extractor: Pendulor system of overall efficiecy 0.5
Breakwater Caisson: Unit length of 20 m
Rating output per unit: 10×20×0.5 = 100 kW for a casson
                            with three pendulum plates

TABLE 2   MARKET COST ESTIMATION OF ELECTRICITY

A: Cost of Caisson = 8 (M¥ /m) × 300 = 2,400 M¥	
B: Cost of Generating System = 5 M¥ × 15 = 750 M¥	
C: Total Construction Cost = 3,150 M¥	
D: Specific Cost = C / Output = 3,150/1500 = 2.1 M¥/kW	
E: Fixed Cost =Total Cost×Equivalent Coefficient	
= 3,150 × 0.07581 = 238.8 M¥/Year	
= 83.7 '' (Without Caisson)	
F: Expenditure = E + F = 276.3 M¥/Year	
= 121.2 '' (Without Caisson)	
G: Amount of Generated Electricity	
= Output ×24hr ×365d ×Working Rate = 1500 ×24 ×365 ×0.7	
= 9.2 ×10^{6}kWh/Year	
H: Cost of Electricity = F / G = 30.0 ¥/kWh	
= 13.2 ''(Without Caisson)	

Applying the result above obtained Pm, we can predict the
true cost Pt as shown in Table 3.
As for oil in the table Pgas is estimated as the cost of CO_2 disposal
(Fulherson et. al, 1990) and Ps is estimated from the Japanese expense for
Persian Galf War.   Meanwhile Ph and PH for wave is estimated as the
consfruction cost of caisson.

TABLE 3   ESTIMATION OF TRUE COST

Primary Energy	Pm(¥/kWh)	Ph	PH	Pgas	Pe	Ps	Pt
Oil	12			12	12	0.1	24.1
Wave	30		-16.8		-16.8		13.2

The result shows that wave energy utilization at breakwaters is undoubtly
one of the major promising renewable energies to the island.

          COCLUSION
With taking account environmental and social costs, renewable energies are
becoming to be prospective since fossil energies have to pay expensive
environmental cost.
Wave energy utilization which brings good effects environmentally and
cheaper true cost compared with oil fired electricity must be utilized in
coastal countries for overcoming the woldwide energy and environmental
crises.

REFERENCES

Fulkerson, W, R. R. Judkins and M. K. Sanghvi(1990). Energy from Fossil Fuels, Scientific American, Sept/90, 83-89.

Kondo, H. and T.Watabe(1991a). Environmental Protection Effect due to Wave Power Extraction, Proc. of 3rd Conf. on Wave Energy Utilization, JAMSTEC, 83-89 (in Japanese).

Kondo, H. and T.Watabe(1991b). Economical Evaluation of Ocean Energy by Considering Enoiromental Effect, Abstract. of Annual Conv., 2, JSCE, 832-833(in Japanese).

Kondo, H., T.Watabe and K.Yano(1985). Wave power extraction at coastal structure by means of moving body in the chamber, Proc. of 19th International Conf. on Coastal Engineering, ASCE, III, 2875-2891.

Kondo, H., T.Watabe and A.Kawamori(1989). Wave power utilization toward the is land community free from oil, Proceedings of IFAC Symposium on Energy Systems, Management and Econimics, Tokyo, 67-71.

Watabe, T. and H.Kondo(1989). Hydraulic technology and utilization of Ocean Wave power, Proc. of JHPS. International Symposium on Fluid power Tokyo, 301-308.

Yano, K., H.Kondo and T.Watabe(1985). A device for wave power extraction in coastal sturctures, Coastal Engineering in Japan, 28, 243-254.

A STUDY OF CROSS-FLOW TURBINE - EFFECTS OF TURBINE
DESIGN PARAMETERS ON ITS PERFORMANCE

H. OLGUN and A. ULKU

Department of Mechanical Engineering,
University of Karadeniz Technical, Trabzon, 61080, TURKEY

ABSTRACT

An experimental rig of cross-flow turbine was designed and
constructed   to investigate its performance.   Tests   were
performed by varying the number of blades, the ratio of inner
to outer diameters and the gate openings of the turbine nozzle
under different heads. It was found that the changes of the
blade number and the inner to outer diameters ratio of the
cross-flow turbine are related with the reduced   turbine
characteristics. The results presented in this paper show that
the maximum efficiency of the cross-flow turbine occurs at 28
blades and the ratio of inner to outer diameter of 0.67.

KEYWORDS

Cross-flow turbine; Banki turbine; Ossberger; runner; nozzle.

INTRODUCTION

The cross-flow turbine is a good alternative for electric power
essentially independent from a grid system. Many researchers
have investigated characteristics of this turbine. Cross-flow
turbine consists mainly of two parts, a nozzle and a turbine
runner. The runner is built up of two parallel circular discs
joined together at the rim with a series of curved blades. The
nozzle with a rectangular cross-sectional area, discharges the
jet over the full width of the runner and enters the runner at
an angle of 16 degrees. The blades can be cut from standard
pipes or can be manufactured from steel plates by a simple
process of bending. This experimental study covers the effects
of two different types of nozzles with different runners on the
efficiency.

2834

EXPERIMENTAL RIG

The experimental rig consists of a water recirculation system, the cross-flow turbine and a loading and measurement system as shown in Fig. 1. The water recirculation system is made of two centrifugal pumps and the delivery piping which provides the necessary head and the flow rate to the cross-flow turbine. The rig is organized in such a way that the pumps can be connected in series, parallel or in by-pass mode. The adjustment of head and flow rate is accomplished by the use of two valves. The nozzle and runner are attached to a large tank. The flow leaving the runner is fed back again into the tank.

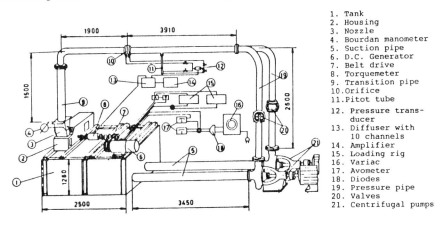

1. Tank
2. Housing
3. Nozzle
4. Bourdan manometer
5. Suction pipe
6. D.C. Generator
7. Belt drive
8. Torquemeter
9. Transition pipe
10. Orifice
11. Pitot tube
12. Pressure trans-
    ducer
13. Diffuser with
    10 channels
14. Amplifier
15. Loading rig
16. Variac
17. Avometer
18. Diodes
19. Pressure pipe
20. Valves
21. Centrifugal pumps

Fig. 1. Experimental rig

Both of the nozzles have a rectangular cross-sectional channel and are built up of iron sheet. Nozzle outlet angles of both solid walls measured from the circumferantial direction are designed 16 degrees as shown in Fig. 2. Nozzle "A" has two doubly evolvent walls. One of these wall is stationary with an angle of 16 degrees and the other one which serves as the guide vane is adjustable. This wall can be adjusted to give inlet jet thicknesses between 0 and S_o as shown in Fig. 2. The width ratio of the nozzle throat ($T_r = 2S_o/\epsilon D_1$) is 0.249. Nozzle nearly has the same length as the blades of the runner. Nozzle "B", also has doubly evolvent walls at the exit as shown in Fig. 2. A guide vane is located inside the flow region where the trailing edge of the guide vane is arranged at 16 degrees from the pressure surface. Its width ratio, T_r is 0.544.

For these experiments 13 different runners were manufactured. All runners have an outer diameter (D_1) of 170 mm. and a runner width of 114mm. The inner to outer diameter ratios of runners, D_2/D_1 are 0.75, 0.67, 0.58 and 0.54. The number of blades of each series of turbine runners are 20,24,28 and 32. Geometrical parameters of the runners are given in Fig. 3. The runner blades were produced as cylindrical and the blade edges were smoothed and rounded as shown in Fig. 3. The side discs of the runners were cut out of 10mm. steel plates. The blades were cut

2835

out of standard pipes with 43, 53.5, 65 and 69 mm. in diameter
and 3mm in thickness for each series. The blades were connected
between the discs by means of soldering. All runners were
balanced before assembling them into the turbine. The runners
have no inside running shaft. The discs with the shafts were
bolted together with the runners. The shafts were mounted
between two ball bearings. The turbine was coupled to a loading
d.c. generator via a belt drive. The generator loading system
was made up 18 resistance bars of 500 Watt each connected in
parallel. The torque generated on the turbine shaft was
measured by a torquemeter. A digital tachometer was used to
measure the rotational speed of the runner. The flow rate was
measured by means of a calibrated orifice located between the
outlet of the centrifugal pumps and the nozzle as shown in
Fig. 1. A diaphram type pressure transducer was used to measure
the pressure difference between the orifice pressure tappings.
The static pressure at the entrance of the nozzle was measured
by means of a Bourdan type manometer. Each measuring device was
thoroughly calibrated.

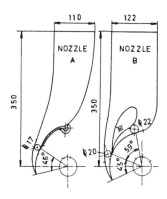

Fig. 2 Nozzles

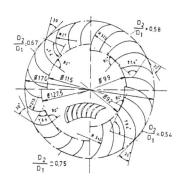

Fig. 3. Runners

EXPERIMENTAL RESULTS AND DISCUSSION

To investigate the effects of the head regions on the turbine
efficiency and the reduced parameters, the experiments are
performed at 10 different heads from 8m. to 30m. for full gate
openings of nozzle "A" and all of the runner combinations. The
obtained results are summarized in Fig 4. From this Fig, it can
clearly be seen that the difference in efficiency between the
8m. and 30m. heads are within 4 % for all runner combinations.
The runners having about 28 blades gives the maximum turbine
efficiency which was about 73 %. However the difference in
efficiency between 20 and 32 blades is within 3 % . These
results are in agreement with the experimental results of
Fukutomi et all.,1985 and Steller and Reymann, 1987. They found
that 30 blades give maximum efficiency. It can also be observed
that the reduced rotational speed and specific speed curves
show almost the same characteristic as the efficiency. But the
reduced flow rate and reduced power were found to be
independent of blade numbers.

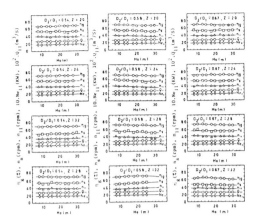

Fig. 4. The effects of the head on the turbine prformance

To investigate the effects of the partial load on the turbine efficiency, the experiments are made at 6 different gate openings of nozzle "A" under 25m. head for all runners. The results are given as Ne-n and η_g-n variations as shown in Fig.5

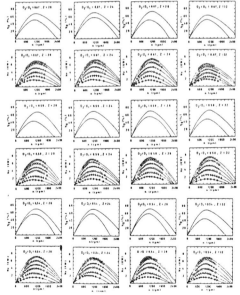

Fig. 5. The effects of the partial load on the turbine performance.

At the η_g - n variations, efficiency curves obtained only for maximum and minumum gate openings are given. To determine the operation of the partial load of the turbine, the efficient factor is defined as follows:

$$\text{The eff. factor} = \frac{\text{effi. at min. gate op.}/\text{effi. at max. gate op.}}{\text{power at min. gate op.}/\text{power at max. gate op.}}$$

The efficient factors of the runners which characterize the maximum efficiency for the partial loading of the turbine are given in Table 1. From this table, it has been dearly seen that the maximum efficient factors are obtained at 28 number of blades for all diameter ratios of runners.

The characteristic curves of the nozzle "B" using the runner with D_2/D_1 of 0.67 and $Z = 28$ under 7 different heads from 4m to 17m is given in Fig 6. The characteristic curves are changed depend on the changing of the head. The maximum efficiency are obtained at 17m heads as 71% . When the head decreases the efficiency also decreases.

Table 1. The variation of the efficient factor

D_2/D_1	Number of blades			
	20	24	28	32
0.75	----	---	3.19	----
0.67	3.10	3.80	3.40	2.40
0.58	2.10	2.60	3.30	2.40
0.54	3.80	4.30	4.40	2.20

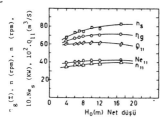

Fig.6. Characteristic curves with nozzle B

NOMENCLATURE

H_o = head, n = rotational speed of runner, $\in$ = nozzle entry arc n_{11} = reduced speed with H_o = 1m, D = 1m, n_s = specific speed Ne = power on the shaft, Ne_{11} = reduced power with H_o = 1m, D=1m S_o = nozzle throat width, η_g = efficiency Q = flow rate, Q_{11} = reduced flow rate with H_o = 1m , D = 1m

REFERENCES

Fukutomi, J., Y. Nakase and S. Hasui (1985). A Study of a Cross Flow Turbine(Effects of the Blade Number and Blade Exit Angle on the Turbine Performance. Bulletin of JSME, 322B, 407-412.
Khosrowpanah, S., A. A. Fiuzat and M. L. Albertson (1988). Experimental Study of Cross - Flow Turbine. Journal of Hydraulic Engineering, 114(3), 299-314.
Macmore, C.A. and F. Merryfield (1949). The Banki Water Turbine Technical Report 25 Engineering Experiment Station, Oregon State College
Olgun, H. (1990). Banki(Cross-Flow) turbini Tasarim Parametrelerinin Incelenmesi Ph.D. Thesis, K.T.U., Department of Mechanical Engineering, Trabzon, Turkey
Steller, K. and Z. Reymann (1987). Some Test Results On Banki Turbines, 8th Conference on Fluid Machinery, Budapest
Toyokura, T. and T. Kanemeto (1984). Studies on Cross - Flow Turbines, Research on Natural Energy, 8, 205-21.

RUNNING CENTRIFUGAL PUMPS AS MICRO-HYDRO TURBINES: PERFORMANCE PREDICTION USING THE AREA RATIO METHOD

J.D. BURTON and A.G. MULUGETA

University of Reading, Department of Engineering, Whiteknights,
P.O. Box 225, Reading, RG6 2AY, U.K.

ABSTRACT

One of the problems of adapting centrifugal pumps to operate as water turbines has been the lack of reliable information about how to predict hydraulic performance in the turbine mode. The present paper examines the area ratio method, used successfully over many years (Anderson 1938, 1955; Worster 1963 and Thorne 1979) for pump operation, and the recently proposed extension for turbine operation. Burton and Williams 1991. It is concluded that the most reliable approach is to select the pump to run with shock free entry in the turbine mode, rather than at peak hydraulic efficiency. The area ratio method would seem to provide a reliable basis for making these estimates.

KEY WORDS

Micro-hydro power; pump as a turbine; shock free entry; no whirl at exit.

NOMENCLATURE

Subscripts

1	refers to impeller eye
2	refers to impeller tip
p	refers to pump operation
t	refers to turbine operation

B^2	area of the volute throat
D	impeller diameter
H	head
K_H	ratio of dimensionless head coefficients
K_Q	ratio of dimensionless flow coefficients
Q	volumetric flow
T	torque
U	impeller tangential velocity

V_m absolte meridional velocity
V_r relative flow velocity
V_u absolute tangential flow velocity
Y area ratio   i.e. $\dfrac{\text{area between impeller blades at exit}}{\text{throat area of volute}}$
b width of impeller flow channel in axial direction

d distance betwen consecutive vanes
g gravitational constant
h_o impeller slip factor - pump operation
y area ratio : $\dfrac{\text{circumferential area between impeller blades tip}}{\text{circumferential area between impeller blades eye}} = \dfrac{V_{m1}}{V_{m2}}$
n shaft rotational speeds c.p.s.
r leakage flow through wearing rings as a fraction of Q
z number of impeller blades
α_v angle of log spiral volute
β impeller blade angle
$\in$ factor of utilisation
Ω shaft angular velocity
v diameter ratio $D_1/D_2 = \dfrac{U_1}{U_2}$
η efficiency
η_v hydraulic efficiency of volute

INTRODUCTION

A standard centrifugal pump can be operated in reverse as a turbine or P.A.T. (pump as turbine), enabling a mass-produced piece of equipment to be used instead of a custom-built turbine. Figure 1 shows a typical system curve for a micro-hydro site. The gross site head is reduced by losses in the penstock which will be more or less proportional to the square of the flow rate. A radial centrifugal P.A.T. will have a performance characteristic as shown, where the constant speed H-Q curve lies between locked rotor (n = 0) and runaway (T = 0) curves; these latter curves also follow an HαQ^2 form.

If the hydraulic machine's speed is fixed by a load controller, then operation will settle down at point P where the P.A.T.'s constant speed characteristic cuts the penstock/site characteristic. It should be observed that for flow-rates below best efficiency point, the efficiency falls rapidly, whereas, at over capacity the efficiency declines only gradually. Burton and Williams 1991 recommend that the pump should be selected so that its characteristic cuts the penstock/site characteristic at a slightly higher capacity than the predicted best efficiency point for the P.A.T.. In this way any discrepancy between the actual running conditions and the predicted head and flow, does not result in a large loss of efficiency.

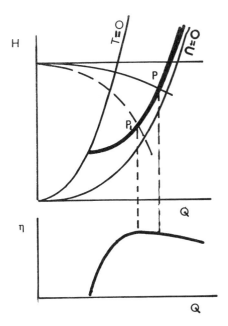

Fig.1 P.A.T. coupled to a typical micro-hydro system

What factors determine the best efficiency point of P.A.T.? A closer look needs to be taken at the internal flow mechanics.

SIMPLIFIED FLOW MECHANICS OF THE CENTRIFUGAL P.A.T.

Figure 2 shows the simpified/idealised velocity diagrams for radial, two dimensional, flow through a centrifugal pump rotor running as a turbine. The absolute velocity V_2 at entry to the rotor will have its <u>direction</u> largely determined by the <u>shape</u>, or spiral angle α_v, of the pump volute. At a certain flow rate, and at a corresponding meridional velocity component V_{m2}, the relative velocity V_{r2} at entry will just match the inlet blade angle β_2. This condition will be called "shock-free" entry, although it is recognised that this may not be exactly the angle of ideal incidence.

The flow at entry to the volute will be irrotational and will largely remain irrotational as it passes through the volute; here assumed to be two dimensional and of log spiral form. This means that once within a blade pocket, the water must possess a strong relative eddy or whirl of equal and opposite rotation to Ω the angular velocity of the rotor. In pump operation the relative eddy is able to cause substantial deviation of the flow from the blade shape at exit owing to the fact that the blades are relatively far apart around the periphery of the rotor. This in turn leads to a loss of momentum and therefore "slip" as the flow leaves the rotor. In contrast, during turbine operation, the flow leaves at the "eye" of the impeller, where the blades are much closer together and offer good guidance to the departing flow. To a good approximation then the relative velocity V_{r1} at exit from the blades, under turbine operation, will follow the blade angle β_1. At some particular flow rate, and corresponding meridional velocity component V_{m1}, the absolute velocity of the flow leaving the rotor will be entirely radial, $V_{u1} = 0$, and the operation will be "without whirl" at exit. This means that the minimum amount of kinetic energy is being released from the turbine with the discharge flow.

2841

In a custom built turbine the design can be arranged so that the two conditions of "shock free entry" and "no whirl at exit" occur at approximately one flow; the design condition. In a P.A.T. the two conditions normally happen at different points on the characteristic and the machines' peak turbine performance will be at some intermediate point between these two flow conditions.

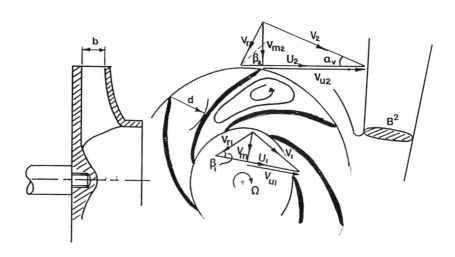

Fig. 2   Velocity diagrams for a P.A.T. operating under turbine conditions: The relative eddy in the blade pocket is also shown.

Figure 3, taken from the work by Mulugeta 1992, clearly shows that the peak turbine efficiency resides part way between the estimated point for "shock-free entry" and the observed "no whirl at exit" conditions, for a P.A.T. tested at Reading University.

Clearly for a relatively low powered utility machine it would make good sense to design the installation to run the P.A.T. close to the "shock-free entry" condition, (point P in Fig.1). This leaves the option in a dry period of partly closing a valve in the penstock thus moving operation to P_1, which although at reduced power would never the less leave the P.A.T. running at a reasonable efficiency - probably close to the " no whirl at exit" point.

With complete information about the pump geometry it would be possible to make reasonable estimates, for unit flow $\dfrac{Q}{\sqrt{H}}$ and unit speed $\dfrac{n}{\sqrt{H}}$ at both "shock free entry" and "no whirl at exit" conditions. Another approach might be to use the widely known area-ratio method for pump performance prediction and extend it to make sensible estimates for the two points, "no whirl at exit", and "shock free entry" under turbine conditions.

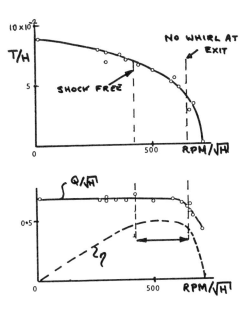

Fig. 3 Unit torque T/H (N); Unit flow Q/H (Litres s^{-1}. m^{-1}); and efficiency η characteristics versus unit speed $\dfrac{n}{\sqrt{H}}$ (RPM,$m^{-1/2}$) for a centrifugal pump run as a turbine. Mulugeta 1992.

THE AREA RATIO METHOD

During pump operation the energy per unit mass added to the flow by the impeller U_2V_{u2} slightly exceeds the energy gH_p delivered at the discharge branch. This is owing to irreversibilities in the volute represented by an efficiency term η_{vp}. The rotor characteristic then may be written as:

$$\frac{2gH_p}{\eta_{vp}.U_2{}^2} = 2h_o - 2\left(\frac{Q_p}{U_2B^2}\right)\frac{\cos\beta_2}{Y} \tag{1}$$

where h_o is the usual slip factor, Worster 1963, to take account of the effect of the relative eddy and Y is the area ratio (Fig. 2) vis;-

$$Y = \frac{\text{area between the impeller blades at exit}}{\text{throat area of volute}}$$

$$Y = \frac{b_2d_2z}{B^2} \tag{2}$$

Anderson (1955) and Worster (1963) make the approximate assumption that

$$Y = \frac{0.95\pi D_2\, b_2\sin\beta_2}{B^2} \tag{3}$$

where the constant 0.95 is an arbitrary allowance for blockage due to impeller blade thickness.

2843

The rotor characteristic (1) will only be valid close to the best efficiency point of pump operation, when the volute shape "fits" the flow pattern leaving the impeller. Recognising this fact Worster (1963) develops a volute characteristic:

$$\frac{2gH_p}{\eta_{vp} U_2^2} = \frac{2(2^B/_D)}{\log_e (1+2^B/_D)} \left(\frac{Q_p}{U_2 B^2}\right) \tag{4}$$

Simultaneous solution of (1) and (4) for different rotor and volute geometries yield best efficiency values of $\left(\dfrac{2gH_p}{\eta_{vp}U_2^2}\right)$ and $\left(\dfrac{Q_p}{U_2 B^2}\right)$. They are found to be strongly dependent upon factor Y and only slightly influenced by B/b; and β_2. These solutions have been reworked here and shown for pump operation in the two uppermost diagrams of Fig. 4. Values of course correspond to those obtained by Worster (1963) and are in agreement with a large amount of test data worked through by Anderson (1955).

Burton and Williams (1991) extend the area ratio method to turbine operation. Now the energy per unit mass gH_t available at entry to the volute will be reduced by irreversibilities in the volute. Only $\eta_{vt}gH_t$ will be available at entry to the rotor. At exit from the rotor there will be a kinetic energy per unit mass $V_1^2/2$ which will not have been used. Usually this is taken care of by incorporating a factor of utilisation $\in$ where:

$$\in = \frac{\text{Energy per unit mass utilised by rotor}}{\text{Energy per unit mass available to rotor}}$$

Thus $\in \eta_{vt}gH_t = U_2 V_{u2} - U_1 V_{u1}$ \hfill (5)

Substituting for the geometrical terms (see Burton and Williams 1991) and assuming log spiral blades ($\beta_1 = \beta_2$) for the impeller, yields:-

$$\frac{2gH_t\eta_{vt}\in}{U_2^2} = 2\left(\frac{Q_t}{U_2 B^2}\right) \left\{\frac{2^B/_D}{\log_e (1+2^B/_D)} + vy\frac{\cos\beta_2}{Y}\right\} - 2v^2 \tag{6}$$

This, the turbine rotor characteristic, will be valid over a wider range of flow conditions than the corresponding pump equation (1) since the pressure distribution around the tips of the blades for a P.A.T. is reasonably constant. Mulugeta (1992).

The volute characteristic (4) with the pump variables $\dfrac{2gH_p}{\eta_{vp}U_2^2}$; $\dfrac{Q_p}{U_2 B^2}$ replaced by the corresponding turbine variables will enable (4) and (6) to be solved simultaneously for $\dfrac{2gH_t\eta_{vt}\in}{U_2^2}$ and $\dfrac{Q_t}{U_2 B^2}$ over a range of Y values. These solutions, representing the "no whirl at exit" condition have been plotted in the lower diagrams of Fig. 4 with a broken line. Immediately it can be seen that the "no whirl at exit" condition is strongly influenced by not only Y but also B/b; y, the area ratio between the periphery and the eye of the impeller; and v the ratio of eye diameter to the impeller tip diameter. In contrast the "shock free entry" condition with:

$$\left(\frac{Q_t}{U_2 B^2}\right) = \left(\frac{v}{y}\right)\frac{Y}{\cos \beta_2} \tag{7}$$

and its corresponding value of $\dfrac{2gH_t\eta_{vt}\in}{U_2^2}$

,calculated from (6), is found to be remarkably independent of all factors except Y. The values have been plotted with a solid line in the lower diagrams of Fig. 4.

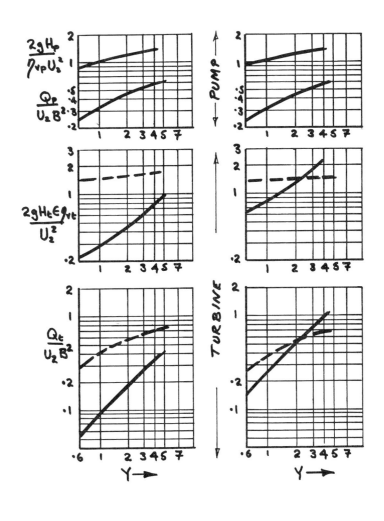

$$\frac{B}{b} = 2.0$$

$$v = 0.25$$

$$\beta = 25°$$

$$y = 3$$

$$\frac{B}{b} = 1.5$$

$$v = 0.45$$

$$\beta = 25°$$

$$y = 2$$

Fig 4 Values of dimensionless head and flow against area ratio Y for pump and turbine operation. Full line "no whirl at exit". Broken line "shock free entry".

P.A.T. PERFORMANCE ESTIMATION: THE RECOMMENDED PROCEDURE.

Sensitivity analysis of the type undertaken in Fig. 4 would suggest that the values of $\dfrac{2gH_t\eta_{vt}\epsilon}{U_2^2}$ and $\dfrac{Q_t}{U_2B_2}$ associated with "shock free entry" are mainly influenced by the area ratio Y; indeed the head parameter only varies by some 7% over a wide range of $4 \geq Y \geq 0.6$. This would suggest that at <u>least one procedure for estimating P.A.T. performance</u> would be to measure the volute throat area B^2, then d, b and z for the impeller (see Fig. 2) so that the area ratio Y could be established. Entering Fig. 4 at the appropriate value of Y it immediately becomes possible to read off the ratios:-

$$\frac{\left(\dfrac{2gH_t\eta_{vt}\epsilon}{U_2^2}\right)}{\left(\dfrac{2gH_p}{\eta_{vp}U_2^2}\right)} = K_H \tag{8}$$

$$\frac{\left(\dfrac{Q_t}{U_2B^2}\right)}{\left(\dfrac{Q_p}{U_2B^2}\right)} = K_Q \tag{9}$$

It is recommended that (8) and (9) be estimated either from the appropriate equations for pump best efficiency point and turbine "shock free entry" or from Fig.4. Sensitivity analysis would suggest the following average values, with a wider spread of $\pm 7\%$ for K_H.

Y	0.6	1.0	2.0	4.0
K_H	1.75	1.57	1.39	1.26
K_Q	1.26	1.26	1.26	1.26

Knowing the area ratio Y and the appropriate values of K_H and K_Q it is a simple matter to estimate the head and flow ratios vis:-

$$\frac{(H_t) \text{ shock free}}{(H_p) \text{ Best efficiency}} = \frac{K_H}{\epsilon\,\eta_{vt}\eta_{vp}} \tag{10}$$

$$\frac{(Q_t) \text{ shock free}}{(Q_p) \text{ Best efficiency}} = K_Q \tag{11}$$

The head ratio calculation involves the estimation of the volute efficiency terms η_{vp} and η_{vt} for pump and turbine operation; moreover the factor of utilisation ϵ will need to be determined. Thorne (1979) discusses the estimation of η_{vp} and as a first step ϵ and η_{vt} can be combined as joint components of the turbine efficiency.

One further correction might seem to be in order, and that is for leakage losses through the wearing rings. Such leakage will tend to <u>reduce</u> the actual pump discharge rate and tend to <u>increase</u> the flow requirement during turbine operation. For a given leakage rate "r" measured as a percentage of average through flow, it would seem appropriate to multiply K_Q in (11) by the ratio $\left(\dfrac{1+r}{1-r}\right)$.

CONCLUSIONS

Using the extended area ratio method proposed by Burton and Williams (1991) it has been possible, through sensitivity analysis, to show that simple ratios K_H and K_Q, which are related to Y, may be used in determining head and flow ratios for pump and turbine operation. Further work obviously needs to be undertaken to test the method and to relate the volute performance coefficients η_{vp} and η_{vt} to machine design.

REFERENCES

Anderson, H.H. (1938). *Mine pumps*. J. Kings's College Min. Soc., University of Durham July.
Anderson, H.H. (1955). *Modern Developments in the Use of Large Single-entry Centrifugal Pumps*. Proc. Inst. Mech. Engrs. London Vol. 169, p141.
Burton, J.D. and Williams, A.A. (1991). *Performance Prediction of Pumps as Turbines (P.A.T.s) Using the Area Ratio Method*. Proc. of 9th Conference on Fluid Machinery, Budapest, Hungary. pp 76 - 83.
Mulageta, A.G. (1992). *Performance Evaluation of an End Suction Centrifugal Pump run as a Turbine*. MSc thesis, Department of Engineering, University of Reading, U.K.
Thorne, E.W. (1979). *Design by the area ratio method*. 6th Technical Conference of B.P.M.A., U.K.
Worster, R.C. (1963). *The flow in volutes and its effect on centrifugal pump performance*. Proc. I.Mech.E. Vol. 177, No.31.

A COMBINED SYSTEM OF RENEWABLE ENERGY FOR GRID-CONNECTED ADVANCED COMMUNITIES

M. A. ALHUSEIN, O. ABU-LEIYAH, & G. INAYATULLAH

Faculty of Engineering, Mu'tah University, Karak, Jordan

ABSTRACT

It is aimed by this research to demonstrate the possibility of utilizing renewable energy resources in Jordan such as solar and wind in a hybrid system, to help in saving energy cost by cutting down on fuel, which is considered to be the main source of energy. Based on data of both energy demands and renewable energy resources, solar and wind, that collected and analyzed during this work for a selected site, a suitable hybrid system comprising solar, wind, and conventional energy sources was then suggested and investigated for this purpose. The utilization of such system could cover a considerable portion of both electrical power and thermal energy demands and can be applied to similar conditions.

KEYWORDS

Hybrid system, Solar energy, Wind energy, Advanced community.

INTRODUCTION

Apart from natural gas and oil that have been recently discovered in small quantities, Jordan has no indigenous energy sources. Thus energy requirements are primarily satisfied by imported oil which represents a substantial fraction of the total national imports, making the energy costs to be the highest in the region. However, energy need is remarkably increasing due to electrification and industrialization schemes of the last two five-year plans in the country (JEA, 1990). Therefore, energy conservation and utilization of alternative resources have been encouraged to reduce the heavy burden of oil import on the general national income and consequently to reduce the country's dependence on imported oil.

Since Jordan is blessed with an abundance of solar insolation which is about 5.5 kwh/m^2 as mean daily insolation, and sunshine hours of more than 3000 per annum (Daghestani et al.,1983), solar energy can be considered technically and economically a feasible alternative with wide spectrum of applications. Solar collectors for water heating are produced locally at

2848

reasonable prices and becoming increasingly popular. More than 20% of homes are employing solar collectors for water heating.

In addition to the above mentioned potential of solar energy, the average wind velocity distribution in various sites in Jordan (El-Hayek & Anani, 1989) indicates that wind energy is worth to be considered for energy planning. But due to the nature of solar and wind energy sources, the utilization is restricted to individual or isolated consumers in rural areas far from national grid. Suitable technologies based on energy needs of the communities and available energy resources must be identified to promote energy self sufficiency. Although there is a large number of small communities scattered in rural areas, there are still no satisfactory efforts made to implement hybrid energy systems which enhance the capability of conventional sources by integrating renewable energy resources with them. Also no single one of either solar or wind energy resource can provide all the energy requirements. Therefore, this study is focused on the feasibility of implementing a system comprising solar, wind, and conventional electric energy sources in a hybrid scheme in small isolated communities which are connected to the national grid, and to investigate the possibility of utilizing such system in providing the energy demand. Such communities are considered to be technically advanced based on the energy consumption per capita. The use of hybrid systems in remote community applications especially for countries in lack of major energy resources have been studied by Helfrich (1979) and Manasse (1980). Various integration schemes of renewable energy resources depending on the community location characteristics and its energy requirements have been discussed before (Griffith & Brandt, 1984; Nigim & Nigim, 1987; Dubois et al., 1987).

SITE SELECTION

Mu'tah University campus has been chosen for this study as a typical representative of a small isolated but technically advanced community. Such communities could be small residential areas or small industrial estates that scattered in the country. The present choice was justified by the high figures of per capita energy consumption of more than 1300 kwh compared with the national figures of 1054 kwh (JEA, 1990). In addition to that the energy flow in this site is controllable and the facilities needed to complete this study are also available.

This site is approximately 140 km south of the capital, Amman, at latitude 31° 30' north and longitude 35° 45' east of Greenwich. Its residential area includes about 160 apartments and housing more than 600 people.

ENERGY CONSUMPTION

The energy consumption pattern of the considered community varies according to the type and utilization of the energy appliances. In general, energy is used for space heating, hot water services, lighting, catering, fans and pumps, and domesting appliances in addition to the computers and laboratories instrumentations. The consumption rates of the main energy sources, oil and electricity from grid, were collected for a period of three years and assessed in this work on an annual basis. The estimated monthly average consumption rates of the above energy supplies are shown in Figs. 1 and 2 respectively. It is obvious from the curves that the maximum load demand of both supplies is during the winter season, while the dip in

2849

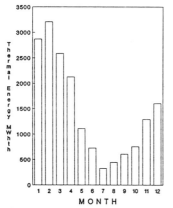

Fig. 1. Monthly average distribution
of equivalent thermal energy.

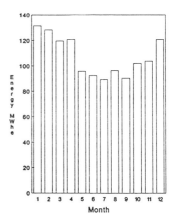

Fig. 2. Monthly average distribution
of consumed electrical energy.

the curves indicates a minimum load demand in the summer season due to the
fact that there is no space heating is required in addition to the summer
vacation.

RENEWABLE ENERGY RESOURCES

In order to explore the possible contribution of renewable energy resources
in covering the energy demand of the considered community, data of both
solar and wind has to be available for enough period of time. Analysis of
three years data for the period 1989-1991 on both solar and wind reveals
that the daily average of solar insolation is about 5.9 kwh/m^2 with a total
sunshine duration of more than 3200 hrs. The average wind speed measured at
10 meters above ground level is about 4.69 m/s. The monthly average
distributions of the available solar and wind energy are shown in Figs. 3
and 4 respectively.

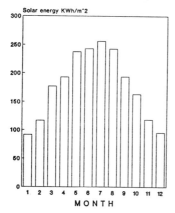

Fig. 3. Monthly average distribution
of solar energy available.

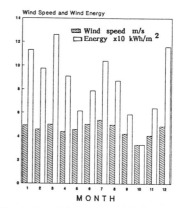

Fig. 4. Monthly average distribution
of wind energy available.

ENERGY CONVERSION SYSTEMS

Taking into account the high availability of solar energy during the period from April to October, (Fig. 3), a considerable substitute to the thermal energy demand can be achieved by using solar water collectors to provide hot water to the community, where no space heating is required during this period. A separate study including details of matching the thermal load by solar energy is currently under investigation by the authors, while this paper is focussed on meeting only the electrical load requirements.

Electricity generation

Wind system. The power output from wind is estimated by installing a 55-kW rated power wind machine with the characteristics shown in Fig. 5, and using the mean annual distribution of wind speed. For this wind machine, the minimum wind speed to operate must be higher than 3.5 m/s. The rated speed is 7.5 m/s and the cut-out speed is 12.5 m/s. Its power curve can be represented by the following equation:

$$P = - 10.06 - 2.2 \, V + 1.45 \, V^2 \qquad \text{for } 3.5 \leqslant V \leqslant 7.5 \qquad (1)$$

where P is the power output in kW and V is the adjusted wind speed in m/s at the rotor height of 24 m.

Using equation 1 together with the recorded wind data at the site, the annual energy production by the 55-kW wind machine was obtained and found to be about 354 MWh. Therefore, it can be seen that four 55-kW wind machines are needed to meet all the electrical demand from wind. The extracted wind energy distribution along with the consumption rates are shown in Fig. 6.

Photovoltaic system. The measured solar radiation for horizontal surface was corrected for a fixed slope of 30° according to the formula given by Overstaeten and Mertens (1986) in order to size the required pv-system. The output energy from the pv-system was calculated assuming a conversion and inversion efficiencies of 8% and 90% respectively. By considering a commercial available pv-module rated at 48 W_p with a surface area of

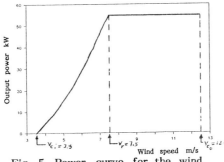

Fig. 5. Power curve for the wind machine

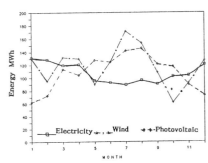

Fig. 6. Solar and wind extracted energy distributions.

$0.438m^2$, it was found that 22463 modules with a peak power of 787 kW are needed in order to meet all the electrical demand. The monthly average distribution of pv-energy is also shown in Fig. 6.

Solar-wind combined system. To provide the required electrical demand, there are three possible scenarios using the commercial available systems of renewable energy. The first is to use four 55-kW wind machines (Fig. 6). It is revealed that there is a lack of supply for nearly 4 months if no storage system is implemented. The second is to use pv-system of 787 kW$_p$. Also here there is a lack of supply for nearly 6 months. The third is to combine three 55-kW wind machines with 4043 pv-modules of a 142 kW$_p$ as shown in Fig. 7. This combination was suggested according to the resources availability. The lack of energy supply can be covered by connecting this system to the utility grid, where the energy surplus can be fed back to the utility (Fig. 7). In this case the need for storage system is eliminated.

CONCLUSIONS

Both solar and wind energy sources prove to be practical and highly available in the considered site. The combination of these resources with the grid should be able to provide the required energy with high reliability. The grid-connection will result in saving of the storage cost. The use of such systems will lead to reduce the fuel consumption and consequently to reduce the country dependency on imported oil.

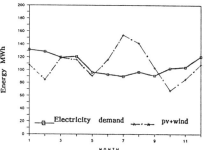

Fig. 7. Combined solar/wind system.

REFERENCES

Daghestani, F.A., Badran, I., El-Mulki, H., Abu Mugli, F. and Qashou, M. (1983). The potential of solar energy application in Jordan. Vol. 1, Royal Sceintefic Society, Amman, Jordan.
Dubois, A., Boyle, G., and Dichler, A. (1987). Design and operation of a solar photovoltaic-wind energy cogeneration system for a group of houses. ISES Solar World Congress, Vol.1, 441-446, Hamburg, Germany.
El-Hayek, S. & Anani, F. (1989). Measurements and evaluation of wind speed data in Jordan performed by Royal Scientific Society. Paper to the Int. Conference on Wind Energy, Amman, Jordan.
Griffith, L.V. & Brandt, H. (1984). Solar-fossil hybrid system analysis: performance and economics. Solar Energy J., Vol.33, No.3/4, 265-276.
Helfrich, C. (1979). Integrated energy systems in New England. UNH Senior Project Report.
Jordan Electricity Authority (1990). Annual Report.
Manasse, F.K. (1980). An integrated approach to energy supply for small communities. American Institute of Aeronautics and Astronautics, Inc. Paper No. 80-0651, 209-217.
Nigim, K.A. & Nigim, H.H. (1987). The need for integration in use of renewable sources. Melecon 87 Conference Proceedings, 631-634, Rome, Italy.
Van Overstraeten, R.J. and Mertens, R.P. (1986). Physics, technology and use of photovoltaics. Adam Hilger Ltd.

YOUNG SCIENTISTS IN RENEWABLE ENERGY:
A REPORT ON CHI-SEE'91 - CHILDREN'S SOLAR ENERGY EXHIBITION

SYDNEY M. CHAGWEDERA

Department of Science and Mathematics Education, University of Zimbabwe,
P.O. Box MP 167, Mount Pleasant, Harare, Zimbabwe

ABSTRACT

A recently held special Young Scientists' Exhibition CHI-SEE'91 - Children's
Solar Energy Exhibition (described) highlights continuing efforts by the
Zimbabwe Young Scientists' Exhibition Society to encourage Renewable Energy
use and awareness in Zimbabwe by appealing to the young. A "summary" of
the exhibition is given in the form of a poem and individual exhibits
described in the context of themes in the poem - giving a dramatic
pedagogical effect.

KEYWORDS
Solar energy; children; young scientists; curriculum; sun; science education
exhibition; Zimbabwe.

INTRODUCTION

The role of young scientists in Renewable Energy has been a subject of
discussion by the author before (Chagwedera, 1990, 1991) mainly in the
wider contexts of co-curricular approaches to science education by the
Zimbabwe Young Scientists' Exhibition Society (ZYSES). At a recent
exhibition - ZYSE'91 (Young Scientists' Exhibition, 2-6 May 1991) governmental
support come in the form of a speech by Minister Fay Chung (1991)
(Education and Culture) who stressed: " --- the economy cannot expand, nay,
cannot even be maintained at its present level, without very important
inputs by scientists."

THE CURRICULUM

Charters (1990), Hague (1990/91), including ZYSES, being aware of the
environmental aspects to development, have highlighted renewable energy
education. The Ministry of Energy and Water Resources and Development and
the Friedrich Ebert Foundation even asked ZYSES to coordinate a Children's
Solar Energy Exhibition when the former organized an International Solar
Energy Conference from 14-17 November 1991 in Harare for the Young
Scientists in Zimbabwe.

This marked a turning point: namely, the first time an exhibition entirely devoted to the subject of Solar Energy in its wider sense was held. A successful Children's Solar Energy Exhibition (CHI-SEE'91) was held from 14-17 November 1991 on the occasion of an International Solar Energy Conference and Exhibition of Solar Energy Technologies.

POEM

Asked to enter exhibits on Solar Energy in its "wider sense" (Renewable Energy) young exhibitors' entries reflected a wide range even to the extent that during processing a pattern emerged enabling the author (the coordinator) to summarize the ideas in the following a poem which also appeared in CHI-SEE'91 prospectus:

SUN

Oh Sun! I shall not laugh
At the ancients again
For worshipping you for long
At an eclipse of you
Of your power we're reminded
For even birds to sleep they go
Without you there is no life
Oh Sun! I shall not laugh
At the ancients again.

Oh Sun! May you forgive
Us the twentieth century generation
For even the coal and oil you made
For over six hundred million years
In a mere five centuries
We are squandering them all
Oh Sun! How greedy we are
Oh Sun! How wasteful we are
No wonder we laughed at the ancients.

Oh Sun! Even the trees which
In our time to us you gave
Without replanting we carelessly cut
But with diligence pollutants we create
To destroy our "life screen": the ozone layer
Just to get your ultraviolet rays
In amounts which you very well knew
Good for us were not Oh Sun!
Laugh at the ancients I shall not.

Oh Sun! Abundant silicon there is
Directly to tap electricity from you
But Oh Sun! the patience may we have
For lasting solutions to seek
To avoid "stopgap": shortcut measures
Of which we are otherwise so fond
For one last chance us you've given
Maybe this time learn we have to
And not laugh at the ancients
Oh Sun!

Inspiration for stanza 1 mainly come from exhibits numbers 5 (Life
and the Sun) and 15 (The Solar System) by Simbarashe Chikanza and
Simbarashe Mupfumira respectively, both of Haig Park School, Harare.

Stanza 2 and stanza 3 were linked to exhibits numbers 29 (Ethanol
from Sugar Cane) by Kayton Mudzengerere of Nyakatsapa High School,
25 (Energy Conversion) by Cosam Maliro of Dzivaresekwa No. 2 High
School, 24 (The Water Cycle and Budget) by Psychedelia Nzara of
Mabelreign Girls High School and 22 (Tropical Rain Forests and the
Desert Plants) by Chenai Mariri also of Mabelreign Girls High School.

David Muchemwa a school leaver from Kadoma entered his "Solar Energy
Traffic Lights" (exhibit No. 28) to demonstrate use of solar cells.
Other projects on photovoltaics were by Molly Nyakudya (Seke No. 7
Primary School) on "Solar Power Television), Tendayi Fusire (Haig
Park School) on "Solar Panels", Marakia Majoni (Seke No. 7 Primary
School) on "Communication Satellite", Clara Maumbarudze and
Marvellous Manyowa (Glen View No. 2 High School) on Solar-Powered
Telephone. These projects formed the basis for stanza 4, the last
stanza.

CONCLUSION

Parallel to the children's exhibits were exhibits by adults and
solar energy companies. The children siezed the opportunity to
widen their horizons. For instance, barely a week after CHI-SEE'91,
Molly Nyakudya an exhibitor from Seke No. 7 Primary School wrote
this author describing her experiences and ideas from CHI-SEE'91.
She was anxious to know when the next exhibition would be held
as she was looking forward to attending it. What more is required
for a positive evaluation of an exhibition?

REFERENCES

Chagwedera, S. M. (1990). Renewable Energy Education in Zimbabwe. In
Energy and the Environment into the 1990s. A. A. M. Sayigh (Ed),
Vol. 5, pp. 3229-3235, Pergamon Press, Oxford.

Chagwedera, S. M. (1991). The Young Scientists' Exhibition: A Co-
curricular Approach to Renewable Energy Education in Zimbabwe. Presented
at ISES Solar World Congress, August 17-24, 1991, Denver, Colorado, USA.

Charters, W. W. S. (1990). Development of Primary and Secondary School
Teaching Packages for Renewable Energy Education. In Energy and the
Environment into the 1990's (A. A. M.. Sayigh, Ed), Vol. 5, pp. 2845-
2860, Pergamon Press, Oxford.

Chung, F. (1991). Young Scientists' Exhibition Society in Hints and
Kinks, Zimbabwe Science News, 25, 72.

Hague, B. (1990/91). Energy in the National Curriculum. Review, 14,
8-9.

Teacher in Zimbabwe (1991). I (6), 8.

THE DOMINICAN REPUBLIC'S RENEWABLE ENERGY:
CURRENT USES AND FUTURE DEVELOPMENT

Dr. Francisco Perelló-Aracena
Research Center for Energy and Natural Resources (CEERN)
Pontificia Universidad Católica Madre y Maestra
Apartado 822, Santiago, Dominican Republic

ABSTRACT

In order to evaluate the potential uses of renewable energy sources in the Dominican Republic, a long-term LP model has been applied. Use of this tool provides the 'optimal' primary-energy profiles, investment requirements and technology mix needed to meet exogenous demand levels. Strategies discussed in this paper are based on results refering to the participation of renewable energy in the whole energy supply structure.

KEYWORDS

Dominican Republic; optimization model; long-term strategies

INTRODUCTION

As a result of deficiencies in the energy sector, the nation has been facing problems of high dependency on imported energy and daily black-outs which currently average 6-12 hours for more than 5 years, among others.

In order to provide answers to the energy and environmental problems in the Dominican Republic, the mathematical-computer model EFOM-ENV (Energy Flow Optimization Model - ENVironment) has been applied to the country's specific conditions. By using this model, total discounted costs of the whole energy structure are minimized under different techno-economic and environmental constraints. These costs are those required to meet exogenous final-energy demand levels.

CURRENT USES OF RENEWABLE ENERGY

Intensive use of oil and oil derivate products has been, and is still, the most specific characteristic of the Dominican energy structure, although diversification of the primary energy profile (e.g. coal and hydropower) has been on the increase in the last decade. At the moment, more than 50% of the primary energy consumption (PEC) corresponds to oil.

In recent years the share of firewood and sugar cane bagasse has remained stable in the range of 30% and 10%, respectively, of the country's PEC. Almost 70% of the Dominican Republic's population depends directly or indirectly on wood-fuels as main energy carriers for cooking purposes.

Due to the important socio-economic and environmental impacts of wood-fuels and bagasse in the country's energy structure, they will be denoted as non-commercial energy sources hereafter, and the term renewable energy will refer to such kinds of energy as solar, wind, hydro and geothermal.

As shown in Table 1, Dominican energy resources consist of around 75 million tons of lignite, a total hydropower potential of 2,000 MWe, corresponding to a technically feasible potential of 8,000 GWh/a, and a yearly production of nearly six million tons of bagasse, which can provide around 30,000 TJ/a. Renewable energy resources, for example 5.43 kWh/m^2/day of solar radiation, are also available. In 14 different locations, wind speeds of 5 m/s have already been found, representing a potential that could reach 0.4 kW/m^2. Geothermal and ocean thermal energy potentials have also been researched, but additional information and data are required for a complete evaluation.

Table 1. Overview of Dominican Republic's energy reserves

RESOURCES	DESCRIPTION
Lignite	75 million tons
Hydropower	2,000 MW$_e$; 8,000 GWh/a
Bagasse	6 million t/a; 30,000 TJ/a
Solar	5.43 kWh/m^2/day (radiation)
Wind	5 m/s (speed); 0.4 kW/m^2
Geothermal	40-50 °C (shallow water)
Ocean thermal	21-24 °C (temperature differences)
Energy farms	536,000 ha; 76 PJ/a

APPLICATION OF THE EFOM-ENV MODEL

The basic difficulties encountered in any attempt to use mathematical tools in the Dominican energy planning process are no more due to a lack of methodology than they are to a lack of quantitative information. Aside from the publications COENER (1987) and Schorgmayer (1989), no regular report on energy data and information is available.

In the case of the Dominican Republic the model has been successfully adapted and applied to country's conditions. Figure 1 shows the model representation of the sub-system, in which renewable resources (left side) are represented. They are linked to other sub-systems (right side) through many transformation processes, for which an extensive data base of technical, economical and environmental information has been compiled. Due to the size restrictions for this paper, it is not possible to describe the whole model. More details can be found in (Perelló-Aracena, 1991; Rentz et al., 1990).

2857

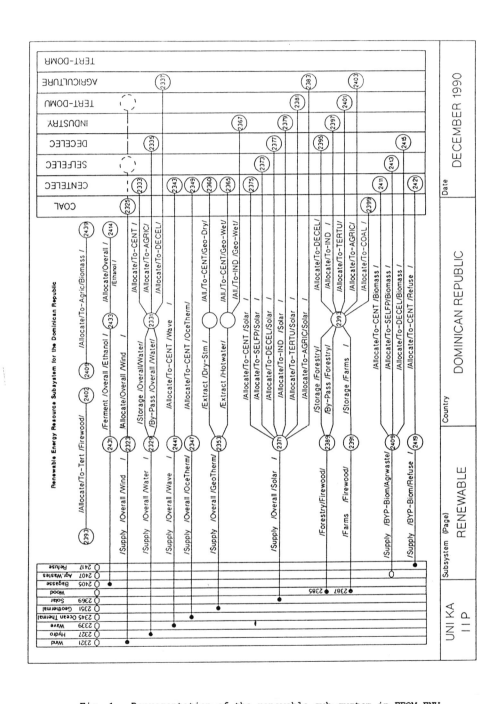

Fig. 1.   Representation of the renewable sub-system in EFOM-ENV
from Perelló-Aracena (1991)

SELECTED MODEL RESULTS

Definition of the scenarios

By calculating the future structure of the Dominican energy sector, correlation factors and specific energy data have been taken into account in the different study cases. Energy scenarios have been prepared based on national and international socio-economic forecasting. In all scenarios, population growth rate is expected to remain around 1.8%/a.

In the reference case (DR1) and in the renewable case (DR4), national economic development (in relation to GDP) is forecasted to increase at annual rates of 4.6% and 3.7% in 1985-1995 and 1995-2015, respectively. By the years 2000 and 2015, total final-energy demands are expected to be 170-300 PJ and 200-600 PJ, respectively. These demand vectors are used as input for optimization of the energy supply structure (Perelló-Aracena et al., 1988).

Renewable scenario: DR4

Projection of primary energy consumption is an important issue for future development strategies. As for the DR4 scenario, PEC will increase quickly at an average annual rate of 6.7%. PEC in the year 2015 (740 PJ) is expected to be 4 times higher than in 1990 (176 PJ).

Based on model reults, the 'optimal' primary energy profile will be as follows, see Fig. 2: (1) Renewable energy sources will increase their share from 4% in 1985 to 24% in year 2015, (2) The share of non-commercial energy resources will decrease from 39% to 16% in the period 1985-2015, (3) The use of oil will decrease from 54% in year 1985 to 45% in year 2015, (4) Lignite will increase from 3% in 1985 to 15% in 2015. These results show that country's oil dependency can technically be reduced considerably.

In the electricity generation sector, the 'optimal' additional capacity programme should be based on traditional units like hydroelectricity as well as new technologies which use solar and wind energy potentials. Firewood and agricultural wastes can be used to generate electricity or heat, but appropiate technologies must be applied. Between 1990 and 2015 an additional capacity of 100 MW will be required within the self-producer sector. This should be implemented in various steps by installing pilot-plants with a capacity in the range of 3 to 30 MW. These small plants using renewable energy resources should be located especially in areas which are not connected to the national grid.

The major consumption of wood-fuels is designated to heat generation in small processes, i.e. in private households and bakeries. Firewood consumption for energy purposes will be substituted between 40 and 70% within a 30-year period in comparison to the 1985 consumption level. These requirements correspond to a wood consumption of around 33 million m^3 in the year 2015, corresponding to around 1.2 thousand ha land needed.

In order to meet such demands, more energy farms have to be created. Measures like subsidies, price preferences, credits and others, are needed so that farmers will see forestry as a rentable business, and not, as at the present time, an "obligatory-illegal" activity.

2859

Another heat demand in the Dominican Republic corresponds to hot water. Results show that by the year 2015 more than 300,000 solar-based water heaters for urban households should be in use. Additionally, more than 20% of all commercial hot-water demand could be met with solar energy.

CONCLUSIONS

By applying the EFOM-ENV model to the Dominican Republic's energy structure as presented here, an appropiate data base management system has been made available. For such a country, with small scientific capacity, the work done here offers a new and excellent solid basis for analysis in this area. The techno-economical options obtained indicate that diversification is urgent and it can be achieved if specific directions are followed.

REFERENCES

COENER (1987). Boletín estadístico No. 18 del sector energía en la República Dominicana. National Commission for Energy Policies, Santo Domingo.

Perelló-Aracena, F. (1991). Energy-environmental long-term strategies for the Dominican Republic. Schulz-Kirchner Verlag, Idstein.

Perelló-Aracena, F., H.-D. Haasis and O. Rentz (1988). Future Development of the Energy Sector in the Dominican Republic. Energy, 15, 1029-1034.

Rentz, O., H.-D. Haasis, Th. Morgenstern, J. Remmers and G. Schons (1990). Optimal control strategies for reducing emissions from energy conversion and use in the European Community. KfK-PEF Series 72, Karlsruhe.

Schorgmayer, H. (1989). Energía 1. PUCMM, Santiago.

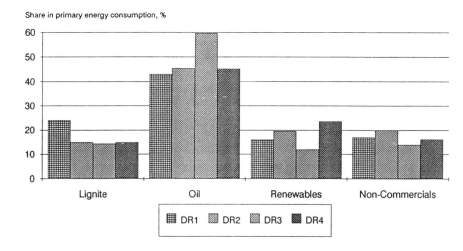

Fig. 2. Primary energy consumption in the Dominican Republic at year 2015 from Perelló-Aracena (1991)

SOLAR FURNACE WITH "FIVE" CONCAVE MIRRORS
AND "VICARIO" CYCLE FOR THE PRODUCTION OF
HYDROGEN BY THERMOLYSIS-ARCHITECTONICALLY-
ACCOMPLISHED BY THERMODYNAMIC CONVERSION
OF SOLAR ENERGY INTO STEAM.

Emilio C.O.VICARIO
Director of : L'INVENTIVA -Solar Hydrogen Centre
Avenue Luigi Settembrini,NR.6
81100 CASERTA (ITALY)

ABSTRACT

Into the field of the SOLAR ENERGY and WATER,the architectural design called PENTA ETA RADIATIONS 1958 identifies the
3-Dimensional structure of the "NUCLEON" having two dodecahedral and hollow volumes.

These volumes are concentric and they are drawn from
complex and similar atomic structure of the molecule of
Helium.

This is examined also by photographic recording of the
dodecahedral and 3-Dimensional rating distribution in
energetic interaction of inter-quark forces of the nuclear
mass of Helium (VICARIO photograph in BRUSSELS 1958 and
called PENTA ETA RADIATIONS 1958 with international trade-
mark registered in 1963).

This "architectural design",proposed with the colour
photography in FIG.1 ,is bounded by news functionality of
the "five" concave and pentagonal mirrors obtained by one
and only semi-dodecahedron hollow volume. -In fact, with
my new geometrical construction, by "one" only semi-dodecahedron hollow volume, they are obtained : "five" concave
and pentagonal mirrors , and "one" plane and pentagonal
mirror.The plane and pentagonal mirror is placed on the
leval ground parallelly at lower base of the dodecahedral
boiler. This plane and pentagonal mirror, by the radiation
reflected, work with orthogonal vector.- This is useful for
to promote the "thermic-intensity" of the FRUSTUM OF CONE,
energetically, constructed with the radiation reflected by
"five" concave and pentagonal mirrors.-With the photographs
FIG.1 and FIG.2 it is proposed only 1/6 of the "SIX"mirrors
necessary for to identify the "power amplifier" concentrated
under the base of the dodecahedral boiler.

2861

The power of the concave and pentagonal "mirror" is :
7 Watt/cm^2 ,and this power is projected and concentrated in
one "focal lenght" that is double comparatively to "focal
lenght of classical and optical construction.

It is very important to consider that this thermical
power, in the "five" different directions,is concentrated
under the base of the dodecahedral-boiler placed on a short
and central tower.

It is, also, very important to consider that for one
concave and pentagonal mirror having 50 m^2 = 500.000 cm^2
we have one thermical power corresponding to :

(1)   7 Watt . 500.000 = 3.500.000 Watt = 3,5 MW

R E S U L T S

VICARIO cycle for the massive production of Hydrogen by
Photolysis of the steam, architectonically, is accomplished
by thermodynamic conversion of Solar Energy and of the.water
into steam.

The steam,in installation presented with photography
FIG.1 ,is dynamically concentrated in concentrical and
dodecahedral volume.-
This small volume is flatted by "five" valves (one for
each pentagonal face of the base of dodecahedron) and this
valves push the steam toward the internal and dodecahedral
boiler into the lower share superheated by the power of
concentration of the projected solar calories.

With this VICARIO cycle ,into the interior of the boiler
(dodecahedral and greater volume !) and upon the "five"
pentagonal and plane surfaces of the lower semi-dodecahedron,
it is obtained the simultaneous functionality of the following nuclear reactions :

(2) H_2O (gas) $\xrightarrow[1000°C]{}$ $\frac{1}{2}$ O_2+ 2H$^+$ + 2e$^-$

(3) SiO_2+2H_2O (gas) $\xrightarrow[1000°C]{}$ SiO_2 + O_2+ 4H + 4e

That because these reactions happen with the "L I G H T "
and " H E A T ".=

For this reason the architectural ,exterior and dodecahedral volume is construed in quatz crystal or in crystal
called Pyrex.

Whereas the interior and concentric volume is construed in stainless steel because it works with the thermodynamic pressure.

This not exclude that also the dodecahedral and greater volume that be manufactured in "iron". In this case it is realized the chemistry reaction :

(4) $3Fe + 4H_2O$ (gas) $\xrightarrow[1000°C]{}$ $Fe_3O_4 + 4H_2$

but this reaction endure the phenomenon called " the firecoat ".

This architectural plan between the "two" concentric volumes involves the consideration of value of the mechanic "equivalent" of the "heat" in "work".

But,in fact,when the steam into the larger volume,raises the temperature it effects one work of expansion of a gas and that reduces the pressure,when the steam falls in torrents on the "five" faces of the "red-hot" lower semi-dodecahedron.

That because,into mechanical relation :

(5) $p = \dfrac{F}{S} = \dfrac{L\,M\,T^{-2}}{L^2}$

with :   p = pressure ; F = force ;   S = plane surface

it will be that greater is the plane expanse of −S− , smaller and insignificant is the value of −F−.

Moreover the "pressure" is inconstant because the Hydrogen produced by " P H O T O L Y S I S " is transferred , through "causa and effect" of the low atomic weight of the mass of Hydrogen (absence of the force of gravity) ,in another container by fit and higher pipes.

One electro-magnetical system of classical knowledge sorts in two different pipes the " Oxygen" and the "Hydrogen", or only "Hydrogen" in case of application of the chemistry reaction (4).

     C O N C L U S I O N

The scientific and photographic discovery of the 3-Dimensional hollow NUCLEON having two hollow-dodecahedral and concentric volumes (mine photographic discovery of the 1958 and 1962 !), scientifically and with technological concepts of applied architecture, realize one contraposition to classical and theoretical concepts of the NUCLEON formed

2863

one "T E T R A H E D R O N " having "five" q u a r k s :
on the contrary,in hollow and dodecahedral volume of the
N U C L E O N, the quarks are -21- and that barycentric
is "maximum" because it realize one work of thermodynamic
interaction between the inter-quark forces.

Then the commutation and the transfer of this discovery,
in architectural technology, realize one new and optical
Boiler having two concentric volumes : in "quartz crystal"
the external and dodecahedral volume , in hard-metal the
dodecahedral and interior volume.

The solar VICARIO cycle, with one similar structure ,
turn water into Steam and, with thermodynamic cyclo,removes
and pumps the Steam in "interior volume" of the OPTICAL-
BOILER for the purpose of to make concrete the chemical and
classical reactions (2) , (3) , and (4).

This thermodynamic cycle,with the architectural plan
called : PENTA ETA RADIATIONS 1958 ,finds its optimum
efficiency and, therefore, allows the production of Hydrogen
at low cost and at high quantities without electrical energy
consumption. The general planimetry is presented in FIG.3

The cost of the "plant" is not prohibitive.

        R E F E R E N C E

1)- Emilio C.O.VICARIO
    The New Architectural Work called PENTA ETA
    RADIATIONS 1958.-
    Copyright by : Emilio VICARIO, 1958
    U.S.A. Certificate Registration of a Claim
    A-B Af 29215.

FIG.1 and FIG.2 - Optical Boiler having two
dodecahedral concentric volumes
and "concave"-pentagonal mirror.

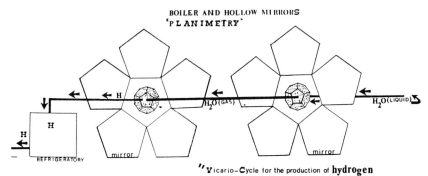

BOILER AND HOLLOW MIRRORS
'P L AN IMETRY'

"Vicario-Cycle for the production of hydrogen
by water thermolysis , architectonically ,accomplished
by thermodynamic conversion of solar energy
into steam.*"

FOR GENERAL INFORMATION WRITE TO AUTHOR :

" Academic Emilio Vicario "
aven luigi settembrini n.6 - 81100 Caserta-Italy
telephone : 0823 -32722 2·

FIG.3 - The general planimetry of the "plant".

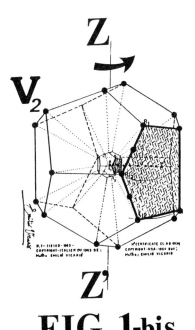

FIG. 1-bis

3-DIMENSIONAL STRUCTURE OF THE "NUCLEON"
HAVING -21- QUARKS.

HYSTERESIS IN LaNi$_5$-H$_2$ SYSTEM DURING ACTIVATION

S.L. ISAACK[*], E.A.KARAKISH[**], H.I SHAABAN[*], and
S.H. KANDIL[**]
[*] Atomic Energy Authority, Cairo, Egypt
[**] Institute of Graduate Studies and Research,
Alexandria University, Alexandria, Egypt

ABSTRACT

In hydrogen storing materials, hysteresis in a hydriding - dehydriding cycle present an efficiency loss which is to be minimized in order to get high performance. In this paper, some of the factors that may influence the hysteresis loop are investigated. Thus, the intermetallic compound, LaNi$_5$ was activated under constant volume conditions using different initial pressures by cooling in hydrogen gas from 523 K down to room temperature, then, thermally delydriding the obtained hydride. It was found that both the degree of hysteresis (DH) and the pressure difference (ΔP) across the two branches of the hysteresis loop were highly affected by the initial pressure used and the hydriding-dehydriding temperature. The greatest values of DH and ΔP were observed in the temperature range 373-473 K with a shifting tendency towards lower temperatures with lower initial pressures.

KEYWORDS

Degree of hysteresis, pressure difference, constant volume hydriding-dehydriding, hysteresis loop, shift, LaNi$_5$

INTRODUCTION

Research work activities in the field of hydrogen storing materials and methods became nowadays one of the important topics of intenset in non-conventional energy studies. Although, the intermetallic compound LaNi$_5$ was the first to be discovered as having hydrogen recharging properties in 1976 (Van Mal 1976) its hydriding-dehydriding hysteresis characteristics was described earlier before in 1971 (Kuijpers and Van Mal 1971). Hysteresis phenomena in rechargeable hydrides presents a loss which is to be minimized in order to get a better performance efficiency from these systems. Published

2868

research papers dealing with the subject are few rare and mostly hypothetical. Some authors (Flanagan et al., 1980, Van Mal, 1976) investigated the effect of annealing on hysteresis and attributed it to dislocation creation. Others, (Akiba et al. 1987) studied the pressure-composition isotherms (PCT) of the $LaNi_5$-H_2 system and found a clear hysteresis loop to develope above 353 K. Dynamic isotherms where the $LaNi_5$ was subjected to continuous hydrogen flow were also given (Goodell et al. 1980) whereby hysteresis was found to increase significantly as compared with that for static isotherms. The phenomena was found to be markedly reduced by using large hydrogen aliquots (Park and Flanagan, 1983). The effect of temperature in the α-β two phase region was also studied (Biris etal, 1976, Tanaka and Flanagan, 1978, Tanaka 1983, Van Mal 1976).

In practical use of metal hydrides, going through constant volume hydriding-dehydriding cycles is a usual necessity in most cases. So, it is obvious to study the hysteresis characteristics under this condition specialy during the early activation cycles.

EXPERIMENTAL

Equal weight samples of $LaNi_5$ were subjected to activation by hydriding in a standard Sievert's apparatus using different initial hydrogen pressures. The process was achieved by show cooling (2 ^{O}C/min.) the sample in pure hydrogen from 523 K down to room temperature. The obtained L_aN_i5 hydride was then reheated slowly to 523 K (2 ^{O}C/min). The above cycle was repeated till two identical hydriding-dehydriding curves were obtained.

RESULTS & DISCUSSION

Figure 1 shows the hysteresis loops that were developed when different $LaNi_5$ samples (of equal weights), were activated by cooling in hydrogen gas having different initial pressures P_i. The values of the degree of hysteresis (DH) at different temperatures were calculated and given in table 1 for the first activation cycle of each sample. The DH was calculated using the equation (Shenghua et al 1989)

$$DH = \frac{1}{2} RT \ln (P_f/P_d)$$

where P_f and P_d denote the hydride formation and desorption absolute pressures respectively, R being the gas constant and T the absolute temperature.

Table 1. Effect of Initial Pressure on DH

| Temperature | D H Values | | | |
| | Initial Pressure P_i, MPa | | | |
	1.1	1.6	2.6	4.2
323	698	319	145	40
373	904	628	391	244
423	615	699	422	297
473	184	446	328	149
523	0	69	85	0

The obtained data show clearly that hysteresis peak values are get mostly in the temperature range 373-

2869

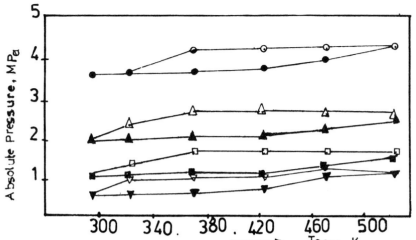

Fig.1 Hysteresis cycles during activation at various initial pressures, (o,●) 4.2 MPc (△,▲) 2.6 MPa, (□,■) 1.6 MPa, and (▽,▼) 1.1 MPa, solid symbols denote desoption.

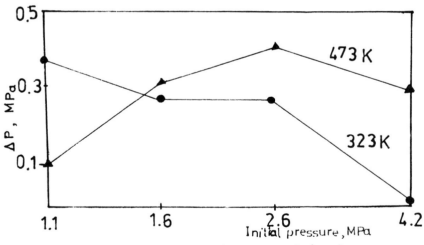

Fig. 2 Relation between initial pressure and △ P

423 K with a shift towards lower temperatures as the initial hydrogen pressure P_i was decreased. Also, the DH increases largely with pressure decrease.

The DH values could be considered as a tentative measure for the dislocations happening during the hydriding - dehydriding cycles (Flanagan and Clewly 1982). Hence, it could be claimed that most of the dislocations developed in the LaNi$_5$ hydride lattice take place in the temperature range 373-423 K when the material is found under a hydrogen gas pressure exceeding 2MPa. This dislocation forming temperatures range tends to lower values by lowering the gas pressure over the L$_a$N$_{i5}$. This means that dislocation forming temperature depends largely on both the hydrogen pressure and the hydriding temperature.

As the DH gives no good description of the hysterisis loop since it includes temperature and pressure ratio terms it is better to consider the pressure differential $\Delta P(= P_f - P_d)$ across the hysteresis loop at any given temperature. Figure 2 presents the relation between P_i and ΔP for two temperature, namely 473 and 323K. These plots show clearly that ΔP(distance between the two arms of hysteresis) increases with temperature decrease as the P_i decreases. This means that the hysteresis loop's position and shape are largely affected by the working hydrogen pressure.

CONCLUSION

Data obtained by the cosntant volume hydriding-dehydriding of LaNi$_5$ intermetallic compound in the temperature range 298-523K under different initial hydrogen pressures (from 1.1 to 4.2 MP$_a$) lead to the following:

1- The shape and position of the developed hysteresis loop is highly affected by both the temperature and the initial hydrogen gas pressure used.

2- The temperature range within which the majority of dislocations takes place in the L$_a$N$_{i5}$ hydride lattice is highly dependent on the hydrogen pressure used.

REFERENCES

Akiba, E. Nomura, K., and ONO,S. (1987). J. Less Common Metals, 129, 159-164.
Biris, A., Bucur, R.V., Ghete, P., Idrea, E., and Lupe, D. (1976), J.Less. Common Metals, 49, 477.
Flanagan, T.B., Bowerman, B.S., and Biehl, G.E. (1980). Hysteresis in Metal Hydrogen Systems., Scripta Metallurgica 14, 443-447.
Goodell, P.D., Sandrock, G.D., and Huston, E.L. (1980). Kinetic and dynamic Aspects of Rechargeable Metal Hydrides. J. Less Common Metals. 73, 135-142.
Kuijpers, F.A., and Van Mal, H.H. (1971). J. Less Common Metals. 23, 395
Shenghua Qian, and Northwood, D.O. (1989). The effect of hydride formation and decomposition cycling on Plateau pressures and hysteresis in Zr(Fe$_x$C6$_{1-x}$)$_2$-Hsystems. J. Mat. Sc. let. 8, 418-420.
Park, C.N., and Flanagan, T.B. (1983). The effect of hydrogen aliquot size on the plateau pressures of LaNi$_{s-H}$J.less Common Metal. 94, L$_1$-L$_4$.
Tanaka, S., and Flanagan, T.B. (1978), J. Less Common Metals 52, 9.
Tanaka, S. (1983). Hysteresis of hydrogen absorption and desorptian isotherms in the $\alpha\beta$ two-phase region of LaNi$_s$, J. Less Common Metals. 89, 169-172 Van Mal, H.H. (1976). Philips Res. Rep. (Suppl.), 18

APPLICATION OF LOW TEMPERATURE GEOTHERMAL ENERGY ON HOT AIR DRYING AGRICULTURAL PRODUCTS AND CHOICE OF SYSTEM DESIGN PARAMETERS

Zhu Long-Huei Xu Jing-Qiu Ou Zhi-Yun

Guangzhou Institute of Energy Conversion,
Chinese Academy of Sciences

ABSTRACT

Application feasibility of geothermal water of 65-90°C on hot air dryiny agricultural products is described. Method of choice geothermal drying system design parameters is suggested.

KEYWORDS

Low temperature geothermal energy; Agricultural products; Hot air drying.

1. INTRODUCTION

Agricultural products are mostly thermosensitive, with the drying temperature ranging between 55°C-70°C. In respect of efficient utilization of thermal energy, low temperature geothermal water of 65°C-90°C is an ideal heat source for drying agricultural products by hot air. Thermal source in the form of geothermal water is featured by stable flowrate and temperature, cleanliness, which are favorable to product quality.

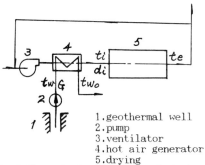

1.geothermal well
2.pump
3.ventilator
4.hot air generator
5.drying
Fig.1.Flow chart of geothermal drying
    system

Flow chart of a geothermal drying system is as shown in Fig.1. The system consists of thermal water well, pump, ventilator, hot air generator and dryer. The proper selection of design parameters has a direct influence on product quality, capital cost and operation costs.

After investigation and experiment on drying process and optimized design of drying system for agricultural products, we found the following five are the key parameters for the system: inlet air temperature to the dryer, ti (°C); absolute humidity, di(gw/kgDA); drying rate of material, S(kgw/h); dryer exhaust temperature, te(ºC); temperature drop of geothermal water accross the system, tw(ºC). Among these, S is the most critical.

2.CHOISING OF SYSTEM DESIGN PARAMETERS

2.1 Drying rate S

Drying rate S depends on the structure, moisture content and temperature of the raw material. Agricultural products are mostly of capillary-porosity type coloid structure, with a moisture content exceeding 60% in most cases. It is generally admitted that for agricultural products the drying process is composed of four stages, i.e. temperature rising stage, constant rate stage, first rate reducing stage and second rate reducing stage. However, as known from our experimental result, constant rate stage appears only when much water is adhered to the surface of material, and is not usually obvious in ordinary circumstances. Fig.2 is our experimental result.

Material: mushroom   Material: chinese cabage
Hot air tem.:~65 ºC   Hot air tem.:~60 ºC
Hot air vel.:× 0.85 m/s   Hot air vel.:× 1.0 m/s
       ⊙ 0.5 m/s   ⊙ 1.5 m/s

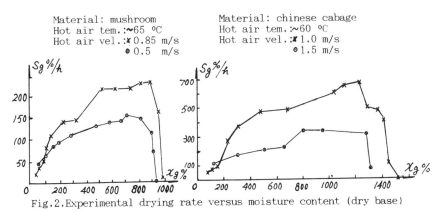

Fig.2.Experimental drying rate versus moisture content (dry base)

In usual, averaged drying rate over the whole drying process is taken as the design value. By practice we found for material of high moisture content, the practical drying rate in the fore stages is too low in the system so designed, resulting in worsened product quality. Moreover, there are microbes (bacteria, yeast, mold) in material which cause the material to deteriorate. Microbes will propagate very fast under the condition of moisture activity Aw≱0.65 and the temperature between 25ºC- 37ºC. For material of initial moisture content over 70%, material temperature in the early drying stages is approximately equal to the wet bulb temperature of the hot air, or between wet bulb and dry bulb temperature (a little nearer to wet bulb temperature), which is very favourable to microbes growth. Low drying rate often provides conditions in humidity, temperature and time for microbes growth, thus resulting in deterioration of material. Hence for material of moisture content over 70% (wet base) and of high protein content like soybean products, it is imperative to reduce moisture content to a safety limit before microbes have grown in large number.Hence, for agricultural products containing high mechanically combined water, taken as design value should be a drying rate by which moisture content will be reduced down to safety value before deterioration of the material is likely to occur. For material which is high in moisture content but with no risk of deterioration, design value may be taken as the rate averaged over that from the initial moisture content to the end of the first rate reducing stage. For material such as wood, which contain principally physiochemical water and in which stresses during drying process have to be limited, average rate of the whole drying process shuold be taken as design value. In practice, excessive drying rate is not preferred because shrinking and crusting will occur on the material surface, thereby blocking the passages through which water diffuses to the surface and preventing water transfer to the surface with the result of "wet interior, dry exterior", a result contrary to desired.

2.2 Dryer air inlet temperature ti

Air inlet temperature has great influence on system thermal efficiency, the speed of water transfer in material, and product quality.

At a given consumption of effective heat, higher inlet air temperature leads to reduced air flow and hence heat loss in exhaust air.

As to water transfer inside and outside a material, higher air temperature leads to correspondingly higher material temperature, and improved water transfer as can be seen from the formulas below.

The relationship between coefficient of internal moisture diffusion and temperature:

$$\alpha_m = \alpha_{mo}[(t+273)/273]^n \qquad m^2/s \qquad (1)$$

where $n=10-14$; α_{mo} is the value at $0°C$. According to experiment, for bread, α_m is 0.45×10^{-5} m^2/s at $40°C$ and up to 12×10^{-5} m^2/s at $120°C$.

The relationship between coefficient of external moisture diffusion and temperature:

$$Dab = 0.22 \times 10^{-4}[(t+273)/273]^{1.8} \qquad m^2/s \qquad (2)$$

Formula (1) and (2) are valid in the condition of 1 atmospheric pressure and temperature range between $0-1220°C$. It may be seen from above formulas that influence of temperature t on m is far greater than on Dab. Therefore, to increase air temperature is maily for increasing diffusion rate of moisture inside the material.

However, excessive temperature would lead to less product quality. For example, at temperature between $55-60°C$, protein will deteriorate; long time high temperature will cause fat to be oxidated. For legume products, high temperature will cause them to become yellowish, thus degrading their quality class.

In summation, inlet air temperature to the dryer should generally be the upper limit tolerable by the material. For material containing much mechanically combined water especially in the case where much water adheres to material surface, in the early stage of temperature up-rising and the beginning of drying period, air temperature may be higher by $5°C$ than the upper limit, because the actual material temperature is only between wet bulb and dry bulb temperature in this time.

2.3 Absolute humidity of inlet air to dryer di

Medium used in drying of agricultural products is usually atmospheric air. Therefore, minimum humidity of inlet air to the dryer is that of the atmospheric air. However, to increase thermal efficiency of the drying system, exhaust air partly recycling is used which is accompanied by greater humidity than inlet humidity. In this case, in choosing di, the primary consideration should be to maintain the projected average drying rate and at the same time to avoid a too larger air flow than in the case that di equals to atmospheric value.

2.4 Exhaust air temperature from the dryer te

To reduce exhaust heat loss, a lower te is favourable. However, it is important that when hot air passes through the dryer, there is sufficient heat transfer intensity and hygroscopicity between the air and the material from the very beginning to the end so as to ensure uniform drying. Through practical operation we found that a difference of 8-10°C between wet bulb and dry bulb temperature of exhaust air is proper. For wood, this difference may be low as 5-8°C.

To sum up, in the four design parameters to be chosen, drying rate depend principally on the material itself, and the parameters of air, i.e ti, di and te have an enhancing effect on drying rate. A higher ti will speed up transfer of moisture from within to without while a reduced di only improves mass transfer exteral of material. An optimized condition would be reached only when internal and external mass transfer rates are equal to each other. Hence, di is not the smaller the better. The influence of te is principally on drying evenness and system thermal efficiency. The lower limit of te should be a little higher than the exhaust dewpoint.

2.5 Temperature drop of geothermal water accross the system (tw-two)

Geothermal water temperature drop accross the system is consisted of two parts: heat loss causing the drop Δt1 in delivery of the water and heat exchange causing the drop Δt2 in hot air generator. Among these Δt2 is usefuly. Assuming Δtw=(tw-two), then:

$$\Delta tw = \Delta t1 + \Delta t2 \qquad (3)$$

Assuming heat delivery efficiency as ηtr, then:

$$\eta tr = \Delta t2 / \Delta tw \qquad (4)$$

The temperature drop accross the system has a direct effect on water flowrate needed, and further on equipment investment and water pump operation cost. Through practice, we found that the operation cost of the pump taken a large percentage in equipment operation costs. Using geothermal water as the heat source for drying, the essence is to use its apparent heat as the apparent and latent heat needed for moisture temperature rising and evaporation. For 1 kg of water (1 kgw) evaporated, thermal balance may be expressed as :

$$Cp \cdot G \cdot \Delta tw \cdot \eta tr = (Cps \cdot te + \gamma) / \eta s \quad (kj/kgw) \qquad (5)$$

where: Cp=4.19 kj/kg·°C, water specific heat at constant pressure.
    Cps=1.97 kj/kg · °C, steam specific heat at constant pressure.
    G - geothermal water flow needed for evaporating 1kg water, kg/kgw.
    te - Exhaust air temperature from the dryer, °C.
    γ=2501 kj/kgw, water latent heat at 0°C.
    ηs- thermal efficiency of dryer in the stage of design drying rate (not the averaged efficiency over the whole drying period).

When above known data are substituted to formula (5), we obtain:

$$G=(0.47te+596.9)/(\eta s \cdot \Delta t2) \qquad (6)$$

Assuming $\eta s=0.6$, $\Delta t1=2°C$, and
exhaust air temperature from the dryer
te=40°C, then formula (6) is becomed:

$$G=1026.2/\Delta t2 \qquad (kg/kgw) \qquad (7)$$

The relatioPship between G and $\Delta t2$ in
formula (7) is hyperbolic, as shown in
Fig.3. The relationship between heat
delivery efficiency ηtr and $\Delta t2$ at
$\Delta t1=2°C$ also is shown in Fig.3. From
the figure it may clearly be seen that
with $\Delta t2$ lower than 10°C, G and ηtr
change rapidly with $\Delta t2$, that with $\Delta t2$
larger than 30°C, the curve becomes
slow, and that with the dimensions of
hot air generator taken into account,
$\Delta t2=5-20°C$ is preferred, with smaller
$\Delta t2$ in correspondence to lower
geothermal water temperature tw.

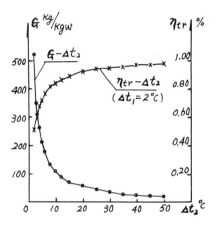

Fig.3.Geothermal water flow
G versus tem. drop $\Delta t2$

If the discharged water from the drying
system is to be used further, $\Delta t2$
should be chosen basing on an overall consideration of the comprehensive
utilization system.

3.EXPERIMENTAL RESEARCH OF LOW TEMPERATURE GEOTHERMAL ENERGY APPLICATION ON
AGRICULTURAL PRODUCT DRYING

Basing on the above design concept, we have designed and built a small hot
air type experimental installation for drying agricultural products.Test
run was performed in geothermal site, with satisfactory result.The dryer has
a batch capacity of 200 kg raw material. 10-15 t/h of geothermal water at
60-66°C from the well was used. Materials for drying were chinese cabage,
sliced sweet potato and unpeeled peanuts. Dried products quality was quite
good.

In winter, water discharged from the system was piped to fish ponds for
overwintering of fish fry or to a bath-water tank in a hotel. Because of
the required thermal load of fish ponds and the temperature requirement of
bathing water, temperature drop accross the hot air generator is selected as
only 2°C in design. To pump the discharged water to the water tank in the
hotel,the lift of the pump is large as 48 meters and the operation cost of
the pump account for 80% of the total equipment .operation costs.

4. CONCLUSION

Geothermal water of 65°C to 90°C is ideal thermal source for drying
agricultural products. The economy of such drying system depends on the
difference of product price before and after dried. Proper choice of design
parameters has great influence on investment of the system, operation costs
and utilization efficiency of geothermal water.

SOME PERTINENT TRADITIONAL ASPECTS IN PRODUCING
BIOGAS AT LEALUI IN ZAMBIA

PREPARED BY:

MR A. BANDA, B Eng., MEIZ
NATIONAL ENERGY COUNCIL
P O BOX 37631
LUSAKA 10101,
ZAMBIA

ABSTRACT

    The National Energy Council installed a biogas digester
at Lealui in Zambia's Western Province. Lealui is the summer capital
of the traditional Lozi people.

The Biogas Demonstration Project at Lealui is the first one
in Western Province. Despite the project being technically
sound it has however, not been without administrative and
traditionally derived problems which should have been addressed
at the project planning stage. Although the project was
commissioned on 16th November, 1990 it has continued to receive
a half-hearted response from the beneficiaries. This attitude,
the Author believes, is most probably due to overlooked traditional
aspects by the project implementors during planning.

WORLD RENEWABLE ENERGY CONGRESS
READING, UNITED KINGDOM 13 - 18 SEPTEMBER 1992

BACKGROUND

In Zambia biogas is produced using a digester which was modified from the Indian design. The Zambian digester is a fill and draw floating roof type (see fig. 1) which has successfully worked well using secondary biomass of either cow-dung or pig manure. Unlike other countries who have worked in the psychrophilic and thermophilic ranges, methanogenesis in this country has been primarily in the mesophilic range where digester heating has been through solar energy.

The National Council for Scientific Research (NCSR) has been the principal research institution in biogas technology in which they have worked for just over a decade. Although research is continuing into alternative raw materials, cow-dung and pig manure have constituted the main feeds used so far. Success with the Chalimbana and Kasisi digesters by NCSR spurred the National Energy Council (NEC) to call for the transfer of biomethanation technology to a rural setting. This it was felt would help in derivation of socio-economic factors that should be taken into account if implementation of anaerobic digestion was to be envisaged on a wide scale. Based on the 1984 and 1986 surveys on energy demand patterns in Zambia conducted by National Energy Council and University of Zambia (UNZA), Western and Southern Provinces were chosen as favourable areas for installation of biogas demonstration projects.

Having selected Western Province, the NEC sent a mission to Mongu, the provincial capital, to study the feasibility of where to locate a biogas demonstration digester. Four areas were identified namely: Namushakende Agricultural Training Centre, Lukulu, Lealui and a Dairy Farm on the outskirts of Mongu. In order to come to a decision as to which of these four areas was most suitable, discussions with the Royal Establishment, the traditional governing authorities, were commenced.

From the discussions it came to light that dry cow-dung and a form of grass known as Makuku and Matetelembwe were already in use as fuel resources in some villages especially in the plain. It was noted furthermore, that in one of these villages, Lealui, the summer capital of the Lozi people, supply of firewood to families which remain in the capital during floods (which are seasonal) was erratic. The foregoing was considered strong ground for NEC to choose Lealui as a place where to build the first anaerobic digester in Western Province.

Procurement of materials for building the fermenter started in 1987 and by July, 1988 the fermenter was to a larger extent completed. The Provincial Natural Resources Officer (PNRO) based in Mongu (12 Km away from Lealui) would act as coordinator of the project and was to be the link between the beneficiaries and project implementors (NEC) who were 585Km away from Mongu. Commissioning of the project was not until 16th November, 1990.

2878

OBSERVATIONS AT LEALUI

Beneficiaries of the Lealui biogas demonstration project
are three families living in three round grass-thatched houses
built in a line but separated by a distance of about 10m from
one house to the next. In the first house there lives an old
couple. The man is eighty years old while his wife may not be
more than twenty years younger than him. In the second house
there is an old woman aged over fifty who lives with her daughter
who is in her early twenties. In the last house there is a
woman in her late thirties who lives with small children below
the age of fifteen.

The Author interacted much with the beneficiaries of the
biogas demonstration Project and noticed a number of things on
his itineraries. Below is a collection of observations. On one
occasion he had gone to test the performance of the 400-500 candle
power mantles. They only lit half way showing that gas pressure
was not adequate. In one of the houses the valve to the lantern
had been left open. When the main valve was opened gas filled
that house making it dangerous to conduct any tests. The outlet
valve was closed and the house was left to be naturally ventilated
before testing was done. The beneficiaries had misconceptions
concerning utilisation of biogas:-

(i)   they thought that when the gas holder was full
      the stoves (gas burners) would not work well;

(ii) they were convinced that simultaneous use of the
      stoves in the three houses was not permissible.

From September to November there is a critical water shortage which
contributes   to digester non-usage. During this period an
insect pupa was discovered to have blocked one of the gas delivery
pipes. When it was removed normal gas flow resumed.

Evident also were organisational problems since some families
reneged to feed the digester when their turn came.

More confusing was the fact that some families often gave
conflicting reports concerning availability of gas in the digester
and when the gas was put to use. It was observed that some of the
reports which were given were not factual but were meant only to
appease the project implementors.

Because of the hollow enthusiasm to the project shown by
the beneficiaries the Author had suspicion that there could be other
reasons that contributed to such an attitude. Other observations
were made during the visits to the site which are recorded in the
first column of table 1. A small survey was conducted (see table 1)
in which people within and out of Lealui were interviewed based on
the questions in the second column of the table. The survey
responses are shown under 1 - 4. Responses 1 and 2 were from men
and 3 and 4 from women.

SURVEY RESULTS AND ANALYSIS

The survey response that showed the highest correlation with site observations was that under 1. The results in this column were closest to the response that the Chief Induna (chief local ruling authority) and another person within Lealui gave as explanations of the observations made. Responses 2 3 and 4 were not very close to observations showing that most people, even if, they use the same dialect with those at Lealui, do not know the cultural taboos that exist there.

The threat of perpetual haemorrhage seems to be the main deterrent to women from entering cattle kraals. Dry cow-dung is used as a fuel resource in general and especially so in brewing local beer known as Sipesu. Smoke from burning dung is sometimes used as a medicine to cure nose bleeding and as a mosquito repellent. Wet cow-dung has competing uses of house plastering, sealing holes in traditional dishes and as a raw material for the biogas digester. House plastering is the main competing alternative use of wet dung to its use as an input to the digester. House plastering is a yearly requirement as most of the houses are flooded during the rainy season and occupied again or reconstructed during the dry season.

The very high day temperatures from September to November averaging 32 degrees celsius force people to work mainly in the mornings. During this period there are shortages of water and people have to prepare their fields before the onset of floods. These activities compete for the people's labour and are rated higher than dung collection for biogas purposes.

There is only one man among the beneficiaries who because of the existing taboos remains the only one to feed the digester. He is able to do this only intermittently because of frequesnt ailments due to old age.

While it has been noted in some research reports that a 10 cubic metre, digester is adequate for three households, having an average of six members per family, for cooking and lighting this may not always be so as food items cooked in the various regions and social cultural habits differ.

Project follow-up by implementors has been unsystematic since it was subject to availability of funds.

CONCLUSION

Some researchers have said that people in Lealui welcomed the biogas demonstration project enthusiastically. Good as this was it should not have been the only criterion used to arrive at a decision of implementing the project. Adequate conscientisation of the people together with an understanding of the people's social cultural values should have been included in the many considerations leading to that decision.

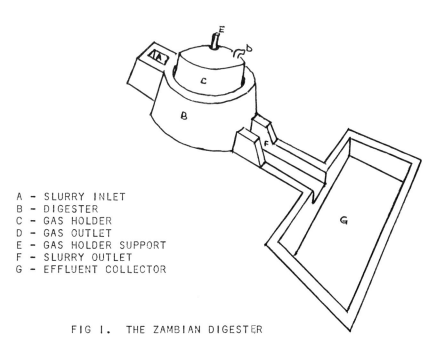

A - SLURRY INLET
B - DIGESTER
C - GAS HOLDER
D - GAS OUTLET
E - GAS HOLDER SUPPORT
F - SLURRY OUTLET
G - EFFLUENT COLLECTOR

FIG I.   THE ZAMBIAN DIGESTER

Observation on Site	Question	Survey Response			
		1	2	3	4
Women never went into cattle Kraal to collect cow-dung	(a) Why are women not allowed in a cattle Kraal?	- it is taboo for women of childbearing age to enter a Kraal.might because they.might neutralise fertility medicine performed on cattle - to avoid perpetual blood flow in women	Women are free to enter a cattle Kraal.	Not all families follow this. Those who do, do it in order to prevent rendering useless medicines performed on cattle	Women are not restricted.
Cow-dung was not collected from the Kraal while cattle were out	(b) Why should'nt cow dung be removed from a Kraal in the absence of cattle?	- it is believed that cattle would die and that; - cattle would not return to the Kraal if they did	I don't know of any such restriction	Not always followed	There's no restriction known
Dry cow dung was used as a fuel resource	(c) What uses of cow-dung do you know?	- fuel resource - fertiliser - house plastering	- fuel resource	- fuel resource - fertiliser - house plastering - sealing holes on traditional dishes	- fuel resource - fertiliser - house plastering - chasing mosquitoes (smoke from burnt cow-dung)
-	(d) Do you know any medicinal use of cow-dung?	No. I don't perhaps wizards and witches might know.	No	Dried cow-dung is burnt to cure nose bleeding	No

Table I Site observations and Survey Responses

Observation on Site	Question	Survey Response			
		1	2	3	4
Looking after cattle involved walking long distances under intense heat from the sun.	(e) Why don't women manage their own cattle?	- Traditionally it is a man's job - it is laborious	- its laborious - women fear to be gored by cattle	Traditionally cattle are clan assets Men keep them in order for them not to go from one clan to another.	it is man's job but where there are no men women manage their own cattle.
-	(f) What are the consequences of contravening the taboos?	A woman of child-bearing age might end up having, perpetual haemorrhage	Apart from unproven threats there are no consequences	Menstruating women may have perpetual blood flow.	There are no consequences. Its just a way of denying women certain things.
-	(g) Do you know of other taboos concerning male-female relationships?	There could be some but I can't remember any at the moment.	Women are not allowed to eat alligators and certain fish as they would be rendered barren.	Women who have just given birth are not allowed in cattle kraals	Women are not supposed to jump over fishing nets and are not allowed to eat certain types of fish.

Table 1 Site observations and Survey Responses

REFERENCES

1. Reports of A Banda's Working visits to Lealui from
   1990 - 1991.

2. M. Macwani and G. Ngulube, National Energy Council Report
   on Completion of Phase 1 of Demonstration of Biogas
   Technology in Western Province - 1987.

3. A. Banda   A Synoptic view of biogas generation in Zambia
   - 1989   FIMA 89 ZARAGOZA, SPAIN

THE GRAVITY OF EARTH IS THE "CLEAN ENERGY" WHICH CAN NOT BE USED UP

With The Example of the Conceptive Train Using Clean Energy

Tuan Guoxing   Senior Engineer, Department of International
Affairs, China Association for Science and
Technology (CAST)
32, Baishiqiao Road, Haidian District,
Beijing 100081, China

Shi   Pengfei   Head of Liaision, Chinese   Wind   Energy
Development Centre(CWEDC)
CWEDC, Huayuan Road 3, Beijing 10083, China

Preface

People often see the following phenomina in their daily life:
---A ball on a horizontal plane will slide down when the plane inclines.
---When you are riding a bicycle on a slope, your bicycle will rush down the slope automatically, and the speed accelerates as it goes down. When the bicycle has dashed over the lowest point, it can rush up the slope for some distance with the inertia, and then slows down gradually. With the inertia and some more hard pedals, you can reach the summit of another slope smoothly.
---According to a newspaper, a train stopped on a slant tracks at a station. Accidentally, the train was out of control and slided down the slope. The speed became faster as the train rushed down, and there was almost an accident.
---We often see the water of a "great waterfall" falls down from a height of several meters or even more than ten meters with a thunderous sound. The clouds and mist circulating around it makes a magnificent view.

There are more examples which I shall not list any more here . People all acknowledge that the above reflects the role of and shows great energy of the earth gravity.

The Gravity Of The Earth Is The Clean Energy
Which Can Not Be Used Up

The author holds that under the situation of the overall development of the modern thecnology of computers, automation control and various new power system, the energy of the earth gravity can be the clean energy which can not be used up by mankind if we exercise control over it and combine it with other new energy and modern power technology.

This article tries to outline the idea of introducing the earth gravity as one of the clean energy of train.

The Basic Conception Of The Earth Gravity
Used By The Train As The Clean Energy

---The train runs on the tracks with up and down slopes instead of the normal ones. It is not(or basically not) the normal energy but the earth gravity that makes the train run on the inclined railways
---To bring"the speed acceleration" which occurrs when the train slopes down under control by installing on the train various energy-converting equipment(speed limitation with load), and store the controllable energy produced by energy-converting equipment for the train to use as the power source when it climbs up a slope.
---Together with other stored clean energy such as the solar and wind engery and so on, the stored energy of earth gravity will be used as the power sources for the train to climb up a slope and runs on normal tracks.
All of these are applicable only when the technology of computer and modern power is adopted.

The Basic Condition For Introducing
The Earth Gravity Power Source

1. The basic conditions for train using clean energy:
a.Tracks (or road) with up and down slopes
b.All kinds of the equipment used to limit the speed(speed limitation with load) and turn earth gravity into controllable energy.
c. The installations for storing the controllable energy.
d. The power system for the train to go up a slope (the new power technology is included)
e. The electric computer control system as well as the automatic control equipment of each sub- system.

2. The sketch map of power system of the conceptive train with clean energy in a unit section of a railway.

Note:
a.(1.6) the area for the train to get in and out of the station.
b.(2) the area of the first stage of sliding without speed limitation.
c.(3) the area of the second stage of sliding--speed acceleration with speed lmitation with load.

d.(4) the area for the train to rush up a slope with inertia.
e.(5)the area to push the train up a slope by the controllable power source stored in the train.
f.(6)Whenever necessary, the train can use the normal power source as a means of assistance in order to get in or out of the station.

3.One railway line for the train using the clean energy should be consisted of several "unit sections". The number of "the unit section" will be decided by the length of the whole line.

The sketch map of a whole line for the conceptive train with gravity source

S	To go down a slope	The region for the train's speed acceleretion and speed limitation with load	N	To go up a slope	Using the controllable power to draw the train up a slope	S
		[1]	2		[3]	

Notes:
1.(1)The area for train to slide down a slope in an unit section.
2. "s" means the area for the train to get in and out of the station.
  (2) "n" means several same unit sections.
3 (3) The area for the train to climb up a slope in an unit section.

The Conversion And Storage Of Energy

In view of the overall development of the power facilities at present ,many new technology can be introduced to limit speed(speed limitation with load) and effectively turn the earth gravity into the controllable energy, and store it.

The methods in most common use are as follows:
1. To put the power generator(speed limitation with load) into operation when the train goes down a slope.
2. To set the wind tunnel power generator(speed limitation with load) into operation when the train goes down a slope.
3. To install the wind power generator on the train (the generator works mainly with the wind produced when the train runs forward).
4. To install solar cells on the train. The electricity is to be stored in high efficient battery.
5. To put the air compressor(speed limitation with load) into operation and store the compressed air when the train goes down a slope.
6. To make the inertia flywheels on the train run at a high speed and store the controllable inertia power energy produced when the train goes down a slope.

The sketch map of changing gravity energy into controllable energy.

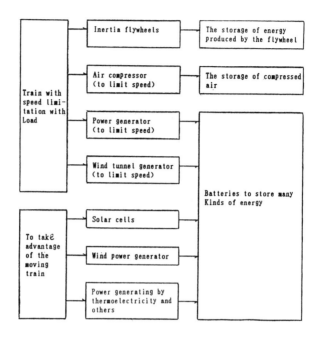

Note:
1.The energy converting ways listed here are based on the examples in this article.
2.The stored controllable power energy is in dynamic state and can be changed randomly.

The Methods Of Using Controllable Energy
For The Train To Go Up A Slope

The following methods should be applied when the train slows down gradually after rushing up a slope for some distance with the help of gravity inertia produced when the train goes down a slope:
1. First to drive the train up a slope with the power source stored by inertia flywheel on the train and other facicilities.
2. To bring the air-compressing engine into operation so as to push the train up a slope.
3. To start electric engine (with power source of electricity stored in battery) to push the train up a slope.

The sketch map showing the application of controllable power energy resource when the train goes up a slope.

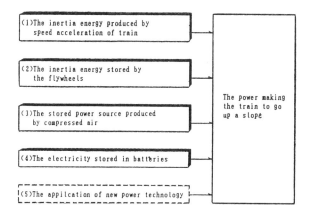

(1)The inertia energy produced by speed acceleration of train

(2)The inertia energy stored by the flywheels

(3)The stored power source produced by compressed air

(4)The electricity stored in batteries

(5)The appilcation of new power technology

The power making the train to go up a slope

In order to effectively settle the problem of energy sources used by the train when it goes up a slope, we should also introduce such new power technology as increas ed bouyancy of the train body, electromagnetic power, air cushion , electromagnetic suspension, etc. into the systems that make the train run and go up a slope.

The Application Of Electronic Computer Technology

The comprehensive application of electronic computer technology is an indispensable component part of the conception of the train with clean energy. The following are several examples:

1. The overall control of the whole train systems.
2. The automatic control of random braking and train's speed when it goes up and down a slope.
3. The automatic control and selection of the start, increase and decrease of various kinds of speed limitation with load; the running and stopping of the train; the converting and storing of energy.
4. The automatic control of the application and counter-convertion of controllable energy ,and the adjustment of speed limitation with load.
5. The automatic control system to keep the train in balance when it goes up a slope or maintain best condition as it runs.
6. The automatic control of the driving system of train and the new power technology system.
7. The communication and automatic signal systems.

Conclusion

The clean energy train discussed in this article involves a wide range of science and technology. It is sincerely expected that persons who are interested in it will futher explore and study it.

(Participants in translation: Tuan Ying and Tuan Jie
Translation proof reading by: Wu Yongguan)

NEW PERSPECTIVES ON INTEGRATED RURAL ENERGY SYSTEMS - A BLUEPRINT FOR RENEWABLE INDIAN SOCIETY

D K Dixit
Department of Mechanical Engineering
Visvesvaraya Regional College of Engineering, Nagpur - 440 011, India

ABSTRACT

The paper discusses the parameters and variables for evolving meaningful development strategies to harness the potential of renewable energy systems, critically examines the complex issues hampering the transition to a renewable energy society, and assesses the existing priorities in the use of resources compatible with the long-term sustainability of society. A case study involving the techno-economic evaluation of an exemplary integrated rural energy systems tailored to the local needs is presented. The concept of integrated rural energy centre is recommended as a role model for the Third World countries.

KEYWORDS

Integrated energy systems; Renewable energy; Rural development; Energy alternatives; Appropriate technology.

INTRODUCTION

The energy crunch has necessitated the exploration and exploitation of renewable energy resources, like solar energy. Nuclear power in India has far too much state financial commitment and importance than is warranted. Environmentally benign sources like solar, wind, biomass etc. are equally significant and perhaps more germane to the Indian cultural ethos and social milieu. In any case it is unwise to put all the fruits in one basket. Solar orange is as vital as the nuclear apple. Sound energy planning must incorporate socio-economic and techno- ecological factors.

Energy plays a catalytic role in any development strategy. Evolving an ingenious strategy for optimum exploitation of renewable energy systems in a country like India with 70%

of population living in villages eventually boils down to promoting primarily the integrated rural energy centres, which can provide for cooking, lighting, pumping water for irrigation, drinking and what have you.

There is a strong case for integrated rural energy system concept. For one thing, it is more economical than the rural electrification proposals and works out more pragmatic in the long run. For another, it generates power from natural resources with minimum waste, provides employment to the youth who thereby contribute to the village economy, and helps the housewife by piping in cooking gas and drinking water. And there is a cleaner and healthier environment - no dung and agricultural waste all over, no smoke - filled kitchen either. A unique and successful experiment in developing an integrated rural energy centre in an Indian village is now described at length.

INTEGRATED RURAL ENERGY SYSTEMS - A CASE STUDY

Khandia Experiment

Khandia, a village in Gujarat, is perhaps the only integrated rural energy centre in India, functioning with significant success. It has provided total energisation rather than bringing electricity to a few affluent homes. The entire capital cost of the Khandia package is Rs.1.5 million, which is Rs.0.3 million less than the estimate made for bringing in equivalent quantum of electricity to the village. The cost covers building of the irrigation network, biogas distribution pipelines criscrossing the village, and burners and electrical connections and fittings in the streets and homes. This works out to a per capita investment of Rs.1,875 which is less than one year's per capita income in rural areas (Dixit, 1990)

The Khandia energy package consists of a solar hot water system, three solar stills for distilled water and soalr photovoltaic refrigerator for the primary health centre, a solar photovoltaic powered TV and radio community sets, two community solar cookers for the free mid-day meal scheme in the village primary school, four 4.5 kW biomass gassifier engine pump-sets on each of the village well for irrigation, a 24 kW gassifier - based power generation system for domestic lighting (two points per household at Rs. 6 a month) and street-lighting as well as for community water supply with surplus spare capacity to run a flour mill, and an 85 m^3 per day hiogas plant run on locally available agricultural wastes and cattle dung. The livestock population of 350 is adequate to sustain plant operation round the year. It will also provide energy for agro-industries set up by the villagers themselves. The plant is expected to replace completely the scarce and expensive firewood traditionally used in Indian villages today.

With the entire energisation package being managed and run by the energy users' co-operative, new systems for cooperation had to be evolved. Twelve hectares of wasteland on the village outskirts has been used for captive energy plantations that will eventually yield 250 tonnes of biomass per year for the gasifier sets. The biomas (fuelwood) is presently being purchased from a private company. Villagers will have to work four days a month in the 'energy farms' for getting free cooking gas for their homes if they are unable to contribute the requisite four buckets of dung per day or pay Rs.30. The cooperative keeps track of dung contributed, the gobar gas drawn per household and at the end of the week, tallies are made. Both biogas and slurry are on sale. The latter combined with protective irrigation is expected to improve harvests considerably.

Under the Khandia scheme, based on recycling locally available waste matter, all the energy requirements of a village are met. The concept, if applied systematically to larger areas, may be a veritable boon for the industry. Electricity Boards, always pesterd by governments to expedite rural electrification, have never enough to spare. Every unit saved could be diverted to the industrial sector, tied down as it is by power cuts and load-shedding. It is in this context that the rural energy package has been conceived, switching from dependence on urban installations to energy autonomy. Another welcome feature is that the quality is not compromised while ensuring equitable distribution to all.

Several official agencies, each with a stake in successful development of the scheme have contributed to a common pool. The funds come from the Government of India's department of non- conventional energy sources; the planning and execution is done by State Government bodies - the Gujarat Energy Development Agency (GEDA) and the Gujarat Electricity Board (GEB).

In the Khandia experiment, the maintenance cost of the system might well be marginally higher than that for a conventional one. But the benefit of energy independence that it confers, in a sense, transcends these narrow economic calculations.

There is one more aspect that needs to be touched upon. A field experiment of this nature must address not merely to the techno- economic feasibility of the energy systems but also the amenability of social and political institutions of the village to the potential innovation. The day to day functioning of the Khandia rural energy centre - the mobilisation of inputs, the distribution of the output according to the optimal pattern, the periodic repair and maintenance of the institutions etc. –is done through the rural energy cooperative which can contain conflicts of interest. Many well-meaning community projects fail for want of popular support. The relatively strong tradition of cooperatives in Gujarat is a welcome factor here.

Intricacies and internecine interests of course can torpedo community rural development schemes. The key question therefore will always be : whether the system has been designed to fit local conditions - physical and biological, as well as socio- economic and political. Success is not ensured solely by techno- economic viability. How the technologies - appropriate or otherwise - are going to be institutionalise in the context of power structures within the village are far more crucial (Reddy and Prasad, 1977).

STRATEGIES AND OPTIONS

It is worth noting that the principal characteristics of solar energy are its abundance, its dilute nature and its variability. In contrast to fossil fuels or nuclear reactors which represent hightly concentrated forms of energy, it is dispersed and environment–friendly. However there are very few locations even in a tropical country like India having cloudless sunny days for more than 80 per cent of the year. This limitation implies that the applications of solar energy have to be largely confined to reasonably small units which serve the needs of families or communities.   A logical outcome is that life in the years to come would probably be more decentralized and life styles more simple. As a consequence, current trends which favour centralization and population shifts to the urban areas may well be reversed.

In the Indian context there is need to marry the solar applications to appropriate or intermediate technology. Thus, for example, the generation of mechanical or electrical power through a thermal cycle seems best adapted to magnitudes of 10 or at most 100 kW through systems like the low- temperature Rankine cycle coupled to a solar pond or a medium- temperature Rankine cycle coupled to cylindrical parabolic concentrators. The same is true for a photovoltaic system. For example, it seems best to provide photovoltaic power modules for a small irrigation system or for supplying the electricity needs of a small isolated community. The generation of larger quantities through the central receiver concept or the satellite power station does not seem to be the right approach for most countries and most certainly not for India (Sukhatme, 1984). Today one is trying to fit the energy source to an existing economic structure which demands a high degree of centralization. In fact, it should be the other way around; the economic structure should adapt itself to the energy source. It is therefore unlikely that a scheme like the central receiver concept would ever be cost effective.

The provision of solar water heating systems can be envisaged for homes or for establishments like hostels, hotels or hospitals. But a centralized water- heating system with its huge collector area requirement and the associated piping even for a community spread over a few square kilometres in perhaps illogical.

As far as indirect applications are concerned, there are possibilities of large-scale power generation in two applications. These are the energy plantation concept and the OTEC (ocean thermal energy conversion) concept. In both these cases, no collectors or storage tanks have to be fabricated. Collection and storage will be done by nature itself. This will help in reducing costs substantially and should make these applications economically attractive. It would certainly be profitable to pursue these applications in a coordinated manner in India. In the case of the energy plantation concept, however, land sites selected should not notrmally be needed for agricultural purposes.

Community biogas plants could reduce the dependence of the rural areas on firewood for burning and on kerosene for domestic application. The dependence of stationary engines on diesel oil could also be appreciably reduced if biogas is used as a substitute fuel.

The most encouraging aspect of the progress made in the last decade is that it has cleared the way for gradual change. Energy conservation has provided breathing room while new technologies are developed that will allow a meshing of renewable and conventional energy sources during the decades of transition. Changes will be continuous and the challenges enormous, but this process of historic change will also provide opportunities for creativity and growth for generations to come.

In the future, differences in climate, natural resources, economic systems, and social outlook will determine which energy sources will be used in which regions. New patterns of employment, new designs for cities, and a revitalized rural sector could all emerge with renewable energy development. For individuals and the environment the changes would be rejuvenating. because 'renewables' are less polluting than traditional fossil fuel resources.

2893

Renewable energy development is a gradual process that unfolds with many small investments. A mistake today does not foreclose another option tomorrow. Banking on renewable energy and energy-efficiency is fundamentally the most conservative energy course one can take. Risks are minimized, options preserved.

FINALE

It is thus evident that given the appropriate economy and infrastructure, the integrated rural energy systems promise to meet local energy requirements in a self- sufficient and self-sustaining manner. There is no pollution fall-out, no problem of depleting fuel reserves and no sophistication in technologies involved either. The 'total energy concept' and 'soft energy paths' should constitute a powerful and meaningful strategy while envisaging a resurgent renewable energy society based on optimum exploitation of non-conventional energy sources. This would not only promote better quality of life of the simple rural folk but prevent the present exodus to urban metropolises by providing suitable employment opportunities in the villages themselves.

A multi-pronged, pragmatic strategy should be to tap the labout-intensive, ecology-friendly renewable energy sources with the power to choose and the promise to reinforce the local economy and autonomy. No energy transition can unfold overnight but the inexorable trend today is towards 'vernacular technologies' - decentralised and dispersed, find-tuned to local climatic conditions. It is hoped that the elite - oriented, capital - intensive and ecologically inimical energy systems of today will be supplanted by location - specific, rural - oriented and environmentally benign smaller scale renewable technologies by the turn of the century. Success already achieved reinforces this expectation.

Despite the overall cost advantage of renewables, the picture is very different for the individual consumer because of his having to make the total investment himself as against merely paying the (usually subsidised) electricity tariff. Appropriate policy measures are therefore urgently needed to ensure that micro-level interests coincide with macro-level opportunities, and to correct the prevailing distortions that could play havoc with investment patterns in the energy sector.

It is high time that we as a nation moved away from discussing questions like the need and relevance of renewable resources. These questions now belong to the past. The need today is to move ahead with a mojor thrust on developing the relevant technologies and on their rapid commercialisation.

REFERENCES

1. Dixit, D.K. (1990). Strategies for the promotion of utilization of renewable energy systems. *Proc. National Solar Energy Convention*, Calcutta, India.

2. Reddy, A.K.N. and K.Krishna Prasad (1977). Technological alternatives and the Indian energy crisis. *Economic and Poitical Weekly*, 1465 - 1502.

3. Sukhatme, S.P. (1984). *Solar Energy : Principles of Thermal Collection and Storage.* Tata Mc Graw Hill, New Delhi.

ENVIRONMENTAL IMPACT OF GEOTHERMAL RESOURCES: EXAMPLES FROM NEW ZEALAND

Keith Nicholson

School of Applied Sciences, The Robert Gordon University, Aberdeen AB1 1HG, Scotland

ABSTRACT

Geothermal energy resources are natural polluters of the soil, atmosphere and aquatic environments. The polluting capacity of the fluid is dependent upon water-rock reactions at depth, the flow-rate of the fluid discharged to the surface and the temperature of the reservoir. Species of most concern include CO_2, H_2S, As, B and Hg. The latter three species can accumulate in soils, geothermal deposits and aquatic sediments to levels far in excess of their dissolved concentrations. These accumulations represent pollution time-sinks which are enhanced by the discharge of large masses of hot, untreated waste geothermal steam and water from geothermal power plants. An awareness of the environmental impact of geothermal fluids enables the design of environmentally-sensitive power plants such as that at Ohaaki, New Zealand.

KEYWORDS: Geothermal energy, arsenic, boron, mercury, pollution time-sinks, environmental impacts

INTRODUCTION

Geothermal fields are found throughout the world in a range of geological settings, and are increasingly being developed as an energy source. Each of the different types of geothermal system has distinct characteristics which are reflected in the chemistry of the geothermal fluids and their potential applications. However, they all have in common a heat source at a few kilometres depth, and it is this which sets water, present in the upper sections of the Earth's crust, into convection. The heat source for the system is a function of the geological or tectonic setting. A magma commonly provides the driving heat flux for many high-temperature systems. However, a magma is not necessary for the generation of a geothermal system: heat can be supplied by the tectonic uplift of hot basement rocks, or water can be heated by unusually deep circulation created by folding of a permeable horizon or faulting. Most geothermal resources can be used for space heating applications (eg. urban district heating schemes, fish farming, greenhouse heating), but it is only the hotter systems (>~180°C) which are used to generate electricity through the production of steam.

Geothermal fields are commonly classified by the nature of the reservoir: equilibrium state, fluid, temperature. This paper focuses the environmental impacts, notably chemical pollution, of dynamic, high-temperature, liquid-dominated systems and the development of these systems for power generation. However, many of the observations are also relevant to geothermal resource development in general.

ENVIRONMENTAL IMPACTS OF GEOTHERMAL DEVELOPMENT

To set the discussion on geothermal chemical pollution in context, this overview briefly describes environmental impacts which are enhanced or a consequence of geothermal developments. To emphasise the geothermal-related impacts, those which would be encountered in any construction development (eg. increased road transport, land clearance) are not considered.

Local hydrology disturbance: reduced flow-rate in steams due to water withdrawal for cooling/drilling; increased levels due to exploitation-induced subsidence.
Disturbance of forest/flora: defoliation due to hydrogen sulphide content of steam, and the boron concentrations of water aerosols in the steam discharge.
Hydrothermal eruptions: increase in incidence of shallow focus eruption due to formation of steam zone on exploitation (Bixley and Brown, 1988).
Unsafe hot ground: increase in area of steaming ground on formation of a steam zone on exploitation with lateral migration of steam.
Noise: of high velocity steam discharge.
Drilling blow-outs: can develop into eruptions in addition to the uncontrolled noise and steam pollution.
Subsidence due to fluid withdrawal: removal of geothermal fluid can cause compaction in some lithologies creating subsidence over the field.
Loss of geothermal discharge features: Lowering of the geothermal water table by withdrawal on exploitation leads to a loss of water discharge from springs and pools, with consequent impact on tourism and local culture.
Disposal of steam: hydrogen sulphide emissions of particular concern for health; carbon dioxide emissions contribute to global warming.
Disposal of steam condensates: potentially enriched in ammonia, boron, and mercury.
Disposal of hot geothermal waste water: pollution by heat, As, B, Hg Li, Rb of concern.
Environmental accumulation of toxic species: in soil, sediment and geothermal deposit time-sinks.

GEOTHERMAL CHEMICAL POLLUTANTS

Geothermal resources are natural polluters of the environment. Water and steam discharges containing potentially harmful constituents are found over most geothermal fields, and the development of geothermal resources enhances this pollution potential through the transmission and disposal of larges masses of waste water and steam extracted from the reservoir. The type and concentration of pollutants carried by the geothermal fluid is dependant upon several factors:
- host-rock - water reactions in the reservoir
   (organic-rich, sedimentary host-rocks provide greatest levels of pollutants)
- flow rate of fluid discharged by the system
   (greatest impact is caused by systems discharging large masses of geothermal water and steam)
- reservoir temperature
   (the concentration of pollutants increases with increasing reservoir temperature)

Geothermal water pollutants: Near-neutral-pH waters represent the deep geothermal fluid. They are composed predominantly of chloride, sodium and potassium and can carry significant concentrations of As, B, Hg and other trace elements (Nicholson, 1992).

Geothermal gas pollutants: Geothermal steam is produced by boiling of the reservoir fluid at depth. Gases typically form about 2% of the steam discharge. Carbon dioxide, comprising 95-98% of the gas content, and hydrogen sulphide (2-3%) are the main gases. Minor and variable amounts of ammonia, nitrogen, hydrogen and methane may also be present, together with the volatile species of boron, arsenic and mercury. Atmospheric pollution by hydrogen sulphide is of greatest concern, although the contribution of carbon dioxide emissions to global warming has also received attention. However, geothermal plants emit about 25% of the sulphur gases and 5% of the carbon dioxide of coal-fired power plants (Bowen, 1979).

Dispersion Processes and Pollution Time-sinks

Geothermal species of environmental concern are dispersed through the environment through geothermal waters and steam by a series of primary and secondary dispersion processes (Fig. 1). While many species entering into the atmosphere or aquatic systems are diluted and dispersed away from the discharge area, others can form significant, potentially toxic, accumulations over time.

Geothermal water commonly enters the local drainage system either by direct discharge or seepages, contributing heat, solutes and particulate matter (silica, carbonates, sulphides). Most solutes remain in solution and are carried from the point of discharge some, however, are taken up onto the river or lake sediments. These species can accumulate on the sediments to form sinks or reservoirs at concentrations far in excess of the soluble concentrations. Remobilization of species in such sinks would release an elevated, potentially toxic, flush of the species into the environment. To emphasise this time-accumulation-remobilization factor, the term "pollution time-sink" is used to describe these sedimentary elemental reservoirs.

Geothermal steam and gases discharge directly into the atmosphere through vents or fumaroles. However, vapours also seep through overlying soil horizons, slowly diffusing into the atmosphere. During this latter process some of the volatile species in the steam, notably ammonia, arsenic, boron and mercury, are taken up in the soil, particularly by organic matter but also by clay minerals, to form zones enriched in these species. To illustrate these pathways and sinks the fate of three elements of environmental concern will be examined: arsenic, boron and mercury from New Zealand geothermal systems. All three elements form time-sinks in soils, sediments and geothermal deposits.

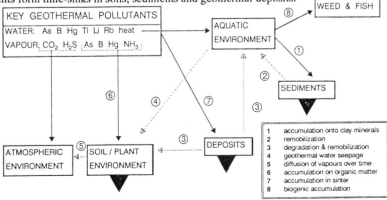

Fig. 1. Dispersion pathways and sinks of geothermal pollutants. ▼ = pollution time-sink.

2897

Arsenic

Aquatic systems and sedimentary sinks: Geothermal waters in New Zealand contain arsenic at concentration up to about 6.5 mg/kg (Parker and Nicholson, 1990). The Wairakei power plant has discharged untreated geothermal waste water at 60-80°C into the Waikato River at a rate of 6500 tonnes/hour for over thirty years. These geothermal waters contain an average of 4mg/kg As and therefore discharge about 190 tonnes/hr of arsenic into the drainage system. Other geothermal fields drain naturally into the river, but the Wairakei power plant effluents accounting for 75% of the dissolved arsenic content of the river (Aggett, *et al.*, 1982). Dissolved arsenic levels along the river range from 8 mg/kg (background) to 95 mg/kg. However, significant levels of arsenic accumulate on river and lake sediments of the Waikato drainage system. Aggett et al. (1982) determined mean sedimentary arsenic levels of 335 mg/kg (dry weight) representing accumulation rates of 8 tonnes As/yr. Moreover, this arsenic is remobilized on lake stratification during the Summer (January) and builds up in the hyperlimnion to levels representing up to 40% of the mean arsenic content of the lake water, until turnover in April-May.

Geothermal deposit sinks: Arsenic-rich sulphide deposits have been recorded from several sites in New Zealand. These invariably form millimetre-size layers within silica sinter deposited around hot springs and pools. These amorphous sulphide deposits can also contain antimony, thallium and mercury at the several hundred or thousand mg/kg level. The concentration of arsenic in New Zealand sinters is typically in the range 50-200mg/kg, but examples up to 12 000 mg/kg are known from the Waiotapu geothermal field (Nicholson and Parker, 1990; Parker and Nicholson, 1990). These deposits are often friable and can be easily weathered. They represent arsenic time-sinks which once eroded can return years of arsenic accumulation into the environment at high levels.

Boron

Soil sinks: Boron concentrations in New Zealand geothermal waters are typically in the range 10-100 mg/kg, although exceptionally high boron levels in excess of 1000 mg/kg are found at Ngawha, a high-temperature, sedimentary-hosted field. Boron is also a trace constituent in geothermal steam discharges. Soil-boron time-sinks can form from the accumulations of boron from vapour diffusing through the near-surface horizons, and from geothermal water seepage. Vapour-produced accumulations of plant-available soil-boron at levels of up to 9.6 mg/kg (dry weight) have been reported over a 3km² area from Naike, a low-temperature, sedimentary-hosted field (Liu and Nicholson, 1990; Nicholson and Liu, 1991). High-temperature fields show even greater soil-boron accumulations; vapour diffusion and water seepage have produced soil-boron levels of up to 874 mg/kg over the Tokaanu field (Bovelander, 1991). All are well in excess of 4 mg/kg, regarded as the maximum level of available soil-boron for optimum plant growth.

Aquatic systems and sedimentary sinks: The Waikato River receives natural discharges of 10-100 mg/kg B from several geothermal fields and. as noted above, effluent from the Wairakei power plant. Collectively these contribute 1400 tonnes B/yr. Boron is adsorbed on the river sediments and may be remobilized during times of flooding and increased turbulence (Nicholson, unpublished data). At Tokaanu, where boron discharges at a rate of about 3 kg/s into the local steam, sedimentary accumulations of up to 45 mg/kg have been recorded (Bovelander, 1991).

Mercury

Aquatic systems and sedimentary sinks: Geothermal discharges into the Waikato River add 50 kg mercury/yr and this is regarded as partly responsible for the high concentrations of mercury (often > 0.5 mg/kg wet flesh) in trout from the river, and sediment-mercury levels in excess of 200 ug/kg (Timperley, 1988). Of the 1500kg of Hg discharged over the lifetime of the Wairakei Power Plant, Timperley (1988) suggests that at least part (possibly up to 50%) is lost from the aquatic environment to the atmosphere by volatilization from the river, particularly at the effluent outfall where the water temperature is 27°C.

Soil sinks: As with boron, vapour-borne mercury can accumulate in the upper soil horizons. A soil survey the Reporoa high-temperature field showed soil-mercury levels up to 1510 ug/kg with anomalous mercury concentrations in excess of 500 ug/kg covering an area of about 20 km² (Sheppard et al., 1990).

CONCLUSIONS

Geothermal energy does have an environmental impact. Geothermal fluids contain potentially toxic species and can pollute aquatic, soil and atmospheric environments if discharged untreated. Condensation of the steam discharge, the use of holding ponds to cool waste water and precipitate solids and the reinjection of waste water and steam condensates at depth, away from the production zone, are some of the measures practised to significantly reduce the impact of geothermal development. Environmentally sensitive development of a geothermal power plant is exemplified by the recently commissioned Ohaaki power plant (see a thematic collection of papers in Brown and Nicholson, 1989). This station boasts New Zealand's only cooling tower - now a tourist attraction.

REFERENCES

Aggett, J., Aspell, A. and O'Brian, G. (1982). Arsenic: An environmental problem associated with operation of the Wairakei and Broadlands power plants. Proc. N.Z. Geothermal Workshop, 4, 157-159.
Bixley, P.F. and Brown, P.R.L. (1988). Hyrdothermal eruption potential in geothermal developments. Proc. N.Z. Geothermal Workshop, 10, 195-198.
Bovelander, M.J.W. (1991). Boron abundance and distribution in the surface geothermal environment at Tokaanu. N.Z. Geochemical Group Conference Abstracts, 17.
Bowen, R.G. (1979). Environmental impact of geothermal development. In: Kruger, P and Otte, C. (Eds) Geothermal Energy, Stanfoord University Press, 197-215.
Brown P.R.L. and Nicholson, K. (Editors) (1989). Proceedings of the 11th New Zealand Geothermal Workshop 1989. University of Auckland Press, 377pp.
Liu, W. and Nicholson, K. (1990). Boron in soil over Naike and Whitford low-enthalpy geothermal fields, New Zealand. Proc. N.Z. Geothermal Workshop, 12, 177-182.
Nicholson, K. (1992). Geothermal fluid chemistry. Springer-Verlag, in press.
Nicholson K. and Liu, W. (1991). Energy resources and the environment: Influence of natural boron emissions on soil chemistry over two geothermal resources in New Zealand. In: Selinus, O. (Ed.) Second international symposium on environmental geochemistry. Swedish Geol. Surv. Report 69, 95.
Nicholson, K. and Parker, R.J. (1990). Geothermal sinter chemistry: towards a diagnostic signature and a sinter geothermometer. Proc. N.Z. Geothermal Workshop, 12, 97-102.
Parker, R.J. and Nicholson, K. (1990). Arsenic in geothermal sinter: determination and implications for mineral exploration. Proc. N.Z. Geothermal Workshop, 12, 35-39.
Sheppard, D.S., Favre-Pierret, Orange, C.J. and Le Guern, F. (1990). Soil-gas surveys: a cheaper alternative to geophysical surveys - three examples from the Taupo Volcanic Zone. Proc. N.Z. Geothermal Workshop, 12, 125-128.
Timperley, M.H. (1988). Mercury in the Waikato River. Proc. N.Z. Geothermal Workshop, 10, 235-238.

ON THE WIDER USE OF RENEWABLE ENERGY SYSTEMS IN INDIA
- SOME SPECIFIC ISSUES

M. SIDDHARTHA BHATT, S. SEETHARAMU and K.R. KRISHNASWAMY
Central Power Research Institute
Sir C.V. Raman Avenue, Bangalore - 560 094, India.

ABSTRACT

This paper discusses some of the issues related to the wider use of renewable energy systems in India. The specific technologies under consideration are solar water heaters, biomass gasifiers for stationary use and autonomous wind electric generators. It is concluded that high capital cost and inadequate maintenance back up have resulted in a slow pace of development of these technologies.

KEYWORDS

Solar collector; biomass gasifier; wind electric generator.

INTRODUCTION

A concentrated national level policy and programme on use of renewable energy systems were evolved in the early 1980s. As a first step capital subsidies (30 to 100%) and exemptions from certain categories of taxes, were offered to users. These gave an initial impetus to the programme and a large number of systems were installed. However, capital subsidies are only incentives to help faster penentration of specific technologies. It is desirable that the subsidy programme yield the following results:

(a) Perfection of technologies for reliable field use (b) Establish technical competance and capabilities of manufacturers for specific products (c) Promotion of public awareness (d) Creation of substantial open market.

Subsidy programmes have produced good results. Yet, several new problems and issues have crept in. These have to be addressed immediately to accelerate the process of bringing new users into the network of renewable energy utilisation.

The common problems of solar, biomass and wind have been high
capital costs, long payback periods, inadequate
infrastructure for repairs and handling trouble shooting of
systems installed at site, etc..

In this paper, an attempt is made to bring out some of the
specific issues which need to be tackled to create a wider
usage of renewable systems.

SOLAR FLAT PLATE COLLECTORS

Solar flat plate collectors are used in India mainly for
obtaining hot water for bathing. The temperature of water
used for bathing is around 42°C and the requirement is about
35 litres per person per bath. This water is generally
needed during the morning (6.00 to 9.00 AM) and is presently
being generated by electric geysers, kerosene stoves or wood
fired geysers.

In Bangalore city alone (population: 4 million), there are
about 0.3 million all-electric-homes (AEH). These consume
nearly 1.2 million kWh/day of electrical energy for water
heating giving rise to a peak demand of 200 MW (Ramprasad,
1992). Every year several thousand new AEH connections are
being given in Bangalore and this is creating demand
management problems. An alternative is the use of solar
energy. The rainy season is limited (June-Sep.) and it is
sunny at other times of the year.

Technological status: Over the last few years the quality of
the collector has been optimised and improved considerably.
The quality of the glass cover, collector coating, sealing,
housing, etc., have been improved. The standard module is of
2 m^2 with an insulated storage tank and booster electric
heaters. In larger units water is circulated by a pump. A
task force for repair and maintenance has yet to be formed.
This is a major requirement as small problems render the
system in-operational. Presently, about 500 domestic
collectors have been installed and problems such as leakage
of water pipes, hot air escape from collector, improper
thermo-syphon action, high temperature drop overnight, etc.,
are being experienced. CPRI is setting up a testing centre
for a wide range of tests to evaluate solar collectors.

Capital cost: An impediment in the adoption of solar heaters
is the very high capital cost (Rs. 120/litre/day or Rs. 12,000
for a collector module of 2 m^2). The payback period on the
investment is around 10 to 12 years in comparison with an
electric geyser. An important area of R&D would be to
develop low cost, long lasting and reliable flat plate
collectors of 2 m^2 size to be used in domestic installations.

Subsidy programme: Subsidies are available for only a
fraction of the total demand (say 10%). This introduces a
differential pricing system and users are not willing to buy
unless subsidy is given. Also, the facility has been used
only by the top income bracket groups.

A welcome development in recent years is that the public
utilities and electricity boards have shown interest in
subsidising (10 to 20%) and/or providing soft loans (interest
rate below 5% and payments spanning over 60 to 80 months).
The authorities providing loans will also be instrumental in
ensuring product quality and service back up. The cost of
installation, generation and transmission is Rs. 15 million/MW
and if solar energy is used as an alternative it is only
Rs. 12.5 million/MW.

BIOMASS GASIFIERS

Technological status: While the technology for updraft steam
gasification is available upto 1 MW (thermal), downdraft
gasifier-engine driven systems are available only in the 1 to
10 kW range. Though systems of 100 kW are in use, their long
term reliability is yet to be proved. The downdraft
technology is available for only wood chips and rice husk.
There is a need for extending the technology for different
biomass fuels and larger sized wood pieces as availability
and sizing of wood to the required size is a major deterent
to the downdraft technology. Updraft reactors with
multi-fuel capability and ability to be operated on a range
of particle sizes and moisture contents has been developed
(CPRI, 1990). Developmental work is also needed in the
elimination of tar from the product gas before it enters the
IC engine. There is also a need for development of fluidised
bed gasifier-engine systems for capacities in the range 100
to 500 kW. A strong maintenance, repair and service back up
facility must be created at the earliest. Also, diagnostic
feedback of system performance to R&D Centres is called for.

Financial factors: The capital investment on gasifiers is
high but the payback period is only around five years.
Hence, the technology is economically attractive.

WIND ELECTRIC GENERATORS

Technological status: The designs successful in the West are
largely unsuccessful in India as the wind regimes are
moderate. Specific design changes are needed. The
technology for units upto 55 kW is indigenously available.
However, some developmental work needs to be done in the area
of electrical components of the system such as terminal
voltage controllers, charge controllers, etc.. Typical
problems faced in small sized autonomous units are covered in
a CPRI paper elsewhere in this Congress.

There is a need for developing units of 250 kW or so for
integration into a wind farm of 2 to 5 kW in selected
locations in the country. Presently most of the efforts in
wind engineering are from the public sector. There are very
few private industries interested in developmental work on
wind systems.

Economics: The capital cost for stand alone units is around
Rs. 30,000/kW and does not break even with conventional

sources of energy. However, the technology is environmentally friendly and independant of the grid. Hence there is a need for development of low cost systems to make it economically attractive. A significant development in the area of low cost blade design is the CPRI design (Kamalakar and Seetharamu, 1989). The long term reliability of this system is being established.

Promotors: The role of the promotor is to get the systems fabricated in large numbers, create the market potential, arrange for subsidies, advertise the product, etc.. Promotors are hesitant of taking up wind technologies for commercialisation even in cases where the technological risks are limited and the economics is attractive. At present there are very few manufacturers making wind electric generators and all efforts are limited to public agencies.

CONCLUSION

High capital costs, inadequate infracstructure for repairs and maintenance, technological risks and poor participation from the private sector are some of the reasons for the slow development of renewable energy systems in India.

ACKNOWLEDGEMENT

The authors wish to thank the management of CPRI for permission to publish this paper.

REFERENCES

CPRI (1990), Development and performance of a novel high rate updraft biomass gasifier, Technical Report No. 209, Central Power Research Institute, Bangalore - 560 094, Sept. 1990, 55 pages.

Kamalakar, R.S. and Seetharamu, S., (1989), Efficient, low cost blades for windmills, Paper No. 899396, Proc. IECEC, Vol. 4, Energy Management and Renewable Resource System, IEEE, New York, pp. 2009-13.

Ramprasad, M.S. (1992), Solar water heating systems - overview, Paper presented at Seminar on Strategies for commercialisation of renewable energy technologies, Hotel Taj Residency, Bangalore, India, 7th Feb. 1992.

A HYBRID POWER SUPPLY CONCEPT OF THE PENDULOR DEVICE

T.Watabe* , M.Fujiwara* , M.Kuroi** , K.Yaguchi*** & H.Kondo*

 * Muroran Inst. of Tech., 27-1, Mizumoto, Muroran, 050, JAPAN
 ** Eng. raboratory of Hitachi Ship Builder Co. Ltd.,
            1-3-22, Sakurajima, Konohana-ku, Osaka, 554, JAPAN
 *** NIIGATA ENG. Co. Ltd., 1-30-20, Kamata-Honcho, Ohta-ku,
                                      Tokyo, 144,JAPAN

ABSTRACT

A wave power system *Pendulor* was designed as an example of hybrid power supply concept which is based on a principle of mutual support. The *Pendulor*
which is an incomplete system in itself, when coupled with a *Diesel power
plant* not only completes the system but also yields benefits to the *Diesel
power plant.*

KEYWORDS

Natural Energy; Hybrid; Pendulor; Diesel Engine; Concept.

INTRODUCTION

A wave power system *Pendulor* was designed as an example of a hybrid power
supply concept which is based on a
principle of mutual support. The concept is to generate steady electricity from natural energy using an incomplete simple system. The *Pendulor*
(Watabe & Kondo 1990) was specially
considered for suitable operation combined with a *Diesel power plant*.
Behaviour of the *Pendulor* driven by
irregular waves has been studied theoretically, and the output was shown
in several cases. This concept can be
applied for other renweable energy
systems.

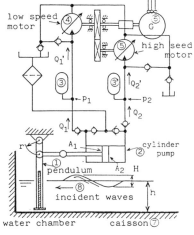

Fig.1 Principle of the Pendulor

FUNDAMENTAL CONCEPT

A hybrid power supply concept was accepted in order to break the barrier
of economy and storage problems of

2904

natural energy. The concept is to generate steady electricity from nature in a simple way. It depends on a principle of mutual support combining two or more systems together. An example is a combination of a *Diesel power plant* and the *Pendulor*. The combination brings the benefit as shown in Table1.

Table 1 The benefit of the combined system

The energy storage problems are free for steady power supply.
The *Pendulor* consists of no speed controller.
The combination makes the output bigger.
The *Diesel power plant* can save fuel.
The *Diesel power plant* needs no additional equipment.

DESIGN CONCEPT OF THE HYBRID SYSTEM

The *Diesel power plant* is suitable for local use and the *Pendulor* as well. Fig.1 shows principle of the *Pendulor*. As the pendulum ① swings to the left, the pump ② delivers oil to the motor ④. The delivery pressure depends on the amount of the oil trapped in the accumulator ③. Similarly, while the pendulum ① swings to the right, the pump ② delivers oil to the motor ⑤. The delivery pressure depends on the amount of the oil trapped in the acc. ③'. The motors ④ & ⑤ drive the generator ⑥ together. Torque of each the motor changes periodically though, the sum of the two torques makes a uniform torque.

When the natural frequency is selected in accordance with the waves' frequency, the pendulum swings violently in resonance. If the pressure p is so controlled that the pendulum comes to an impedance matching, the *Pendulor* works best to generate electricity. The best condition can be kept automatically even when the waves change height H and period T (Watabe & Kondo 1989).

The *Pendulor* deeply depends on the technology of oil hydraulics which has been widely used in many fields. However, the pendulum motion excited by the waves is quite slow but accompanies huge force. The motion is very suitable for driving a

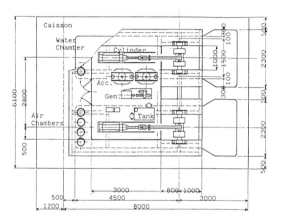

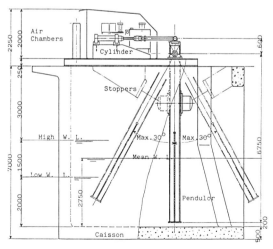

Fig.2 The Test Pendulor (a conceptual design)

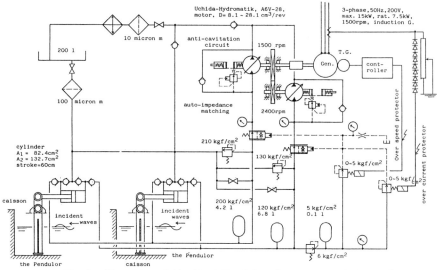

Fig.3 Circuits of the Test Pendulor (a conceptual design)

cylinder which acts as a pump of a HST(Hydro-Static Transmission). HST is robust for marine use and matches for driving a generator which is governed by a *Diesel power plant*.

DYNAMIC SIMULATION

Fig.2 shows the *test Pendulor*(Watabe & et al 1991) adopted for an object of this study. The *Pendulor* consists of the two pendulums which drive a 15kW generator. Fig.3 shows the circuits. Equa. of motion of the pendulum is written in the next form.

$$(I_0 + I) \ \ddot{\theta} + N \dot{\theta} + (K_0 + K) \ \theta + M_p = M_w \qquad (1)$$

Where, I_0 : moment of inertia of the pendulum, I : moment of inertia by adding water, θ : angular displacement, N : damping factor by wave generation, K_0 : restoring moment factor by gravity, K : restoring moment factor by changing water level behind the pendulum, M_p : damping by the pump, M_w : exciting moment by the waves

$$\left. \begin{array}{l} M_p = r \, A_1 p_1 \quad \text{while} \ \ \dot{\theta} < 0 , \\[2mm] M_p = r \, A_2 p_2 \quad \text{while} \ \ \dot{\theta} > 0 \end{array} \right\} \qquad (2)$$

when suction pressure is atmospheric, r : length of the arm(see Fig.1), A : piston area of the pump, p : delivery pressure to the motor

$$p_1 = (k_a / A_a{}^2) \, V_1 \quad , \quad p_2 = (k_a / A_a{}^2) \, V_2 \qquad (3)$$

where, k_a : equivalent spring const. of the acc., A_a : piston area of the acc., V : volume of oil in the acc.

$$V = V_0 + \int (Q - Q') \, dt \qquad (4)$$

2906

where, V_0 : vulume at the initial condition, Q : delivery flow of the pump, Q' : supplying flow to the motor

$$Q' = n_m D_m / \eta_v \qquad (5)$$

n_m : motor speed, D_m : displacement of the motor, η_v : volumetric eff.

The incident power E entering into width B of a pendulum is calculated by Equa.(6).

$$E = \rho g B \int_0^\infty S(f) C_g \, df \qquad (6)$$

where, ρ : density of water, g : acceleration of gravity, $S(f)$: wave spectrum, C_g : group speed of the waves, f : frequency of the waves

Wave spectrum is defined by the next relationship.

$$\int_0^\infty S(f) \, df = \sum_{n=1}^\infty \frac{1}{2} \left(\frac{H_n}{2} \right)^2 \qquad (7)$$

here, H_n : wave height of nth wave
Therefore, the exciting moment M_w is written in Equa.(8).

$$M_w = \sum_{n=1}^\infty A_n B \frac{H_n}{2} \sin 2\pi f_n t \qquad (8)$$

here, A_n : constant related on f_n and water depth h

The output of the *test Pendulor* was investigated by this simulation which was considered under the same wave condition; — significant wave height $H_{1/3} = 1.5$ m, its period $T_{1/3} = 4$ sec. Fig.4 shows an accumulator designed for the *test Pendulor*(Watabe & et al 1991).

The typical results are shown in Fig.5 which were derived by applying one of the best combinations of pump and motor. In Fig.5, since the size of the accumulator is different in each case, the output appears in several features in terms of time history, but concerning the average output, there is no significant difference. The bigger the accumulator is, the smoother the output and the other hand, the more expensive. On the contrary, out of the best combination brings a lower output accompanied by an unusual (lower or higher) pressure appearance. It is expected that the total output can be flattened even in the worst case in Fig.5 by coupling the *Pendulor* with a *Diesel plant*.

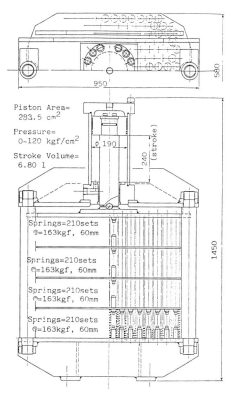

Piston Area= 283.5 cm^2
Pressure= 0~120 kgf/cm^2
Stroke Volume= 6.80 l

Springs=210sets @=163kgf, 60mm

Springs=210sets @=163kgf, 60mm

Springs=210sets @=163kgf, 60mm

Springs=210sets @=163kgf, 60mm

Fig.4 Accumulator designed for the Test Pendulor

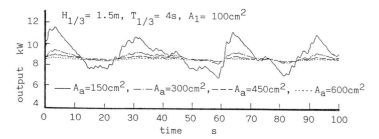

Fig.5   The output of the Test Pendulor

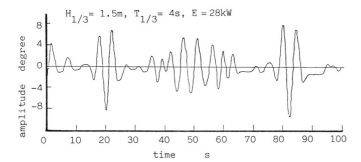

Fig.6   Response of the Pendulum

Fig.6 shows a typical motion of the pendulum studied by the simulation (Watabe & et al 1989)

CONCLUSION

This hybrid power supply concept based on a mutual principle is a key idea in order to solve the problems of natural energy utilization. A hybrid system coupled with the *Pendulor* and a *Diesel power plant* is one of the examples and its total output can be flattened even in the worst case in Fig.5. Applying an optimal combination of pump and motor, the *Pendulor* can generate a steady output maintaining best efficiency.

REFRENCES

T.Watabe & H.Kondo(1989). Hydraulic technology & utilization of ocean wave power, Proc.JHPS,301-308.
T.Watabe, H.Kondo & M.Kobiyama(1989). A case study on the utilization of ocean wave energy for fish farming in Hokkaido, Proc.CRHT, 159-164.
T.Watabe & H.Kondo(1990). Progress on Pendulor-Type wave energy converter device, Proc.1st PACOMS, 161-169.
T.Watabe, H.kondo, H.Yamagishi & M.Kudo(1991). Design of a wave plant of the Pendulor (in Japanese), Proc.3rd symp. OWEU-JAMSTEC, 3-13.

CURRENT TRENDS IN RENEWABLE ENERGY SYSTEMS IN INDIA

M.ARUMUGAM and S.NATARAJAN,

Department of Physics, College of Engineering,
Anna University, Madras-600 025, India.

ABSTRACT

India is looking forward for the use of new and renewable sources of energy to face the great demand of energy. The industrial growth in India during 1990-91 is reduced to 0.3 percent due to shortage of electrical power. Economic Progress of India, to come on par with developed countries requires the availability of large quantities of renewable energy as a primary input. The Department of Non-conventional Energy Sources, India in collaboration with different statewide Energy Development Agencies and Electricity Boards has taken up national programmes relating to demonstration and extension of various new and renewable sources of energy systems. The above programmes include the setting up of windmills for water pumping and power generations, solar water heating systems, solar photovoltaic powered lighting and water pumping systems and bio-gas plants for cooking and electrical power generation.

KEYWORDS

Renewable energy; Unit; Solar pond; Briquette;

POWER NEEDS OF INDIA DURING 2000 A.D.:

The energy resources in India are finite and exhaustible. There is no assurance of regular or periodical replenishment or renewal. We have commercial forms of energy like coal, hydro and thermal electric power, oil, atomic energy and natural gas. Similarly we have also non-commercial forms of energy like fuel wood, vegetables and animal wastes. Table 1 shows the present Indian status of energy availability and requirement during 2000 A.D. The expectations shown in table 1 are based on the fact that the expected population of India during 2000 A.D. is about one billion. It is believed that the world reserves of coal and oil will be used within 130 years. Therefore we are in a position to use alternative energy sources.

Table 1: Indian Energy Status

Forms of Energy	Present Availability	Requirement during 2000 A.D.
A. COMMERCIAL		
Electricity (in Billion kW)	264	424
Coal (in Million Tonnes)	211	288
Oil (in Million Tonnes)	33	196
B. NON-COMMERCIAL		
Fuel Wood (in Million Tonnes)	49	192
Dung Cake (in Million Tonnes)	96	105
Plant Waste (in Million Tonnes)	38	59

UTILIZATION OF SOLAR ENERGY IN INDIA

Actually India receives large amount of solar energy such that 6000 billion MWh per year. If we utilize 0.1% of this energy, the created energy amounts 35 times the present generation of electricity in India. Solar energy is usefully exploited through solar photovoltaic and thermal systems. Upto March 1991, nearly 7500 solar photovoltaic operated street lights have been installed in India. Solar photovoltaic power plants are producing the electrical energy of 400 kWh. The Central Electronics Limited, Sahibabad in Utter Pradesh, is manufacturing solar photovoltaic deep well water pumping systems suitable to any place in India. The daily water pumping capacity of these solar powered pumps ranges from 10,000 litres at a depth of 100 metres to 25,000 litres at a depth of 20 metres. The solar thermal water heating systems are popular even in the metropolitan areas. The solar panels are available to deliver hot water upto 90°C. Each panel output is ranging from 100 litres per day to 100000 litres per day. Hot water is also made available during non-sunny hours by storing it in the insulated steel tanks. The solar thermal panels having total area of 172000 square metres are installed so far according to the data given in 'Renewable Energy for all' (1991) published by the Department of Non-conventional Energy Sources, New Delhi.

Solar ponds

Researches are going on, in the area of tapping electrical energy from the nonconvective salt gradient solar ponds. Solar ponds are the artificial ponds. These are constructed such that the bottom layer of water is made more saline than that at the surface and thus the solar pond is rendered non-convective. The salinity gradient in the pond gives a vertical density difference so that the heated water remains at the bottom. Temperatures as high as 103°C have been recorded at the bottom of such solar ponds. The solar pond having area of 1000m^2 can produce a power of 5000 MW. The Central Salt and Marine Chemicals Research Institute in Bhavanagar has erected a 1600 m^2 solar pond which has been in operation for the past 5 years. The production cost of electricity is about 4 rupees per unit (kWh). The erection works of solar ponds of 5000 m^2 amd 2000 m^2 are under progress at Pune and Pondicherry respectively.

ENERGY FROM TIDES AND WINDS

In India, some experimental projects are carried out in a successful manner to tap the energy from tides. Near Bombay, the tides having spring range of 3.6 metres can be exploited for power generation purposes. We can expect some commercial turbines very soon along the coastal basins of Tamil Nadu, Gujarat, Maharashtra and Orissa states to generate electricity from the tides. Initially the windmills and solar powered water pumping systems are installed in the places where there is no electrical distribution system. Windmills are now installed at the locations where the average wind speed is above 6m/sec. The wind farms are operating at various places in Gujarat, Tamil Nadu, Orissa and Maharashtra states. The capacity of these wind farms ranges from 0.55 MW to 11.2 MW. There are over 1500 windmills installed across the country with 10,000 more envisaged for the eighth plan. The Asia's largest wind farm is located at Lamba (Gujarat) having a total capacity of 11.2 MW. The research and development wing of Bharat Heavy Electricals Limited, Hyderabad and Natural Energy Processing Company, Madras are manufacturing commercially water pumping windmills having power generation from 5kW to 400 kW. The Natural Energy Processing Company has made a valuable contribution in the development and utilization of wind energy for power generation. It undertakes the installation of wind turbine generators upto 250 kW in collaboration with MICON of Denmark. Based on the annual report (1991) of National Energy Processing Company and the data given in 'Utilization of Wind Energy' (1991), published by the Tamil Nadu State Electricity Board, it is found that it has completed a project of 10 MW wind farm consisting of 30 machines of 250 kW at Kayathar and 20 machines of 250 kW at Muppandal in Tamil Nadu for Tamil Nadu State Electricity Board apart from 14 machines of 55 kW, 2 machines of 110 kW and 2 machines of 250 kW at Gujarat. Further it is going to install 1.5 MW wind farm for M/s. Metal Powder Company Limited, Madurai to generate electrical power for their industrial needs. India has a wind potential of 20,000 MW per annum. In wind energy utilization amongst all the states in the country, Tamil Nadu and Gujarat share major production. The wind farms are generating a total power of 15.4 MW in Tamil Nadu and 15.7 MW in Gujarat respectively. Meanwhile, the total power generation from wind in India is about 38 MW. It is estimated that the wind energy is supposed to yield 5,000 MW of power during 2000 A.D. Substantial subsidies are made by Government of India for erection of these windmills. The D.C. output from each windmill having 3 blades of glass fibre reinforced polyster is 1 kW and using an inverter, the A.C. voltage of 220 volts, 50 Hz is provided. Specially made wind turbines which pump water from a depth of 300 metres and generate the electrical power upto 400 kW are also available.

BIOGAS PLANTS

India is very much interested in installing the biogas plants for the production of cooking fuel and electricity for the rural areas. Upto March 1991, the biogas plants produced an electrical energy of 1402000 units. Agrowastes availability in India is 295 million tonnes per year. In this amount, the amount of rice straw availability per year is 110 million tonnes. The cost of the electricity per unit from various sources is given in Table 2.

Table 2 : Cost of Electricity per unit (kWh) in India according to the data given by the Tamil Nadu Energy Developmental Agency, Madras (1991).

Source	Cost per unit Rs.P.
Hydroelectric	0.40
Biogas	0.30
Coal	0.65
Nuclear	0.65
Wind	0.90
Diesel	1.40
Urban wastes	1.25
Gas from petroleum products	1.75

It is seen that in India, the production of electricity from biogas is very cheap, compared with other sources. Further according to the Entrepreneurship Development Programme Proceedings (1991) published by the United States Agencies for International Development, New Delhi. Stirling Dynamics, a private limited company in Madras commercially manufactures 5 H.P. Stirling Engines which are externally heated engines. Utilizing saw dust, rice husk and waste wood as fuels they convert the heat from combustion into mechanical shaft power. Indian Institute of Science, Punjab Agricultural University and Tamil Nadu Agricultural University are making gasifiers with the power of 5 kW to 50 kW using agrowastes such as sunflower earhead/stalk and redgram stalk. Fuel replacement of 60-85% has been achieved when the biogas is used in 5 H.P. diesel engine pumpset. Large biogas plants are manufactured by Lotus Energy Systems Private Limited, Bangalore and Indian Council of Agricultural Research Laboratories, New Delhi. Briquetting machines are available for the commercial manufacture of briquettes and are marketed by Alternative Fuel Private Limited, New Delhi and Spectoms Engineering Private Limited, Baroda. At present the biogas is produced at the rate of $0.19m^3$ of gas per kilogram of manure (animal waste). Eventhough only 15% of the manure is converted into biogas, the remaining amount of dry sludge is used as a fertilizer having 2% nitrogen per kilogram of manure.

<u>Biogas data</u>

The energy value of gas	=	20 MJ/m^3
The burning efficiency of the biogas	=	60%
The effective (useful) heat obtained	=	12MJ/m^3 of gas
Total no.of biogas plants installed during 1974 - 81	=	101213
Total no. of biogas plants installed during 1981 - 88	=	908644

Now there are so many villages which are fully utilizing biogas (Gobar gas) for lighting, pumping water, cooking and operation of other equipments like radio, fan and television.

Table 3 :   Cost of pumping 1 Million litres of water through a head of 12 metres with alternative devices.

Pumping Soruce	Cost Rs.P.
Diesel pumpset (using 5 H.P. engine)	252.99
Wood gasifier driven engine (Duel fuel - 20% diesel + 80% wood gas)	258.24
Windmill	202.00
Biogas engine (Duel fuel - 20% diesel + 80% biogas)	120.00
Electric pumpset	74.79

Table 3 shows the amount of expenditure involved for pumping 1 million litres of water through a pressure head of 12 metres using different pumping machines. It is seen that among the non-conventional energy sources, the biogas engine is the cheapest one. It is found that using biogas alone as the fuel, an output power of 2 kW is obtained from the 3.7 kW engine. The cogeneration systems are experimentally tried in many hotels and hospitals. They generate electricity using gas or diesel. Meanwhile the developed waste heat is utilised for space heating and water heating. Ministry of Energy, Government of India also promotes the development of new renewable energy sources by means of giving financial help for the research projects. One such project finished during 1990 is the high efficiency wood burning stove which has high thermal efficiency and no smoke. It is reported that the saving of fire wood is more than 25% using this stove. Now it is marketed by M/s. Heat Systems, Bangalore in a large scale. In Anna University, Madras effective researches are going on to use Methanol and Hydrogen as automobile fuels. Methanol and Hydrogen are available as byproducts from the Chemical Industries situated around Madras City. It is proved that the duel fuel system using Methanol or Hydrogen along with diesel is a successful one to run the automobiles and to reduce the cost of the fuel. Further researches are going on for the effective storage of Hydrogen and the development of engines using Hydrogen fuel alone. Government of India has taken keen interest on production of energy from non conventional energy sources. In the present eighth plan, the allotted amount for the development of non conventional energy sources is about rupees 6,930 crores; but it was about rupees 100 crores in the sixth plan.

In conclusion it would appear that the energy scenario for India in the next century is a matter of great concern. Proper attention and more facilities in research and aids from Indian Government would make rapid progress in the development of renewable energy systems. Time is not far when one can visualise a future with renewable energy systems like wind farms and solar farms shouldering the major share of the power generation in India. This will not only solve India's energy crisis but also provides a clean, radiation free and unpolluted environment maintaining the natural setup adapted to its eco-system.

REFERENCES

Annual Report for the year 1990-91 (1991), Natural Energy Processing Company, Madras.

Proc. of the Entrepreneurship Development Programme in New and Renewable Energy Technologies (1991), United States Agency for International Development, New Delhi.

Progress on New and Renewable Energy Systems during 1985-1990 (1991), Tamil Nadu Energy Developmental Agency, Madras.

Renewable Energy For All (1991), Information and Publicity Wing, Department of Non Conventional Energy Sources, Government of India, New Delhi.

Utilization of Wind Energy (1991), Tamil Nadu State Electricity Board, Madras.

ROLLING ELEMENT BEARINGS FOR RENEWABLE ENERGY AND INTERMEDIATE TECHNOLOGIES

CE Oram

Development Technology Unit, University of Warwick, Coventry, CV4 7AL UK.

ABSTRACT

Manufacturers of machinery in less developed countries, are commonly troubled by the problems of obtaining and fitting bearings. Wooden bearings suffer from problems of wear and friction; and conventional rolling element bearings cost foreign exchange and are difficult to get. There are also the difficulties of mounting bearings on shafts and chassis of sufficient accuracy and alignment to exploit their performance. Going back to the essential features of the anti-friction bearing, that two relatively moving parts may be separated by rolling elements made of readily available materials, may provide a new answer to these problems. Suitable materials for bearings may include, mild steel round bar, malleable iron water pipe and even plywood. Bearings manufactured along the lines suggested in the present paper could be manufactured in the country, and very likely in the locality, of use. This paper presents some results of the work undertaken at Warwick over the last nine years or so, and suggests that re-inventing the wheel, or at least the rolling element bearing, might result in significant cost reductions in many renewable energy machines.

KEYWORDS

Anti-friction bearings; rolling element bearings; wooden bearings; renewable energy; water pumping wind turbines; animal engines; appropriate technology; intermediate technology.

INTRODUCTION

In third world countries, problems with the provision and installation of conventional rolling element and plain bearings are well known. Unfortunately the diffuse nature and slow speed of many renewable energies requires machinery with large shafts, which in turn require large bearings, and these are expensive. Such machinery is also susceptible to significant deformation under load and may require careful design to avoid early bearing failure through misalignment and bearing fight. Another problem concerns the accuracy of rolling element bearing seats and housings. Conventional bearings are lifed on the basis that they are mounted on shafts and in housings that control their running clearances. These tolerances are on the limit of what can be achieved with a single point tool in a good lathe and with a good operator.

What is required is a bearing technology that is not much harder to manufacture than wooden bearings, and yet is able to tolerate bad misalignment and poor lubrication for long periods. There are some compensations: most loads are modest, speeds are usually low, much higher wear can be tolerated, and the criterion of bearing failure can be relaxed; in the present work failure has been taken to mean jamming that cannot be rectified by cleaning and re-lubrication.

Working in Botswana in 1982, as a student, the author became aware of the difficulties of using conventional rolling bearings and carried out a short research project upon his return to the UK (Oram, 1983). The work was based on the supposition that rolling-element bearings made of less than ideal materials, might have useful performance. He had made ball bearings with races of mild steel and there were other precedents for soft rollers: industrial castors, thermoplastic rolling element bearings, nineteenth century English farm machinery and mining trucks in Germany up until the 1950s, for example. The author's tests were made on bearings with an inner race diameter of 160mm and 20mm thickness, carrying loads up to 7kN at about 150rpm.

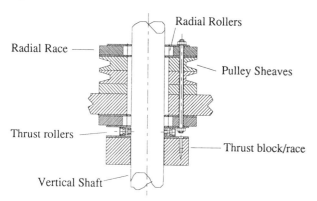

Figure 1: low technology rolling element bearing arrangement in an animal engine.

Other students: Godden (1986), Austin (1987) and Mascia (1989), have subsequently worked on alternative rolling element bearings, including crossed roller bearings, needle roller bearings, a cylindrical roller thrust bearing, bearings consisting of steel strip tyres on wooden backings and bearings with wooden races only. Fig.1 shows a bearing installation used on an animal engine being developed at Warwick and illustrates the freedom that bearings made in-house provides. The shaft here, was not available for machining. Both the steel races shown, and on other occasions the raw inside diameters of the plywood pulleys, were used as bearing races. In both modes the bearings functioned perfectly carrying loads around 200kg.

Low technology bearings mark a change from the conventional approach to bearing provision; namely from using highly selected materials, to one of using only those materials that are more readily available and accepting that performance will be adversly affected. Although it would be possible to use hardening processes, (a course adopted by one African manufacturer, Mujemula 1992), this pressure has been resisted from considerations of quality control.

STRESS ANALYSIS OF ROLLING ELEMENT BEARINGS

Hertz developed expressions for both compressive stress and shear stress which show that shear stress reaches a maximum of 0.304 of the maximum compressive stresses at a depth of 0.78 times the contact strip half width (for parallel cylindrical bodies). Relevant formulae are:

maximum compressive stress is given by $\sigma = \sqrt{(FK/\pi L)}$

contact strip half width $b = \sqrt{(4F/\pi LK)}$

load at yield $F_y = \pi L \sigma_y^2 / K$

where $K = (1/r_1 + 1/r_2)/((1-v_1^2)/E_1 + (1-v_2^2)/E_2)$ and σ_y is compressive yield stress, v is Poisson's ratio, F is roller load, L is roller length, r is roller radius and E is Young's modulus. Subscripts $_1$ and $_2$ denote the two materials in contact.

Thus roller load at the point of plasticity rises with the square of material hardness (or yield stress) and inversely as material stiffness. Table 1 below shows the load to just produce plasticity, together with the contact strip half width, for a range of materials. Permissible loads of low technology roller bearings are very low by comparison with those of conventional bearing steel, but reasonable loads can still be carried if enough race area can be provided. Provision of enough area is not necessarily too difficult; a range of vehicle wheel hubs is under development at Warwick and these, rated at the maximum load capability of the tyre, occupy a volume only slightly larger than a comparable hub using conventional bearings.

Table 1: Alternative bearing materials and bearing loads.

material	Young's modulus [MPa]	compressive yield stress [MPa]	Poisson's ratio ¶	roller load at 3 times yield stress [N] §	contact strip half width at yield [mm]
bearing steel	210 000	1 850	0.29	10 049	0.10
mild steel	210 000	350	0.29	360	0.02
galv water pipe	210 000	195	0.28	112	0.01
aluminium	70 000	200	0.36	335	0.03
brass	100 000	400	0.33	960	0.04
wood	20 000	50	0.45	67	0.02
polyacetyl	3 000	65	0.35	832	0.24

¶ for wood: heavily dependant on orientation: average figure taken from Beaver (1986)
§ for 25mm diameter shaft and 10mm diameter rollers each 10mm long.

INTERMEDIATE TECHNOLOGY ROLLING ELEMENT BEARINGS

Work in the present programme at Warwick has centered on the use of cylindrical roller bearings both from considerations of manufacture and of the likely low load carrying capacity of intermediate technology ball bearings. An important issue is how many rollers carry load. Stribeck, assumed that all rolling elements in a thrust bearing, or half of them in a radially loaded bearing, carried load; Allan (1945) and Barwell (1979). This situation is unlikely with intermediate technology rolling element bearings because of low manufacturing accuracy: a roller would be completely unloaded by making it smaller than its neighbours by a small fraction of the contact strip width. Nevertheless it is probable that more than one element is under load at any instant, simply from considerations of kinematic stability: observations suggest that three rollers normally carry load. A certain amount of 'running in' or plastic deformation could take place so that the outer race might become a better fit to the roller profile and might allow for misaligned operation also.

Hawkins (1992) has raised the possibility of an outer race made of a helix of small square or round section bar. The technique holds out the possibility of bearing manufacture without machine tools, and consequently of much more local manufacture. High contact stresses might be avoided by allowing elastic deformation of the outer race so that it may conform better to the required diameter. Very early tests of these bearings are encouraging. The issue of the place of manufacture is important and has implications for the effective cost of many intermediate technologies. Many authors, for example Starkey (1988), have detailed the difficulties experienced by emergent farmers with inappropriately designed equipment manufactured in some distant location.

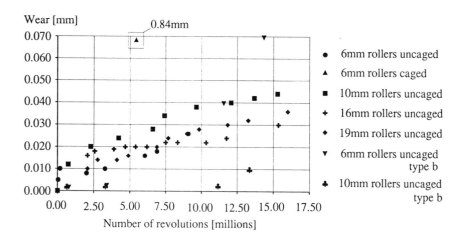

Fig. 2: low technology rolling bearing wear; taken from Mascia 1989.

Figure 2 taken from Mascia (1989) shows that wear at the nominal loading of about 2 MPa (load/projected area) is about 2 microns/million revolutions when using UK obtained free machining mild steel. Bearings in these tests were 10mm thick, had 26mm shafts and carried 500 N at 500 or 1000rpm. Type 'b' bearings had brass washers positioned so as to prevent damage to the bearings by contamination.

CONCLUSIONS

Results of tests on low technology rolling element bearings have been encouraging. Even at high loadings on the edge of plasticity, bearings have survived for periods adequate for animal power renewable energy applications for example. Tests have, in most cases, been suspended because of the time taken for testing, rather than failure of the bearing. Looking further afield there are other technologies in the South that could benefit from alternative bearings: transport and crop processing machinery are obvious candidates and the scale of the benefit to some countries may be judged from the case of one manufacturer in Bangladesh who claims to fit about 20 000 bearings per year to rice threshers alone, (Bose 1990). Work is well under way at Warwick to develop a range of axles, vehicles and pieces of equipment to exploit these bearings, but much further investigative work is required in the areas of lubrication, effect of dirt, sealing and other materials; and into high lifetime performance. The author is of the opinion that these bearings merit further attention and may allow a breakthrough in many bearing problems of the 'South'.

REFERENCES

Austin M. (1987). *The development of a simply manufactured crossed roller bearing for animal drawn carts in Zambia.* Third Year Undergraduate Project Report, Dept. Engineering, University of Warwick, Coventry, UK, 110p.

Allan R.K. (1945). *Rolling bearings.* Pitman, London, UK. 401p.

Barwell F.T. (1979). *Bearing systems, principles and practice.* Oxford University Press, Oxford, UK. 565p.

Beaver M.B. (1986). *Encyclopedia of materials science and engineering. 7,* p5417-8. Pergamon Press, Oxford, UK.

Bose A.N. (1990). Personal communication. Mr AN Bose, Chairman, Comilla Cooperative Karkhana Ltd, PO Box 12, Ranir Bazar, Comilla, Bangladesh.

Godden P. (1986). *Simply manufactured crossed roller bearings - their design for use in animal drawn carts in Zambia.* Third Year Undergraduate Project Report, Dept. of Engineering, University of Warwick, Coventry, UK, 110p.

Hawkins A. (1992). *Animal power units for developing countries.* MSc thesis, Dept. of Engineering, University of Warwick, Coventry, UK, 150p.

Mascia A. (1989). *The analysis and testing of low carbon steel roller bearings for use and manufacture in developing countries.* Third Year Undergraduate Project Report, Dept. of Engineering, University of Warwick, Coventry, UK, 100p.

Oram C.E. (1983). *Mild steel cylindrical roller bearings for developing countries and alternative technologies.* Third Year Undergraduate Project Report, Dept. of Engineering, University of Warwick, Coventry, UK, 202p.

Starkey P.H. (1988). *Animal Drawn Wheeled Toolcarriers: Perfected Yet Rejected.* Vieweg and Sohn, Braunschweig, Wiesbaden, Germany. 159p.

Umara B. (1992). *Transportation in Nigeria and the development of intermediate technology rolling element bearings;* MSc Thesis, Dept. of Engineering, University of Warwick, Coventry, UK. 200p.

Research of Geothermal Flowing Warm Water   High Density
Anguilla Culturing Technology

Zhu De-yan

Guangzhou Institute of Energy Conversion,
Chinese Academy of Sciences
81 Central Martyrs'Road,Guangzhou,China

ABSTRACT

Using geothermal resources, an open system of flowing water was
set up,which maitained the fish pond water temperature between
24 and 28 degrees C for successful anguilla culture at high
released density.

KWYWORDS

Geothermal; Flowing Warm Water ; High Density; Anguilla;Culture

PREFACE

As a warm water animal, anguilla will have poor appitite when
water temperature is below 15 degrees C and will eat nothing if
water temperature drops below 10 degrees C under which it will
hibernate on the pond bottom. In natural waters, it has to take
5 years as long to be ready for market. By artificial culture,
the period is also long as 3 years.to shorten culture cycle,
reduce energy consumed and heighten production in anguilla culture,the author undertook a research of anguilla culture technology in an exposed type flowing warm water system using geothermal water.by this technology,water temperature in anguilla
ponds were controlled between 24-28 degrees C.Breeding of
anguilla fry and high density culture of commodity anguilla
were carried out with high yield.

EXPERIMENTAL SITE AND CONDITIONS THEREOF

The experimental site is at the Wenquan Anquilla Farm, Jiexi
county, Guangdong province. Where, northwest wind prevails in
winter, with a maximum wind speed of 11 m/s and an averaged
wind speed of 2.3 m/s. Average temperature in January is   12.5

degrees C. The area is annually hit by 4 to 5 cold airs lasting
about 10 days, when the ambient temperature drops by 13 to 19
degrees C. In March and April, there are about 40 days in which
the weather is overcloud, warm,rainy , wet and with little
sun. Hence, in each year,there are about 5 months in which
water temperature is not good for anguilla growing.

The anguilla farm is by the side of Tanghu Hot Spring near the
water reservior of Longjing. For experiment,geothermal water
(33.5 °C) of 800 t/day from No.1 geothermal well is used.In
summer, water temperature at the area of Longjing water reservior is only about 19 °C,and this is favourable to the use of
flowing warm water for aquaculture.the geothermal water and the
reservior water are both clear and meet the standards of water
used for aquaculture.

EXPERIMENTAL PONDS

Basing on the features of anguilla living, the fry has to be
tended on a graded basis according to their size.Therefore, 4
adjoining pounds of 860 square meters in total area were chosen
as anguilla culture ponds.The ponds for commodity anguilla was
1.1m in height, 80cm in water depth. The pond of 169m^2 for
anguilla fry breeding was 80cm in height.All ponds were brick-
stone and cement in construction,with facilities for escape
prevention, draining and overflow.

ANALYSIS AND CALCULATION OF THERMAL SYSTEM

Thermal system of the exposed type culture ponds is composed of
pipe networks. A centrifugal pump draws water from the geothermal well and delivers it through a 2" manifold to the pondside.
And then the water is branched into 4 perforated pipes over the
ponds.From the perforation, 1 mm in diameter,positioned basing
on calculation, the water is sprayed onto the ponds at an elevation angle of 10. The branch pipes are controlled by valves.A
flow field of warm water is created by use of impeller type
oxygen adding machines which stir the pond water. Maximum quantity of water sprayed is 21.83 tons per hour,but for each pond,
3 tons per hour or less is enough.When water level goes above
preset level, the excess water will be discharged via the overflow.

The thermal system is designed basing on the thermal load of
exposed culture ponds calculated according to meteorological
elements. Thermal losses from the pond water body usually in-
clude:losses due to thermal conduction of pond bottom,which may
be neglected; losses due to convection, radiation and evaporation which take place concurrently and across the interface
between water and air.Among heat losses from water surface,
those attributed to evaporation are the pricipal. For evapora-
ion, water takes heat from the pond rather than from the air
over the pond to change its state, thus reducing water surface
temperature. If air temperature is lower than pond surface temperature, convection will lower surface water temperature,and
vice versa.During operation, the thermal system creates a local

2921

minor climate by hightening temperature and humidity of air over the pond(upon practical measurement,air temperature in the space 1 to 2 meters above the pond was higher by 1 to 3 degrees C than the atmospheric temperature), thus restricting evaporation of pond water and reducing thermal losses.In this way,water temperature of the exposed ponds is put under control.

Evaporation rate E is calculated according to Dalton principle,and then heat dissipation G due to evaporation(in general, takes about 80% of the total heat losses) is calculated.

Evaporation rate:

$$E=17.5(1.15e_s-e) \text{ mm/month}$$

where E-evaporation rate,e_s-saturated water vapor pressure at evaporation surface,and e -actual vapor pressure of air.

Quantity of water evaporated:

$$G=E \cdot S \cdot C \text{ kg/hour}$$

where S-evaporation area,m ,and C-water specific heat.

Heat dissipation:

$$Q=r \cdot G$$

where r-heat required for evaporation,kj/kg.

Hence, total heat dissipation Q_1(kj/hour) from the pond surface can be calculated.

Heat provided by the geothermal well refers to the averaged heat at the outlet of perforation during operation:

$$Q_2 =W \cdot (T_1 -T_2) \cdot C_1$$

where,W-heat supplied by geothermal water,kj/hour:T_1-temperature of water sprayed,$^\circ$C;T_2- pond water temperature, C;and $^\circ C_1$- specific heat of geothermal water.

Basing on the average atmospheric temperature in January and 24 $^\circ$C of pond water temperature,$Q_2=8.49 \times 10^8$kj/hour and $Q_1 =7.40 \times 10^8$ kj/hour are obtained.Obviously,$Q_2 > Q_1$,i.e.total heat supplied is greater than total heat dissipated.Therefore,during experiment, pond water temperature was fundamentally maintained between 24 and 28 degrees C even under the condition of strong cold air.

EXPERIMENT

The experiment was carried out in a continuous process from anguilla fry acquisition, young anguilla breeding and high density commodity anguilla culture.Young anguillae were released beginning on March,11,1988, ending on May,8,1988. 17kg of fry was released in one batch. The experiment of commodity anguilla bagan on January,14,1988, ended on March,31,1989, with the young anguillae released in two batches. After carefut tending, commodity anguillae were harveted in batches. Acceptance of the experiment (all anguillae were harvested) was made on April,2, 1989.(see table of experiment)

Table of Experiments

Content of experiment		Time	Pond No.	No.per Ang.	Weight kg	Size No. per kg	Density No. per m^3	Vita-lity %
Commodity anguilla culture, 1st batch	Releatch	88.1.14 88.1.28 Total	2.4 1.3 1.3mus	25791 44201 69992	894.35 562.9 457.25	28.84 78.52 48.03	78.636 122.78 av. 101.73	92
	Harvest	88.5.23 88.9.4 88.10.7 88.10.7	1 2 3 4	670 14000 9400 40278	120 3032 1794 1654.4	5.6 4.6 5.3 24.34	av 93.53	
	Totel		1.3mus	64348	6600.4	Net Wt. 5143.15 kg		
Commodity anguilla culture, 2nd batch	Releatch	88.6.22	1	63014	273.5	230.4	342.47	**
	Harvest	89.4.2	1 2 3 4	20657 19151 10903 8882	834 2833 1613 1314	24.3 6.76 6.76 6.76	112.26 104.08 61.95 61.68	94.6
	Total		1.3mus	59611	6600	Net Wt. 6326.5 kg		

** Released into pond no.2,3 and 4 progressively basing on size

FEEDING AND TENDING

Graded culturing in different ponds

During the stage of young anguilla breeding, young anguillae were sorted out with grading screen each 10 to 15 days, and young anguillae which were up to standard were collected and put into the respective anguilla pond.In culturing of commodity anguilla,the animals were sorted and moved to next pond each 20 to 30 days. Before moving, they had to be put under fast. After sorting, 2 to 3 days were necessay for them to adjust appitite and restore from fatique before casting of feed normally.

Cast of feed

Within about 10 days before young anguilla culture, small earthworms were cast in times per day,with the total quantity equal to the total weight of anguillae.The process from spreading of

2923

feed all over the pond to the centralized casting of feed at
the feeding stand, i.e., the training process for the animal,
took about 5 to 6 days. After the 7th days, synthetic feed for
young anguilla was gradually used. When casting of feed, oxygen
adding machines had to be in operation. For cammodity anguilla
culture. Japan-made synthetic feed was cast at 9 am. and 4 pm.

Others

Other thing to be done in feeding and tending include water
temperature control,water quality control, disease prevention,
pond monitoring, etc., all being important matters. During the
experiment, all these four aspects were given special importance, thus rendering the experiment successful.

Experiment result and conclusions

Breeding of anguilla fry:This took 58 days from white coloured
fry of 7114 individuals/kg to young anguilla of 465.5 individ-
uals/kg. This size of young anguilla was double as big, as compared with that of 1000 individuals/kg in the previous years
when no geothermal was used. Vitality of young anguilla reached
95% and the feed factor was 1.63.

Culture of commodity anguilla:Annual production for ponds:N0.1,
2.29 tons; No.2,14.07 tons; No.3,8.55 tons; No.4,8.79 tons. The
experiment resulted in a net yield of 11469.65kgs with a
maximum production per mu of 14.07 tons and an averaged per mu
production of 7.3 tons annually. The unit production was twice
that of the previous years i.e.3.83 tons, with a commodity rate
of 74.4% and a feed factor of 1.8.

The technology is featured by less investment required, quicker
harvest ,energy saving, culture cycle shortened by 1.5 years,
higher vitality up to 93.3% and an obvious production increase.
In the past three years, the technology has been used in 35 mus
of water area in the farm,resulting in a production increase of
116.8 tons of commodity anguilla and an obvious socio-economic
effect. At present, the technology has been disseminated to
Zhangzhou of Fujian etc.in high density culture of anguilla,
colossoma brachypomum,tilapia,etc.with successful results.

REFERENCES

Chen Shi-xun and Zuo Bin-bai(june.1982), Meteorology and Chima-
    tology,199-221,People's Education Press.(in Chinese)
Chu Bing(aug.1989), Overwinter culturing of crab by use of Geo-
    thermal,New Energy Source,No.8.1989,28-36.(in Chinese)
Chu Bing(editor)(Feb.1991),Handbook of Agricultureal Geothermal
    Use,China Machine Press,166-189.(in Chinese)
Pan Hua-lin,Guangdong Aquaculture Technology Popularization
    General station(sept.1985),Technical report: Anguilla Cultu-
    reing,1-12.(in Chinese)

OIL SPILL RECOVERY TECHNOLOGY

JAY NASH, WILLIAM COOPER, VICTOR NEE, H. NIGIM

University of Notre Dame
Department of Aerospace and Mechanical Engineering
Notre Dame, Indiana 46556
United States of America

ABSTRACT

Current deficiencies in oil spill cleanup processes have resulted in research and development of new cleanup technologies at the University of Notre Dame. Emphasis on reducing, reusing and recycling equipment and waste at a cleanup site has prompted advances in oil recovery technology as well as improvement in sorbent materials.

KEYWORDS

Cleanup; environment; oil; recovery; recycle; reduce; reuse; skimmer; sorbent; spill.

OIL SPILL RECOVERY TECHNOLOGY

In recent years, the global community has become increasingly sensitized to mankind's effect on the environment. Driven by the fresh images of environmental disasters such as the Exxon Valdez oil spill and the current state of the Persian Gulf waters, environmental conscienceness is growing world wide. It is unfortunate that it took tragedies of this magnitude to ignite such a response. The immediate solution to the problem of oil pollution is, of course, prevention of the accident. However, no preventative measures can ever be completely fail-safe as it is impossible to absolutely control all outside influences. In this case we must be prepared to cleanup the environment as fast and efficiently as possible to minimize any lasting effect. The University of Notre Dame Applied Engineering Research Laboratory is currently conducting research to do just that.

In the initial phases of research, priority was given to determining what were the state of the art practices in a modern day cleanup operation. Interviews were conducted, deployment schematics studied, and equipment was evaluated. From this research three major problem areas were defined; the speed at which oil may be recovered, the effects on the environment of the processes used in cleaning up the spill, and the economic considerations of cleaning a spill of that magnitude. These three issues are the focus of the research being conducted at Notre Dame.

A common piece of mechanical equipment used on oil spills is the oil skimmer. Even with today's advanced technology base, the philosophy of skimmer technology is still brute force over finesse. Weir skimmers are dragged along the water at a spill sight and try to collect oil off the surface. This process is not considered to be efficient, and is severely limited in its application on anything but calm seas. Typical recovery percentages for these type of skimmers range from 30 to 60 percent oil content (Meyers et. al., 1989), and most will not operate in slicks less than 1 centimeter thick (Meyers et. al.,1989). This trait narrowly constrains the usefulness of many skimmers. In this range of efficiency, skimmer operators are collecting roughly one gallon of water to every gallon of oil they recover. This low efficiency manifests itself in problems with storage space, emulsification of the recovered product, and increase time and expense in treating the oil back on land.

The most common tool used for cleaning up oil spills of all sizes is sorbent materials. In an operation like the Exxon Valdez cleanup, the vast majority of these sorbents are hydrophobic, oleophillic polymers. These synthetics, contrasted with the skimmer technology, are quite efficient. Depending upon the specific polymer used, and the processed used in it manufacture, and the viscosity of the oil encountered, these sorbents can adsorb between 14, and 25 pounds of oil per pound of material. On the Exxon Valdez spill alone, all of the available stockpiles of sorbent materials in the United States were used in the cleanup effort. The problem with the indiscriminate use of booms and pads made of this material is the tremendous amount of waste that is generated. In most cases incineration is not a desirable option, and these materials require much energy to burn. Therefore, most of the used sorbents are secured in containers and landfilled. This option does not sit well with the contractors in charge of cleaning up the spill as it is very costly, but burying this material is effectively just moving the waste from one place to another. Another problem with the massive use of this material occurs when the booms containing it break. These broken booms then release millions of tiny pieces of synthetic material into the ocean environment, which can have adverse effects all their own.

All of these operations cost vast sums of money. The geographical location of the spill, hiring of qualified manpower, and the cost of the specialized equipment all contribute to the large monetary expenditures that are incurred during a cleanup process. To address this topic, the University of Notre Dame Applied Engineering Research Laboratory has adopted the philosophy of REDUCE,REUSE and RECYCLE in developing new oil spill cleanup technologies. REDUCING the amount of equipment, material and manpower needed, REUSING materials on-site and possibly on future sites, and RECYCLING recovered oils will decrease the amount of capital needed to effectively clean a spill sight. Money saved in these areas my then be spent on development of alternative energy sources, or more advanced spill prevention measures.

Combining all of these considerations, at The University of Notre Dame Applied Engineering Research Laboratory, efforts are underway to research and develop new oil recovery systems and techniques. The primary design considerations are high oil-to-water recovery ratios, high recovery rate-to-unit size ratios, and low waste-to-oil recovery ratios. Again, focusing on low maintenance, high efficiency, and low cost systems.

As mentioned previously, the belief that the future of oil recovery lies in quick, efficient, and flexible recovery capabilities drives the research. The emphasis is on improving those technologies by increasing system efficiencies and reducing the resulting waste. In the same way that the aerospace industry used to be worried about "higher and faster", but is now worried about system efficiencies. Oil recovery methods must be improved to take advantage of current technology and reduce the environmental impact of spills both during and after the spill cleanup.

Research efforts are concentrated on containment booms, sorbent pads, and industrial quality oil skimmers as these components are central in the great majority of oil recovery operations. These tools have also shown the greatest need for improvement. In response to research on current skimmer efficiencies, we are currently developing systems and techniques which utilize newly developed sorbent materials. In conjunction with advances in skimmer design these systems are being designed to collect greater than 95% oil by volume. Water recovered by these techniques is in droplet form (excluding emulsions) which can easily be separated using gravity separation tanks prior to pumping. Such systems have recovered upwards of 98% oil to water in laboratory tests.

Conservative projections based upon developed skimmer designs indicate that over 28 liters per minute of 95% oil can be collected by one modular unit. The high recovery efficiencies translate into more efficient use of recovered-oil storage facilities, which are always in short supply near spills. Prototype skimmers have been undergoing laboratory testing over the last ten months with excellent results. Due to interest from several corporations, plans are being drawn up to test full scale skimmers in a variety of applications in the near future. The waste to oil-recovered ratios using these skimmers is extremely low making them an attractive recovery device. Modular designs allow multiple units to be piggy-backed for flexibility. In addition, our skimmers are capable of operating in any slick more sizable than a sheen on the surface of the water. This last characteristic opens door to applications in land-based industrial sites. Sumps, bilges, and small boat harbors who traditionally have ignored small spill due to the cost and effectiveness of cleanup equipment, may find these systems useable.

Another facet of our current research is the development of "active" containment devices. These devices will transform "passive" nonaggressive containment booms into systems for continuous oil recovery. The "active" containment booms utilize hydrophobic materials to filter oil from the surrounding water and internal ducting networks to convey the oil to storage sites. The influence of bad weather, waves, and turbulence which makes conventional devices inefficient can be minimized by tailored utilization of the hydrophobic materials. The booms can be designed to be linked together, thus containing and cleaning up the spill. This starkly contrasts conventional booms which only contain the spill, or absorb a finite amount of oil, creating a high waste-to-oil recovered ratio. The problem of booms breaking during the cleanup is also being addressed by researching different materials, their configurations and physical strengths.

A complementary (estimated half billion dollar a year problem in the United States) problem facing nearly all industries is how to dispose of sorbent materials once they have been used to cleanup oil spills on water or land. In sticking with the REDUCE, REUSE, RECYCLE philosophy, our answer is that these materials should be reused and recycled if possible. Towards this end, several new techniques have been developed in conjunction with the above research for recycling sorbent materials.. These advances may greatly reduce the solid waste generated by water and land spill cleanup.

The end goal is the development of fast, efficient, flexible and inexpensive tools for the cleanup of oil spills. The University of Notre Dame Applied Engineering Research Laboratory has taken a few steps in the development of these systems, and hopes that the increased environmental awareness that exists today will help propel future advances. It is never too late to be responsible for our environment.

REFERENCES

Meyers, Robert J. & Associates Inc., Research Planning Institute, Inc. (1989) Oil Spill Response Guide, pp. 257-289, Noyes Data Corporation, New Jersey

INFORMATIC SYSTEM FOR GESTION OF METEOROLOGICAL DATA

J.J. Rivera and J.E.Segarra

Departament of Meteorology and Soil Science
Polytechnical University of Catalonia
Av. Bases de Manresa, 61-73
08240-Manresa (Spain)

ABSTRACT

This work present one informatic system of low cost, based on PC, for analyzer, on integrated way, meteorological data provinent of diverses instruments . Their aspect most important is that the hardware is very simple and the software is made-to-measure, according the needs of the user.

KEY WORDS

- Satellite images, pronostic charts, informatic network, digitizer tablet, scanner, multiuser system.

It is important for the entities wich work in explotation of renewables energies, especially hidraulic, solar or eolic energies, to have one automatic system for gestion of meteorological data in real-time: precipitations, temperature, wind, etc.

It exists diversity of meteorological instruments, each one produces information of variables related with some aspects of weather. These informations do not, frequently, interrelated; each instrument produces their data that are treated on independent way, therefore the information is disperse.

This work presents one informatic system of low cost, to treat meteorological data provinent of diverses instruments in real-time and integrated way. From among other data, they are available satellite images, prognostic charts, presion, temperature, wind, humidity, etc.

The satellite image may arrive to system across the informatic network or directly from receiver. In the last case, the system digitalize the image and introduces it in the computer. The system is syncronized with the exterior for receiver succesives images and treat its without human intervention.

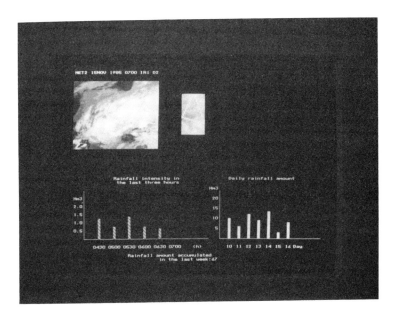

Fig. 1. Some results of the program

The prognostic charts that bring up by fax, they are reading in digital form and puts its in the computer as one image for treatment. Equally, the system digitalizes charts or documents with digitizer tablet or scanner and introduces its in the computer. The data provinents of other conventional meteorological instruments: termometer, barometer, anemomenter, etc, are introduced in the computer directly.

With the lot of information in the computer, the system treat its and send results to screen or to printer (fig.1). The results may be, from among others: full satellite images, zoom of a region desirable, graphics for intensity of rainfall, rainfall accumulated, temperatures, wind, prevision of frost and intensive rainfall, alarms according with the caracteristics of each installation, etc.. One important aspect for the display information is the posibility to work in the multiuser mode; in this way may be displayed simultaniously some results in differents screens connected to a single computer.

It is necessary to mark that some of this results are calculated, of course, with theoretical models, the authors of this work will update the software with the improvements that appear in the scientific literature.

All results are stored in files for treatment "a posteriori", if it is desirable. The images may be stored in compact form in order to occupy few space.

REFERENCES

Peter Norton and Richard Wilton (1990). El IBM PC y PS/2. Microsoft Press, Washington.

Mark Goodwin (1990). Graphical User Interfaces in C++. Mis. Press, Portland, Oregon.

Chris H. Pappas and William H. Murray, III (1990). Turbo C++ professional. McGraw-Hill, Berkeley, California.

Frank Richards and Phil Arkin (1981). On the Relationship between Satellite-Observed Cloud Cover and Precipitation. Monthly Weather Review, 109, pp. 1081-1093.

Olli M. Turpeinen, Azzouz Abidi and Wahid Belhouane (1987). Determination of Rainfall with the ESOC Precipitation Index. Monthly Weather Review, 115, pp. 2699-2706.

HIERARCHICAL CONTROL OF MULTI-AREA NONLINEAR
STOCHASTIC SYSTEM WITH APPLICATION TO SOLAR

Dr. FATEN H. FAHMY

NATIONAL RESEARCH CENTER, ELECTRONICS RESEARCH INSTITUTE,
CAIRO, EGYPT.

ABSTRACT

This paper presents two different optimization techniques to
solve the nonlinear stochastic problem associated with
renewable energy sources, a conventional technique and
hierarchical technique. A comparision has been carried out
between the results of both techniques.

KEYWORDS

Hierarchical; Stochastic; renweable energy system; nonlinear
system.

Introduction

Nowadays interest has taken a more important dimension, that
is renewable energy sources are clean sources with no
harmfull effect on the enviroment.
The purpose of this work is to obtain an optimal combination
of the renewable energy sources,when connected to the proposed load.a hybrid system is composed of PV solar system, and
wind generator source to feed a D.C load (Faraghal,1988,Said
1991,Groumpos, 1983). This optimization problem has a random
variables in nature. Therefore, it has to be formulated as a
stochastic nonlinear programming problem. In this paper, two
different techniques are proposed to solve the problem at
hand.the first is based on conventional control theory techniques whilst the second uses a two level decoposition-coor-
dination approach (Faten, 1987, Singh, 1977) .
A comparison has been carried out between the two methods.

System Description

The system under investigation is a hybrid photovoltaic and
wind energy sources to supply remote application.this hybrid
system is coupled to a separately excited D.C motor loaded by

2933

a centrifugal pump as the main load(Said,1991),Fig(1).

Definitions

A   completation factor (q/AKT = 13.6 1/V).
B   Ideality factor = 1.92
E_h The hybrid energy generated during interval $\triangle$ t.
E_s The solar energy generated during interval $\triangle$ t.
E_w The wind energy generated during the same period $\triangle$ t.
E_L The required load energy during the same period.
E_{Go} Band gap for Silicon = 1.11 ev.
I_o Cell saturation current.
I_{or} Saturation current at reference temperature.
I_{LG} Light generated current.
I_s Solar cell output current.
K/q   Boltzmann's constant/electronic charge=8.62x10 e.v/K.
I_{SCR} Short circuit current.
T   temprature coefficient at I_{SCR} .
R_s Cell series resistance.
T   Cell temprature in Kelven.
T_r reference temprature.
V   cell output voltage.
W_m motor angular speed.
W_w wind angular speed.
ϕ cell illumination (mw/cm).
λ Lagrange multiplier.
u_i Kuhn-Tuker multipliers.
v   the cell voltage during the interval   t (volt).
$\triangle$ t_ the interval time taking in consideration.
E_s,E_w the   max   energy   limits   of   solar   &   wind   sources.

Model Formulation (Said, 1991)

objective function

Maximize $E = E_S + E_W$ (1)
where the solar system equation can be affected by the cell
illumination as shown in the following:

$$E_S = I_S \, V_S \quad \triangle t \qquad (2)$$

$$I_s = I_{LG} - I_o \{\exp[\frac{q}{AKT} (V_s + I_s R_A)]-1\} \qquad (3)$$

$$I_{LG} = [I_{SCR} + K_T (T_c - 28)]\phi / 100 \qquad (4)$$

$$I_o = I_{or} (\frac{T^3}{Tr}) \exp[\frac{E_{Go} q}{BK} \{\frac{1}{Tr} - \frac{1}{T}\}] \qquad (5)$$

while the wind system is function of the wind angular speed w

$$E_w = \frac{1}{2} \sum_T \sum_g V^3 \, a \, \triangle t \qquad (6)$$

as

$$v = w.r \qquad (7)$$

$$E_W = \eta_g K_W w^3 \qquad (8)$$

η_T is the turbin efficiency
η_g is the wind generator efficiency

r   is the turbin radiance
σ is the air density = 1.225 Kg/m

K_W is the proportional constant
v   is the wind velocity.
w   is the angular turbin speed

Constraints

Equality constraints

The mass balance equation of the system represents the
relation between the hybrid energy generated & the energy
required to feed the load.

$$E_S + E_W = E_L \qquad (9)$$

as $E_L = T_L \ w_L = 0.5 \ w_L + 0.002387 \ w_L^2 + 0.00039 \ w_L^{2.8}$ (10)

while the coupling equation can be written as:

$$E_W = Z$$

i.e $\wedge_g k_W \ w_W^3 = Z$ (11)

Inequality constraints
----

i) $E_S \leqslant \bar{E}_S$ (12)

The solar energy generated must not exceed its boundry limits.

ii) $E_W \leqslant \bar{E}_W$ (13)

i.eThe wind energy generated must be within its boundry limits

Control Problem

The object of this problem is to maximize the combination
energy generated from the hybrid system.Depending on the
solar insulation which has no specific value, i.e it is
random in nature. Thus we shall study here a stochastic
nonlinear optimization problem.

The Stochastic Nonlinear Programming Technique

When the stochastic nature of this problem has been taken
into consideration,a very complicated stochastic optimization
problem has been obtained which can not be solved easily using conventional estimation theories & stochastic control
theory. To overcome this difficulty we apply the stochastic
programming approach which is mainly the chance constraint
theory.The new objective function and constraints depend on
the mean values and the standard deviations respectively :

i.e $F = K_1 \, \mathcal{Y} + K_2 \, \mathcal{Q}$

$$=\{V \ \{-\frac{\Phi}{100}(I \ +C \ (T-303))-[I \ (\frac{1}{SCR})\ \frac{3}{or \ Tr} \ exp[\frac{qE_{GO}}{BK}(\frac{1}{Tr} - \frac{1}{T})]]$$

$$[exp(\frac{qv}{AKT} - 1)]\}\wedge_g K_w \ w_w\}\Delta t=\{\{\frac{1}{100}(I \ +C \ (T-303)\Delta t\}^2 \ \mathcal{Q}_{\phi}^2$$

$$+ \{3 \wedge_g K_w \ w_w^2 \ \Delta t\}^2 \ \mathcal{Q}^2 \ \}^{1/2}_{w_w}$$

where $K_1 \geqslant 0$, $K_2 \geqslant 0$ & their numeric values indicate the
relative importance of and respectively.

The new formula of the equality constraints can be
formulated as function of its mean & variance.

$$\bar{g}_j - \Phi_j(p_j) \ [\sum_{i=1}^{N} \frac{\partial g_j}{\partial x_j})^2 \ \mathcal{Q}_{x_i}^2]^{1/2} = 0$$ (15)

where $x_j = [\Phi, w_w, w_L]$ & $\Phi_j(p_j)$ is the value of the standard
normal variates corresponding to the probability P_j .

As a result, the transformed deterministic nonlinear

2935

optimization problem is to maximize the objective function
(14) subject to new constraint (15) & the inequality
constraints (12,13) the corresponding optimum coordinates
(ϕ,w,w) must be defind. The proposed multilevel stochastic
optimization Approach is shown in Fig(2).
The global stochastic nonlinear techniques suffer from the
drawback for increasing of the number of state variables
representing the system, other approaches like hierarchical
system theory are readily available. Two fundamental notions
constitute the core of hierarchical control theory,the partitioning of the problem and the coordination . We consider
that the system is composed of j subsystems interconnected
together as shown in Fig.(2). Then,it is possible to apply a
two level decomposition- coordination technique .

Problem Decomposition

The Lagrangian is modefied by fixing certain variables in
order to obtain a separable form of L, i.e:

$$L = \sum_{j}^{-} \; L_{j} \; [I_{j} \; (i), \; V_{\backslash} \; (i)]$$

i.e $\quad L=F+\lambda(E_{S} +Z - E_{L})+\lambda(E_{w} -Z) +\int_{1}(E_{S} -\bar{E}_{S})+\int_{2}(E_{w} -\bar{E}_{w})$

where L is the Lagrangian of the j subproblems which has the
local variables T (i),i.e"ϕ ,w & w",and the coordinator variables V(i),i.e" & Z".The latter are considered to be fixed
as far as the lower level is concerned,but they evolve at the
higher level until the global solution is obtained, Fig(4).

Concluding Remarks

This paper treats two main points.Firstly,depending on the
nature of equations involving random variables,the stochastic
optimization problem reduces to a stoclastic nonlinear programming problem.Using the propability theory,the stochastic
optimization problem is converted to an equivalent determin-
istic.Secondly,as the number of variables of the control problem increases,the problem becomes more difficult to solve.
In this work control problem of a multiarea power system has
been studied using decomposition- coordination technique.In
such cases,one should seek simplifying procedures that effectively reduce the computational storage and distribute the
associated control effort,Table(1).

REFERENCES
----

1. S.A.Farghal, M.Roshdy abdel Aziz, "Generation Expansion
   Planning Including The Renewable Energy Sources"IEEE
   Transaction on Power Systems vol.3, No.3, August 1988.
2. M.Said Abdel M.,H.El K.&Faten H.F.,"Optimization of PV-
   Wind Stand-alone Systems IEE JAPAN-IAS'91,Sapporo,Japan,
   july,1991.
3. P.P.Groumpose,PV power systems:Basic Theoretical Consid-
   eration,REC-R-06,Energy Research Center,Department of
   Electrical Engineering, Cleveland, State University,
   Cleveland Ohio, March, 1983.
4. Faten H.Fahmy,"Hierarchical Stochastic Optimization with
   APPlication to Hybrid Power Plants Chain Operation",Ph.D.
   Dissertation , Faculty of Engineering , Cairo University,
   Egypt,1987.

5. M.G.Singh,Dynamic Hierarchical Control. North Holland
    Publishing Co., Amesterdam, 1977.

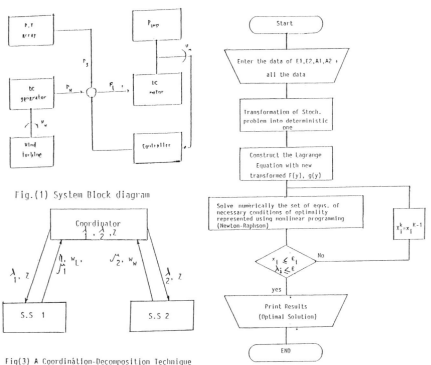

Fig.(1) System Block diagram

Fig(3) A Coordination-Decomposition Technique
       to Solve An Optimization Control Problem

Fig(4) The Flow chart depicting the conventional Technique
       to Solve globally stochastic System.

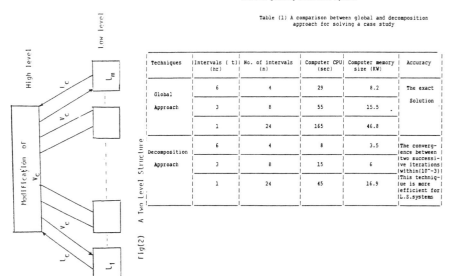

Fig(2) A Two Level Structure

Table (1) A comparison between global and decomposition
          approach for solving a case study

Techniques	Intervals (t) (hr)	No. of intervals (n)	Computer CPU (sec)	Computer memory size (KW)	Accuracy
Global Approach	6	4	29	8.2	The exact Solution
	3	8	55	15.5	
	1	24	165	46.8	
Decomposition Approach	6	4	8	3.5	The converg-ence between two successi-ve iterations within(10^-3)
	3	8	15	6	This techniq-ue is more efficient for L.S.systems
	1	24	45	16.9	

ENERGY REQUIREMENTS FOR NEW TECHNOLOGIES SUPPLYING FINAL ENERGY

N.PUCKER[*] and W.SCHAPPACHER[**]
UNIVERSITY of GRAZ
[*]Institute for Theoretical Physics, Universitätsplatz 5/I,
[**]Institute for Mathematics, Heinrichstraße 36,
A 8010 GRAZ/AUSTRIA

ABSTRACT

The installation of a new energy system (wind energy, solar energy a.s.o.) requires the investment of construction energy. The whole system is characterized by the energy harvest ratio of a typical power station. Energy pay-back time, asymptotic power level and necessary foreign energy are investigated for different installation strategies. Specific results are given in the case of a photovoltaic system.

KEYWORDS

Energy systems, Energy harvest ratio, Energy pay-back time, Photovoltaics.

INTRODUCTION

If we start to install a new energy system, energy is also needed to produce the necessary components. This energy is taken from an existing power system: the installation of a photovoltaic plant for example needs construction energy supplied by fossil fuels. In the course of time the new system has to become selfsustaining: it yields power for the public and also the necessary power for new plants. Such a progress to selfsustainment can be described by means of various quantities like the energy harvest ratio R or the energy pay-back time τ_E (Seifritz, 1978; Milasinovic et al., 1988). The harvest ratio contains all relevant information about the energy system:

$$R = (p_0 \cdot \tau_L)/Q_0 \qquad (1)$$

(Q_0 = construction energy of a single plant, p_0 = power produced by the plant per year, τ_L = life time of the plant).

The construction energy Q_0 of a single power plant is the key quantity. Its

2938

determination may be very difficult because many contributions are condensed to a single number. But as a consequence we can deduce from it the main features of the new energy system we start to set up. These are the energy pay back time of the single power plant $\tau_{E0} = Q_0/P_0$ as well as the energy pay back time of the whole system τ_E, which depends also on the strategy how we set up the necessary number of power stations constituting our new energy system. That system finally will work at a stationary power level. The model calculations show how much energy has to be invested for the achievement of a certain power output for the public. We give realistic numbers for the energy pay-back time in relation to the harvest ratio R and the life time of a single plant. It is also demonstrated how the possible technological progress changes the situation. Finally the results are applied to photovoltaic modul installations.

STRATEGY OF POWER INVESTMENT

Let q(t) describe the strategy of power investment for constructing the power stations of the new technology. Among a large number of possible investment functions we have carefully investigated three specific forms (PUCKER et al.,1991):

$$
\begin{array}{ll}
\text{a)} & q(t) = N = \text{constant} \\[2mm]
\text{b)} & q(t) = \left\{ \begin{array}{ll} \alpha \cdot t & \text{for } 0 \le t < t_\alpha = N/\alpha \\ N & \text{for } t_\alpha \le t. \end{array} \right. \qquad (2) \\[4mm]
\text{c)} & q(t) = \dfrac{N+1}{1 + N.\exp(-a_1.t)} - 1.
\end{array}
$$

(The parameter a_1 measures the velocity with which q(t) approaches N)

concerning their influence on the pay-back time τ_E of the system, on the energy $E_{foreign}$ which has to be borrowed from already established technologies and on the achievement of a stationary asymptotic power level. The investigation assumes a power function

$$
A(t) = - q(t) + p_0 . \sum_{i=1}^{K} H(t-t_{0,i}) - p_0 . \sum_{i=1}^{\hat{K}} H(t-t_{0,i}-\tau_L) \qquad (3)
$$

(The number K of operating stations differs from the number $\hat{K}$ of already closed ones), H(t) being the Heaviside function. At every time $t_{0,i}$ a new plant starts operation until the time $t_{0,i}+\tau_L$. The energy pay-back time can now be computed by

$$
0 = \int_0^{t_{0,1}+\tau_E} [- q(t) + p_0 . \sum_{i}^{K} H(t-t_{0,i}) - p_0 . \sum_{i}^{\hat{K}} H(t-t_{0,i}-\tau_L)] dt \qquad (4)
$$

At time $t_{0,1}+\tau_E$ we have equality between the energy invested by the strategy q(t) und the energy produced by the power stations of the new system. Note that the energy pay-back time is correlated with the startup time of the first power plant.

Comparing cases with R=2/5/10 for the three investment strategies we can see that there is no dramatic difference between the three cases of formula (2) concerning the pay-back time τ_E(Pucker et al., 1991). Finally they achieve an asymptotic power level of

$$L_{max} = N.(R - 1). \tag{5}$$

Figure 1 shows the energy and power development for three systems with R= 2/5/10 according to the investment strategy q(t)=N=30(suitably chosen power units). The respective values for τ_E and $E_{foreign}$ are τ_E= 25/10/5(years) and $E_{foreign}$= 250/100/50 (suitably chosen energy units). For the other two cases of q(t) τ_E is larger and $E_{foreign}$ is somewhat smaller: if we have more time to wait for a positive netto output of power we can borrow a fewer amount of energy (Table 1). L_{max} in formula (5) is the stationary final power output for the public corresponding a finally achieved investment power N. A harvest ratio R=1 provides only enough power to reproduce the plants which are at the end of their life time. Therefore we have to require R>1 theoretically. From the viewpoint of a reasonable energy economy we probably should have values of at least R=2 or 3. The time when L_{max} is achieved depends on q(t). The fastest approach is along version a).

Table 1. Results for τ_E and $E_{foreign}$

R		2	5	10
a)		25	10	5
b)	τ_E	25,22	11,02	5,37
c)		25,25	10,46	5,5
a)		50	100	50
b)	$E_{foreign}$	245,1	75,8	29,3
c)		247	89,43	33,78

THE EFFECT OF TECHNOLOGICAL PROGRESS

We can take into account a technological progress by changing the harvest ratio R from a constant quantity to a time dependent one. Making R(t) smaller and smaller means that fewer energy is needed for the construction of the respective power plants: the technology has been improved. We use a logistic time function

$$R(t_{0i}) = R_0 . \frac{a_2}{b_2 + (a_2-b_2).exp(-a_2.t_{0i})} . \tag{6}$$

which leads to an asymptotic value of $R(t)=R_\infty$. This simulates a mature technology. The parameters a_2=0,2 and b_2=0,08/0,04 in combination with q(t)=N=30 lead to R_∞=5/10 if we start with R_0=2. The resulting changes in τ_E and $E_{foreign}$ as well as the change from R_0=5 to R_∞=10 are given in Table 2. After some time the system can provide the power L_{max} corresponding to the asymptotic R_∞-values. We see that the progress in technology yields an appreciable shortening of the pay-back time. The savings in $E_{foreign}$ vary between 25 and 10 per cent.

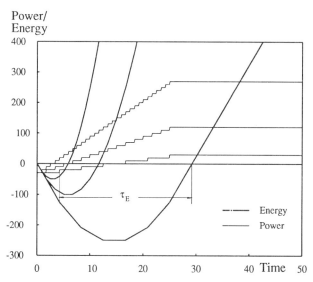

Fig. 1. Time development of three energy transforming systems $(R=2/5/10, \quad N=30, \quad \tau_L=25, \quad p_0=10)$. Each step in the power curve shows the installation of a new power station. The energy curve shows the amount of borrowed or produced energy.

Table 2. Results for τ_E and $E_{foreign}$ taking into account technological progress

R	τ_E	$E_{foreign}$
2 $\longrightarrow$ 5	14,9	203
2 $\longrightarrow$ 10	11,6	188
5 $\longrightarrow$ 10	7,8	91,4

APPLICATION TO PHOTOVOLTAICS

During the last years the harvest ratio of photovoltaic moduls has grown remarkably. That is the result of considerable improvements in Silicon technology (HAGEDORN et al.,1989). A careful examination of all steps of the production of photovoltaic cells and moduls yields a harvest ratio around R=3 in connection with a life time $\tau_L=20$ years. The strategy a) of constant power investment leads to a pay-back time of the system $\tau_E=13,3$ years. A further important question is the amount of necessary $E_{foreign}$ for the installation of a certain amount of L_{max}. We assume an average value of 1200 kWh/m^{2}y of incoming solar radiation (which is typical for Austria). A modul efficiency of

2941

10 per cent gives 120 $kWh_{el}/m^2 y$ or 300 $kWh/m^2 y$ equivalent of primary fossil energy. So we have also $p_0 = 300$ $kWh/m^2 y$ according to formula (1) which leads us to $Q_0 = 2000$ kWh/m^2 (HAGEDORN et al., 1989). Assuming a modul area of 1000 m^2 as one single power plant means $1,2.10^5$ kWh_{el}/y or $p_0 = 3.10^5$ kWh/y in units of equivalent primary energy. This means likewise $1,5.10^7$ kWh/y investment power N and approximately $E_{foreign} = 5,1.10^7$ kWh. The lifelong of one power station is 6.10^6 kWh. To provide $E_{foreign}$ needs therefore the lifelong work of 8,5 power stations. That number has to be compared with the total number of 150 units. The whole system provides a power $L_{max} = 1,2.10^7$ kWh/y.

CONCLUSION

The energy balance of energy transforming systems can be studied very successfully with the help of the energy harvest ratio of the system. Important quantities like asymptotic power output or "foreign" construction energy can be related to investment strategies. Photovoltaic energy transformation is discussed as a specific example. The analysis shows a remarkable progress concerning the energy balance and the energy pay-back.

We close by noting that we can generalise the above considerations to any (technically reasonable) investment strategy. Again formula (3) and (4) are valid and can be used to calculate numerically the startup times t_{0i} and the energy pay-back time τ_E. Moreover it turns out that if q(t) converges to some limit q_∞ for $t \longrightarrow \infty$, then the system tends to a power level which is in general however not a constant but varies between two neighbouring levels (Pucker et al., 1991).

REFERENCES

Hagedorn, G., St. Lichtenberger and H. Kuhn (1989). Kumulierter Energie-verbrauch für die Herstellung von Solarzellen und photovoltaischen Kraftwerken. Forschungsstelle für Energiewirtschaft, München.
Milasinovic, T. and R.Sizmann (1988). Energie-Rückflußzeiten in modularen Energiesystemen. Universität München, Preprint. München.
Pucker, N. and W.Schappacher (1991). N e w energy technologies: installation strategies, energy pay-back and achieveable power level. UNIGRAZ UTP-20-10-91. Graz.
Seifritz, W.(1978). Zur Dynamik von Substitutionsprozessen. In: Reaktortagung 1978, p.1063 - 1066. Deutsches Atomforum eV, Zentralstelle für Atomkernenergiedokumentation. Eggenstein-Leopoldshafen.

CLEAN ENERGY APPLICATIONS IN DEVELOPING
COUNTRIES

JACOB KOW MENSAH

Energy Research Group, University of Science &
Technology, C/O. Box 111, Tema.

ABSTRACT

Deforestation and in some areas Dessertification and hence Global warming is
very much being contributed in Africa since the commonest source of energy
is Biomass ie fuelwood & Charcoal. Tropical moist forest are believed to
contain up to half of estimated 10 million species of plants and animals in
the world. In Ghana (area 92,098 sq km population 12,205,594 in 1984, it
has been estimated that current removal of forest is 11 to 12 million cubic
meters per annum of this amount 1.02 is for commercial use while 10.10
million is used as woodfuel. This story could be repeated for other develop
ing countries as regards in the rate of depletion of available forest resources.

Abundant solar energy source in Africa is very little harnessed. In Ghana,
tree planting activities, the use of L.P. Gas, improved Coal Pot, improved
firewood stoves, and hydro-electricity are being encouraged. Hydroelectricity is at the moment the major source of Clean Energy in Ghana and
is being extended to areas within 20 Kilometers radius from the national
grid.

Solar conversion and applications technology to harness this abundant
source of energy is needed to provide electricity to the rest of the country
ie the rural areas by the next 30 years..

KEY WORDS
Large scale heat energy Irrigation system, Clean energy Cooking pots,
Electrical energy saving cookers, Effective use of Solar and Hydro-electric
potential in the Tropics.

INTRODUCTION
The MENSAH'S ELECTRIC AND CLEAN ENERGY COOKERS WILL disourage and gradually
phase out the use of pollutant and outmoded fuel in developing countries. I
counceived this idea in November 1989 after presenting the paper entitled
("maximizing the use of Electricity and L.P.G. potential in Ghana at
University of Science & Technology Kumasi, Ghana

A Grazed Ceramic electric heater, designed to facilitate our traditional cooking method not only provided a sustainable electric shock
free heating but also conserved considerable amount of heat energy
when it was placed in a ceramic flower pot and covered. The conserved
heat in the flower pot continued to do the cooking after the food to be
cooked's temperature had been raised to 100°C and the Power switched off.
This clean energy saving cooker could reduce the cooking energy cost by
80% and therefore need to be promoted for use in developing countries.

In Africa the greatest energy potential is Solar hydro Power. It is
estimated that the continent has 20% of the world's potential for hydro-
electric power generating. A good number of the major hydro-electric
power scheme has been established and several more are planned.

In Ghana, hydro-power is the main sources for electricity production.
The two hydro-electric plants produce a total 1,072 MW, that is 912 MW
with 15% overload capacity from the Akosombo dam and 160MW from Kpong
dam. There is more hydro potential for a third Dam. This electric
power potential has been extended to our neighbouring countries namely:
Republic of Bennin, Togo, La Cote D'Voir and is being extended
to Bukina-Faso. Nigeria is also taping her Hydro-power potential and
so forth.

ENVIRONMENTALLY SUSTAINABLE DEVELOPMENT MODELS

Large Scale Mensah Heat Energy Irrigation System

The biomass water pump presented in the volum 3 page 1953-7 of the
proceeding of the 1st World Renewable energy congress, has been
developed for Large Scale Irrigation. The improved system now use
electric (clean) Energy or L.P.Gas as the source of heating. It can
therefore be used at the desert region of the world for both irrigation
and rural water supply. The major advantage of this pump is that it
is affordable to the ordinary rural dwellers since it also operate on
Biomass Waste and is cheap to Manufacture.

Clean Energy Cooking in Africa.
In Ghana, as other developing countries, even though there is high poten
tial for the Electiricy it is not fully utilized for fear of being elec-
tricuted. A sustainable electric shock free heater has been developed
and adapted to the traditional cooking methods in Africa (see fig 1)
This will encourage the use of it in addition to the use of natural Gas
and L.P. Gas which is non-renewable and therefore not very reliable
for very long term projects. Improved coal pot and improved firewood
(cookstove) may reduce fuel consumption but they do not eliminate
deforestation, dessertification and environmental pollution. By
tradition, most families prefer charcoal, Kerosine and firewood to
electricity for cooking. The few durable and efficient electric
stoves and cookers are imported and quiet expensive for an average
Ghanaian both to purchase and to use. They have metal bodies and
because metal are good conductor of electricity, the users receive
electric shock whenever faults are developed. It is also observed that
the metals used become rusty quickly leading to the rapid breakdown of
the stoves especially when the food boils over and the brine affects
the heating elements.

From firewood & charcoal to electriity

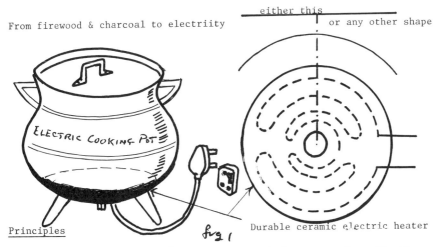

Principles   *fig 1*   Durable ceramic electric heater

The principles of the Mensah's Energy Saving Cooker state that effective cooking occurs at temperture of 100°C ± 5%, any additional heating becomes a waste of energy.

(Fig2) shows the energy saving cooker. The sustainable electric shock free heater was placed in the earthenware flower pot. The food to be cooked was placed on the heater and the flower pot covered. The initial cooking time and temperature were noted. The boiling time of the water was also noted. The electric power was then switch off and the final cooking time was noted, after the food has been well cooked. It was observed that the total energy cost for cooking the same quantity of food with charcoal and with firewood was 2:40 and 2:20 respectively, as compared with normal traditional cooking method.

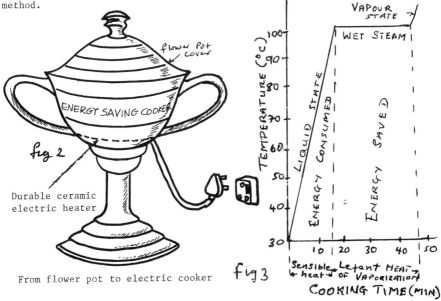

fig 2

Durable ceramic
electric heater

From flower pot to electric cooker   *fig 3*

COOKING TIME (MIN)

CALCULATIONS & RESULTS

Workdone Using the Energy Saving Cooker

Heating 1kg of water from 30°c ambient to 100°c ± 5% cooking temperature. During this stage, heat was applied to raise the temperature of the water from 30°c ambient to 100°c saturation temperature t_s corresponding to pressure p. The volume of 1kg of water was increased from V_{30} to V_{100} (see fig 3). Since the heat added produced a temperature change, sensible heat was added. The specific enthalpy of water donated by hw at 100°c is giving by

$$hw = Uw + PVw/J \dots\dots\dots\dots\dots\dots\dots (1)$$

and the change in enthalpy is given by

$$(h_{100} - h_{30}) = Uw + P(V100 - V_{30}/J \dots\dots\dots\dots (2)$$

Workdone During Complete Cooking of the Food not under the Energy Saving Cooker (ie by the electric cooking Pot)

For the external work done by the boiling water

$$hs = P (Vs - Vw) \text{ Kgfm.}$$
$$= P(Vs - Vw)/K \text{ kcal}$$

But the product P.Vw is regligible quantity.
External workdone per Kg of steam

$$= PVs/J \dots\dots\dots\dots\dots\dots\dots\dots\dots (3)$$

and for wet steam with x driness fraction, the external workdone

$$W = PX \ Vs/J \dots\dots\dots\dots\dots\dots\dots\dots (4)$$

Actual latent heat of evaporation is more than this workdone. Therefore heat required to overcome the intra-mslecular resistance to change the phase of the temperature and the difference between L and PVs/J ie. internal latent heat.

$$\text{Thus L = Internal latent heat + PVs/J} \dots\dots\dots (5)$$

Therefore total workdone W = internal workdone + External workdone.

The Internal Energy

Hence the value of Internal energy may be calculated from the definition of enothaply given by

$$h = U + PV/J \dots\dots\dots\dots\dots\dots\dots\dots (6)$$

Therefore, for saturated liquid, we have

$$hw = Uw + PV/J \dots\dots\dots\dots\dots\dots\dots (7)$$

or the internal energy of the liquid is given by

$$Uw = hw - Ps \ Vw/J \dots\dots\dots\dots\dots\dots (8)$$

Similarly the internal Energy for the wet steam is given by

$$Us = hs - Ps \ x \ Vs/J \dots\dots\dots\dots\dots\dots (9)$$

and the saturated dry steam is given by.

$$Us = hs - PsVs/J \dots\dots\dots\dots\dots\dots\dots (10)$$

Nomenclature

V_{30} = (ie. initial volum at 30°c ambient.
V_{100} = Valum at 100°c
XV_S = Volume of 1 kg of wet steam.
UW = Initial energy of water at temp. t
hw = specific enthalpy of water
h = Enthalpy of 1 kg of wet steam
L = Latent heat of vaporization per kg.
hs = Enthalpy of 1 kg of dry steam.
Kgf/cm = absolute Pressure.
J = 427 (Joule's constant or mechnical equivalent of heat).

Table 1 **TYPICAL ELECTRICITY SCHEDULE TARIFFS EFFECTIVE OCTOBER, 1990**

UTILITY	TARIFF CATEGORY	RATES
V.R.A E.C.G. N.E.D.	H.V. Supplies	M.D. ₵660.00/KVA per month Energy ₵6.50 KWH Service Charge 1,850/Per Month
E.C.G.	L.V. Supplies 100 KVA to 800 KVA	M.D. ₵730 KVA/per Month Energy ₵7.00 KWH Service Charge ₵1,850/Per Month
E.C.G.	Mon Residential	Energy ₵20.40 KWH Service Charge ₵140.00/Per Month
E.C.G. N.E.D.	Residential 0 - 50 Kwh 51 - 200 Kwh * 201 - 600 Kwh 601 +	Block Charge ₵300.00/Month Energy ₵4.90/Per Kwh Energy ₵7.70/Per Kwh Energy ₵17.00/Per Kwh

VRA - Volta River Authority
ECG - Electiricy Corporation of Ghana
NED - Northern Energy Department.

Economic Valuation

To know your bills when using the energy saving cooker,multiply the boiling time by one unit (see above table) 1 * to calculate your bill for cooking 1 kg of beans.

1000 watts for 1 hour = 1 unit = 70.70 cedis
15 minutes boiling time = $\frac{15}{60}$ x 7.70

$$= \frac{1}{4} \text{ x } 7.70$$

= 1.925 Cedis Approx. 2 Cedis
Equivalent cost of firewood = 20 Cedis
Equivalent cost of charcoal = 40 Cedis.

Conclusion

It was therefore recommended that the use of modern clean energy devices such as the electric cooking port, the Mensah's heat energy saving cooker, Electric driving cars, and Hydrogen driven vehicles are promoted in developing countries.

Acknowledgement

I shall therefore take this opportunity to thank Council for Scientific & Industrial Reserch (CSIR) and Ghana Regional Appropriate Technology and Industrial Service (GRATIS) for their enumeraus assistance.

REFERENCE

WACLAW PLICUTA, Bellerive Foundation (1985) Modern stoves for all,Intermediate Technology Publishers ltd. 9 King Street, London Pages 58-68.
P.L. BALLANEY (1980) Applied thermodynamics,Khame Publishers, Delhi pp. 92-126, 2ND CONGRESS OF AFRICAN SCIENTISTS JAN, 1990 Proposals for a minimum programme, Pan-African Union of Science & Technology, Accra pp 27-31.

ESTIMATION OF THE SOLAR RADIATION AT THE EARTH'S SURFACE FROM METEOSAT SATELLITE DATA

A .MECHAQRANE*, M. CHAOUI-ROQUAI**, J . BURET-BAHRAOUI*

*Laboratoire d'énergie solaire, Faculté des Sciences de Rabat-B.P 1014
**Ecole Nationale de l'Industrie Minérale-B.P 753-RABAT

ABSTRACT

A simple physical method to estimate the hourly and daily global solar radiation from METEOSAT-2 satellite data is tested for two maroccan sites. The root mean square error is of 116 Wh.m-2 (21% of the mean measured value) for hourly values and 737 Wh.m-2 (15%) for daily values.

KEYWORDS

METEOSAT-2; digital count; calibration factor; atmospheric effects; global solar radiation

INTRODUCTION

Last works which use satellite data to estimate global solar radiation at the earth's surface showed that the physical models are more interesting than statistical models because of spatial and temporal variations of the regression coefficients.

In this paper a simple physical method is proposed to estimate global solar radiation from METEOSAT-2 satellite data. This method is based on a physical radiative transfer model taking into account only the Rayleigh scattering and the ozone and water vapor absorption. We have then neglected the aerosols effects and the clouds absorption.

SATELLITE AND GROUND DATA

The digital counts (8 bit) in the shortwave channel (0.4-1.1 μm) of the geostationary satellite METEOSAT-2 have been collected for the two maroccan sites: Rabat (34°N,6.75°W) and Marrakech (31.62°N,8.03°W) during the year 1985. These data are provided by numerical pictures which have been reduced to obtain a resolution of 30 km in space and 3 hours in time, according to the B2

2948

format of the International Satellite Cloud Climatology Project procedures(ISCCP). Only three slots (09, 12, 15 T.U) are exploited to estimate the global solar radiation.

The pyranometer data are the three hourly values of 8-9h, 11-12h and 14-15h solar time (TSV) and the daily values of the solar global radiation.

METHODOLOGY

Radiative Transfer Model:

In order to estimate the global solar radiation at the ground (E_s), we use a simple and well known equation taking into account the solar radiative transfer in the system earth-atmosphère:

$$E_s = E_{oh} \frac{1 - \alpha - A}{1 - A_s} \qquad (1)$$

Where α is the total atmospheric absorption, A the top of atmosphere albedo and A_s the ground albedo. E_{oh} is the extraterrestrial solar radiation on a horizontal surface.

The top of atmosphere albedo :

The shortwave channel of Meteosat-2 is not calibrated. To relate noncalibrated digital counts to the solar irradiance leaving the top of the atmosphere, we use a calibration method developed by Kopke,1983. This relationship between the digital count C and the solar irradiance in the solar spectum (Lsol) is written as:

$$Lsol = k (C - 2) \qquad (2)$$

k is a calibration factor (0.575 ± 10 % $W.m^{-2}.sr^{-1}.count(8bits)^{-1}$) weighted by a conversion factor which depends on the atmosphere and the ground optical properties. This factor varies around average values of 2.8 for vegetated surfaces and 2.4 for bare surfaces. So, its accurate knowledge is very difficult. We use for our work a constant conversion factor equal to 2.6. Thus, k is equal to 1.495.

The top of atmosphere albedo A is computed from the radiance Lsol, assuming isotropy of radiation field:

$$A = \frac{\pi \ Lsol}{E_{oh}} \qquad (3)$$

The ground Albedo :

Deschamps et al,1983 have established a simple equation to correct the satellite signal from the atmospheric effects. The top of atmosphere albedo is writted as

$$A = [A_a + \frac{T(\theta_s)\,T(\theta_v)\,A_s}{1 - S\,A_s}] \prod_i t_{gi} \qquad (4)$$

A_a is the intrinsic atmospheric albedo, S the spherical albedo, A_s the ground albedo, $T(\theta_s)$ is the diffuse transmittance of the atmosphere for the sun-earth path and $T(\theta_v)$ for the ground-satellite path. t_{gi} is the atmospheric transmittance after absorption by gases in both incident and reflected path. Ground albedo A_s can be calculated from equation (4).

For our work, we need the atmospheric parameters integrated over the whole band pass of pyranometer 0.3 - 3 µm. So, we assume that the spectral formulation of Deschamps et al (1983) is averagely valid in this whole band, and each atmospheric fonction $(A_a, T(\theta_s), T(\theta_v)$ and S) is adjusted by use of the exact computation obtained from the "5S" program (Simulation of the Satellite Signal in the Solar Spectrum) (Tanré et al,1985). We take for A_s the minimal value observed in a temporal serie of one month. Then we assume that this value is observed for a cloud free sky whose aerosol content is very weak. So, we neglect the aerosol effect and we take in account only the Rayleigh scattering and the water vapor and ozone absorption.

The atmospheric absorptance

The aerosol absorption is relatively weak, especially in the visible band. We have neglected it, and according to Dedieu et al,1987, we also neglect cloud absorption. So, the atmospheric absorption is only due to the atmospheric gases. In this conditions and assuming that,for cloudy sky, the gaseous effects remain identical to those for a clear sky, we can express the total atmospheric absorption as :

$$\alpha = 1 - \prod_i t_{gi} \qquad (5)$$

t_{gi} is the transmittance after gaseous absorption in both incident and reflected path.

On a first step, we have used value of the content of water vapor given, for each hour, by the meteorological stations. Ozone content has been calculated from Heuklon,1978, parametrisation. We have then noticed that the results are not very affected by using only the constant content of the midlatitude summer standard atmosphere ($U_{O_3}=0.319$ cm atm, $U_{H_2O}=2.93$ g.cm^{-2}). We have then adopted the constant contents.

RESULTS

Hourly Values

Figure 1 gives, for the 3 hours and the two stations (2055 cases), the hourly satellite estimates plotted against the pyranometer measurements.The correlation coefficient is 0.90, the root mean square error (RMSE) is 116 Wh.m-2, which is

21% of the mean measured value; and the mean bias error (MBE) is 31 Wh.m-2 (6%).

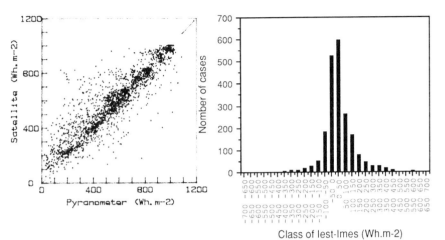

Figure.1: Comparison of measured and estimated hourly global radiation

Figure.2: Histogram of the difference between estimated and measured hourly global radiation

The large scatter of data around the line of slope 1, illustrates the difficulty to compare the two measurements having different time and space sampling (the satellite data is an instantaneous measurement over a small solid viewing angle, while the pyranometer measurement is integrated over an hour and a solid angle of 2π). A second problem is linked to the difficulty of the localisation the pyranometer site on the satellite reduced picture (B2 format).

Figure 2 gives the histogram of the difference between calculated and measured hourly values. We note that more than 86% of the cases presented a difference between -150 and +150 Wh.m-2.

Monthly Average of Hourly Values

Figure 3 gives the comparison between predicted monthly averages and measured values. The RMSE is 47 Wh.m-2 (8% of mean measured value). A surestimation of the satellite predicted values is observed for 8-9h and 14-15h. This can be attributed to the large solar zenith angle for which the atmospheric effect is important but also to the larger azimut angle between satellite and sun when the shadow effects and spatial and temporal differences are consistant. We notice the best results obtained for the 11-12h hour for which the RMSE is 28 Wh.m-2 (4%) and the MBE is 4 Wh.m-2 (0.5%).

Daily Values

The trapeze's method is used to calculate the daily global radiation from the hourly values for 9 h, 12 h and 15 h. Figure 4 gives the comparison of the estimated values plotted against the pyranometer measurements (631 cases). The correlation coefficient is 0.93, the RMSE is 791Wh.m-2 (15% of the mean measured value) and the MBE is 285 Wh.m-2 (5%).

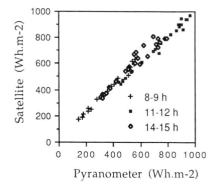

Pyranometer (Wh.m-2)

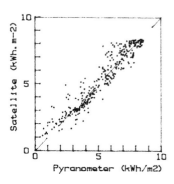

Figure.3: Comparison of measured and calculed monthly average hourly values

Figure.4: Comparison of mesured and calculed daily values

For Rabat (323 cases) the correlation coefficient is 0.95, the RMSE is 668 Wh.m-2 (13%) and the MBE is 61 Wh.m-2 (1%). For Marrakech (308 cases) this results are 0.93, 903 Wh.m-2 (18%) and 519 Wh.m-2 (6%) respectively. The results obtained for Rabat are better than those for Marrakech.

CONCLUSION

The spatial and temporal differences between satellite and pyranometer measurements are the principal problem and affect cosiderably the precision. But there are also same errors induced by the simplifications in our determination of the atmospheric parameters. A more accurate calculation of this parameters should be made to improve the results.

REFERENCES

Deschamps, P.Y., M. Herman, D. Tanré (1983). Modélisation du rayonnement solaire réfléchi par l'atmosphère et la terre entre 0.35 et 4 micromètre. Contrat ESA 4394/80/F/DD/SC

Tanré, D., C. Deroo, P. Dahaut, M.Herman, J.J. Mocrette,J. Perbos, P.Y.Deschamps (1985). Effets atmosphériques en télédétection - Logiciel de simulation du signal satellitaire dans le spectre solaire. Proceedings of the 3rd International Colloquium on Spectral Signatures of Objects in remote sensing, Les Arcs, France (ESA SP-247)

Kopke, P. (1983).Calibration of the Vis channel of METEOSAT-2. Advanced Space Research 2, 93-96

Dedieu, G., P.Y. Deschamps, Y.H. KERR (1987). Satellite estimation irradiance at the surface of earth and of surface albédo using a physical model applied to METEOSAT data. J. of Climate and Applied Meteorology 26, 79-87

T.K.Van Heuklon, T.K. (1978) Estimation atmospheric ozone for solar radiation models. Solar Energy 22, 63-68

THE ROLE OF ITALIAN AGRO-FORESTRY SYSTEM IN CONTROLLING THE CARBON DIOXIDE AND METHANE BALANCE IN ATMOSPHERE

G. Barbera*, T. La Mantia*, G. Silvestrini**

* Istituto di Coltivazioni Arboree, Università degli Studi, Viale delle Scienze
Palermo, Italy

** CNR-IEREN, National Research Council,
Via Rampolla 8 - 90142 Palermo, Italy

ABSTRACT

After the EEC decision to stabilize the carbon dioxide emissions by year 2000 at the 1990 level, a study has been financed by the Italian Ministry of Environment in order to define what targets could be set by the year 2005 and what strategies could be implemented in Italy in order to achieve consistent carbon dioxide reductions. The results of the research indicate the possiblity for Italy to reduce by 25% the CO_2 emissions compared to the 1990 level. In this paper the options to use the biomass in order to increase the sink of carbon in Italy are analyzed. The role of forestry, agricultural wastes and residues, urban wastes, energy crops and organic soil matter has been considered. In a climate stabilization scenario it could be possible to avoid the emissions (or to capture) an yearly quantity of carbon of 18 millions of tons. The potential reduction of methane emissions from the agro-forestry sector on the urban wastes disposal is also presented.

KEYWORDS

Global warming, carbon dioxide, biomass, urban wastes, energy crops

INTRODUCTION

The Italian agriculture can play a significant role against the increase of greenhouse gases in the atmosphere, also in relation to the new agricultural policy of EEC. Recently large areas (850.000 ha during the 80s) were abandoned by agricultural crops and are now available for forestry and landscaping. In the future (2000), according to the EEC policy against agricultural surpluses, others 1.750.000 ha of good agronomical value will be available for energy crops and short rotation forestry. In such areas many measures can be envisaged in order to reduce greenhouse gases releases or increase carbon fixation in forests and soils. A contribution can derive also from the traditional crops, a sector where economic and environmental constraints facilitate the introduction of measures concerning energy saving, waste and residues recycling and improvement of soil organic matter levels.

CARBON DIOXIDE

Forests

The Italian forest sector stocks today 253 millions of tons (Mt) of carbon. This value has been obtained by the revision of the data presented in the National Forestry Inventory and considers the productivity of the existing forests and of new plantations (IR, 1992)

In the "Business as Usual" (BAU) scenario the national forestry sector will stock, in 2005, 331 Mt of carbon. According to the model used in our research (IR, 1992), by the year 2005 there will be an annual increase of 6 Mt. But the role of trees can be enhanced considering a new forestry policy based on an afforestation campaign and a reduction of the area damaged by fire and pests: in this case the carbon storage will reach 379 Mt with an increase, compared to the BAU scenario, of 48 Mt from 1991 to 2005 and an annual carbon increase of 11 Mt at the end of the period. It is also possible to give a specific emphasis on forests as an energy source: in this case the annual increase of carbon fixed will be in 2005 of only 5 Mt. This is less than in the BAU scenario, but we must consider that in the "energy scenario" there will be a reduction of annual carbon dioxide releases of 7,5 Mt (2 Mt of carbon) due to the substitution of fossil fuels by biomass (RI, 1992). In fig. 1 the carbon stored in different scenarios is presented.

Soil Organic Matter

Considering that the Italian agricultural soils have an average of 1.5% of organic matter, they store in the upper layer 394 Mt of carbon. A technological reform of current agricultural systems according to organic farming practices (green manure, conservation tillage, planting legumes, use of organic fertilizers, etc.) could increase this value. In accord with the estimates reported by Lovins (1991), such increase could reach 0,02% per year. In this case and considering an area of 1,5 million of ha (the actual agricultural land used is of 15 millions ha), in the year 2005 the annual storage could be of 1 Mt of carbon. Moreover we have to consider that also in the soils interested by afforestation programs the amount of carbon will increase.

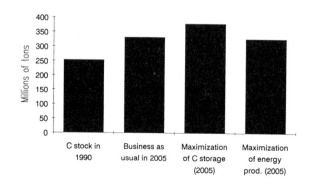

Fig. 1   Quantity of carbon stored in the italian forests in 1991 and in 2005 according to different scenarios

CARBON DIOXIDE

Forests

The Italian forest sector stocks today 253 millions of tons (Mt) of carbon. This value has been obtained by the revision of the data presented in the National Forestry Inventory and considers the productivity of the existing forests and of new plantations (IR, 1992)

In the "Business as Usual" (BAU) scenario the national forestry sector will stock, in 2005, 331 Mt of carbon. According to the model used in our research (IR, 1992), by the year 2005 there will be an annual increase of 6 Mt. But the role of trees can be enhanced considering a new forestry policy based on an afforestation campaign and a reduction of the area damaged by fire and pests: in this case the carbon storage will reach 379 Mt with an increase, compared to the BAU scenario, of 48 Mt from 1991 to 2005 and an annual carbon increase of 11 Mt at the end of the period.   It is also possible to give a specific emphasis on forests as an energy source: in this case the annual increase of carbon fixed will be in 2005 of only 5 Mt. This is less than in the BAU scenario, but we must consider that in the "energy scenario" there will be a reduction of annual carbon dioxide releases of 7,5 Mt (2 Mt of carbon) due to the substitution of fossil fuels by biomass (RI, 1992). In fig. 1 the carbon stored in different scenarios is presented.

Soil Organic Matter

Considering that the Italian agricultural soils have an average of 1.5% of organic matter, they store in the upper layer 394 Mt of carbon. A technological reform of current agricultural systems according to organic farming practices (green manure, conservation tillage, planting legumes, use of organic fertilizers, etc.) could increase this value. In accord with the estimates reported by Lovins (1991), such increase could reach 0,02% per year. In this case and considering an area of 1,5 million of ha (the actual agricultural land used is of 15 millions ha), in the year 2005 the annual storage could be of 1 Mt of carbon. Moreover we have to consider that also in the soils interested by afforestation programs the amount of carbon will increase.

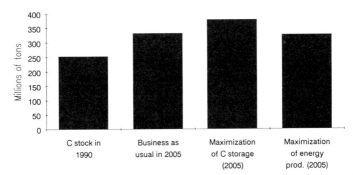

Fig. 1   Quantity of carbon stored in the italian forests in 1991 and in 2005 according to different scenarios

Agricultural wastes and residues

Just a small portion of residues and wastes of the agricultural crops is used today. Considering the non utilized part, an energy content of 3,85 Mtep/year could be used to produce, ethanol, biogas or electricity (CESTAAT, 1988). It is difficult to evaluate the fraction of this potential that could be converted by 2005.
A prudential evaluation, based on the transformation of 10% of the potential amount, lead to a reduction of 11,1 Mt of CO_2 per year in consequence of the saving of fossil fuels (Tab.1).

Source	Quantity	Energy recoverable	CO_2
	Mt/year	Mtoe/year	Mt/year
Agriculture	17,14	1,6	4,65
Livestock	13,17	1,93	5,60
Agro-industry	*	0,30	0,87
Total		3,85	11,12

*   the quantities considered in this sector are not in homogeneus units

Tab. 1 Agricultural wastes and residues produced annualy in Italy, energy potentially obtainable, and carbon dioxide potential reduction

Urban wastes

The annual production of urban wastes in Italy is of 20,5 millions of tons, that in large majority is dumped in landfills.
We analyze in this paper only the possible contribution of the organic part (that represents nearly 40% of the total garbage) and of the cellulosic fraction (20-25%).

Three options are considered, inceneration with heat recovery, composting and recycling of paper.
The production of electricity from steam produced by incenerators is of 1 kWh for 2-3 kg of wastes. In terms of CO_2 this means that for each kg of wastes incinerated it is possible to avoid the emission of 0,25 kg of carbon dioxide (in reality to strictly calculate the contribution of the organic part of the wastes, the plastic, almost 10% of the total share, should be subtracted).
In the case of composting an evaluation has been made on the energy consumption necessary for the production of the substitute chemical fertilizers, that can be estimated of 40 kg CO_2/ton of waste.
Finally for the paper recycled, considering that the energy saving connected with the use of recycled material is of 3600 Kcal for each kg of new paper, a potential reduction of 0,22 kg CO_2 per kg of urban wastes has been estimated.

Considering the possibility to incinerate 40% of the total urban wastes and to compost the organic part of another 20% of the wastes, an annual reduction of 2,2 millions of tons of carbon dioxide could be achieved in Italy.

Energy Conservation

Italian agriculture consumes 9,85 Mtep/y, including indirect energy consumption for fertilizers (CNR, 1989). If saving measures will be adopted, as in fertilizations, pest control, soil cultivation, a remarkable reduction - up to 20% in accord with many international reports (Turlow, 1990) - could be achieved. With this assumption, in 2005 the CO_2 emissions could be reduced by 5,71 Mt/year.

Energy Crops

The diffusion of energy crops for electricity or renewable liquid fuels production seems to be very interesting for Italian agriculture. Many crops can be grown in productive systems characterized by low inputs, in order to minimize energy costs.
With regard to the most promising one, sweet sorghum, Grassi (1991) affirms that, through a process of gassification and the use of gas turbines, 1-2 million hectares would be sufficient to eliminate the national electric deficit (32 billions of kWh/y). To obtain this goal the crop's yield should be of 10 t/ha of sugar and 25 t/ha of bagasse. The benefit in term of avoided carbon dioxide emissions would be of 22 Mt/y. In our climate stabilizations scenario we have considered achievable by 2005 a production of 6,5 TWh/y, a fifth of the figure reported by Grassi.
In the same scenario we have considered the goal of a 5% reduction of the national consumption of gasoline and diesel oil used for transportation. If a yield of 6000 and 1100 l/ha will be reached, respectively by sugar (as sweet sorghum, Jerusalem artichoke, sugar beet) and oil crops (rape, sunflower, soya), 834.000 ha should be planted. The benefit in term of avoided carbon dioxide emissions would be in this case of 4,7 Mt/y.

Product	crop	hectars	CO_2
		thousands	Mt/year
ethanol	sweet sorghum; Jerusalem artichoke; sugar beet	149	2,1
biodiesel	rape; sunflower; soya	685	2,6

Tab. 2 Agricultural area necessary to produce a quantity of biofuels equal to 5% of actual italian gasoline consumption and related tons of carbon dioxide avoided emissions

The global potential for carbon dioxide reduction from the agro-forestry sector in our scenario is of 66 MT CO_2/y, an amount that represents 16% of total italian emissions in 1990.

METHANE

Livestock production, rice cultivation and urban wastes are the national sectors involved in methane emissions. The contribution of the first sector is of 536.000 t/y if only the national animal husbandry is considered; it grows up to 1,75 Mt if we consider the national beef consumption. Reducing beef consumption up to 50%, in accord with a balanced diet, could allow to reduce global methane emissions by 760.000 t/y (270.000

t/y considering only the italian meat production). Italian rice paddy fields are a source of 111.000 t/y. Considering the possibility to increase by 20% dryland cultivation, it would be possible to reduce methane emissions by 22.000 t/y. Finally, considering that in a landfill for each ton of garbage 80/90 m^3 of methane are produced due to anaerobic digestion, the collection of this gas and its use to produce electricity could avoid the emissions of 1,7 tons of carbon dioxide equivalents per ton of garbage (considering that methane has a Global Warming Potential 21 times greater than CO_2).
The possible complementary options for waste treatment (recycling, incineration, composting, landfill disposal with methane recovery) applied to 80% of total urban garbage could avoid the emission of a quantity of methane equal to 25 Mt CO_2.

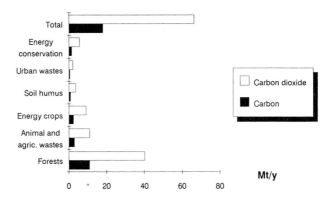

Fig. 2 *Potential of carbon dioxide emissions reduction from the italian agro-forestry sector by the year 2005*

REFERENCES

CESTAAT (1988), Impieghi dei sottoprodotti agricoli ed agroindustriali, Roma

CNR - ENEA (Biondi P., Panaro V., Pellizzi G., eds.), (1989). Le richieste d'energia del settore agricolo italiano - Sottoprogetto biomasse ed agricoltura. Roma.

RI (1992) Barbera G., La Mantia T., Pettenella D., Picciotto F., Silvestrini G., Strategie di riduzione dei gas climalteranti in Italia, Rapp. intermedio, Ministero per l'Ambiente, Roma

Lovins A., Lovins H. (1991), Least -cost climatic stabilization, Annual Revew of Energy

Grassi G. (1991), Produzione di elettricita' da risorse di biomassa, Agricoltura e Innovazione n. 19, Roma

Thurlow(ed.) (1990), Technological responses to the grenhouse effect, Watt Committee on Energy Report n. 23. Elsevier Applied Science.

Coppola S., (1991), Trattamento di fanghi di depurazione e residui solidi urbani in vista del loro impiego in agricoltura come materiali utili alla produzione, contribuendo nel contempo al problema del loro smaltimento, Agricoltura e Ambiente ,Edagricole, Ed., Bologna.

AN APPROXIMATE EXPRESSION FOR
THE UTILIZABILITY FUNCTION

Jadranka Vuletin, Petar Kulišić*, Ivan Zulim,
and Nenad Sikimić
Faculty of Electrical and Mechanical Engineering and Naval
Architecture, University of Split, Split, Croatia
*Faculty of Electrical Engineering, University of Zagreb,
Zagreb, Croatia

ABSTRACT

Monthly average values of the utilizability function were calculated for a photovoltaic system using the hourly data of global and diffuse solar radiation incident on a horizontal surface for two different locations. The results obtained have been compared with data calculated according to the approximate expression given by Klein. Values of the coefficients in the ϕ–equation are modified using the measured data for the locations being considered.

KEYWORDS

Photovoltaic system; solar radiation; utilizability function.

INTRODUCTION

The concept of utilizability function, introduced by Whillier (1953) and generalized by Liu and Jordan (1963), has been developed and applied as a design method by Klein (1978). The approximate expression for the average value of the utilizability function (Klein, 1978) correlated to the monthly average clearness index and critical radiation ratio has provided the possibility to determine the ϕ–charts.

THE UTILIZABILITY FUNCTION

The hourly values of the utilizability function ϕ can be determined (Liu and Jordan,

1963) according to the relation:

$$\phi = \frac{1}{N} \frac{\sum_N (H_t - H_{t,c})^+}{\bar{H}_t'} \tag{1}$$

where H_t is the hourly total radiation on the tilted array, $H_{t,c}$, the critical level, $\bar{H}_t'$, the monthly average hourly radiation for a given hour of the day and N, the number of days in the month.

The monthly average utilizability $\bar{\phi}$ can be calculated (Klein, 1978) using the monthly average daily total radiation $\bar{H}_t$ instead of the monthly average hourly values $\bar{H}_t'$ as:

$$\bar{\phi} = \frac{\sum_{days} \sum_{hours} (H_t - H_{t,c})^+}{\bar{H}_t N} \tag{2}$$

The superscript "+" in the relations (1) and (2) indicates that only positive values are considered.

The critical level $H_{t,c}$, defined as the radiation level at which the ratio of the electrical energy production equals the load, is given by:

$$H_{t,c} = \frac{E_L}{A \eta_c \bar{\eta}_f \bar{\tau}} \tag{3}$$

where E_L is the load, A, the array area, $\bar{\tau}$, the monthly average transmissivity of array cover, $\bar{\eta}_f$, the monthly average array efficiency and η_c, the efficiency of the control subsystem.

The array efficiency can be expressed (Siegel et al., 1981) by the equation:

$$\eta_f = \eta_R[1 - \gamma(T_a - T_R) - (\gamma \tau \alpha H_t)/U] \tag{4}$$

where γ is the temperature coefficient of efficiency, η_R, the reference array efficiency, T_a, the ambient temperature, T_R, the reference cell temperature, $\tau \alpha$, the transmissivity-absorptivity product of the solar panel and U, the overall loss coefficient of the array.

In this work the monthly average values of the utilizability function are determined using the hourly global and diffuse radiation data on a horizontal surface of eight years' period and the data of ambient temperature for the *Bar* ($\varphi = 42°06'N, \lambda = 19°06'E\,Gr$) and *Bitola* ($\varphi = 41°03'N, \lambda = 21°22'E\,Gr$) regions. The calculations concern the south-facing photovoltaic array for the tilt angles from $0°$ to $90°$, the array area of $0.7m^2$, the load from 10 to 50 W and the following parametar values:

$\eta_R = 0.12, \eta_c = 0.9, \gamma = 0.004 K^{-1}, T_R = 0°C, U = 21 W/Km^2, \tau \alpha = 0.95, \bar{\tau} = 1.$

ϕ–EQUATION

It has been found by Klein (1978) that the monthly average daily value of the utilizability function can be obtained by the relation

$$\bar{\phi} = \exp\{[A + B(\frac{R_n}{\bar{R}})](\bar{X}_c + C\bar{X}_c^2)\} \tag{5}$$

The coefficients A, B and C given by Mitchell *et al.* (1981) are:

$$\begin{aligned}
A &= 7.10 - 20.20\bar{K}_T + 12.08\bar{K}_T^2 \\
B &= -8.02 + 18.16\bar{K}_T - 10.68\bar{K}_T^2 \\
C &= -1.02 + 4.10\bar{K}_T - 1.96\bar{K}_T^2
\end{aligned} \qquad (6)$$

where $\bar{K}_T$ is the monthly average clearness index, $\bar{R}$ is the monthly mean total radiation tilt factor and $\bar{X}_c$ is the monthly average critical radiation ratio.

The monthly average critical radiation ratio $\bar{X}_c$ is related (Klein, 1978) to the critical radiation level $H_{t,c}$ by expression:

$$\bar{X}_c = \frac{H_{t,c}}{r_n R_n \bar{H}} \qquad (7)$$

where $\bar{H}$ is the monthly average daily radiation on the horizontal surface, R_n is the ratio of radiation at noon on the tilted surface and that on a horizontal surface for an average day , while r_n is the ratio of the total radiation at noon and the daily total radiation.

The monthly average daily values of the utilizability function have been calculated using the expressions (5) and (6) in order to test the validity of its applicability to the Bar and Bitola regions. The comparison of these values with the values obtained from the relation (2) shows certain descrepancy for particular months. While trying to reduce the differences between the experimental and calculated values the coefficients A, B and C were modified according to the measured data of the climatic region observed. The modifications were made using the gradient method (with steepest descent procedure). It has been found that the suitable values of the coefficients are:

$$\begin{aligned}
A &= 7.76 - 20.66\bar{K}_T + 10.81\bar{K}_T^2 \\
B &= -8.45 + 19.01\bar{K}_T - 9.99\bar{K}_T^2 \\
C &= -1.80 + 5.24\bar{K}_T - 0.21\bar{K}_T^2
\end{aligned} \qquad (8)$$

The monthly average values of the utilizability function were calculated for the south-facing photovoltaic array of $0.7m^2$ area varying the tilt angles from $0°$ to $90°$ and the load from 10 to $50W$ in the period of twelve months for Bar and Bitola regions. In Fig. 1 and 2 the monthly average values of the utilizability ϕ as a function of the critical radiation ratio X_c for the tilt angle of $55°$ are shown for January and July (Bar region). The mean square deviation between the calculated values of the utilizability function from the equation (5) and the values obtained from the equation (2) equals 0.033 for the case when the coefficients (6) are used and 0.024 when the modified coefficients (8) are applied.

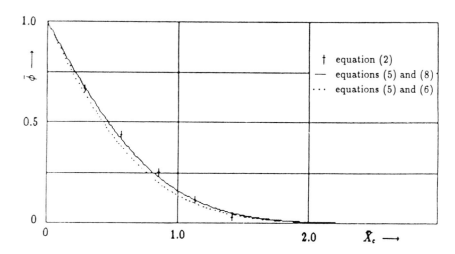

Fig. 1. The monthly average values of the utilizability function
in dependence of the critical radiation ratio for January.

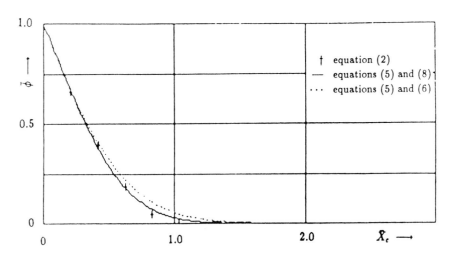

Fig. 2. The monthly average values of the utilizability function
in dependence of the critical radiation ratio for July.

CONCLUSION

It has been found that the approximate expression (5) for the utilizability function can be applied with a sufficient accuracy to the region being observed. The differences appearing between the values obtained from equation (5) with coefficients (6) or with modified coefficients (8) and experimental results are in the range of tolerances for all other photovoltaic system parameters.

REFERENCES

Klein, S. A. (1978). Calculation of flat-plate collector utilizability. *Solar Energy*, 21, 393–402.
Liu, B.Y.H. and R.C. Jordan (1963). A rational procedure for predicting the long-term average performance of flat-plate solar-energy collectors. *Solar Energy*, 7, 53–74.
Mitchell, J.C., J.C. Theilacker and S.A. Klein (1981) Calculation of monthly average collector operating time and parasitic energy requirements. *Solar Energy*, 26, 555–558.
Siegel, M.D., S.A. Klein and W.A. Beckman (1981). A simplified method for estimating the monthly-average performance of photovoltaic systems. *Solar Energy*, 26, 413–418.
Whillier, A. (1953). Solar Energy Collection and Its Utilization for House Heating. *Ph.D Thesis in Mechanical Engineering*, M.I.T., Cambridge, Massachusetts.

MANAGEMENT OF RENEWABLE ENERGY SYSTEMS AT VILLAGE LEVEL

D. SINGH

Rural, Energy and Environmental Development Society
1, Sheetal Apartment, Parvati Nagar
Subhanpura, VADODARA - 390 007 (INDIA)

ABSTRACT

This paper discribes various management problems faced in operating and maintaining of the renewable energy divices and systems at village level. It assesses the effectiveness of some of the measures taken in this direction & highlights a few success stories. Successful management of a community biogas plant catering cooking energy needs of whole of a village community by village women is found a mile stone in managing such systems in villages. Role of the nongovernmental voluntary organisations in management is inevitable, while it is noticed that the village level Energy Cooperatives are the best choice.

KEYWORDS

Renewable Energy System, Village Level Management.

INTRODUCTION

Energy plays a vital role in the development of a country. Delivery of efficient form of energy like electricity for lighting, irrigation, industrial applications and natural gas for cooking especially in developing countries, where there are many such potential consumers scattered over a large area, is a major concerned of the policy makers and energy planners. Conventional methods of electricity transmission to villages through high voltage distribution network is not economical vialble and most of the village population is still uses Kerosene for lighting and due to non-availability of natural gas, firewood, cow-dung and agricultural residues remain the cooking fuels.

2964

Some of the renewable energy technologies have proven their suitability for meeting the rural energy needs. Boigas Technology has pervaded and accepted as an efficient cooking medium besides other advantages of providing clean environment and producing enriched manure. Solar photovoltaic technology has been welcomed due to its instantly power generation characteristic. Boimass gasification technology offers saving of diesel. Improved models of the cooking stoves give almost double efficiency. However, solar water heaters, solar cookers and solar dryers are yet to penetrate in villages.

Most of these systems and devices are being supplied and installed under various governments' subsidy schemes. The programmes are being implemented by the respective state government with the help of central and state governments funds. Though, initial investment on these systems is highly subsidised, neither respective state government nor central government provide funds for post installation operation and maintenance. Further due to the lack of infrastractural facilities,it is difficult for the implementing organisations to operate and maintain these systems especially in villages.

A TYPICAL INDIAN VILLAGE

In general a village is described as a cluster of houses.Most of the houses are constructed with mud and stones. Only a few houses are of bricks and cement, those belong to the upper class rich families. Decision making power in respect of any new thing is rested with these rich families. Majority of the villagers are either small farmers or landless labourers. As most of villages do not have even primary schools, the level of literacy is very low. Village women are amlost illiterate. Poor road links, irregular transportation, minimal credit facilities, uncertain power supply, illiteracy, non-availability of skilled labour etc. make all village based occupations less competitive. Political influence, social acceptibility and technological ignorance in general comes forward in penetrating any new divice in villages. India is a country of such 560000 villages where its more than 70% population lives.

RURAL ENERGY SCENE

Indian rural energy scene is characterised by the use of firewood, cow-dung and agricultural residues for cooking, edible oil and Kerosene for lighting, diesel mix electricity for water pumping for irrigation and animal power for ploughing and transportation. Against an annual production of 49 million tonnes, the consumption of wood is 133.10 million tonnes, while consumption of animal dung and agricultural residues is 73 million tonnes & 41 million tonnes against their production of 324 million tonnes and 203 million tonnes respectively. Therefore, it is evident that

2965

more than two and half times consumption of the fuelwood leads for deforestation and above 20% comsumption of animal dung and agricultural wastes stops a huge part of these to be returned to agricultural land.

RENEWABLE ENERGY TECHNOLOGIES

Since ancient days, solar energy is being utilised in the Indian villages for drying purposes however, it was not utilised for water heating and cooking etc. Production of biomass and its burning for cooking is still continued. Therefore, renewable sources of energy are of course not the new discoveries, these are the devices and the systems, which give a second look at the age old technologies of the sun. With the development of various renewable energy technologies, villages are the potential beneficiaries where some of the renewable energy technologies can play a crucial role in meeting the rural energy needs and uplift the standard of villagers. Therefore, shifting to these sources such as the sun, wind, biogas, biomass etc. can bequeath them a limitless energy treasurer.

URJAGRAM (ENERGY VILLAGE) CONCEPT

Urjagram concept is nothing but it aims at introducing decentralised energy supply options in villages and to bring about energy self sufficiency by harnessing locally available renewable energy resources. Integrated Rural Energy Programme also has the similar objectives being implemented in the country. Under these projects, different devices & systems are installed in villages and some of them are for community use such as community biogas plants, small capacity solar photovoltaic power plants, gasifiers etc. in order to fulfill the maximum energy demands from renewable energy sources. A significant progress has been made in the installation of various renewable energy devices & systems. In Table 1 status of these devices installed upto 1991 has been presented.

Table 1 Status of Renewable Energy Devices Installed in India

Sr. No.	Device/System	For Individual Use (Nos.)	For Community Use (Nos.)
1.	Biogas Plants	1.402 million	657
2.	Improved Cookstoves	.900 million	---
3.	Gasifiers	760	---
4.	Windpumps	2711	---
5.	Solar Cookers	.177 million	*
6.	Solar Timber Kiln	*	---
7.	Solar Water Heaters	0.172 m M^2 Area	---
8.	Solar PV Lights	.035	*
9.	Solar PV Pumps	1171	---
10.	Solar Power Plants	---	50 (244.64 KW)
11.	Wind Farms	---	37MWGrid Connected

* Data not available

2966

Out of the above listed devices, a number of devices' have
been installed under 140 Urjagram Projects.

MANAGEMENT OF RENEWABLE ENERGY SYSTEMS

Most of the community type systems have been installed under
heavy subsidy. The subsidy amount ranges between 70% and
100% for years old systems. Individual devices get a little
less subsidy which is ranging 30% to 50% depending upon the
type of the device and the beneficiary. If the device is
installed for the use of an individual where his contribution
is more than the subsidy amount, the device is used and
managed properly. On the other hand, if it is a community
system and the contribution of the individuals is very
little, the operation and maintenance becomes a crucial task
and as per the experience gained so far, it is noticed that
these systems becomes meager junks after a period of one or
two years of installation. Therefore, it is not only the
money which goes waste, it also gives the negative
demonstration value. Several routine inputs which seem
matter of fact to a city manager take on herculean
proportions when viewed in the village context: collection of
dung, water shortage, quatity and quality of dung, finance
management, distribution of slurry, maintenance, technical
failure and so on. If these problems are not attended to
immediately, they can lead to decreased gas supply, internal
differences, dissatisfied and angry beneficiaries and
ultimately plant shut down. Lack of skilled manpower and
funds to replace a bulb, battery etc. are responsible for non
working of a SPV Power Plant, Community Television etc.

State nodal agencies owned the responsibility of site
selection, supply and execution however, poor roads,
irregular public transport etc. comes in the way of these
agencies to operate and maintain renewable energy systems
life long. So, transferring of the responsibility to
villagers is found a reasonable and acceptable solution of
the management problem. The decision regarding scheduling ,
pricing, scale of operation etc. can thus be taken
judiciously and offer flexibility since the priorities of the
rural users are known fully by the villagers themselves. As
control and management of energy demand and supply will be
virtually in the hands of the users, it is likely to lead to
a more responsible use of energy. Moreover, collective
thinking, decision making and clout can prevent exploitation
of illiterate villagers. Thus the formation of a
co-operative and its infrastructure at the village level is
likely to ease the troubles of the city based programme
implementing agencies and the government.

ENERGY CO-OPERATIVE MODEL

Co-operation in the form of association aimed at securing
mutual economic benefits. Co-operatives have been experimented in India to protect small farmers and other villagers

from exploitation by money lenders and middlemen buying farm products. The concept grew out of the innate human instinct for self-helf and is based on the universal characteristic of equality, fraternity and dignity of man. A village energy co-operative is formed by the users of a particular system to make them the members of the society.

ENERGY CO-OPERATIVES : A FEW SUCCESS STORIES

India's first energy co-operative society was formed in village Khandia (Gujarat) to look after the management of the country's first Solar Village. As on today there are so many energy co-operatives exist in the state of Gujarat and other states of the country to look after the management of biogas plants & SPV power plants. Some successful management stories through energy co-operatives are operation of a community biogas plant of 255 Cu.m capacity catering the gas need of 120 families, at village Motipura and another community biogas plant of 630 Cu.m capacity catering the gas need of 350 families at village Maithan (Gujarat). SPV power plants of 8KW, 4KW and 2KW at villages Kalyanpura, Choradungri and Ralyati Gurjar respectively are also being managed by the village co-operatives since last a number of years successfully.

INVOLVEMENT OF WOMEN IN MANAGEMENT

Village Motipura Biogas Co-operative Society presents a unique feature because it now has 5 women members on its management committee. The post of Chairperson since last several years has for the most part been held by a woman. One year of plant operation, the crunch for dung arose-people were not supplying enough dung to the plant. Therefore, it was realised that involvement of women could this problem be overcome. With the induction of women, the management committee began to acquire the profit it now has. Thus Motipura heralds a new chapter in the Energy Co-operative Movement. In village Maithan, after the involvement of two women members in the Management committee, the Committee had declared dividends in its second year of operation.

CONCLUSION

Management of these systems through Village Level Energy Co-operatives seems to be a solution to overcome numerous problems otherwise faced by the city based implementing organisations. Formation of Energy Co-operatives in the state of Gujarat has been successful. Involvement of women in the management committee of these co-operatives has further proved a milestone in this direction.

REFERENCES
Annual Report (90-91),DNES, Government of India.

EVALUATING CAPACITY VALUE FOR RENEWABLE ENERGY TECHNOLOGIES

M.R. Crosetti

RCG/Hagler, Bailly, Inc., 50 California, Suite 400,
San Francisco, California 94111, U.S.A.

ABSTRACT

Capacity value can be an important component of the overall value of new power generation technologies, particularly in countries which face chronic generation shortages. This paper presents a relatively simple method for estimating the firm capacity equivalent of a new generation unit by assessing the ability of the unit to reduce total unserved energy on the utility system. The method requires data on only three operating parameters: plant output over time, and the corresponding system load and frequency. The methodology depends upon frequency variations, such as those associated with generation shortages, to indicate the relative levels of expected unserved energy over time. This approach is particularly appropriate for assessing renewable energy technologies in developing countries, which in many cases experience chronic load shedding, and which often have neither the data nor the resources to conduct detailed production costing simulations. The paper illustrates the application of this methodology to the economic evaluation of windfarms planned for India. In this particular case, it is shown that the capacity value of the windfarms in question is about one-quarter that of a conventional power plant on the basis of a nominal kilowatt of plant capacity.

KEYWORDS

Power economics; economics of renewable energy; utility capacity credits; capacity value; wind power; India.

INTRODUCTION

In developing countries such as India, China, and Indonesia, power demand is growing rapidly and capital resources are at a premium. Thorough evaluation of generation options is particularly critical in these countries to ensure the maximum benefit from the resources available for power system investment. These countries are also often identified as potentially huge markets for renewable energy technologies (e.g. SERI, 1987), but realization of this market potential first depends on whether these technologies are seen by utility planners to be more economically attractive than conventional alternatives.

The decision to add a new generation unit to a power system depends on the energy value, capacity value, and system operating impacts of that new unit relative to other generation options. A concise overview of these factors is given in Grubb (1989). Capacity value is a particularly important component of value in many developing countries because of the chronic generation deficits these countries face. Capacity value refers to the capital cost of capacity additions which is foregone as a result of the addition of the unit under consideration, assuming an equivalent impact on system reliability.

Comprehensive economic models of utility system operation, such as production costing models, can be used to assess capacity value. However, these models are often data-intensive, costly to run, and typically not well-suited to evaluation of renewable energy technologies in terms of their ability to model the operation of these technologies. These models typically assume that generation units can be dispatched subject only to forced and planned outages. The operation of many renewable generation technologies, however, also depends on the uncertain availability of natural resources such as wind or sunlight.

This paper presents a simplified technique for assessing the capacity value of grid-connected renewable energy systems which is suitable for systems with chronic generation deficits. The approach proposed here explicitly recognizes the variations in generation output from a renewable technology, and relies on data typically available for any utility that operates under a Supervisory Control and Data Acquisition (SCADA) system.

It has been suggested that for low levels of penetration the contribution of any power plant to firm capacity is approximately equal to the average power it generates (Swift-Hook, 1988). Despite the theoretical soundness of such claims and the results of *ex-ante* utility system simulations which confirm them (e.g. VanKuiken *et al.*, 1980), the conventional engineering and finance community remains suspicious of capacity value claims for renewable technologies. This methodology provides yet another perspective on capacity value which will hopefully reinforce past findings, and which some may find intuitively appealing. Furthermore, review of windfarm experience in California suggests that capacity value may be much higher than than these theoretical approximations, depending on the characteristics of the utility and the renewable technology (Smith and Ilyin, 1989). The approach presented here specifically takes into account the utility and plant characteristics which may lead to higher or lower capacity value than suggested by theory.

A METHODOLOGY FOR CAPACITY VALUE ASSESSMENT

Capacity value derives from the ability of the new generation unit to improve system reliability or, alternatively, to reduce the amount of expected unserved energy on the system. Capacity value is sometimes thought to arise from the correlation between output of the generation unit and system load. On the contrary, higher load does not necessarily imply an increase in expected unserved energy. If a proportionally greater share of total capacity is available during peak periods than during non-peak periods (i.e. if peak period reserve margins are greater than in non-peak periods), then additional capacity may be most valuable (i.e. may result in the greatest reduction in expected unserved energy) if it is available during non-peak periods. For instance, a system which depends heavily on seasonal hydroelectric resources may face a greater loss of load risk during a dry season with lower peak demands than during a wet season with higher peak demands.

The approach proposed here examines the correlation between plant output and indicators of expected unserved energy. On systems with chronic generation deficits, frequency sometimes remains below design levels for extended periods of time. This affects the operation of induction motors and other end-use equipment. Should frequency sag enough, generators themselves may be damaged. It is at that point that blackouts are initiated. These variations in system frequency can provide an indicator of the relative incidence of expected unserved energy.

System frequency falls when load exceeds generation output. Changes in system load and frequency are related as follows:

$$\Delta P + K \Delta f = 0 \qquad (1)$$

where:
ΔP = *the change in system load*
K = *a system constant which depends on the characteristics of system generators*
Δf = *the change in system frequency*

Frequency-corrected load is the load which would have been met if system frequency were to have been at its design level, i.e. either 50 or 60 Hz, depending on the system. Since the constant K and the design frequency is known along with the actual frequency and load (retrieved from the utility's SCADA system), one can calculate the frequency-corrected load.

Instead of attempting to calculate the actual level of unexpected unserved energy in each period, the methodology evaluates the *proportion* of total annual expected unserved energy attributable to each period. The basic assumption is that periods with a larger difference between frequency-corrected and actual load relative to other periods will be characterized by proportionally higher expected unserved energy. Specifically, it is assumed that the fraction of total annual expected unserved energy occurring in any hour of any month is proportional to the difference between the frequency-corrected system load and the actual load for that same hour on the peak day of that month, weighted by the contribution of that month to total annual expected unserved energy.

Firm capacity equivalent (FCE) of a generation addition is defined here as the reduction in expected unserved energy given actual operation of the unit, divided by the reduction in expected unserved energy that would occur if the generation addition could operate throughout the period in question with a 100% capacity factor. For dispatchable technologies, the FCE would be equal to the availability of the unit, assuming that forced and planned outages occur independently of the incidence of unserved energy over that period. FCE can be greater than the capacity factor since it is assumed that even if a unit is not operating at the time a loss of load occurs, it can be brought immediately on-line, so long as it is available, i.e. not out of service as a result of a forced or planned outage. If the unit is operating, but not at its nominal rating, when a loss of load occurs, it is assumed that it can be brought up to its full nominal output immediately.

For non-dispatchable technologies such as wind power or photovoltaics, FCE depends upon the match between the temporal distributions of plant output and expected unserved energy. During certain times of the day or season, perhaps during peak periods, expected unserved energy is greater than at other times. If plant output coincides with these periods, then its firm capacity equivalent, and hence capacity value, is greater.

The methodology first derives the monthly FCE of a plant on the basis of hourly profiles of unserved energy, i.e. the difference between frequency-corrected and actual load, and plant output. The FCE for that month is the sum over 24 hours of the products of the hourly ratios of actual plant output to nominal capacity and the normalized difference between actual and frequency-corrected load. The model then derives annual FCE by multiplying monthly FCE's by the respective monthly weights, which are calculated in a similar way as the hourly weights. The mathematical formulation for evaluating FCE is given below:

(1) Compute the contribution of each hour in a month to that month's expected unserved energy by calculating the normalized frequency-corrected load minus actual load for all hours in the month. (A month is often characterized by a single representative day of 24 hours).

$$ND_{m,h} = \frac{FCL_{m,h} - AL_{m,h}}{\sum_{h}(FCL_{m,h} - AL_{m,h})} \tag{2}$$

where:
m = month of the year
h = hour of the day
$FCL_{m,h}$ = frequency-corrected load in month m, hour h
$AL_{m,h}$ = actual load in month m, hour h
$ND_{m,h}$ = the monthly normalized difference between frequency-corrected
and actual load in month m, hour h

(2) Compute the contribution of each month in a year to that year's total expected unserved energy by calculating the normalized maximum difference between frequency-corrected and actual peak load for each month.

$$W_m = \frac{\max_{h}(FCL_{m,h} - AL_{m,h})}{\sum_{m}(\max_{h}(FCL_{m,h} - AL_{m,h}))} \tag{3}$$

where the new variable is:
W_m = the weight for month m

N.B. If the difference between frequency-corrected and actual load is less than or equal to 0, then it is assumed to be 1.

(3) Compute plant FCE in MW per MW of installed capacity.

$$FCE = \sum_{m}(\sum_{h}(ND_{m,h} \times PO_{m,h}) \times W_m) \tag{4}$$

where the new variable is:
$PO_{m,h}$ = Plant output per MW of capacity in month m, hour h

Once the FCE is known, capacity value can be calculated by multiplying FCE by the cost of the least expensive capacity alternative, such as a gas turbine, adjusted for its availability.

A CASE STUDY OF A WINDFARM EVALUATION IN INDIA

This technique was used to help assess the economic merits of commercial-scale windfarms in the Indian state of Tamil Nadu (ESMAP, 1991). Wind logger data was used to estimate hourly windfarm output, and corresponding data on system load and frequency was collected from the Tamil Nadu Electricity Board (TNEB), the local utility. Relevant data for the day representing August, 1989, is shown in Figure 1. In this case, the hours with the greatest contribution to expected unserved energy (and, coincidentally, the hours of greatest windfarm output) are

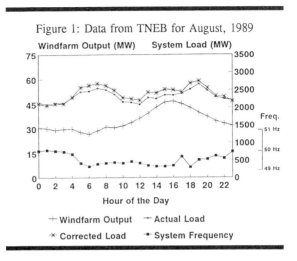

Figure 1: Data from TNEB for August, 1989

not peak load periods. The analysis concluded that the FCE of the proposed windfarms was approximately 0.16. Incidentally, the capacity factor for these windfarms was about 0.18, consistent with the theoretical claim that FCE should be equivalent to average power output. Since the availability of conventional power plants is India is about 0.65, the capacity value of the proposed windfarms is 0.65/0.16, or about one-quarter, of a conventional plant.

CONCLUSIONS

A technique has been presented which enables capacity value to be calculated based only on the plant's hourly output, and the corresponding system load and frequency. Although this approach is most suitable for the evaluation of generation additions to systems with generation deficits, these are typically the cases in which calculation of capacity value is most critical. Case study results are consistent with the findings expected using other methodologies.

REFERENCES

ESMAP (Joint UNDP/World Bank Energy Sector Management Assistance Program) (1991). *India: Windfarm Pre-Investment Study*. The World Bank, Washington, D.C.

Grubb, M. (1989). How much is wind energy worth?. *Wind Energy Weekly*, vol. 8, nos. 350 to 353. The American Wind Energy Association, Washington, D.C.

SERI (Solar Energy Research Institute) (1987), *Study of the Potential for Wind Turbines in Developing Countries: Phase I Report*. SERI/STR-217-3219. Golden, Colorado.

Smith, D.R., and Ilyin, M.A. (1989). Wind Energy Evaluation by PG&E. Pacific Gas & Electric Research and Development, San Ramon, California.

Swift-Hook, D.T. (1988). Firm power from the wind. Swift-Hook Associates, U.K.

VanKuiken, J.C., Buehring, W.A., Huber, C.C., and Hub, K.A. (1980). *Reliability, Energy, and Cost Effects of Wind-Powered Generation Integrated with a Conventional Generating System*. ANL/AL-17. Argonne National Laboratory, Argonne, Illinois.

NEW OUTLOOK OF THE BASIC STRATEGIES FOR
SOLAR ENERGY USE

D.C.BADEA

Department of New Proceed.and Tech.,Inst.of
Food Chemistry, Str.Gîrlei nr.1, Sector 1,
71523 Bucharest, ROMANIA

ABSTRACT

New outlook of the basic strategies for solar energy use, on
specific experiments, which is refering to the solar plant
integration by new suitable solar technological proceedings
to lead to minimum energy consumption. According to these
proceedings, new solar equipment structures are offered as
alternative priorities for solar energy technological use.

KEYWORDS
Solar energy; solar plant integration; new proceedings and
equipment structures.

THE PRESENT OUTLOOK
The present outlook, concerning the construction and working
of technological solar plants, implies the fact that their
efficiency decreases with the temperature increase, as it is
required by the technological process.

The energy measurements, carried out by us, underlined the
fact that, for a series of big technological solar plants,the
fuel saving is not achieved at the level of the estimated
value from the execution projects.Even if, in the case of
these projects, the solar energy supply estimations might be
too optimistic, the main cause of not reaching the predicted
parameters is the technological and power supplyunintegration
of solar plants, which were developed without taking into
account the risk of the coupling energy systems that have
different ages.
The solar energy use and, generally, the use of the renewable
energy is, rather, a technological problem, than an energetic
one.But, in the actual conception, the solar plant is viewed
rather an energetic system and rarely like a technological
one.This leads to solar energy use strategies which have not
a technological base and a suitable application.

SOLAR PLANT INTEGRATION

The research works, performed at the Institute of Food Chemistry from Bucharest(Romania), in the field of solar energy technological use, aimed at a revaluation of the basic concept for the design of solar plants.

The coupling of a solar energy source to a technological system implies the elaboration of the suitable technological proceeding to lead to minimum energy consumption and to an optimum technological integration. The technological integration of a solar plant means the inclusion of the solar plant functions in the circuit of the proper technological plant, on the basis of a technology having specific solar features,where the operations are carried out by proceedings which are suited for the energy consumption minimalising and to ensure the continuity and the quality of the technological process.According to these proceedings, equipment structures are being shaped in view of ensuring maximum technological yields, at minimal energy levels, close to the possible level obtained in a solar plant, which is, thus, energetically integrated. At present , the priorities for solar energy technological use refer to solar drying, refrigeration and liquid media heating.As far as solar drying is concerned, our experiments have been carried out with malt, a sensitive product from a technological point of view. These experiments illustrated that, in six hours , a moisture extraction of 21% can be achieved. From the viewpoint of the solar plant performance, such a moisture extraction is important but, from a technological standpoint, due to the rapid process, the changes occuring in the product are disturbed (physical, chemical, biological changes). These experiments reveal clearly that the exclusive use of solar energy is not sufficient for providing the thermal conditions for malt drying. The essential feature of technological and energy integration, resulting from the above-mentioned experiments, is that only a proceeding allowing the dilatation in time of malt solar drying curves could provide the first drying stage, fact that implies a new principle dryer design. The experiments concerning the solar obtainment and use of refrigeration have a prospective character and envisage the possibility of applying such a technique at the milk collecting centres. The performed experiments revealed that cooling of about 15 l milk/m^2 is achieved from 35 ^{o}C to 10 ^{o}C; new investigations are necessary to develop new equipment, technologically adapted to solar specific features, such as, for example, the solar cold milk generator. In the domain of liquid media heating by means of solar energy, a technique has been experimented in view of achieving the concentration of a solution at low temperature and atmospheric pressure. The results illustrated that, by an optimum adaptation of the technological equipment,on the basis of highly specific solar techniques, solar plants reach performances, compatible with the technological requirements and with the the necessary energy savings. In this case, the adaptation criterion was the low temperature of the necessary air (40 ^{o}C), leading to the development of a technique based on the fact that water evaporation from the solution is achieved by passing a hot air flow over a rotation surface, semi-immersed in the liquid that has to be concentrated. In the case of the glucose-fructose syrup, an evaporated water flow of 1.45 l/hr,m² has been obtained on this surface. Now, the concept of solar plant integration is applied, by the Institute of Food Chemistry, to the agricultural products conditioning. The main problem of the agricultural products conditioning processes is the maximum degree of conservation obtaining, by the using of minimum ecological effects technologies, with minimum energy consumption and without useful product losses. The agricultural products conditioning implies the following two operations: purification and drying. The fundamental concept, of a new proposed research project, is the ecological integrated solar-electrostatic system. Such a system is based upon proceedings which permit a compact rural or industrial construction with the following functions: technological function, energy function and ecological function. So, such a system is capable to perform the projected technological operations, to supply the necessary energy for the processes and to ensure an ecological protection. The technological function of the system comprises: purification, drying, microorganisms destroying. This function is realized by a general proceeding which, first, ensure the separation of vegetable impurities from products and, then, the drying of the useful matter. In this manner, is avoided the drying of the impurities and an unnecessary energy consumption is eliminated. The separation of impurities is realized by using an electrostatic proceeding, which leads to separation yields of 95-98%. The drying is realized, by solar proceeding, in a compact system. The energy function is accomplished by including of the energy source in the system, by using the solar energy, converted in thermal energy by a compact solar plant. The compact solar plant includes, in its structure, the dryer and the solar energy conversion surface. The compact solar plant has an unconventional geometry, which leads to minimum degree of terrain use. To realize the other operations (separations, microorganisms destroying) the electrostatic plants are used, in the condition of a low energy consumption, which is supplied, also, by solar energy. In our concept, the optimum of the solar plant integration is obtained, mostly , by coupling of the solar proceedings with other new proceedings (electrostatic, ultrasounds, microwaves, etc.), by using, so called, combined proceedings. Without a basic readjustment of current concepts, an optimistic technological perspective for solar energy application cannot be envisaged. It is necessary to unblock the limit of 80 kg c.f./m²year, established for the fuel saving achieved by the existing solar collectors, by operating fundamental technological changes and, thus, surpassing today's generalized outlook based on the purely extensive solar applications, such as the inefficient construction of solar plants independent of the technological plant. From this point of view, as above-mentioned, our institute has initiated research works for the creation of integrated plant where the solar function is included in the functionality of the whole technological assembly, leading to spectacular changes in the solar plant investment values which, for the time being, are very great. In this respect, we could mention the "Fluidized-bed solar dryer with effect of concentrating the solar radiation", which has been tested, as experimental equipment, providing, during cold weather, a drying rate of 18 % moisture/hour. Among the techniques for drying process intensification, it is worth mentioning the material bed vibration process, by means of passing hot air through the material, at a speed that is lower than that initiated at the beginning of fluidization. This is the main principle for developing another type of solar dryer, under the name of "Vibrated -bed solar dryer", having the specific feature that the technological equipment as such is integrated in the construction of the solar plant, in a pyramidal form, the sides having an optimum angle slope, corresponding to the latitude of the application site and to the operation period.

Such intensive solutions offer new perspectives to solar energy use in agricultural processes, in food industry, as well as in other industries.

REFERENCES

Proctor, D. and Morse, R.N.(1977).Solar energy for the australian food processing industry.Solar Energy, 19, 1.
Ozisik, M.N.(1980).Solar grain drying.Solar Energy,24,4.
Dehon, P. and Bera, F.(1981).Evaluation des possibilités du séchage des produit agricole par énergie solaire.Bull.Rech. Agron.-Gembloux, 16(2).
x x x First World Renewable Energy Congres proceedings.(1990) Reading,UK.
Moore, A.D.(1973).Electrostatics and its applications.John Wiley & Sons, New York.

SELF-ACTING SYSTEM TRACKING
FOR PYRHELIOMETERS

* J. R. NAVAS DE LA TORRE; J. SERRANO; G. PEDROS; R. POSADILLO;
+ J. LUNA and A. ALVAREZ DE SOTOMAYOR.

* Departamento de Física Aplicada,
+ Departamento de Mecánica,
  Electrotecnia y Electrónica,

Escuela Universitaria Politécnica,
Avda. Menendez Pidal s/n, 14004 Córdoba,
SPAIN

ABSTRACT

The work to which the title refers constitutes the design and development of a universal solar tracking system, reliable and inexpensive. For this it has been supplied with higher accuracy, great flexibility and great autonomy. To fulfil these goals we propose a system based on a tracking made through solar coordinates by means of a microprocessor, this tracking is completed with a fine optoelectronic tracking. The tracking system is currently being tested with a pyrheliometer.

KEYWORDS

Solar Radiation; tracking; pyrheliometer; optoelectronic; microprocessor.

INTRODUCTION

The dimensioning of solar energy collecting systems, especially those which are based on concentration, requires the most detailed information possible on the values which exist of direct and global solar radiation in one given place. The instrument which is needed to determine direct solar radiation is the pyrheliometer, it is supplied with a polar mode tracking which poses two great inconveniences. On the one hand, the daily manual adjustment of the instrument, and on the other, the fact that it has a continous movement in the same direction (even at night), means the twisting and breakage of the wire by which the readings are transmitted. Very few measures of this type are carried out throughout the world due to the necessary manual operation on a daily basis of these sensors.

The aim of this work has been to design and build a tracking system which enables the automated operation of the pyrheliometer. This system can be applied both to rural solar instalations as well as to industry. Given its wide ranging possibilities, the system's design must perform three main objectives: great flexibility and autonomy, as well as maximum accuracy.

We have chosen a mixed solution, a system based on a tracking by coordinates, calculated by means of a microprocessor, and completes with a tracking device of the opto-electronic type. With this type of development we make use of the advantages of the opto-electronic system (Lynch et. al., 1990), as well as the accuracy of the increasingly inexpensive microprocessors.

The optoelectronic system will work whenever the sun shines, and will carry out tracking "in a real way". As soon as the problem of lack of light occurs for the optoelectronic tracking (clouds, day-break, and sun-set) there will come into operation the programme which calculates the position of the sun in relation to the time, the day, longitude and latitude, in the form of cenital and acimutal angles.

Stepper motors will be used, which introduce the important characteristic of reversibility, and which controlled by the microprocessor, at the end of the day, return to the day-break position of the following day, thus going back in its tracks and avoiding the structural problems of spacing as well as the rupturing of connections which may occur during night tracking.

DETAILED DESCRIPTION

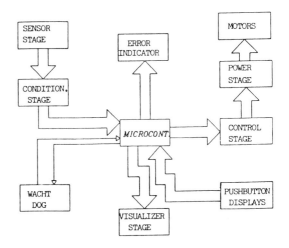

Fig.1. Block Diagram.

The block diagram is represented in Fig. 1. From the functional point of view we can distinguish between three blocks: sensor block, control block and motor block.

The sensor is of the photosensitive type, consisting of photodetecting pairs of elements set up in such a way as to give a null output signal, when the direction of the system coincides with that of the sun, and one signal positive or negative, depending on the deviation. The photosensor is solidly joined to the element which must be aligned with the sun and moves with it.

The topography chosen is to spread the photocells on converging opposite planes on an inclination. In this case there does not exist any fine and coarse appointment since with one of the two pairs of cells any type of angle or deviation in any given direction can be detected. This, compared to other topologies (Villamar, 1983) means economizing on half of the components and it simplifies the construction of the support box.The structure where the photocells are stored is a hollow trunk in the shape of a pyramid, on a cilindrical base, the diametre of which is the same as the diagonal of the main base. The lateral sides have an angle of 60º, each one of which is drilled to quarter to the cells in.

According to Fig. 2a, if the cells were an even with the sides, there would be an overlapping angle in which the sun would light the two cells at the same time, this angle would be 60º. But due to constructive reasons of the photoresistencies, the surfaces of these are not on a level with the sides, but 1 mm. underneath. This means, as can be seen in Fig. 2b, a diminishing of the overlapping angle to 50º. Through all this, the greatest possible lighted surface is attained for very small "relative movements" of the sun (heightened sensitivity).

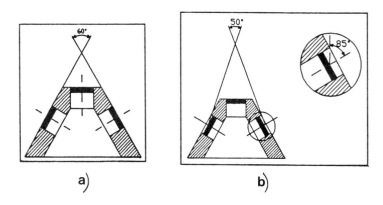

Fig. 2. Overlapping angle.

In order to protect the system of any undesired reflections the pyramid is stored in a hollow cilindre, in avoiding the reflections of the pyrheliometer's metal surface in our case, will not produce a significant decrease in the maximun detection angle.

The photosensitive elements used are photoresistants (LDR). In the outline of the electronic circuit in Fig 3, we can see that the tracking sensor is made up of a Wheatstone bridge LDR1, LDR2, R1, R2. On one of its diagonals the supply voltage is applied, and on the other, the inputs of the instrumentation amplifier. It the photoresistencies are equally lit, the tensions which will appear in the imput of the amplifier will be the same and thus, its output will be null. The moment one of them is more iluminated than the other, a difference in the potential of the amplifier inputs will appear which will give off a positive or negative signal depending on which one is more iluminated. The signal of the amplifier output will show us the direction of the sun's movement.

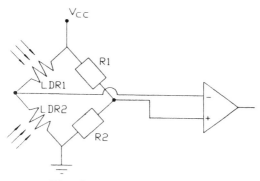

Fig. 3. Sensor Diagram.

In the control block we can differentiate two sub-stages. The one in charge of receiving the signals from the sensor block and of transmitting the signals which govern the motor block; the other sub-stage is the one which generates and transmits the signals which are needed for communication on a user level.

A microcontroller has been chosen as a processing unit due to the fact that a series of doors and functions are integrated in to the same encapsulation, which are accessible through its pins, and these mean a considerable simplification of the hardware required. As to the numbers bits, no specific requirementes exist, since the information which is used does not have the need for high field of values. The nucleus of the CPU is made up of the microcontroller 8052 AH Basic, which belongs to the MCS51 family by INTEL. They possess an internal ROM of 8 k bytes charged with a powerful Basic interpreter. It includes control operations which would otherwise be programmed machine language. Another key point for the election of this microcontroller has been its calculation power. Thanks to this facility we have been able to develop a programme in BASIC using the most accurate formulas recommended in the literature, (Iqbal, 1983) to determine the cenital and acimutal angle of the sun at every moment. Figure 4 shows the diagram of the flow which corresponds to the interactive tracking.

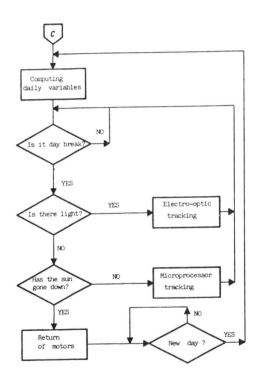

Fig. 4. Diagram of flow.

We must choose the motor block most suited to the existing needs. The most determining factor in choosing the most suitable engine is the accuracy with which the angle turned by the axis can be controlled, as well as its speed. With this in mind, a stepper motor has been chosen. These engines act with impulse voltage of all or nothing, so that every impulse introduced by the winding or windings, corresponds to one fixed turn or "step" of the engine. They also possess great flexibility, being able to function with a fixed or variable speed. The fact that they can be operated directly (with the adequate amplification) by a microcontroller makes them suitable for the most complex control systems.

The stepper motor has been chosen of the hybrid type which allows important pairs of accioning and high work frequencies. It also presents a wide variability in the angle turned by step (a value which oscillates between 90º step y 1.8º step). The use of reductions allows the frictional pair introduced by the charge to be divided by the value of the reduction and that the inertia pair be divided by the square value of the reduction. Because the charge is basically inertial in these types of applications, having an insignificant friction, the engines will operate pratically empty.

We must add to the above, the accuracy of positioning, since the angle turned by the rotor of the engine will appear on the exit axis of the reduction divided by its value, that is, the accuracy will be heightened in the value of the reduction.For the application to follow the sun we will need a configuration of two axis. We have opted for an elevation-acimut configuration.This type allows great mechanical robustness without big complications of the structure and of the turning mechanisms. So, we will need two stepper engines, each one in charge of the movement of one axis.

EXPERIMENTAL RESULTS

A small prototype model was used for testing the tracking system. On a clear day, the system tracks the sun well. When clouds pass in front of the sun, the tracker moves in calculated coordinates. The moment the sun reappears, the photosensors have carried out the necessary fine adjustments. After sunset, the system calculates the suns day-break position and goes back until it is fixed in this position. From this position on, the system will always move westwards. Interaction with the user has also been catered for, in the design of a pushbotton which allows the correction of faults in the power supply and in the microcontroller, as well as to restart the movement.

We have a CN-5 pyrheliometer of Synchotac-Middleton Instrument. We have modified it, designing and building a new metalic structure which a) would permit the acimut-elevation movement in relation to the polar which it brought and b) storing the engines and transmissions in new boxes. We are currently finishing the setting-up of the pyrheliometer with its new stucture and the electronics which governs it.

CONCLUSIONS

This tracking system is simple, moderately priced and with great adaptability. This sun tracker is relatively easy to install. Althought the system requires initial orientation, the accuracy of the set-up procedure is not critical. We have developped it for a very useful aplication such as the tracking of a pyrheliometer. It should be pointed out that we have found no reference in the literature, to the use of automatized tracking systems for this type of sensors.

REFERENCES

Iqbal, M. (1983). An Introduction to solar radiation. Pergamon Press, Toronto.

Lopez, G. (1983). Sistema de seguimiento del sol. In: Energía Solar Fotovoltaica, 123-127. Marcombo, Barcelona.

Lynch, W. A. and Salameh, Z. M. (1990). Simple electro-optically controlled dual axis sun tracker. Solar Energy, 45, 65-69.

Villamar, F. (1983). Sistemas de seguimiento del sol, fotosensor de la orientación de los rayos solares. In: Energía Solar Fotovoltaica, 129-132. Marcombo, Barcelona.

ELECTRICAL POWER FOR ISOLATED COMMUNITIES

LOAD SPECIFICATION AND CONTROL STRATEGIES

by

P.D. Dunn and S. Gómez

Introduction

This paper considers the specification of load demand of an isolated community and the optimisation of the wind power system to match these requirements at minimum cost.

Requirements

Small stand-alone electricity generation systems for isolated communities have to fulfil many requirements. In order to set out a specification it is necessary to consider factors such as cultural, social and economic patterns which affect directly the demand of electricity. This evaluation should take into account the geographical situation of the community, its economic and energy resources and its socio-cultural background.

It is particularly difficult to estimate future electricity demand for a community which does not have a source of electricity available now. Even where such a source exists and the project is one of the replacement of one source of energy by another, renewable source, all the factors mentioned previously should be considered. Additionally the demand for energy also depends on the availability of the resource. For example when wind turbine systems have been integrated with previous existing systems such as diesel generators, the patterns of consumption have changed. The original hours of supply of electricity (by diesel) were limited due mainly to high cost of fuel. Increased hours and in fact increased total consumption of energy has been observed. (See Fig. 1 giving an example from Fair Isle [1]).

2983

Where the load is restricted to lighting and is concentrated in a small area, low voltage d.c. power may be acceptable. In general, this will not be the case and it is desirable that the supply should be compatible with locally available electrical equipment. This defines the voltage and frequency choice, e.g. in the United Kingdom most domestic appliances are designed for connexion to the mains: 240 V 50 Hz.

System Optimisation

The matching of load demand and availability of wind energy is a crucial point in the general efficiency of a Wind Energy Conversion System (WECS). In this paper we will concentrate in the treatment of the load demand. It is important to optimise the end use, in particular to reduce the total size of the equipment, i.e. minimise installed capacity. In order to achieve this, several aspects should be taken into account:

- Recent technological advances such as new developments in efficient fluorescent lamps (assuming local availability) or the use of more efficient domestic equipment. Table 1 compares the energy consumption of some domestic equipment in 1973 and 1988 for Denmark [2] and Japan [3] and the expected energy consumption of these appliances in a "Low Energy House" in Tahiti [4].

- User education so that advantage can be taken of the devices inherent characteristics e.g. sensible use of a deep freezer (open only once or twice a week, preference for top lid types).

- Design/selection of equipment to incorporate storage e.g. thermal storage in refrigerators (by ice storage and also increased insulation) or in heating systems (bigger and better insulated tank).

- Increase number of houses fed from the same supply, effectively reducing the peak demand. For an individual household the magnitude of the peak to mean load demand ratio can be as high as 20. A very high ratio implies that a large capacity is required to supply the peak demand, although full capacity will only be used for short periods each day. If more houses are fed from the same generator the peaks will not be simultaneous so the total capacity is lower than the sum of the individual peaks. (See Fig. 2). Apart from seasonal variations in load level, the demand profile will tend to vary quite considerably from day to day for individual houses, especially at weekends, though as the size of the community increases the overall profile will show more consistency from day to day. According to Bullock (1990) the shape of the load profile of an isolated community is similar to that of a grid connected village, with the timing of the peak load demand depending on the working patterns of the local community.

- Load prioritisation. Development of solid state circuitry and its increased availability will permit the prioritisation of loads. Certain loads such as lighting will require full availability. Others such as refrigerators require only to operate over a sufficient period to maintain temperature. The avoidance of simultaneous switching of induction motors is particularly important.

Wind Speed Spectral Distribution

The combination of the various components of the wind, differing in amplitude, frequency and phase results in the complex traces of Fig. 3.

The wind engineer is interested in two aspects:

1. The statistical distribution of mean wind speed. This information is normally presented as a probability density function which enables the total annual wind energy available to be estimated. The factors which affect the optimal choice of λd (the value of λ at which Cp is a maximum) for a particular site and wind turbine design are discussed by the authors in Ref. 7.

2. The real time fluctuations of wind speed. The probability density function do not give information on these changes. However, this wind behaviour determines storage requirements and is discussed below.

The wind variability in real time can be more conveniently described in terms of spectral density. Wind velocity can be modelled by regarding wind data as a ergodic stationary process. It is then possible to derive a power spectrum which provides information on the individual frequency components contributing to the measured velocity/time data.

The power spectrum essentially expresses the kinetic energy associated with the different frequencies of the spectrum. It is found that the wind velocity spectrum can be broken down into various domains, for example that related to diurnal variations. The very high frequency components can be assumed to be filtered out by the inertia of the turbine (low pass filter in filter terminology).

More significant are the fluctuations in the low frequency section since these require to be averaged by a storage mechanism. As stated by Beyer et al. (1989), the power output fluctuations in the frequency domain of 1/1 Sec down to 1/10 min are of importance for the operation of small (island) grids and performance improvements may be achieved by implementation of short term (some minutes') storage. If we consider the frequency domain related to diurnal variations only, energy storage requirements can be identified and solutions specified. This approach is described in more detail in a later paper.

Very long term swings (days/weeks) must be covered by either prioritisation, long term storage or an alternative source of energy.

Strategy for Load Control and Prioritisation

Two main approaches to overcome power fluctuations in the wind are being developed. Load management systems which accommodate wind calms by cutting low priority demand and storage/resupply systems which flatten out short term fluctuations whilst continuing to meet minimum demand.

The combination of load prioritisation and storage enables the load demand to be modified to provide a close match to the wind energy source.

The difficulty in achieving the ideal situation mentioned above is the lack of information about the short term load demand variations and the subjective characteristics of any prioritisation system. The adequate prediction of system behaviour would enable an optimum sizing to be made not only of the wind turbine and the generator but also of the storage system(s). Decisions related with the relative priority of each appliance or group of appliances will determine the load plan necessary to design and build the electronic control system. Because of the complexity (and hence high cost) of a flexible load demand management system which allows each individual house to switch on/off the different loads, a more rigid approach has been assumed in this study. The different equipment in each household is connected directly to the WECS without the use of sockets or plugs but through electronic devices which allow the use of ripple control to turn on/off the equipment from a central control room. This may sound too rigid but for many isolated communities whose needs are met by a local utility, the supply is currently guaranteed only for a few hours each day and the members of the community become used to a fairly rigid pattern of use of each appliance, realising that demanding too much power may lead to a total cut in supply. For communities without electricity supply the high cost of a more flexible control system could imply the lack of financial feasibility of a WECS. In other words, the decision is between a restrictive freedom of use or non-electricity at all.

Definition of Priorities

Priorities change from individual to individual and among different communities but there are general trends which can be used to determine the relative sizes of the WECS and the storage devices as well as the control strategies. At this stage it is important to distinguish two different sets of priorities; the first one is related to the "convenience" of having electricity, e.g. the use of television sets, radios, etc.; the second is related with the "need" of having a source of electricity for at least some time to avoid losses, e.g. if the refrigerator does not receive enough energy to maintain the food fresh there will be a material loss, similarly if there is a small industry which depends for its production mainly on electricity.

Once a set of priorities is defined, the priority one is assumed to be met all the time, the priority two is included in the dimensioning of the system, but further priorities will only be considered in the design and sizing of the WECS if there is economic justification. To illustrate this, Table 2 shows the preliminary results of a study for a small community in Colombia (the appliances considered are the currently used and not the efficient ones available). As the data is given only for one household, the ownership factor is not included in the table. The diversity factor of 100% means that we allow all appliances to be on at any one time (an extreme case). To supply the priority number 1 we need a system of rated power 138.5 W on average (3.32 kWh/day) but to supply both priorities 1 and 2 this figure increases to about 540 W and to about 2300 W if all the load is supplied from the WECS.

Conclusions

In order to design an optimal wind system to provide electrical power to an isolated community it is necessary carefully to match the power and frequency wind spectra to the load demand.

By recognising the inherent energy storage capacities in the load together with its prioritisation, it is possible to flatten out the demand peaks to some extend. By combining this data with the speed distribution and power spectrum of the wind, an optimised design for the conversion/storage system can be determined. A novel rotating electrical machine design for this purpose is discussed in Ref. 8.

In addition to the system optimisation requirements described above, other more general factors also require attention. These include:

- The provision of public information and education in energy use.

- A legislation framework supportive of energy conservation, e.g. for the production of energy efficient equipment.

- Design of practical installation guides e.g. luminaires and lighting design.

- Development of national standards and testing facilities.

ELECTRICITY CONSUMPTION ON
FAIR ISLE

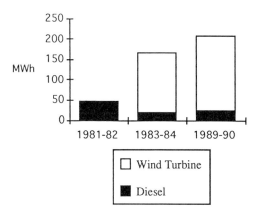

Fig. 1 Evolution of Electricity Consumption in a small Isle

Table 1

Energy consumption of some domestic equipment (kWh/a)

	Denmark		Japan		Tahiti
Equipment	1973	1988	1973	1988	1983
Clothes Washer	500	400	N/A	N/A	96
Color Television Set	250	150	204	121	72
Refrigerator	550	350	955	312	300
Room Air Conditioner	N/A	N/A	752	428	108

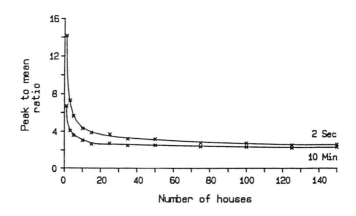

Fig. 2 Variation of peak to mean ratio with number of houses [5]

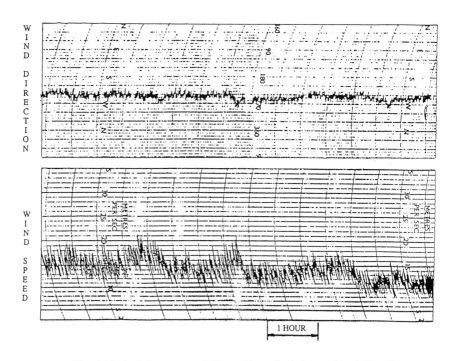

Fig. 3 Typical Anemograph Recording of wind speed and direction

Table 2
LOAD DEMAND per HOUSEHOLD

TYPE	SIZE (W)	TIME/DAY (HOURS)	PRIORITY	ELECTRICITY /DAY (kWh)	ELECTRICITY (%)	DEMAND DIVERSITY FACTOR	ELECTRICITY /YEAR (kWh)
Lighting	360	5.00	1	1.80	3.28%	100.00%	657.00
Television set	300	5.00	1	1.50	2.73%	100.00%	547.50
Radio receiver	2.4	10.00	1	0.02	0.04%	100.00%	8.76
Tape recorder/record player	25	2.00	2	0.05	0.09%	100.00%	18.25
Iron	1250	0.29	4	0.36	0.65%	100.00%	130.36
Resistance (cooker)	1000	4.00	2	4.00	7.28%	100.00%	1460.00
Resistance (water heater)	2000	4.00	3	8.00	14.56%	100.00%	2920.00
Resistance (space heater)	2000	4.00	5	8.00	14.56%	100.00%	2920.00
Refrigerator (cold)	360	3.00	2	1.08	1.97%	100.00%	394.20
Freezer (deep cold)	500	2.00	3	1.00	1.82%	100.00%	365.00
Pump (gravity)	2238	11.00	3	24.62	44.80%	100.00%	8985.57
Motor (mechanical)	746	6.00	2	4.48	8.15%	100.00%	1633.74
Battery charger	10	4.29	3	0.04	0.08%	100.00%	15.64
TOTAL			1	3.32	6.05%	100.00%	1213.26
			2	9.61	17.48%	100.00%	3506.19
			3	33.66	61.26%	100.00%	12286.21
			4	0.36	0.65%	100.00%	130.36
			5	8.00	14.56%	100.00%	2920.00
			6	0.00	0.00%	100.00%	0.00
			7	0.00	0.00%	100.00%	0.00
TOTAL (ALL TYPES)				54.95	100.00%		20056.02
TOTAL (ALL PRIORITIES)				54.95	100.00%		20056.02

REFERENCES

1. SINCLAIR, Barry. (1990). **FAIR WINDS OVER FAIR ISLE.** RE-VIEW, Issue 14, Winter 1990/91. Department of Energy, United Kingdom.

2. NØRGÅRD, Jørgen S., (1989). **LOW ELECTRICITY APPLIANCES - OPTIONS FOR THE FUTURE.** Energy Group, Technical University of Denmark, Lyngby, Denmark.

3. ISHIDA, H. (1991). **ENERGY CONSERVATION IN JAPAN AND INTERNATIONAL COOPERATION.** JETRO, Japan.

4. CEA (Commissariat à l'Energie Atomique), AFME (Agence Française pour la Maitrise de l'Energie). (1983). **ENERGIE PACIFIQUE: L'ENERGIE, L'HABITAT ET LE SOLEIL. A TAHITI ET SOUS LES TROPIQUES.** Programme territoire de la Polynésie Française - CEA - AFME. SOL E.R. (Groupement d'Intérêt Économique SOLaire Énergies Renouvelables), Papeete, Polynésie Française.

5. BULLOCK, Alan (1990). **THE OPTIMISATION OF AN HYDRAULIC ACCUMULATOR ENERGY BUFFER AND WIND/ DIESEL SYSTEM FOR REMOTE COMMUNITY ELECTRICITY GENERATION.** Ph.D. Thesis. Engineering Department, University of Reading, United Kingdom.

6. BEYER, H. G., LUTHER, J., STEINBERGER-WILLMS, R. (1989). **POWER FLUCTUATIONS FROM GEOGRAPHICALLY DIVERSE, GRID COUPLED WIND ENERGY CONVERSION SYSTEMS.** Proceedings of the European Wind Energy Conference, EWEC' 89, Glasgow, United Kingdom, 10-13 July. Part I, pp 306-310, Peter Peregrinus Ltd.

7. DUNN P.D., GOMEZ, S. (1992). **FIXED vs VARIABLE FREQUENCY WIND TURBINE OPERATION.** Proceedings of the 2nd World Renewable Energy Congress, Reading, United Kingdom. 13-18 September Pergamon Press.

8. DUNN P.D., GOMEZ, S. (1992). **ASYNCHRONOUS MACHINES FOR USE WITH VARIABLE SPEED WIND TURBINES FOR ISOLATED ELECTRICAL GRIDS.** Proceedings of the 2nd World Renewable Energy Congress, Reading, United Kingdom. 13-18 September. Pergamon Press.

ASYNCHRONOUS MACHINES FOR USE WITH VARIABLE SPEED WIND TURBINES FOR ISOLATED ELECTRICAL GRIDS

by

P.D. Dunn and S. Gómez

Introduction

Usually, isolated grid systems operate with constant speed turbines, using pitch control, and are coupled to synchronous generators. However, for relatively small isolated grid systems (10's of kW), the cost of the pitch control becomes disproportionately expensive. There is also an economic advantage in the use of asynchronous machines instead of synchronous generators in this power range.

This paper considers the use of a variable speed wind turbine in association with an asynchronous generator. Frequency control is achieved by resonating the leakage inductance of the generator with an external capacitor bank. Medium term voltage control is achieved by load control, however, short term energy storage is also incorporated. The system is shown schematically in Fig. 1

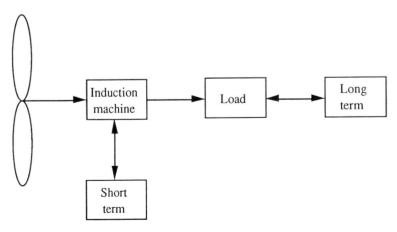

Fig. 1 System Block Diagram

2993

Induction motors and generators

Today the induction motor provides more than 90% of the electrical/mechanical power conversion needs. Usually the motors are connected to the constant voltage, constant frequency grid supply (single or three phase) and rotate at approximately constant speed. The speed is slightly less than that of the grid related synchronous frequency by an amount $1-s$ where s is called the slip. As the load increases so slip increases up to, typically, 5%. The power loss appears as resistive dissipation in the rotor and increases with slip.

There was much interest in the first half of this century in methods of increasing the frequency range of induction motors (i.e. different output speeds) and many solutions were tried, including that of pole changing. The obvious method of speed reduction is to provide more rotor dissipation (increase slip) but this results in low, often inacceptably low, efficiency.

Two different approaches were proposed to improve efficiency. In one method the slip power was removed, by slip rings, and after conversion to grid frequency fed back into the grid. In the second method the slip power was removed and used to drive an electric motor which in turn was coupled to the main motor shaft. The characteristics of the set may be varied according to whether the rotor power is put back into the supply or added mechanically to the shaft. The former case corresponds to approximately constant torque, and the latter to constant power, the motor primary current being assumed constant. In both cases, auxiliary machines were needed, increasing the initial cost of the equipment. Fig. 2 shows examples of these two methods. (Adapted from Ref. 1).

If an induction motor is driven into the negative slip region it will act as a generator and feed power into the grid. This generator, the induction generator, is widely used with constant rotation frequency, grid connected, wind turbines. The grid provides the reference frequency against which slip is measured. As the driving speed increases the magnitude of the slip will increase and hence the losses, resulting in low efficiency.

In the present application the electrical machine is not connected to a strong grid and the standard of frequency is provided by a capacitor bank resonating with the leakage inductance of the motor.

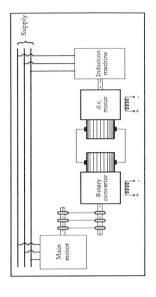

Kramer System (Constant torque)

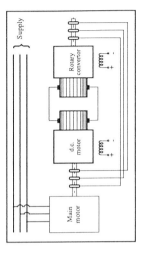

Kramer System (Constant power)

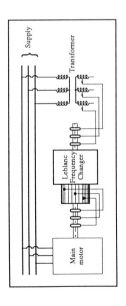

Leblanc System (Constant torque)

Leblanc System (Constant power)

Fig. 2

2995

A number of variable speed configurations have been examined to see if any were applicable to variable speed wind turbine input.

Two new configurations have been proposed and are outlined in the next two sections.

Double Rotor System

The first system incorporates a non-standard asynchronous machine, shown in Fig.3, in which both the '*rotor*' and the '*stator*' are mounted on bearings and are allowed to rotate. The stator, containing the field winding is directly driven by the wind turbine; output power is taken to the grid via slip-rings. The field winding is such that the magnetic field rotates in an opposite direction to the mechanical rotation hence at synchronism the *rotor* will be stationary (slip = 0). As the field winding increases in speed the slip increases and the rotor begins to rotate generating power which is stored in the short term and return as required to flatten out the fluctuations in the wind speed. The short term storage is provided by one of several systems such as a battery, water tank or a flywheel. The hydrodynamic pump coupled to a water tank would appear to match the system best.

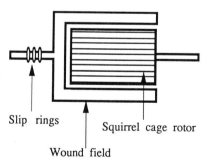

Slip rings

Squirrel cage rotor

Wound field

Fig. 3 Schematic diagram of the proposed induction machine

Divided Mechanical Power System

In the second system a standard induction generator is driven in the conventional manner and a mechanical transmission system is used to divide the input power between the generator and short term storage system. The arrows in the Fig. 4 show the possible flow of energy.

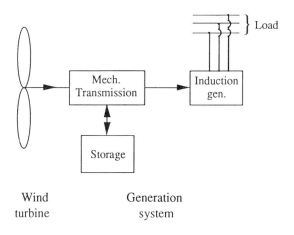

Fig. 4

The first arrangement could be regarded as 'series' and the second 'parallel'.

Conclusions

Two rotating electrical machine systems are described which enable variable speed shaft power to be converted to constant frequency electrical power. Short term storage is incorporated in each.

REFERENCES

1. BROSAN, G. S., HAYDEN, J. T. (1966). **ADVANCED ELECTRICAL POWER AND MACHINES.** Pitman Publishing Limited, United Kingdom.

Late Papers

A COMPARATIVE STUDY
OF THE
TOTAL ENVIRONMENTAL COSTS
ASSOCIATED WITH ELECTRICAL GENERATION SYSTEMS

W.B. Goddard, Ph.D.[1] and C.B. Goddard, M.A.[2]

(1) Research Engineer and Principal (2) Environmental Planner and Principal
Goddard & Goddard Engineering - Environmental Studies
6870 Frontage Rd., Lucerne, California 95458-8504   707-274-2171

Abstract

A study identifying and comparing the Total Environmental Costs (TECs) associated with electrical generation systems (EGS) is presented. The concept of defining TECs is explained as an attempt to develop a total holistic approach to assessing the environmental impacts associated with land disturbance, resource use and/or loss, and pollution emissions during the development and operation of EGS. Each EGS includes the TECs of the impacts of the immediate power plant installation and all associated systems.

The study discusses environmental impacts of EGS to air, water and land and evaluates the magnitude of the impacts in terms of resource use and/or loss and their associated dollar costs where available.

Electrical generation systems considered are limited to those systems with a proven technological record. Systems discussed include power plants using fossil fuels of coal, oil, natural gas and methanol; bio fuels, solar thermal and photovoltaic systems; wind; hydro and geothermal.

Parameters discussed include: air pollutant emissions, the emissions of greenhouse gasses such as carbon dioxide, water use and liquid waste disposal; wastes from air emission control systems (AECS) and fuel processing and the production of noxious and hazardous wastes.

Biobibliography

Wilson B. Goddard, Ph.D. Research Engineer, Principal is a member of the Geothermal Resources Council and his other affiliations include:

Association of Environmental Professionals
Air Pollution Control and Waste Management Association
American Society of Agricultural Engineers
Tau Beta Pi Engineering Honor Society
Sigma Xi Agricultural Honor Society

Prior to 18 years as an independent environmental consultant and principal of Goddard & Goddard Engineering - Environmental Studies, Dr. Goddard was an Assistant Professor at the University of California at Davis with joint appointments in Engineering, Atmospheric Science and Environmental Studies. He conducted research and teaching at UCD in Micrometeorology, Atmospheric Thermodynamics and Statics, Meteorological Instrumentation and Observations, and Environmental Studies.

His air resources specialty in environmental studies includes development of air emissions inventories, analysis of Air Emissions Control Systems (AECS), Best Available Control Technologies (BACT), Lowest Achievable Emission Rates (LAER), and air quality impact analyses including health risk assessments.

His partner and colleague in G&GE, Christine B. Goddard, M.A., Environmental Planner, Principal, graduated with honors from the University of Edinburgh in Physical Geography and has been Advanced to Candidacy for a Ph.D. in the Graduate Group in Ecology at the University of California at Davis. Prior to her 18 years as an independent environmental consultant and principal of Goddard & Goddard Engineering - Environmental Studies, Mrs. Goddard was a Research Associate at UCD. Her areas of specialty include Environmental Planning, Environmental Impact Assessments, mitigations and compliance monitoring, resource management, and regulations analysis. Professional affiliations include:

      Association of Environmental Professionals
      American Planning Association
      California Lake Management Society
      Clear Lake Basin Resource Management Committee
      General Council of the University of Edinburgh

As a team in G&GE, the Goddard's bring together specialties in Engineering, Physical Geography, Atmospheric Science, Ecology, Micrometeorology, Physical Oceanography, and Environmental Studies. Their engineering and planning approach to environmental studies has resulted in the publication of over 100 reports on over 65 projects.

THE EFFECT OF ATMOSPHERIC VARIABLES ON THE PERFORMANCE OF SILICON
CELLS IN A NIGERIAN ENVIRONMENT

M.A.C. CHENDO and MADUEKWE A.A.L.

Department of Physics, University of Lagos
Lagos, Nigeria.

ABSTRACT

The atmospheric effects on the performance of two types of silicon solar
cells namely amorphous silicon, (a-Si), and crystalline silicon, (Si),
cells have been studied for the Lagos environment. The effects of atmospheric water vapor, turbidity, and airmass on the two type of solar cells
have been studied.

The results show : (i) existence of high amount of precipitable water
vapour and (ii) high diffuse content of the total solar irradiance in Lagos.
Based on the above results one can infer that amorphous silicon solar cells
may perform better than crystalline silicon solar cells since the response
of amorphous silicon cells peak off in the visible region of the solar
spectrum, where water vapor effects are very insignificant. Furthermore,
concentrating solar cells may not be ideal for the Lagos environment.

KEYWORDS

Atmospheric, Variables, Silicon Cells, Environment Spectral Characteristics.

INTRODUCTION

The existing atmospheric conditions at a given location affects the solar
spectrum in a very complex manner. Photovoltaic devices are spectrally
selective and as such would respond to these atmospheric conditions.
Spectral distributions for different times of the day and seasons of the
year have to be known in other to predict the performance of Pv devices for
optimum energy conversion efficiency. The earliest work done in Lagos on
the spectral distribution of incident solar radiation by Chendo and
Maduekwe (1) used values of atmospheric variables measured for the month
of February by C.O. Oluwafemi (2). All year round measurement of the
necessary atmospheric variables needed in a computer program like SPECTRAL2
had never been carried out.

An all season data acquisition has been carried. The data obtained has been used in SPCTRAL2 computer model to produce clear sky spectral profiles for the Lagos area. The quantum efficiencies of the two cells were obtained and were used along with the spectral to study the photon fluxes that would be received and the number of photons that would be collected by the cells. The short-circuit current density J_{sc} of the cells can then be obtained from the total number of photons collected and from this the efficiency of the cells can be obtained.

DATA ACQUISITION AND ANALYSIS

Data acquisition was carried out with the following instruments:

(i)   A four channel volz sunphotometer was used to estimate the atmospheric turbidity and water vapor.

(ii)   An Epply Pyrheliometer was used to measure the direct radiation reaching the earth surface.

(iii) Two Middleton Pyranometers along with a shadow band were used to measure the global and diffuse radiation. The data obtained gave an indication of the clearness of the sky through the clearness index.

(iv)   Pressure, temperature and humidity were measured using a barograph and thermo-hygrometer respectively.

(v)   Cloud cover was assessed by observation.

The data obtained were processed for quality check and only those for clear sky conditions were used in the SERI produced SPCTRAL2 computer code to generate spectral irradiance for the various wavelengths. SPCTRAL2 uses the AM1.5 spectra as revised by Neckel and Labs in 1981 and modified by SERI (3) for a resolution of 10 nm.

RESULTS AND DISCUSSION

Fig.1 shows the monthly variations of the clearness index, K_t for a year. The monthly means of the diffuse ratio, K_d, and the beam irradiance ratio, K_b, are also shown in the same figure. The fraction of the extra-terrestrial irradiance reaching the surface in Lagos is less than 50% most of the time and even then over 50% of the total radiation that reaches the surface most of the time is diffuse as can be seen from Fig.1.

Fig.2 shows the monthly means of the hourly turbidity at 0.5um measured for the same period. The monthly means of humidity and wind speed are shown in figures 3 and 4. They play important roles in determining the atmospheric turbidity in Lagos. The nearness of Lagos to the ocean means that the turbidity would be influenced by the sprays of fine droplets containing salts and dissolved organic compounds.

Fig.5 shows the monthly means of the hourly measured precipitable water vapor for the same period. With the exception of July the rainy season months of May to September have higher water vapor content. The water vapor content of the atmosphere in Lagos are generally higher than 2.5cm most of the time. The mean for the period is 4.456cm.

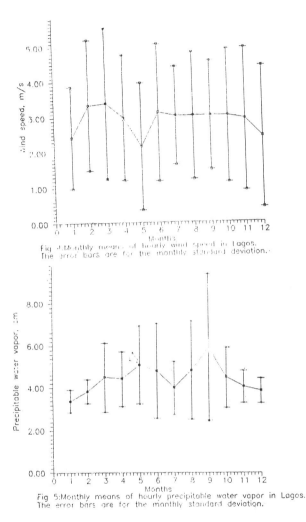

Fig 4:Monthly means of hourly wind speed in Lagos.
The error bars are for the monthly standard deviation.

Fig 5:Monthly means of hourly precipitable water vapor in Lagos.
The error bars are for the monthly standard deviation.

The quantum efficiencies of a-Si and Si are shown in Figure 6. The result
of integrating the product the photon flux densities with the quantum
efficiencies to obtain the number of photons collected are shown for the
two cells in Figures 7 and 8 for the spectra obtained for the months of
April and October which had more clear sky conditions than the other months.
The plots show that the a-Si suffers very little due to water vapor effects
which appear as deep cuts in the infrared region in the plot for crystalline
silicon will not operate at the expected efficiency due to this reduction
in photon numbers generated in the infrared region. Since the efficiency
of a PV device is obtained by converting total photons collected to short-
circuit current density (J_{sc}), multiplying J_{sc} by modeled or measured values
for open circuit voltage (V_{oc}) and fill factor (FF) to obtain power

output, and dividing power output by total integrated irradiance (power
input) it implies that the efficiency as may be specified by the

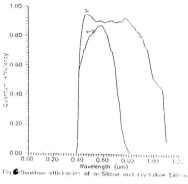

Fig 6:Quantum efficiencies of a-Silicon and crystalline Silicon

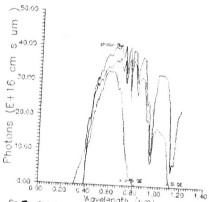

Fig. 7 : Global-Normal spectral solar radiation converted to photon flux and multiplied by the quantum efficiency of an a-Si and a crystalline Si solar cell using data for a clear day in April.

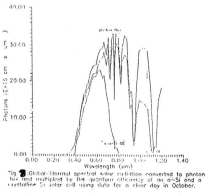

Fig 8 :Global-Normal spectral solar radiation converted to photon flux and multiplied by the quantum efficiency of an a-Si and a crystalline Si solar cell using data for a clear day in October.

manufacturers or as may be obtained under controlled conditions such as in simulations with artificial light sources will be very different from that obtained under outdoor tests.

Furthermore the water vapor content is generally high as seen in Figure 5 and as such under cloudy sky conditions which are mostly prevalent in Lagos the crystalline silicon cell would experience higher losses in efficiency.

CONCLUSIONS

From the above results presented a-Si will perform very well with little loss in efficiency under clear sky conditions in Lagos. The crystallin silicon will suffer loss in efficiency due to effects of water vapor absorption of the incident spectra in parts of the infrared region of the spectrum. Due to the high cloud cover in Lagos resulting in large amounts of diffuse spectral irradiance it is expected that crystalline silicon solar cells may never give optimum performance in this location. It is therefore recommended that a-Si be used in Lagos since its spectral response lies in the region where there is little chance of loss in efficiency due to water vapor absorption.

REFERENCES

Chendo M.A.C. and A.A.L. Maduekwe (1989): A solar Spectral Distribution for Lagos: A Preliminary Study. Nigerian Journal of Solar Energy, 8, 355-363.

Oluwafemi C.O. (1983): Solar Almucantar Radiance and aerosol scattering. Comparative data from one tropic station and one Mid-Latitude station. PAGEOPH, 121, 516.

Bird R.E. and C.Riordan (1986): Simple solar spectral Model for Direct and Diffuse Irradiance on Horizontal and Tilted Planes at the Earth's surface for Cloudless atmosphere. J.Climate Appl.Meteor. 25, 87-97.

W.T. Roach (1961): The absorption of solar radiation by water vapor and carbon dioxide in a cloudless atmosphere. Quart.J.R. Meteor.Soc.,87, 364-373.

Giichi Yamamoto (1962): Direct absorption of solar radiation by atmospheric water vapor, carbon dioxide and molecular oxygen. J.Atmos.Sci., 19, 182-189.

N.Braslau and J.V.Dave (1973): Effect of aerosols on the transfer of solar energy through realistic model atmospheres.

Part 1: Non-absorbing aerosols. J. Appl. Meteor., 12, 601-615.

Part 2: Partly-Absorbing Aerosols. J. Appl. Meteor., 12, 616-619.

W.Wang and G.A.Domoto (1974): The radiative effect of aerosols in the earth's atmosphere. J. Appl. Meteor., 5, 521-534.

A.A. Lacis and J.L. Hansen (1974): A parameterization for the absorption of solar radiation in the earth's atmosphere.J.Atmos.Sci,83,118-133.

K.Liou and T.Sasamori (1975): On the transfer of solar radiation in aerosol atmospheres. J. Atmos. Sci., 32, 2166-2177.

Matson.R,R. Bird, and K.Emery (1981): Terrestrial Solar Spectra, Solar simulation, and Solar Cell Efficiency Measurement. SERI/TR-612-964. UC CATEGORY: UC-63.

Riordan C.J. (1985): Spectral Solar Irradiance Models and Data Sets. Proceedings of Photovoltaic and Insolation Measurements Workshop. SERI/CP-215-2773. DE85012145, 55-67.

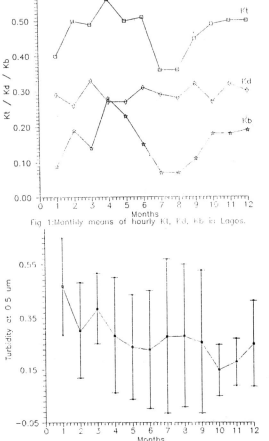

Fig 1:Monthly means of hourly Kt, Kd, Kb in Lagos.

Fig 2:Monthly means of hourly turbidity at 0.5um in Lagos.
The error bars are for the monthly standard deviation.

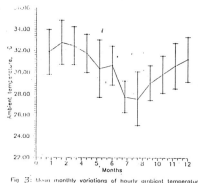

Fig 3: Mean monthly variations of hourly ambient temperature in
Lagos. The standard deviations are shown as error bars.

ENVIRONMENT AS AN ARCHITECTURAL GENERATOR

VAL D'OISE

ANDREW WRIGHT - **RICHARD ROGERS PARTNERSHIP**

GUY BATTLE - **OVE ARUP & PARTNERS**

VAL D'OISE - J101 - 7651

DRAFT PAPER

ABSTRACT

The climate that we inhabit changes constantly, hour by hour, day by day, season by season. Despite this, modern urban form is dumb and does not respond and adapt to changes in weather and take advantage of climatic benefits.

Urban form is governed by the motor car.

The comfort of people is not the prime concern.

Ideally urban form should be built around the needs of people.

We will illustrate how people and their environment can be used to inform an urban strategy and how this strategy can be followed through to a built consequence.

We will focus on Val d'Oise, a region of 1000 hectares on the outskirts of Paris for which we have proposed a new urban structure designed around people and based on simple environmental principles.

Val D'oise has been considered as a series of problems which were isolated and then overlaid to form a clear and cohesive urban matrix.

.The existing urban fabric and environmental climate forms the starting point.

. A transport strategy form the desire to link isolated villages and the town. Allow the area to develop as a community. The strategy is orchestrated around increasing public transport and reducing the need for cars encouraging people to walk in comfort.

. A blue plan sets out how water is collected naturally and stored in reservoirs which are utilised to shape public space.

. A green plan sets how wooded areas can be used to protect development from surrounding noise and pollution. Green and blue corridors wiggle through the new town, providing contrast and creating a variety of aspect, whilst maintaining a cool external environment.

. An orange plan sets policies for integrated planning to facilitate careful energy management and advises on appropriate building types as well as setting clear energy targets.

. The project will be put forward by architect, engineer and ecologist, to set out the necessity for urban design to broaden its discipline and aims.

We will conclude that urban structure which is designed around people based on ecological principles can be totally self sufficient.

- Val d'Oise, a contemporary townscape celebrating quality of life; a great peoples' place.

- A new, sustainable eco-city for people, designed on fundamental environmental principles.

- A new and unique eco-community for the future that sets a European precedent.

- A living plan, a structured approach rather than a rigidly defined solution.

- A catalyst, enhancing physical, social, and economic links between the Val d'Oise communities.

- Creating distinct and individual communities rather than continuous urbanisation; "a network of villages within the city".

- An integrated urban system: a coherent network of green space, waterways and transportation shaping the urban fabric.

- An energy planned region; an energy conserving, low consumption, low pollution urban structure.

- An avant guard approach to current and future EC environmental guidelines.

- A new centre for the developing science of urban ecology.

A New Urban Ecostructure

The Design Approach

Our multi-disciplinary design team, in participation with local communities, is tackling a key issue of the '90s, the creation of a new urban infrastructure based upon ecological principles.

Environmental and Social Parameters

- People are central to the proposal.

- Val d'Oise integrates and stimulates the existing communities physically and economically with enhanced communication throughout the region.

- Providing a beautiful place for living, working and leisure activities; the stage for a rich and diverse social and economic mix.

- Pedestrian and cycle routes weave through hard and soft landscape independent of the road network.

- The heart of town is a human scale, pedestrian prioritised area based upon comfortable walking distances and car free urban routes, squares, spaces and parks.

- Cars may approach the town centre via landscaped routes which terminate in parking nodes and service areas within easy walking distance of the central areas.

- Access time restrictions for deliveries, services, and refuse collection apply.

- Planning responds to the climate, sun path, wind direction, topography and views.

- Greenspace and water influence urban form.

- The local micro-climate is modified for human comfort, so as to maximise environmental advantage.

- Natural resources are used rationally and economically. Recycling is an inherent element of the urban infrastructure concept

- Energy use is minimised through efficient urban planning, building design and transportation strategies

- Compatibility of overlapping uses and activities is planned encouraging a 24hr activity cycle and a balanced year round town life.

- A diverse balance of employment, services, accommodation, leisure, cultural and educational facilities is planned and maintained during the incremental growth of the area.

Context

L'Ile de France

- Val d'Oise is crucially located to benefit from its position within Europe at the hub of a road rail and air transportation network and at the centre of an international telecommunications web.

- Both in a designated growth tributary and as part of the green zone around L'Ile de France, the site must rise to the challenge of reconciling urban development with preservation and enhancement of a rural area.

- The Val d'Oise area restores a valuable link in the green belt, reinforcing the 'Great Green Plan' around L'Ile de France.

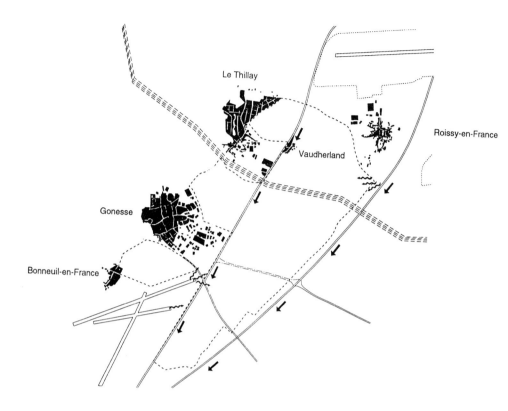

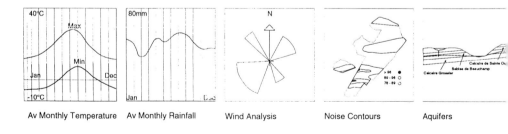

Av Monthly Temperature   Av Monthly Rainfall   Wind Analysis   Noise Contours   Aquifers

Context

Site Analysis

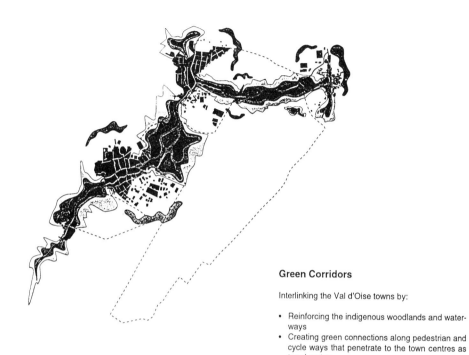

Green Corridors

Interlinking the Val d'Oise towns by:

- Reinforcing the indigenous woodlands and waterways
- Creating green connections along pedestrian and cycle ways that penetrate to the town centres as treed routes.

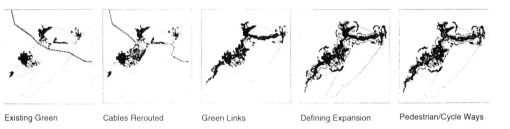

| Existing Green | Cables Rerouted | Green Links | Defining Expansion | Pedestrian/Cycle Ways |

Strategy

Green Corridor

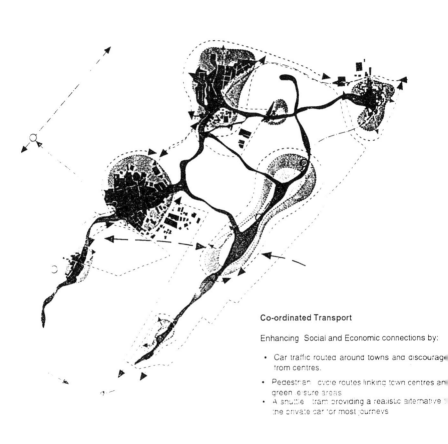

55% Saving in Energy

Co-ordinated Transport

Enhancing Social and Economic connections by:

- Car traffic routed around towns and discouraged from centres.
- Pedestrian cycle routes linking town centres and green leisure areas
- A shuttle tram providing a realistic alternative to the private car for most journeys

Strategy

Co-ordinated Transport

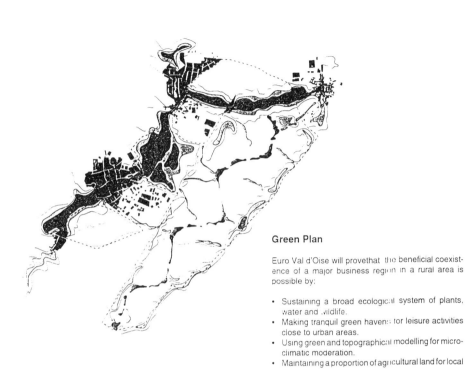

Green Plan

Euro Val d'Oise will prove that the beneficial coexistence of a major business region in a rural area is possible by:

- Sustaining a broad ecological system of plants, water and wildlife.
- Making tranquil green havens for leisure activities close to urban areas.
- Using green and topographical modelling for microclimatic moderation.
- Maintaining a proportion of agricultural land for local food production.

| Green Tributary | Noise Buffer | Agriculture | Central Walk | Pedestrian Links |

Strategy

Green Plan

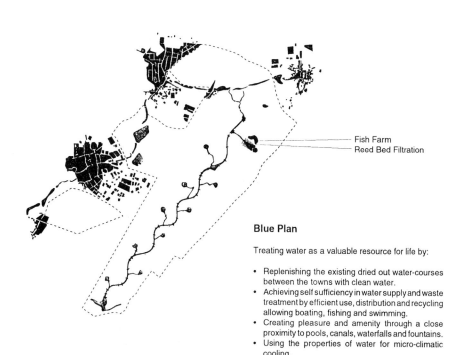

— Fish Farm
— Reed Bed Filtration

Blue Plan

Treating water as a valuable resource for life by:

- Replenishing the existing dried out water-courses between the towns with clean water.
- Achieving self sufficiency in water supply and waste treatment by efficient use, distribution and recycling allowing boating, fishing and swimming.
- Creating pleasure and amenity through a close proximity to pools, canals, waterfalls and fountains.
- Using the properties of water for micro-climatic cooling.

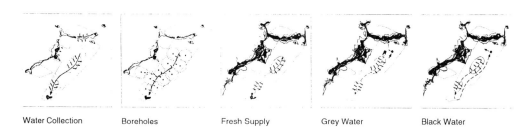

Water Collection　　Boreholes　　Fresh Supply　　Grey Water　　Black Water

Strategy

Blue Plan

CALCULATED USING A POPULATION OF 40,000

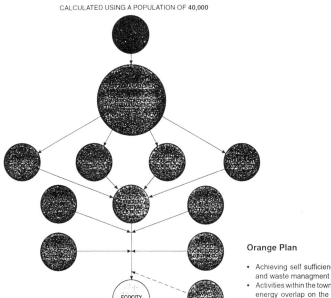

Orange Plan

- Achieving self sufficiency through careful energy and waste managment
- Activities within the town are selected for maximum energy overlap on the 24hr and Summer/Winter cycle.
- Buildings are designed to maximize environmental resources; for natural ventillation, for maximum daylight and thermal exchange.

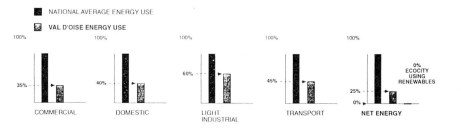

NATIONAL AVERAGE ENERGY USE

VAL D'OISE ENERGY USE

Strategy

Orange Plan

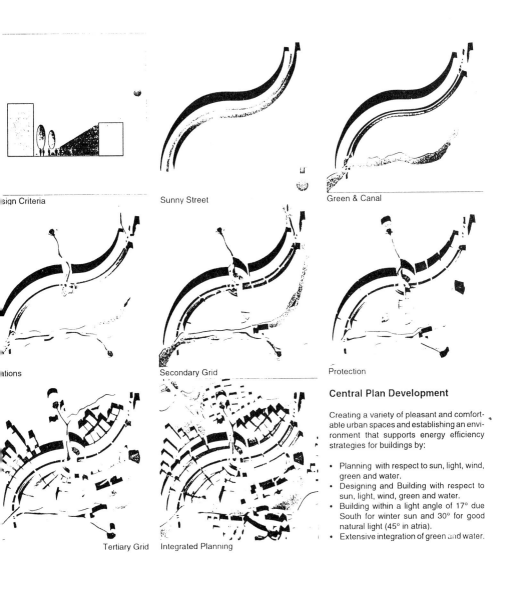

Sign Criteria   Sunny Street   Green & Canal

ations   Secondary Grid   Protection

Tertiary Grid   Integrated Planning

Central Plan Development

Creating a variety of pleasant and comfortable urban spaces and establishing an environment that supports energy efficiency
strategies for buildings by:

- Planning with respect to sun, light, wind,
  green and water.
- Designing and Building with respect to
  sun, light, wind, green and water.
- Building within a light angle of 17° due
  South for winter sun and 30° for good
  natural light (45° in atria).
- Extensive integration of green and water.

Design   **Central Plan Development**

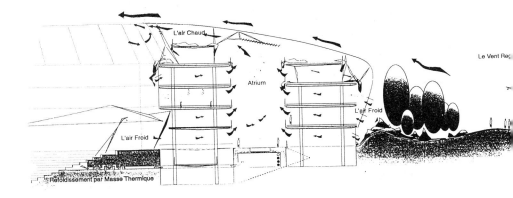

Bâtiment Commercial

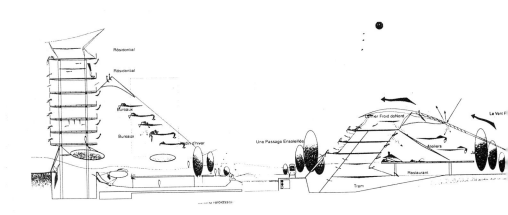

Epine Dorsale

La Conception

Micro-climate

A pleasant and comfortable urban micro-climate, with a variety of qualities benefiting both pedestrians and occupants of buildings will be achieved and an environment that establishes the energy efficiency strategies for buildings.

Strategies
The main strategies involve human comfort considerations with respect to sun, light, greenspace and water.

Sun: The main urban streets and spaces should be designed to allow sun to penetrate at some time during a winter's day. The mid-winter noon sun altitude (17°C) provides a reference for the maximum obstruction angle. Buildings can also benefit from winter sun as a source of free heat when useful. The tertiary urban network will provide a range of spaces, some cool and shady, others more sunny.

Daylight: To ensure that buildings receive good daylighting, an important energy efficient measure, a maximum angle of obstruction is set (30°C). Higher angles of obstruction will rapidly incur energy penalties. In atria, this angle can be increased to 45°C as the internal finishes of glazed spaces can be light coloured.

Greenspace: The use of vegetation in an urban context has a range of benefits from psychological to environmental. Vegetation will provide shade, evaporative cooling, wind shelter, noise attenuation, dust filtering as well as flora and fauna microcosms. Nature in and near urban areas has a particular value and relevance.

Water: In combination with vegetation, the use of water will reduce summer peak temperatures through evaporative cooling. The use of fountains and waterfalls will provide background noise to limit the sound of air, rail and road traffic.

The micro-climate of main streets and squares are thus controlled to allow effective natural ventilation of buildings, using urban spaces as quiet, green, fresh air 'lungs' for the development. Summer ambient temperature reductions of 3°C can be expected; more if the urban landscape, vegetation and water are carefully considered. This can give reductions in potential cooling loads, and facilitate the avoidance of air-conditioning. The most advantage is gained from the first 30% of green coverage in the urban areas, beyond which the thermal advantages of vegetation rapidly diminish.

Winter considerations include providing shelter from winds and rain, whilst allowing sun to penetrate. The design of public space should provide shelter from prevailing (SW) and cold winter winds (NE). An amenable urban microclimate will encourage pedestrian and bicycle movement, minimising the use of cars.

Design Response
The micro-climatic strategies set up geometric rules for spacing and building heights in relation to orientation of the major streets and squares. The results of adopting sun and daylighting criteria generates a range of urban space enclosure configurations.

A wide range of building forms is envisaged which responds to the human micro-climatic considerations and provides lively variety. These may include arcades, covered streets, atria and water gardens.

The design response is thus one that considers people first, ensuring a variety of environmental qualities for various urban activities.

Carbon Dioxide

The aim is to strive for an environmentally balanced development where CO_2 absorption is maximised and emissions are minimised.

Strategies
An energy efficient community will generate little CO_2, mainly through private transport. New EC directives will reduce these levels still further. The general approach for reducing CO_2 emission is the same as that for energy efficiency.

CO_2 emission per fuel type:

Fuel type	CO_2 emission kg/kWh of delivered energy
Electricity	0.187
Gas	0.198
Coal	0.331
Coke	0.378
Other solid	0.421
Petroleum	0.302

Life cycle costing:

Fuel type	CO_2 emission kg/kWh of delivered energy
Nuclear	0.026
Gas	0.409
Coal	0.658
Hydro	0.0089
Wind	0.0062
Tidal	0.0059

The strategy for CO_2 absorption is to ensure sufficient plant growth. Differing species have varying carbon fixing abilities:

Plant species	CO_2 absorption
Trees (average)	1.0 kg/m^2/y
Hawthorn	
(Crataegus macrocarpa)	1.9
Blackthorn (Prunus spinosa)	1.4
Field Maple (Acer campestre)	1.2
Beech (Fagus sylvatica)	0.4

The green strategies, allowing for large parts of Val d'Oise to be planted with a rich diversity of vegetation, reinforce the aim of maximum carbon fixing.

Energy

A sensible and integrated approach to energy generation, and urban energy management, using an appropriate combination of alternative sources described above, makes the development energy largely self sufficient.

Strategies

Buildings: energy efficient design will be achieved by avoidance of air-conditioning, use of natural light and ventilation, control of solar gain, use of thermal mass, night time ventilation, thermal insulation, efficient heating, lighting and ventilation systems and controls.

Potential energy savings (compared national average):

Commercial	65%
Domestic	60%
Light industrial	40%

Transport: efficient and integrated public transport (quick, cheap, comfortable, regular), will be achieved by reclaiming the street for pedestrians, incorporation of bicycle routes, traffic management, integrated access planning and the introduction of light rail/tram systems, linking all central areas.

Potential savings (compared to current standards):

Public transport	35%
Integrated planning, cycling and pedestrian	+ 8%
Energy efficient transport	+ 12%
Overall savings	55%

Overall Energy Comparison

The average overall energy use for France is approximately **160 GJ/year/person.** Breakdown of energy use for France and for an area such as Val d'Oise (omitting freight transport and heavy industry):

Sector	France	Val d'Oise
Buildings:		
Domestic	22%	22%
Commercial	13%	13%
Other	15%	15%
Transport:		
Passenger	10%	10%
Freight	9%	-
Industry	31%	-
	100%	60%

An area such as Val d'Oise consumes 60% of the national average energy, approximately **95 GJ/yr/person**, excluding heavy industry and freight transport.

Savings: Assuming the energy savings defined above, this average figure is reduced as follows:

Sector	Energy	Savings	New Use
Buildings:			
Domestic	22%	60%	8.8%
Commercial	13%	65%	4.6%
Other	15%	50%	7.5%
Transport:			
Passenger	10%	55%	4.5%
	60%		25.4%

The new figure for energy consumption per person is thus **40 GJ/yr/person**, or about 25% of the national average. Of this 83% is consumed by buildings.

For an energy efficient community of 40,000 people the overall energy use is in the order of 1.6x10⁶ GJ/Year.

The Ecocity will utilise free energy to achieve a self-sufficiency

Alternative sources: The remaining energy demand can be supplied by a range of alternative sources.

For Example:
Waste incineration can yield approximately 4 GJ/yr/person when properly sorted.

Sewage digestion can produce 0.4 GJ/yr/person. If the effluent from the surrounding communities and farms is digested, a considerable contribution can be made.

Coppice wood burning can yield 240 GJ/yr/ha. 350 ha with a 20% turn around yields 16800 GJ.

Solar Energy: Apart from the passive use of solar energy in design, active systems will reduce energy demand. Solar water heating can effectively deliver domestic hot water. Photovoltaic panels will be used to generate electrical energy directly.

Wind Energy: Assuming an average wind velocity (V50) equal to 7 m/sec, a 15 meter diameter wind turbine can generate in the order of 1250 GJ/yr. A wind farm with 1,000 windmills will produce 1.2x10⁶GJ. Windmill farms would be placed under flight cones in high noise contour zones.

Cables

The overhead 225kV powerlines, will be buried underground using XLPE cables (cross bonded polythene cables). At this voltage it will not be necessary to cool the cables. The alternative is to divert the cables around the site. However this is likely to be a more expensive solution.

Water

Supply:
Surface water is collected from urban areas in canals and lakes, supplemented from ground water abstracted from boreholes. The central canal effectively forms the fresh water reservoir for the whole development. The principal aquifers beneath the site are:

Limestone - Calcaire de Saint Ouen which has a limited perched water table subject to a high risk of contamination.

Sands - Sables de Beauchamp which is potentially useful for up to about 50,000 m²/ year. This could be extracted from six wells on the plateau with each well supplying 2,000 - 20,000 litres per hour.

Limestone - Calcaire Grossier, Lower Tertiary Age Sands and Cretaceous Chalk. This needs further surveying and more information about present and future policy on extraction.

Demand:
Demand is minimised by the use of water saving appliances. Before consumption, the water is treated and stored in covered reservoirs and fed to individual properties

Waste Management:
Grey and black gravity waste water systems will transfer waste to sewage treatment works. Black waste from WC's will be treated and used for irrigation. Water from washing etc will be treated and recycled in the 'grey water' system to be used for WC flushing. The sludge and methane produced as a by-product of sewage treatment will be used for soil fertilisation and energy recovery. Protein recovery from sewage sludge will provide a source of feed for fish farms.

Noise

Due to its location, the site has particular acoustic planning restrictions. The French noise regulations have been adhered to, and the following criteria have been met.

Isophonic noise level	Development strategy	Comments
>96	No development except agriculture and green space	
89-96	Industrial, commercial, agricultural internal noise	Acoustic treatment where low levels are required.
78-89	Industrial, commercial, agriculture, some domestic	Special treatment for commercial noise insulation for domestic buildings
<78	All types permitted	

Strategy
The planning strategy is to place buildings in the appropriate noise zones, where noise levels are acceptable to the activity, thus making expensive noise attenuation treatment unnecessary.

External spaces: Urban spaces are not deemed to be at risk from noise pollution. Transport is excluded from the main centres, and the development avoids the air flight paths. However, a series of strategies are implemented to improve the external noise environment:
• use of soft planting, vegetation, earth berming, etc, to absorb noise
• use of diffusing surfaces to prevent noise concentration
• avoidance of high narrow streets creating 'noise canyons'

Internal environment: Internal spaces are designed and acoustically configured to allow natural ventilation and opening windows.

Appendix

ENVIRONMENT AS AN ARCHITECTURAL GENERATOR

TOMIGAYA

ANDREW WRIGHT - RICHARD ROGERS PARTNERSHIP

GUY BATTLE - OVE ARUP & PARTNERS

ABSTRACT

The climate that we inhabit changes constantly, hour by hour, day by day, season by season. Despite this, the buildings that we design are insensitive. They do not respond to changes in weather or take advantage of climatic benefits.

Buildings should not be static and dumb, but should interact dynamically with the environment in order to take advantage of free energy and provide comfortable living and working space.

We will illustrate how the environment can be used as an architectural generator and how the building form can be moulded in response to environmental pressures.

The focus will be Tomigaya, a low energy building in Tokyo.

Tomigaya is designed as a total energy system within which every aspect is tuned to achieve low running costs, reduce the necessary plant, thereby increasing its lettable area and to accelerate the client's revenue.

Tomigaya has intelligent reactive facades that change like a chameleon to provide a cool comfortable internal environment.

The northern facade is heavier and permanently pattererned with clear glass, diffused glass and opaque panels to allow views and light to enter where needed whilst insulating elsewhere.

The heavy concrete structure is exposed to absorb heat gain and moderate the internal environment.

Water around the deep basment is used to provide peak summer cooling and to warm cool air during the winter.

The tower works as a chimney under the action of sun and wind to help extract hot stale air.

The building form is modelled to compress and accelerate the prevailing winds between the main building and its service core to drive a vertical axis wind turbine to provide power.

The project is preented by the architectural engineering team not so much for the product itself, although this is significant, but in order to demonstrate a process of design, that leads to a more wholistic architectural solution.

1. **INTRODUCTION**

The Tomigaya project stems from a brief set by one of our Japanese clients.

The **brief** asked for a high profile, visually strong showroom building.

The **site** is triangular and on a small hill overlooking Yoyogi Park in Central Tokyo.

The **client** asked for:

1. A tall building which rose above the surrounding urban form to make use of the unusually high planning restriction.

2. A very deep basement to increase the lettable space.

3. High and open office space within which mezzanine floors can be hung.

4. A visually transparent southern facade.

Our **objective** was to - whilst complying with a tight client brief - look at the building as a 'total energy system' within which every aspect must be considered to achieve substantially reduced energy consumption, reducing plant area, and to increase the lettable area.

Our initial **response** was to separate the core and main building to:

1. To let light enter the deeper areas of the building.

2. To provide clear usable office space isolating the plant noise and maintenance requirement.

3. Visually emphasise height and elegance to emphasise the drama and impact of the building.

2. BUILDING ENGINEERING RESPONSE

THE CLIMATE

In Tokyo, the summers are hot and humid (maximum temperature and humidity 33 °C, 80%), the winters are cold and dry (minimum temperature -6.1°-10° C), there is a distinct midseason period (spring and autumn). Rains are especially heavy in the summer.

Prevailing winds are from the NW in winter and SE in summer, on the same primary axis as the buidng.

COMFORT

Key to the whole response of low energy buildings is a discussion with respect to comfort. Comfort is the complex interaction of the four human senses:

1.	Heat	-	Thermal Environment
2.	Light	-	Visual Environment
3.	Sound	-	Acoustic Environment
4.	Smell	-	Olfactory Environment

Contrary to the traditional view, we believe that comfort can only be broadly defined by an international standard. Cultural and varying climatic conditions do effect people's perception of what is satisfactory, which is primarily in response to adaption rate, time experienced influences and expectations. Thus the peak internal temperature permitted in London is lower than that deemed satisfactory in Tokyo. This does not mean that the quality of the space is any less, in fact it recognises the fundamental connection between culture climate and comfort. Figure 3.1 indicates the thermal comfort zone for a Tokyo office.

BUILDING FORM

The building form and shape is a response to the site, and the building use. However, the form has been particularly influenced by the desire to utilise the potential energy within the prevailing winds. Thus the core is separated from the main building block, and the building is smooth and shaped to encourage both the summer and winter winds to pass through the gap between the core of the building. This creates a venturi effect leading to increased air speed and hence increased energy potential, this is discussed further.

BUILDING FABRIC

In broad terms assuming a typical office use the following prognosis will be true for Tokyo.

In summer the building will generate too much heat and so the building enclosure must exclude additional heat gains (eg. solar), and minimise internal gains (eg. lights and equipment).

In winter, this building will be warmer than the external environment, and hence the building enclosure must keep heat in. Due to the high humidity and high heat gain in summer it will not be possible to provide comfort by natural ventilation alone. Furthermore, the noisy and polluted traffic routes that surround the building prohibit the direct use of fresh air without using some form of filtering system.

The building fabric is the key to moderating the adverse climate conditions, but should be designed to monopolise on its benefits.

It must provide the following:

Even cooling load

The fabric is tuned to respond to the shape of the site, the orientation and the use within. It comprises two basic elements. The rear passive façade has fixed apertures that vary along the length of the building. These apertures are set so that they permit a fixed solar gain (20 W/m^2) across the whole floor. Thus they must be smaller the narrower the width of building. The Southern façade, which is fully glazed as a client requirement has a variable shading system, that is turned according to the time of day, the season and whether the sun is shining. The shades are contained within a double solar flue, that not only effectively control the solar gain, but also act as an efficient acoustic barrier (fig 3.2).

Daylit office

The lighting energy consumption would normaly be 30% of the total energy consumption, it is therefore vital that the building is daylit and has a control system that utilises daylight. However, due to the exteremely narrow plan glaze control is a major factor influencing the design of the facade. Thus the requirements for providing an even cooling load also tie in with the requirements for providing good daylighting. The deeper the office the larger the window aperture required. This was one of the reasons to separate the core from the main building to allow daylight to penetrate into the deepest part of the floor plan, and light back into the main block (fig 3.3).

THE BUILDING SYSTEM

The building system is designed to compliment the moderating effects of the fabric and building form and, to utilise free environment energies whilst providing a comfortable and flexible working space. The system section is drawn in figure 3.4 - 3.5, typical summer and winter schematics are drawn. The schematic indicates the following elements.

Thermal wheel

The use of a hydroscopic thermal wheel will allow transfer o heat, cooling and moisture. This is especially relevant in the high humidity of summer. When it will allow substantial drying of incoming air. During winter it will transfer both heat and moisture from exhaust air to supply air. Output modulation is controlled by speed.

Outside air

By using constant 100% outside air in combination with ar upward air supply system the maximum use can be made o outside air free cooling when it is available. Low far pressures avoid the loss of the cooling due to excessive far temperature gain, and will reduce running costs.

Ground water

Ground water is of limited use in conventional chilled air conditioning because it rarely achieves low enough temperature. However, if it is used in conjunction with higher supply air temperatures, as is associated with a displacement system supplied via the floor, it can make a good cooling source. Careful checking is necessary to ensure the temperature approach between ground water and supply air dew point is compatible. It is therefore proposed to use a small amount of ground water (approx. 3 l/s).

Thermal store

A thermal store allows the peak cooling requirement to be delayed or averaged out over a 24 hour period. For this particular project, water appears to provide most benefits as a storage medium. It allows good heat transfer to the ground water using plate heat exchanges and is easily distributed throughout the building to where it is needed. The water thermal store requires no thermal insulation to the surrounding basement (except to the occupied floor above). With a minimum of treatment, the water can be put straight into the raft foundation.

Our outline investigation indicates a thermal store of approximately 380 m^3 would serve 2400 m^2 of building (at 560 m3 for 3600 m^2). This scenario would require no mechanical cooling. Our schematic illustrates that the heat pump provided for heating could easily provide a cooling source back up.

Thermal capacity

The use of high thermal capacity materials exposed to the occupied room is an integral part of the system. As the supply air must be reduced in temperature to remove moisture at peak loads, it can be supplied to the room via the hollow concrete floor structure where it allows the ceiling to operate as a cool radiant surface (with a long time delay). This results in reduced air volumes because the air has almost 14°C temperature difference between supply and extract instead of 10°C. The reduced supply of volume of 3.5 l/s m² can only provide adequate room air movement because it is put into the room inside the occupied lower zone.

Air handling unit

A single central air handling unit is proposed essentially to minimise cost and area take up.

A single central air handling unit tends to be slightly smaller in capacity than the sum of the separate of floor plant because it can accept the diversity available between floors.

In addition, the better efficiency heat recovery and variable speed fans are unlikely to be cost effective in separate on floor plant.

By providing motorised dampers and variable speed fans individual floors can be shut of and billed accordingly. Indeed, on reduced air velocity the coils and fans operate more efficiently.

The provision of central plant ground water cooling is substantially cheaper for the tenant than separate self contained on floor plant using air cooled mechanical cooling.

The AHU uses low air pressure drop components. This greatly reduces the required fan pressure and electrical consumption. This is important for current research shows that the fan running costs can frequently be greater than those of mechanical cooling.

The plant will be hung in the services tower for if the plant is within the building, it is counted as floor area. If it is hung from the tower it is not. included in the lettable floor area.

WIND

Tomigaya is formed and moulded around accelerating the wind between the main building and its core to drive a turbine and generate electricity. This energy is used directly when possible, any excess is either stored on site or supplied into the main grid.

Wind tunnel tests have been carried out on sectional models of the Tomigaya building.

Measurements of wind velocity in the gap between the core

and the main building for the full 360 degrees of wind direction and the range of variations of building form have shown that an increase in wind speed of over two times can be expected for the more favourable wind directions (east and west wind sectors). Considerable enhancement occurs over the rest of the wind directions except for a 90 degree sector of south wind in which the wind velocities are reduced. Allowing for the probability of wind direction on the site, average increases in captured wind energy per square meter of turbine of about 5 could be expected.

Calculations based on a reasonable turbine efficiency of 50% and estimates of the mean wind speed and shear at the side indicate that turbines occupying the building/core gap would generate in the order of 130 kilowatts on average.

Flow visualisation using smoke supports the velocity measurements showing that enhanced flow passes through the gap for at least two thirds of the total 360 degree wind.

The energy that is generated is more than sufficient to service the building aswell as providing all its small power.

3.   **ENERGY**

An initial estimate of the consumption has been made and a comparison with a typical air conditioned (VAV) solution has been carried out. The assumptions are contained in the appendix.

ENERGY CONSUMPTION

PASSIVE DESIGN   ($2400m^2$)

1.	Lighting	Peak	48kw
		Annual	43,200 kwh

2.   Heating   Peak   20 kw (Assumes Heat
      (electrical)   48,000pump
cop=3.5kw)

3.   Fans & Putups &   31 kw
      Thermal Wheels   Peak   140,616

                       Total
                       Consumption   231,816

ENERGY FROM WIND   Average speed =   7m/s
                        Turbine Area =   1.5 + 31
                        Turbine Efficiency =70%

energy = 43.43 $(V50)^3$ kwh/m^2/yr

———
3.6

TOTAL ENERGY   134,689 kwh
FROM WIND   (58% contribution)

TOTAL ENERGY CONSUMPTION
FROM BUILDING   =   .97127 kwh

FULLY AIRCONDITIONED BUILDING WITH VARIABLE
AIR VOLUME SYSTEM AND PERIMETER HEATING
(monitored data)
                        248 kwh PEAK
        Lighting   120,000 kwh
        Fans & Pumps   153,600 kwh
        Heating   168,000 kwh
                        ——————
        Total   441,600 kwh
COMPARISON

PASSIVE DESIGN   =   97,127 kwh
FULL AIRCONDITIONING   =   441,600 kwh
% SAVING   =   80%
* N.B.   FIGRUES TO BE UPDATED

4   CONCLUSION

Our initial results show that it is possible to make the   Tomigaya Project 100% self sufficient in terms of energy production.

However, it is apparent, that <u>significant</u> savings in energy can only be achieved if the design team works with a clear understanding of the environment and the oppertunity it presents from the onset of a project.

This -when held alongside the advances that have and are being made in energy transfer- will mean that environmentally good sites will be able to feed those sites that are not balanced.

As legislation and need focus attention on the requirement to save energy, buildings will be forced to respond to suit. As this need increases so will the influence of the environment on the form and shape of architecture in the future.

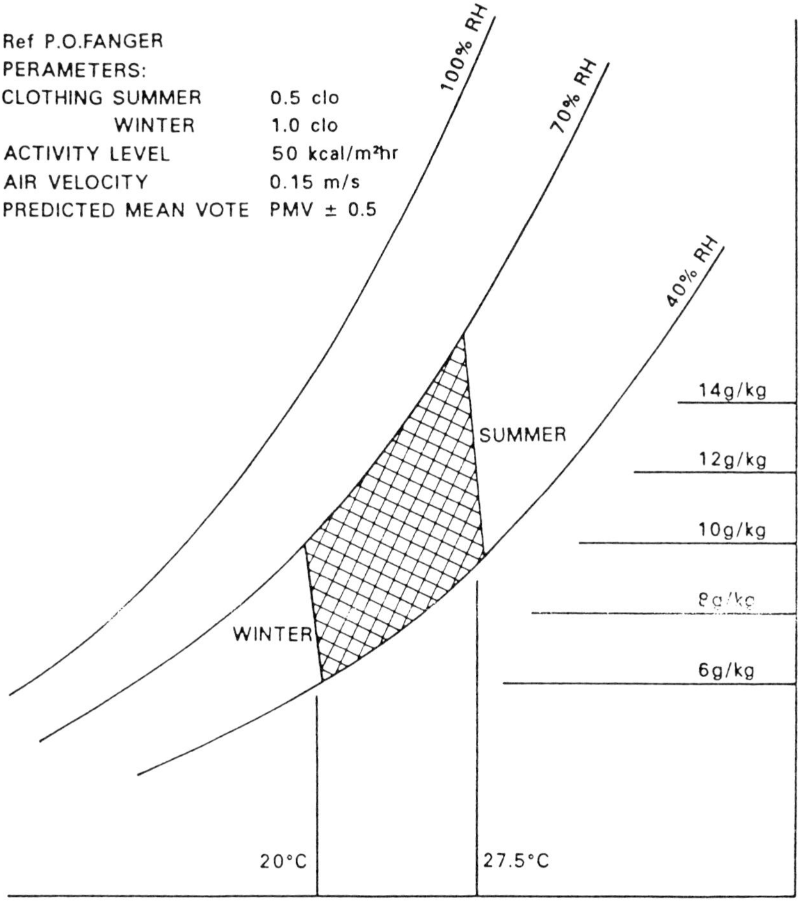

Ref P.O.FANGER
PERAMETERS:

CLOTHING SUMMER	0.5 clo
WINTER	1.0 clo
ACTIVITY LEVEL	50 kcal/m²hr
AIR VELOCITY	0.15 m/s
PREDICTED MEAN VOTE	PMV ± 0.5

100% RH

70% RH

40% RH

14g/kg

SUMMER

12g/kg

10g/kg

8g/kg

WINTER

6g/kg

20°C   27.5°C

PSYCHROMETRIC CHART COMFORT ZONE

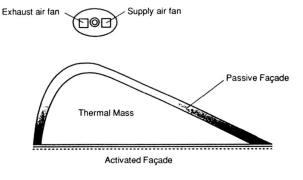

Exhaust air fan Supply air fan

Passive Façade

Thermal Mass

Activated Façade

P L A N

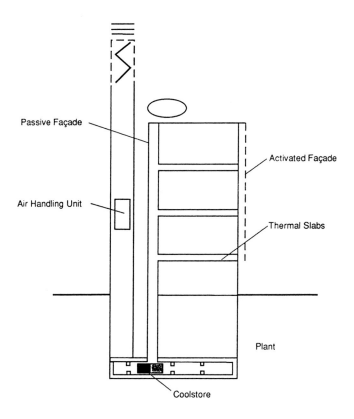

Passive Façade

Activated Façade

Air Handling Unit

Thermal Slabs

Plant

Coolstore

S E C T I O N

PASSIVE & ACTIVE
ENVIRONMENTAL PRINCIPLES

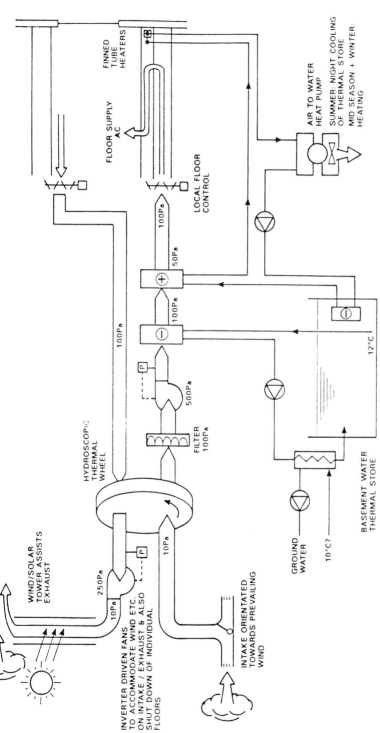

MECHANICAL SERVICES SCHEMATIC

INTENT: TO REDUCE THE HOT SPOTS AND PROVIDE AS
UNIFORM AS POSSIBLE COOLING LOADS WHILE PROVIDING
REASONABLE NATURAL LIGHT TO ALL THE USABLE SPACE

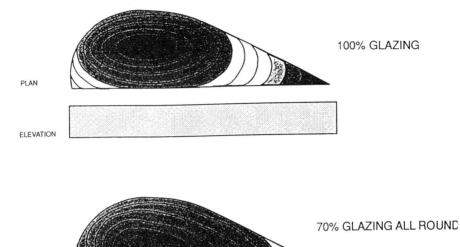

100% GLAZING

PLAN

ELEVATION

70% GLAZING ALL ROUND

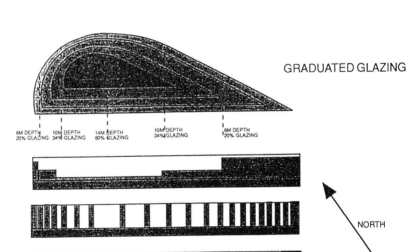

GRADUATED GLAZING

6M DEPTH 10M DEPTH 14M DEPTH 10M DEPTH 6M DEPTH
20% GLAZING 34% GLAZING 60% GLAZING 34% GLAZING 20% GLAZING

NORTH

POSSIBLE ARRANGMENTS OF TRANSPARENT:SOLID TO ACHIEVE "OPTIMUM COOLING LOAD.

COOLING LOAD CONTOURS
ASSUMING 100% DIFFUSE LIGHT

IF THE CORE IS NOT SEPARATE LARGE AREAS OF USEABLE SPACE WILL BE DARK, HAVE REDUCED VIEWS AND NEED ARTIFICIAL LIGHT FOR MOST OF THE DAY.

INTENT :- TO ENSURE AS MUCH NATURAL DAYLIGHT AS POSSIBLE

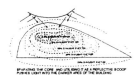

SEPARATING THE CORE AND UTILISING IT AS A REFLECTIVE SCOOP PUSHES LIGHT INTO THE DARKER ARES OF THE BUILDING

USING NATURAL LIGHT PROVIDES A COOL COMFORTABLE LIGHT

USING NATURAL DAYLIGHTING SAVES ENERGY TWICE OVER:
ENERGY TO LUMINAIRES, ENERGY TO COOL LUMINAIRES

ALSO PLANT SIZE IS REDUCED AS MAXIMUM DAYLIGHT IS AVAILABLE AT TIMES OF MAXIMUM SOLAR HEAT GAIN

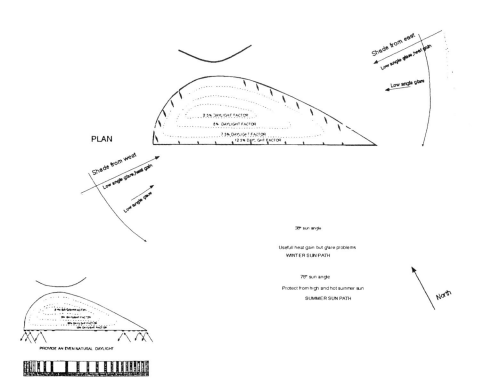

PLAN

2.5% DAYLIGHT FACTOR
5% DAYLIGHT FACTOR
7.5% DAYLIGHT FACTOR
12.5% DAYLIGHT FACTOR

Shade from east
Low angle glare, heat gain
Low angle glare

Shade from west
Low angle glare, heat gain
Low angle glare

38° sun angle

Usefull heat gain but glare problems
WINTER SUN PATH

78° sun angle

Protect from high and hot summer sun

SUMMER SUN PATH

North

PROVIDE AN EVEN NATURAL DAYLIGHT

Allow in 50° winter sun
Usefull heat gain

Direct light

North
Diffuse North light
Permanent floors
Reflected diffuse light

South

Cool North light
Diffuse North light

Permanent floors

SECTION

DIFFUSE COOL LIGHT = USEFUL DAYLIGHTING

NORTH - PROVIDES USEFUL LIGHT WITH MINIMUM HEAT GAIN

DIRECT LOW LIGHT = GLARE

EAST AND WEST - LOW ANGLE SUN IS DIFFICULT TO UTILISE BECAUSE OF THE GLARE PROBLEMS. AVOID WINDOWS FACING EAST OR WEST

SOUTH - HIGH ANGLE SUN CAN BE SHADED EASILY AND REFLECTED TO PROVIDE USEFUL LIGHTING

DAYLIGHT

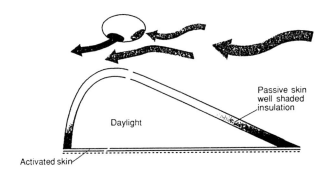

Passive skin
well shaded
insulation

Daylight

Activated skin

Solar radiation + wind
drives the stack effect

P L A N

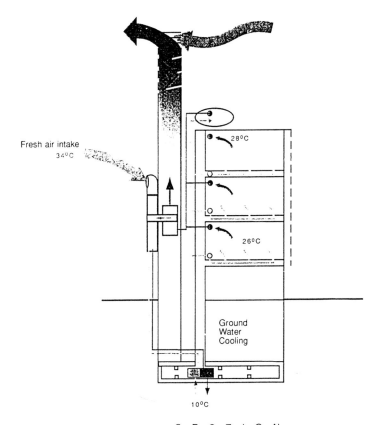

Fresh air intake
34°C

28°C

26°C

Ground
Water
Cooling

10°C

S E C T I O N

M I D - S E A S O N   S U M M E R   D A Y

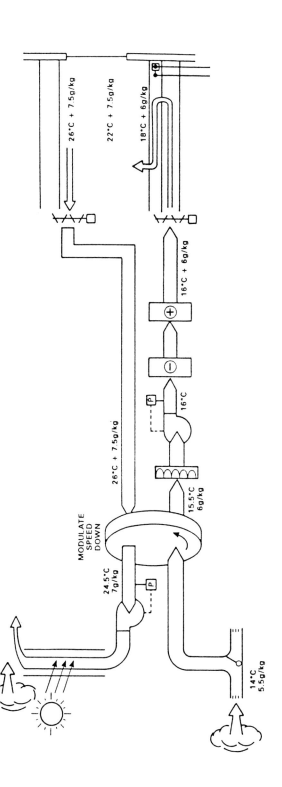

MID SEASON DAY

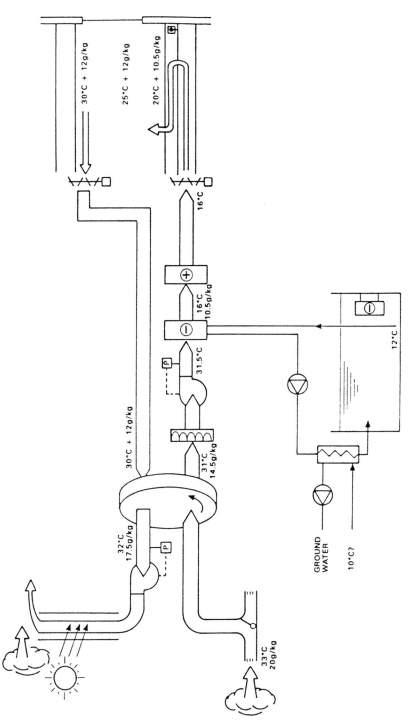

30°C + 12g/kg

25°C + 12g/kg

20°C + 10.5g/kg

16°C

16°C
10.5g/kg

12°C

31.5°C

30°C + 12g/kg

31°C
14.5g/kg

32°C
17.5g/kg

33°C
20g/kg

GROUND
WATER

10°C?

PEAK SUMMER DAY COOLING

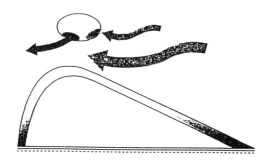

P L A N

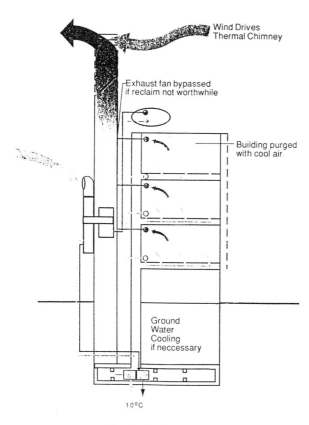

Wind Drives
Thermal Chimney

Exhaust fan bypassed
if reclaim not worthwhile

Building purged
with cool air

Ground
Water
Cooling
if neccessary

10°C

S E C T I O N

M I D - S E A S O N   S U M M E R   N I G H T

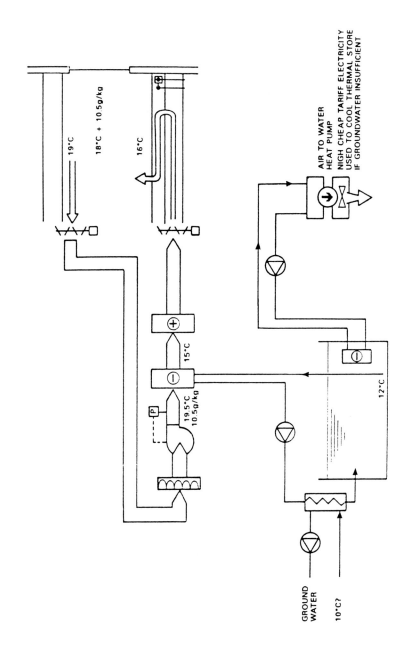

19°C

18°C + 10.5g/kg

16°C

15°C

19.5°C
10.5g/kg

P

12°C

AIR TO WATER
HEAT PUMP

NIGH CHEAP TARIFF ELECTRICITY
USED TO COOL THERMAL STORE
IF GROUNDWATER INSUFFICIENT

GROUND
WATER

10°C?

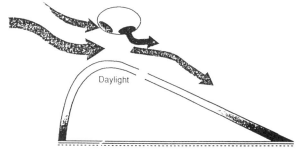

Controlled penetration of
solar radiation for passive
heating at perimeter

P L A N

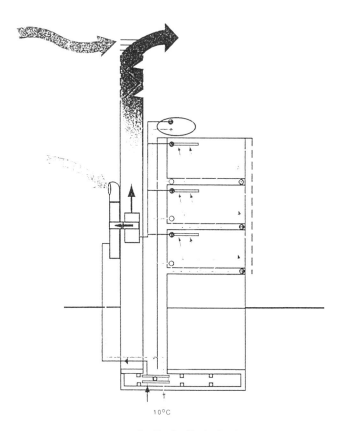

10°C

S E C T I O N

W I N T E R   D A Y

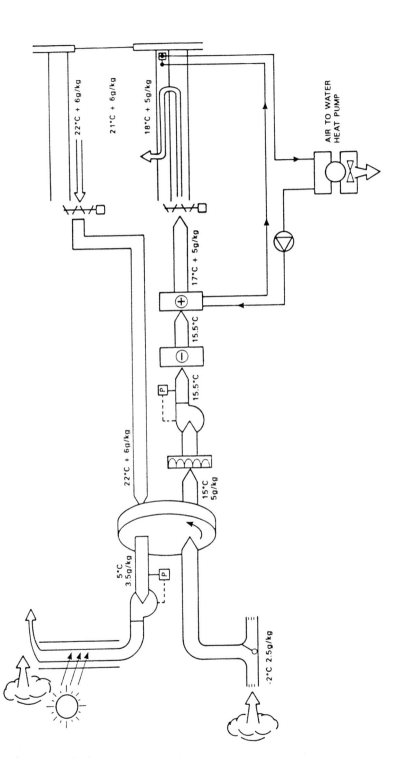

WINTER D

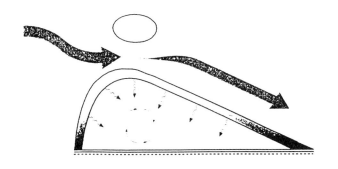

P L A N

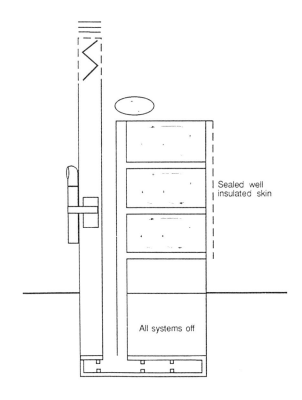

Sealed well
insulated skin

All systems off

S E C T I O N

WINTER NIGHT

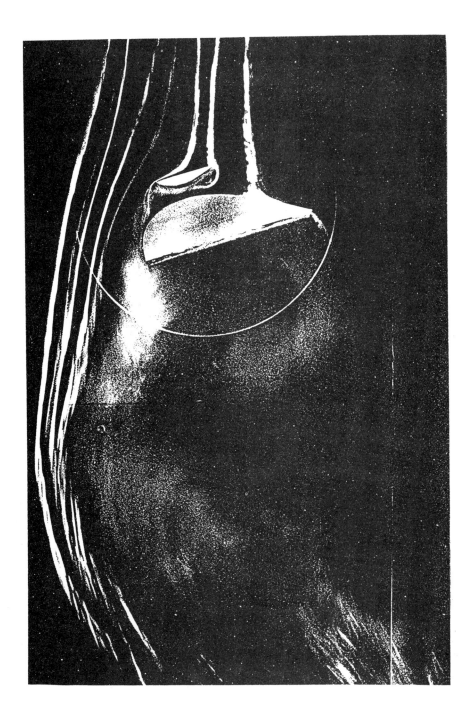

BUILDING WIND METHODOLOGY FOR
ITALIAN CLIMATIC AREAS

Prof.V.CALDERARO,Arch.A.CIOLFI

Faculty of Architecture,University"La Sapienza"
Roma,via Gramsci 53,ITALY

Dr. M.MORONI
Geoex sas,geologia e geofisica applicata
Roma,via Colli del Vivaro

ABSTRACT

The present work,carried out through informations of the
Metereological Service of the Italian Military Aeronautic,is
the first step to carry out a bioclimatic design methodology
based on monitored statistical dates of wind characteristics
in Italian climatic areas.the first dates obtined are relative
to climatic areas of littoral latium and middle interland
and they are:Pratica di Mare,Fiumicino,Ciampino,Guidonia.

KEYWORDS

Wind;Bioclimatic;Direction;Intensity;WMO.

OBSERVING AND PROCESSING METHOD

The present work is a part of a programme on long term which
will identify,for a substantial number of localities in
Italy,dates about wind direction,in per cent,wind velocity,
general characteristics like warm and cold breeze.Other
climatic characteristics will regard temperature,relative
humidity,solar radiation,precipitations.Other phisic parameter
is the thermal inertia of the building.Orografy and exposition
are two remarkable parameter that will affect the operating
system of the design methodology.
The dates iussue from elaboration on informations of the
Metereological Service of the Italian Military Aeronautic.The
investigated localities are around Rome and they are:Pratica
di Mare,Fiumicino,Ciampino,Guidonia.
The various table and graphics in this report,containing

actually yearly frequency distribution of surface wind,agree,
as closely as possible,with WMO requirements.

Each observation has been carried out by taking into account
both wind speed and direction mean velues during the ten
minutes preceding the observation time.
Wind direction,in degrees clokwise from the north,has been
subdivided into 12 equal classes.
Wind speed has been subdivided into 13 classes according
to the anemometric Beaufort scale.

Montly,seasonal and yearly frequency distribution of surface
wind for each synoptic hour(00,03,06,09,12,15,18,21 GMT)is
shown.Each entry in the tables gives the mean number of
occasion in the specified category during the period of
record(month,season or year).
the entry is

$$\frac{n \; x \; k}{s}$$

where

n = total number of observed occurrences in the category
k = number of days in the period
s = total number of available observation

Frequency distribution of surface wind,in per cent,for   each
month,season and whole year,is taking into account.
Each entry in the tables gives the frequency,in per cent,
in the specified category during the period independently
of recording synoptic hour.The entry is

$$\frac{n \; x \; 100}{s}$$

For each investigated locality had been gather dates on
intensity and wind direction relatively to last 20 years.
The dates tranfer on graphic,observed in the eight synoptic
hourshow prevailing wind direction and intensity in per
cent.
The graphic is organized in four concentric circle and
twelve sector from the north.The concentric circle regard
the intensity relate to 11-21 knotes,7-16 knotes,1-6 knotes.
The earth of the circle represent the percentage of calm.The
sector mark wind direction.

RESULTS

The obtined results will be the guide for the design methodology founded on wind characteristics.this is the first
step of the search.further work will regard other informations
about orografy,temperature,relative humidity,precipitations
for each locality investigated.from the obtined results
we can locate the right form,orientation,building characteristics of the building(materials)in function of wind and
other climatic characteristics for summer(passive cooling)
and winter(wind protection).

The same work will be carried out for several Italian localities.

REFERENCES

Aeronautica Militare,CNMCA,1979.Sommario Climatologico
Aeronautico:distribuzione di frequenza del vento al suolo

Aeronautica Militare,CNMCA,1985.Sommario Climatologico Aero-
nautico:distribuzione di frequenza del vento al suolo.
ROMA-FIUMICINO

Aeronautica Militare,CNMCA,1986.Sommario Climatologico Aero-
nautico:distribuzione di frequenza del vento al suolo.
ROMA-PRATICA DI MARE

Aeronautica Militare,CNMCA,1989.Sommario Climatologico Aero-
nautico:distribuzione di frequenza del vento al suolo.
ROMA-CIAMPINO

Aeronautica Militare,CNMCA.Informazioni climatiche su alcune
località italiane.

V.CALDERARO,A.CIOLFI,1991.Architectural Design Methodology
for Environmental Comfort,Proceedings of ISES Solar World
Congress,Denver(CO),1991.

V.CALDERARO,A.CIOLFI,1990.Evaluation Methodology of Environmental Comfort in Bioclimatic Architecture,Proceedings of
the ISES International Conference,Milano,1990.

P.O.FANGER,1986.The PMV Index and the International Comfort
Standard,Practical Application,Proceedings AICARR,Seminar
on thermal Comfort and the Controlled Environment,Napoli,1986.

V.CALDERARO,A.CIOLFI,1988.Evaluation of Thermohigrometric
Behaviour of building with semplified calculation method
for passive cooling.ISES Conference on Bioclimatic Architecture Energy Saving and Environmental Qality,Roma,1988.

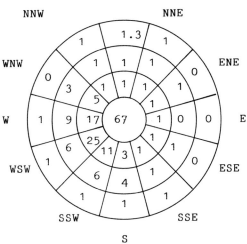

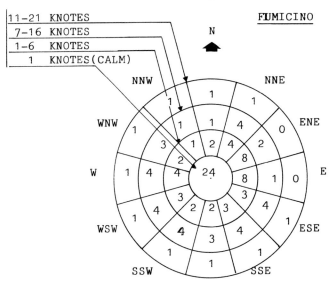

11-21 KNOTES
7-16 KNOTES
1-6 KNOTES
1 KNOTES (CALM)

FUMICINO

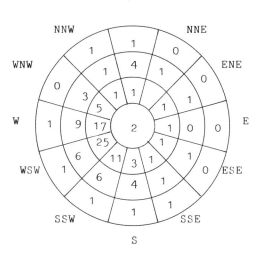

N

N

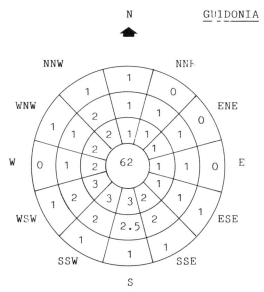

AUTHOR INDEX

* Papers received late are positioned at the end of Volume 5